Watershed Management for Potable Water Supply

ASSESSING THE
NEW YORK CITY STRATEGY

Committee to Review the New York City Watershed Management Strategy

Water Science and Technology Board

Commission on Geosciences, Environment, and Resources

National Research Council

NATIONAL ACADEMY PRESS
Washington, D.C.

NATIONAL ACADEMY PRESS • 2101 Constitution Avenue, NW • Washington, DC 20418

NOTICE: The project that is the subject of this report was approved by the Governing Board of the National Research Council, whose members are drawn from the councils of the National Academy of Sciences, the National Academy of Engineering, and the Institute of Medicine. The members of the committee responsible for the report were chosen for their special competences and with regard for appropriate balance.

Support for this project was provided by the Comptroller of the City of New York.

International Standard Book Number 0-309-06777-4

Library of Congress Catalog Card Number 00-100784

Watershed Management for Potable Water Supply: Assessing the New York City Strategy is available from the National Academy Press, 2101 Constitution Avenue, N.W., Lockbox 285, Washington, DC 20055; (800) 624-6242 or (202) 334-3313 (in the Washington metropolitan area); Internet <http://www.nap.edu>.

THE NATIONAL ACADEMIES

National Academy of Sciences
National Academy of Engineering
Institute of Medicine
National Research Council

The **National Academy of Sciences** is a private, nonprofit, self-perpetuating society of distinguished scholars engaged in scientific and engineering research, dedicated to the furtherance of science and technology and to their use for the general welfare. Upon the authority of the charter granted to it by the Congress in 1863, the Academy has a mandate that requires it to advise the federal government on scientific and technical matters. Dr. Bruce M. Alberts is president of the National Academy of Sciences.

The **National Academy of Engineering** was established in 1964, under the charter of the National Academy of Sciences, as a parallel organization of outstanding engineers. It is autonomous in its administration and in the selection of its members, sharing with the National Academy of Sciences the responsibility for advising the federal government. The National Academy of Engineering also sponsors engineering programs aimed at meeting national needs, encourages education and research, and recognizes the superior achievements of engineers. Dr. William A. Wulf is president of the National Academy of Engineering.

The **Institute of Medicine** was established in 1970 by the National Academy of Sciences to secure the services of eminent members of appropriate professions in the examination of policy matters pertaining to the health of the public. The Institute acts under the responsibility given to the National Academy of Sciences by its congressional charter to be an adviser to the federal government and, upon its own initiative, to identify issues of medical care, research, and education. Dr. Kenneth I. Shine is president of the Institute of Medicine.

The **National Research Council** was organized by the National Academy of Sciences in 1916 to associate the broad community of science and technology with the Academy's purposes of furthering knowledge and advising the federal government. Functioning in accordance with general policies determined by the Academy, the Council has become the principal operating agency of both the National Academy of Sciences and the National Academy of Engineering in providing services to the government, the public, and the scientific and engineering communities. The Council is administered jointly by both Academies and the Institute of Medicine. Dr. Bruce M. Alberts and Dr. William A. Wulf are chairman and vice chairman, respectively, of the National Research Council.

Preface

Watersheds are nature's boundaries for surface water supplies. Natural processes combined with human activities in watersheds determine the inherent quality of these supplies and the treatment they need for potable use. The quality of the drinking water at a consumer's tap depends on the source of that water and, in turn, on the approaches used in managing the activities and processes in the watershed from which the water originates. The effectiveness of these approaches depends to a significant extent on their scientific underpinnings. This report examines watershed management for the water supply of the City of New York. In doing so, it can serve as a prototype for other urban surface water supplies.

In 1997 the National Research Council (NRC), under the auspices of the Water Science and Technology Board (WSTB), established the Committee to Review the New York City Watershed Management Strategy. The NRC chose 15 experts to serve on the committee for the purpose of studying scientific issues associated with the 1997 New York City Watershed Memorandum of Agreement (MOA), a document that outlines strategies for maintaining high-quality drinking water for the nine million residents in New York City and neighboring Westchester County. The MOA has allowed New York City to obtain a filtration waiver under EPA's Surface Water Treatment Rule for its Catskill/Delaware supply system until April 2002. In exchange, the City must (among other things) comply with water quality standards for contaminants in drinking water, it must maintain a watershed protection program (embodied in the MOA), and it must not experience a waterborne disease outbreak.

The MOA is a monumental document in the history of water resources in New York City. There are three main provisions of the agreement: (1) land purchase in the Catskill/Delaware watershed, which supplies 90 percent of the city's drinking water and is predominantly in private ownership, (2) regulations on activities in the watershed, and (3) payments to watershed communities to support local economic development and to promote watershed protection. The MOA owes its existence to multiple parties in New York City, New York State, and the watershed communities, each of which made concessions to protect water quality in the Catskill/Delaware supply and also to meet the needs of the watershed citizens.

To carry out this study, committee members were chosen from a broad range of backgrounds including hydrology, watershed management, environmental engineering, ecology, microbiology, public health and epidemiology, urban planning, economics, and environmental law. The committee met six times over the course of two years, commencing in New York City on September 25, 1997. The third committee meeting was held in the watershed region (Oliverea, NY) to allow the members to visit several of the upstate reservoirs, observe New York City Department of Environmental Protection (NYC DEP) operations, visit a farm enrolled in the Watershed Agricultural Program, and learn about ongoing basic science research in the watershed. At all of the committee's open meetings, the public took great interest in committee deliberations and guest presentations. In all, the committee heard from over 40 invited speakers. In addition, liaisons were established with multiple stakeholders who supplied considerable written information during the course of the study.

Beyond regular committee meetings, four additional activities were undertaken to broaden the knowledge base of the committee. First, a small field trip to the Kensico Reservoir was made in November 1997. In April 1998, six prominent scientists convened with a subgroup of the committee to discuss active disease surveillance in New York City and the prospects for conducting a microbial risk assessment for the Catskill/Delaware supply. In the summer of 1998, committee members met with citizens and farmers in Delaware County and toured the headquarters of the Catskill Watershed Corporation. Finally, in January 1999 the expertise of six buffer zone scientists was used to supplement the committee in its assessment of the adequacy of setbacks in the watershed. The following individuals are thanked for their participation and contribution to these efforts. The members of the Panel on Microbial Risk Assessment are Gunther Craun, Craun and Associates; James Miller, New York City Department of Health; Mark Sobsey, University of North Carolina; William MacKenzie, Centers for Disease Control; and Joan Rose, University of South Florida. The buffer zone experts are David Correll, Smithsonian Environmental Research Center; Art Gold, Rhode Island Department of Natural Resources; Peter Groffman, Institute of Ecosystem Studies; Nick Haycock, Quest Environmental; James Hornbeck, USDA Forest Service; and Lyn Townsend, University of Washington.

In conducting this study, the committee encountered numerous issues that inspired controversy and debate, two of which are noted here. First, out of consideration for time and overall effectiveness, the Croton water supply is not a primary focus of this report. Given the large number of tasks it was addressing, the committee concluded that it could not provide a comprehensive evaluation of Croton watershed issues, many of which differentiate it considerably from the Catskill/Delaware watershed. In particular, New York City is currently under order to filter the Croton system. Kensico and West Branch reservoirs, which are physically located in the Croton watershed, are included in the report to the fullest extent possible. Second, because the filtration avoidance determination for New York City includes the dual-track approach of conducting watershed management and designing a water filtration plant, the report discusses filtration and its benefits. However, in the committee's opinion, this report, which focuses on watershed management, was not the appropriate place to evaluate the global issue of whether all surface waters need treatment beyond disinfection.

Although watershed management is important for any surface water supply, it is critical for an unfiltered supply. The MOA is a remarkable document and a significant milestone in the City's water supply and the region's development. Successful implementation of the MOA is the most important challenge facing the City's water supply managers. It is also important to stress that watershed management is an iterative, living process. This report and its recommendations are provided to support the implementation of the MOA and to contribute to the maturation of the watershed management program.

This report has been reviewed, in accordance with NRC procedures, by individuals chosen for their expertise and their broad perspectives on the issues addressed herein. This independent review provided candid and critical comments that assisted the authors and the NRC in making the published report as sound as possible, and it ensured that the report meets institutional standards for objectivity, evidence, and responsiveness to the study charge. The content of the review comments and the draft manuscript remain confidential to protect the integrity of the deliberative process. The committee thanks the following individuals for their participation in the review of this report and for their many instructive comments: Derek Booth, University of Washington; Kenneth Brooks, University of Minnesota; Gunther Craun, Craun and Associates; Joseph Delfino, University of Florida; Theo Dillaha, Virginia Polytechnic Institute and State University; Walter Lynn, Cornell University; Patricia Miller, Virginia Department of Natural Resources; Jerry Ongerth, University of New South Wales; Joan Rose, University of South Florida; Milton Russell, University of Tennessee; Rhodes Trussell, Montgomery Watson; and Robert Tucker, Rutgers University. These individuals have provided many constructive comments and suggestions. Responsibility for the final content of this report, however, rests solely with the authoring committee and the NRC.

Special thanks must go to several individuals who contributed to the committee's overall effort in so many ways. First, the committee thanks Nancy Anderson and Steve Newman of the New York City Office of the Comptroller.

The committee benefited from their insight into the problems and needs of the City with respect to watershed management. Second, considerable thanks must go to our liaisons at the New York City Department of Environmental Protection: Bill Stasiuk, David Warne, Mike Principe, Salome Freud, Kimberly Kane, David Stern, Ira Stern, Jim Miller, Arthur Ashendorff, and Jim Benson. Third, I appreciate the extensive input from the environmental community in New York City, particularly Eric Goldstein, Robin Marx, and Mark Izeman of the Natural Resources Defense Council and David Gordon of the Hudson Riverkeeper. Finally, I thank our upstate contacts: Alan Rosa of the Catskill Watershed Corporation and Gail Sheridan and Dick Coombe of the Watershed Agricultural Council. All these individuals spent considerable time and energy compiling and passing to the committee the information it needed to complete this study.

Finally, I would like to thank vice chair Max Pfeffer, the members of this committee, study director Laura Ehlers, and project assistant Ellen de Guzman, all of whom devoted many long hours to this project. I have enjoyed immensely the opportunity to work with such a talented and articulate group of professionals. They provided a stimulating environment for addressing the study issues. I especially appreciate their willingness to spend time researching, writing, and revising their contributions. I believe the results of their efforts provide useful guidance for watershed management in the New York City watersheds. This guidance should also be relevant to a broader universe of watershed management programs.

Charles R. O'Melia, *Chair*

Contents

EXECUTIVE SUMMARY

January 21, 1997, marked an important event in the history of American water management: the signing of the mammoth New York City Watershed Memorandum of Agreement (MOA) which provides a legal framework for protecting the drinking water supply of nine million people. The culmination of years of negotiation between upstate and downstate interests, the MOA commits New York City to a long-term watershed management program that combines land acquisition, new watershed rules and regulations, and financial assistance to watershed communities to promote environmental quality and their local economies. Most important for New York City, the agreement currently satisfies provisions of the Environmental Protection Agency's (EPA) Surface Water Treatment Rule (SWTR) that will allow the City to avoid filtering its upstate Catskill/Delaware water supply until at least 2002.

Immediately following the signing of the MOA, the National Research Council (NRC) was asked by the New York City Comptroller's Office to provide a scientific evaluation of the watershed management program. The goal of the NRC study was to determine whether the MOA is based on sound science and to recommend improvements to strengthen watershed management for this large unfiltered supply (see Chapter 1 for the complete Statement of Task). This report is intended to inform New York City and other public water suppliers that are trying to maintain the purity of their existing water sources through proactive watershed management (regardless of whether they presently utilize filtration).

The NRC committee was specifically asked to address the following provisions of the MOA: (1) the use of setback distances to protect bodies of water from nonpoint source pollution, (2) the Total Maximum Daily Load (TMDL) program, (3) siting and technology requirements for wastewater treatment plants and septic systems, (4) the phosphorus offset pilot program, (5) the enhanced monitoring program conducted by the New York City Department of Environmental Protection (NYC DEP), (6) New York State's antidegradation policy, and (7) NYC DEP's Geographic Information System (GIS). In addition to these tasks, the committee also evaluated the Land Acquisition Program and the MOA's requirement for comprehensive land use planning. The role of active disease surveillance in watershed management was explored, and a microbial risk assessment was conducted. Finally, the committee considered the potential impact of future changes in federal regulations regarding safe drinking water. As a result of its studies, the committee determined that the following issues should be taken into consideration as New York City carries out watershed management. More detailed conclusions and recommendations are found in this executive summary and throughout the report.

The watershed management program of New York City should be prioritized to place importance first on microbial pathogens, second on organic precursors of disinfection byproducts, third on phosphorus, and fourth on turbidity and sediment. The prime focus of New York City's efforts in protecting the health of its consumers should be on pathogens in the water supply and on developing means for their control. Because pathogenic microorganisms are the primary target of the SWTR, they should be the focus of any watershed management program designed to gain filtration avoidance. Currently, the main focus of New York City's watershed management strategy is phosphorus, because of its role in eutrophication and its contribution to the creation of disinfection byproducts (DBPs). Considerably less effort has been expended developing monitoring and modeling tools for microbial pathogens, which pose a more significant and direct threat to public health. After pathogens, New York City should focus on the reduction of DBPs, both existing and emerging. Because disinfection via chlorination is planned to continue, control of DBPs will require precursor (organic carbon) reduction. Precursor control can be partially accomplished by the current strong efforts to control phosphorus loading to the reservoirs, which should continue. Nonpoint sources should be the primary focus of future phosphorus control. Finally, greater effort should be made to control sources of inorganic turbidity and sediment at their points of origin rather than in downstream aqueducts and reservoirs. Controlling inorganic turbidity at its source has multiple benefits in relation to other priority pollutants.

The concept of balancing watershed rules and regulations with targeted support of watershed community development is a reasonable strategy for New York City and possibly other water supplies. The City's commitment of financial resources to promote environmentally sensitive development in watershed communities is a fair quid pro quo for enhanced land use planning and regulation intended to protect the drinking water supply. It is possible that resulting economic development will affect water quality. However, existing information convinced the committee that population growth in the Catskill/

Delaware watershed is limited. The potentially adverse effects of population growth and increased economic activity can be offset by careful planning, directed development, more extensive environmental regulation, and improved waste-water management, as provided in the MOA. Such measures will help to maintain high water quality in the Catskill/Delaware reservoirs over the next several years, assuming growth rates do not increase substantially.

The committee encourages New York City and all other water supplies to be receptive to the possibility of additional treatment options. The need for additional treatment may arise because of uncertainties about pollutant sources, better scientific understanding of the impacts of pollutants on human health, and changing regulations. At some point in the future, implementation of the MOA may not be sufficient to avoid additional treatment beyond disinfection. For example, New York City may be compelled to treat its water by coagulation/filtration or equivalent technologies to comply with more stringent regulations for DBPs and microbial pathogens in spite of its best efforts under the MOA. Management of these constituents may be infeasible with current infrastructure simply because increasing disinfectant concentrations to improve pathogen removal *simultaneously* increases the quantity of DBPs.

New York City should lead in efforts to quantify the contribution of watershed management to overall reduction of risk from waterborne pollutants. Watershed management is an essential component of a modern water supply system, but its direct contribution to risk reduction is particularly difficult to quantify. It is much less difficult to estimate the risk-reducing effect of specific treatment options, such as coagulation/filtration, membrane technologies, and alternative disinfection. Because of the comprehensive approach embodied in the MOA and because it has gained national prominence, New York City is in a unique position to demonstrate the overall risk-reducing effects of watershed management. This requires developing the concept of risk reduction as a metric of overall program success and conducting regular risk assessments.

BACKGROUND

New York City obtains its drinking water from three upstate watersheds—the Catskill, Delaware, and Croton systems. These watersheds and a complex infrastructure of reservoirs, aqueducts, and tunnels encompass 1,970 square miles, contain 600 billion gallons of usable storage, and provide as much as 2 billion gallons of water per day (Figure ES-1). Operational flexibility is afforded by interconnections among the systems that allow bypass of reservoirs and blending of water. In particular, water from the Catskill and Delaware systems is generally combined in the Kensico Reservoir, making this terminal reservoir critical to maintaining high-quality water. The combined system serves nine million people in New York City and adjacent Westchester County, the second largest service area of a single water supplier in the United States.

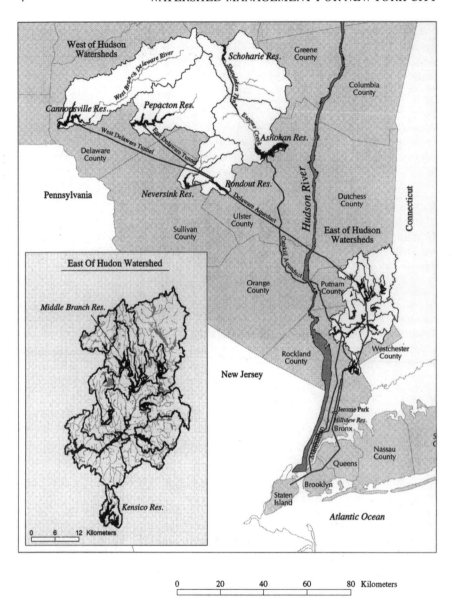

FIGURE ES-1 New York City water supply system. Courtesy of the NYC DEP.

Since the mid-nineteenth century when it turned to the large and sparsely populated Croton River watershed, New York City has recognized the importance of high-quality source water in providing drinking water and protecting the public health. Because the water from the upstate systems historically has met drinking water standards, it has been delivered to consumers without coagulation/ filtration, using only disinfection by chlorination. In recent years, however, the need for filtration of surface water supplies has been increasingly emphasized by regulators and public health experts. Since 1989, the SWTR has required every water supplier to filter its surface water sources prior to disinfection unless its source water meets specific water quality criteria and it establishes a watershed management program. In accordance with this regulation, New York City is preparing to filter water from the more heavily developed Croton watershed, which delivers approximately 10 percent of the City's total water supply. However, thus far the City has been able to satisfy regulatory criteria for water quality and gain a waiver from filtration for the combined Catskill/Delaware system, which delivers the remaining 90 percent of the supply. To maintain its waiver, New York City has embarked on an ambitious strategy of watershed management unprecedented in scope, scale, and cost.

Seventy-four percent of the Catskill/Delaware watershed is in private ownership. With a population that varies from 50,000 to 200,000 depending on the season, there is a fear that human activities, such as increased residential and commercial development, will degrade water quality. National concerns about risks of microbial pathogens such as *Cryptosporidium*, *Giardia*, and viruses in drinking water make the issue of water quality all the more sensitive. Balancing New York City's interest in protecting the watershed for high-quality water with the economic, recreational, and other interests of watershed residents has proven to be a significant challenge. Conflict between the City and upstate residents has been generated by efforts to protect the upland water supply source through land purchase and extensive regulation in the watersheds. The conflicts have largely been resolved by the 1997 signing of the MOA.

KEY ENVIRONMENTAL REGULATIONS

Multiple concerns about drinking water quality have spawned a plethora of federal, state, and local laws, regulations, and policies. Perhaps the most prominent concern is that drinking water will endanger human health, either through acute infections caused by microbial agents or through chronic exposure to chemicals. Although a wide variety of microbes have been linked to waterborne illness, at the present time the microbial pathogens of greatest concern are the protozoa *Giardia* and *Cryptosporidium* because of their resistance to chlorine disinfection. Chemical contamination of drinking water can also produce adverse health effects in humans. The most notable contaminants are DBPs such as trihalomethanes and haloacetic acids, formed in drinking water when chlorine interacts with dissolved organic matter in raw water. Exposure to DBPs has been shown to be related to certain cancers and to miscarriages (Chapter 3).

These concerns, as well as aesthetic considerations, have prompted the creation of environmental laws that directly and indirectly govern the supply of drinking water and watershed management in New York City and across the country. The Safe Drinking Water Act of 1974 (SDWA) sets allowable maximum contaminant levels (MCLs) for multiple contaminants in finished water. The SWTR, pursuant to the SDWA, was created to address the health risks of drinking water derived from surface supplies, particularly those risks associated with microbial pathogens. It describes the criteria that must be met by surface water supplies that want to forgo filtration, including the creation of a watershed management program, meeting standards for turbidity and for fecal and total coliforms, providing adequate disinfection, and avoiding any waterborne disease outbreaks.

The Clean Water Act (CWA) requires the nation's waters to meet water quality levels that are fishable and swimmable. In accordance with the Act, states must designate waters according to use classifications and set specific water quality criteria for those classifications. For those waters not meeting water quality standards, the Act requires that TMDLs be developed to identify sources of pollution and describe mechanisms for reducing them. The National Pollutant Discharge Elimination System (NPDES) issues permits to wastewater treatment plants (WWTPs) and stormwater discharges, ensuring that effluent standards are maintained and best available control technologies are used. Thus, while the SDWA targets public health and the quality of drinking water at or near the point of use, the CWA focuses more on the quality of source waters and discharges into those waters.

State and local environmental regulations also play a role in providing high-quality drinking water to New York City. The Watershed Rules and Regulations of the MOA fill gaps between the CWA and the SDWA by placing specific restrictions on watershed activities. Additional oversight of watershed activities is provided by the New York State Environmental Quality Review Act (SEQR), which creates a role for public decision-making regarding proposed projects through its requirement for an environmental impact statement.

SOURCE WATER PROTECTION

Watershed management is recognized as a key part of the multiple-barrier approach to providing safe drinking water (see Chapter 4). Source water protection specifically refers to management of water supply watersheds, both surface and groundwater. An ideal source water protection program contains several components. First, it should have defined program goals and objectives. An inventory of the watershed to assess existing and potential contamination is also necessary. Once sources of contamination have been identified, protection strategies must be chosen to reduce contamination and prevent future pollution. Protection strategies require proper implementation, a component of source water protection that is often overlooked and sometimes responsible for program failure. A final component of any source water protection program is effectiveness monitoring and evaluation of monitoring data. To be successful, stakeholders must be involved in all stages of a source water protection program. These components exist to varying degrees within the MOA and other watershed management programs.

CURRENT CONDITIONS IN THE
CATSKILL/DELAWARE WATERSHED

Common pollutants in the New York City watersheds include microbial, chemical, and physical parameters. The pathogenic protozoa *Giardia* and *Cryptosporidium* have been detected at low levels in the Kensico Reservoir (see Figure 6-6) and in stream sites throughout the Catskill and Delaware watersheds. Although both nitrogen and phosphorus can cause eutrophication of waters, phosphorus is the more important nutrient in the New York City reservoirs because it is limiting to algal growth during most of the year. Natural organic carbon compounds, which serve as precursors for many undesirable DBPs, are found in all the reservoirs. Sediment transport to reservoirs is primarily a problem in the Schoharie and Ashokan watersheds, where steeper slopes and fine-textured soils are more prevalent. Finally, toxic compounds are used in specific regions of the Catskill/Delaware watershed.

The pollutants described above are derived from both point and nonpoint sources. Point sources, consisting primarily of 41 WWTPs, occur throughout the Catskill/Delaware watershed. Nonpoint sources of pollution are more widespread and abundant in the watershed. Undisturbed forests, comprising almost 70 percent of the Catskill/Delaware watershed, are sources of nutrients, natural organic carbon compounds, sediment, and pathogens. Agriculture, the second largest land use in the Catskill/Delaware watershed, is a potentially significant contributor of microbial pathogens, nutrients, and pesticides. Although urban areas comprise a relatively small percentage of land in the Catskill/Delaware watershed, they can be significant sources of nutrients, pathogens, and pesticides, as well as of heavy metals and other toxic compounds. The Catskill/Delaware watershed includes over 30,000 septic systems (on-site sewage treatment and disposal systems, or OSTDS) that can be sources of both microbial pathogens and nutrients. Finally, atmospheric deposition over land and water is a probable source of nitrogen and mercury to the watersheds that is extremely difficult to quantify and control.

Measures of reservoir water quality indicate a chronic problem with the eutrophic health of the water supply reservoirs, particularly the Cannonsville Reservoir and the Croton system reservoirs. All 19 reservoirs are classified as either mesotrophic or eutrophic, and some have variably high concentrations of phosphorus, chlorophyll *a*, and turbidity. The Kensico Reservoir had elevated levels of fecal coliforms between 1990 and 1993, which were attributed to bird populations. In addition, new commercial development and road expansion in the Kensico watershed may threaten water quality in this reservoir in the near future. Despite these conditions, source water and drinking water in New York City have historically been and continue to be in compliance with the SDWA. Catskill/Delaware source water and associated tap water are meeting requirements for total and fecal coliforms and trihalomethanes, and there have been no documented outbreaks of waterborne disease in New York City since 1974.

WATER QUALITY MONITORING

Water quality monitoring has four general objectives. *Compliance monitoring* evaluates physical, chemical, and pathogenic biological parameters for compliance with local, state, and federal regulations. In almost all cases, the water supply has met the requirements for physical, chemical, and biological parameters. *Operational monitoring* is conducted on a broad set of parameters needed to assess the ambient quality of water and reservoir dynamics, and to determine the sources of pollution that influence water quality. New York City's monitoring program measures a wide range of parameters in streams, subsurface flow, reservoirs, aqueducts, and the distribution system. *Performance monitoring* is used to evaluate the effectiveness of watershed management practices and policies and to isolate design factors that influence pollutant removal. This type of monitoring in the New York City watersheds is confined to a few specific studies. Finally, monitoring data can be used *to support modeling* of projected changes in water quality under different conditions related to land use change or watershed management actions. Monitoring for model calibration and validation involves both intensive and extensive sampling of a reservoir and its watershed to define parameter values, set initial conditions, and physically characterize the watershed.

In general, the committee finds the monitoring program to be informed, extensive, and of high quality, as is appropriate for a water supply of this size. An extensive review of the enhanced monitoring program led to the following suggestions for improvement.

Monitoring should be conducted on the basis of discharge-mediated volume rather than on fixed intervals for stream flow, shallow subsurface flow, WWTP effluent, and precipitation analyses. Event-based or flow-proportional sampling is needed to capture rapid variation in water flow and water quality constituents, including microbes, total and dissolved organic carbon, sediment, and nutrients. Although some flow-proportional monitoring is under way, more efforts and resources should be directed toward converting fixed-frequency sampling to storm-event sampling.

Shallow subsurface and groundwater parameters should be monitored regularly throughout the reservoir watersheds. This is particularly important given the prominent role of agriculture, the high density of OSTDS, and the potential for leaking sewer lines. At the present time, groundwater sampling occurs only in the Kensico watershed. New routine groundwater sampling should be integrated with direct experiments on the efficacy of OSTDS and riparian buffer zone management.

With the exception of dissolved organic carbon, the analytical methods employed for physical, chemical, and biological parameter monitoring are generally adequate. Pathogen-monitoring techniques must be improved to achieve significantly higher numbers of nonzero measurements for (oo)cysts, to assess overall recovery efficiency on a weekly basis, and to build split samples into the routine sampling plan. Additional microbial parameters should be considered for routine monitoring, including *E. coli*, F^+ coliphage, *Clostridium perfringens,* and cyanobacteria that produce toxins.

Monitoring and other special pollution prevention activities in the Kensico Reservoir and its watershed should continue at their present high intensity given the importance of this watershed in controlling Catskill/ Delaware drinking water quality. Similar measures should be considered for the West Branch Reservoir and watershed, which is undergoing more rapid population growth.

ACTIVE DISEASE SURVEILLANCE

Active disease surveillance for giardiasis and cryptosporidiosis began in New York City in the mid-1990s. This activity involves the compilation of case rates from City medical laboratories followed by case interviews to obtain information about suspected risk factors. Current endemic rates of these infections are significantly lower than rates in previous years, primarily because of fewer infections among immunocompromised persons.

Active disease surveillance is generally too slow and insensitive to detect an ongoing disease outbreak, unless the outbreak is very large. To better detect outbreaks in real time, New York City is monitoring sales of antidiarrheal medications, surveying health of nursing home patients, and monitoring laboratories for the number of stool samples submitted. These complementary systems can indicate sudden increases in gastrointestinal illness, which may be indicative of a waterborne disease outbreak.

Linking observed disease to drinking water as the route of transmission can only be accomplished with specific epidemiologic studies. New York City has conducted a case-control study for giardiasis and a cross-sectional study of cryptosporidiosis, in which no relationships between drinking water habits and infection were observed, possibly because the study populations were not large enough. In general, New York City is making concerted efforts to conduct active surveillance and outbreak detection for giardiasis and cryptosporidiosis.

To improve the sensitivity of active disease surveillance, health care providers and laboratories should make *Cryptosporidium* testing part of all routine stool examinations. Active disease surveillance as practiced by New York City and other communities is too insensitive to detect waterborne disease outbreaks on a real-time basis. Data from other communities suggest that only 1 in 22,000 cases are diagnosed and reported via disease surveillance during waterborne disease outbreaks.

New York City should determine the lowest incidence of disease that can be detected by the current outbreak detection program and increase the sensitivity by studying specific populations. It is not clear what size of outbreak can be detected using the combined system of monitoring sentinel nursing home populations, sales of antidiarrheal medications, and submission of stool specimens. To increase the sensitivity and population base covered by these systems, New York City should consider monitoring school absenteeism related to gastroenteritis in sentinel schools, monitoring Health Maintenance Organization nurse hotline calls to track increased incidence of gastroenteritis, monitoring hospital emergency room visits for gastroenteritis, and monitoring gastrointestinal symptoms in a network of sentinel families.

To determine the role of tap water as a vehicle of infection, New York City should conduct additional epidemiological studies. New York City is in an excellent position to conduct a randomized household intervention trial using some type of in-home water treatment device. Although such studies are complex and expensive, they are a rigorous and definitive way to quantify the potential for waterborne disease, and they provide valuable documentation of the safety of the water supply.

MICROBIAL RISK ASSESSMENT

Active disease surveillance is one analytical tool to estimate potential disease impacts from pathogens in water. A complementary tool is quantitative microbial risk assessment. Although risk assessment is not being done by New York City on a regular basis, there are sufficient data being collected to use this technique to estimate potential disease impacts.

The risk assessment conducted by the committee used data on *Cryptosporidium* oocyst concentrations in the Kensico Reservoir from 1993 through July 1998. The daily risk estimate considering all years of data and all sources of uncertainty was 3.4×10^{-5}, with a 95 percent confidence interval ranging from 3.4×10^{-7} to 21.9×10^{-5}. For an exposed population of 7.5 million, this would translate into an estimated 255 infections per day. See Chapter 6 for details and a description of several caveats that *must* be kept in mind when reviewing and interpreting these results. The following recommendations regarding risk assessment are made.

It is recommended that a *Cryptosporidium* risk assessment be performed on a periodic basis for New York City. The goal of these efforts should be to help determine the contribution of watershed management (vs. other treatment options and management strategies) to overall risk reduction. Data that are sufficient for these purposes are currently collected as part of the NYC DEP Pathogen Studies. Prior to commencing this regular effort, a decision must be made as to what level of risk is deemed to be acceptable to the regulatory agencies, the City, and the affected parties. This level should be arrived at after full and open discussion with the various stakeholders.

An ongoing program of risk assessment should be used as a complement to active disease surveillance. Risk assessment allows one to ascertain the level of infection implied by a very low level of exposure that would go undetected by active surveillance, thus acting as a complementary source of information about public health. In general, periodic risk estimates should be examined for concordance with prior computed risks and observed illness rates in formulating subsequent water treatment and watershed management decisions.

LAND ACQUISITION AND COMPREHENSIVE PLANNING

The goal of the Land Acquisition Program is to solicit up to 355,000 acres of sensitive lands in the Catskill/Delaware watershed. Land must be obtained on a willing buyer/willing seller basis, and conservation easements are an option in lieu of direct purchase. To date, the program has over 15,000 acres under contract, and it has made offers on over 100,000 acres. Other program elements to promote water quality include a flood damage buyout program, agricultural easements, and stewardship initiatives. Areas for acquisition within the Catskill/Delaware watershed have been prioritized based entirely on their proximity to bodies of water and the distribution system. In addition, the program sets minimum size limits from one to ten acres for all priority areas.

The committee recommends that greater use be made of the GIS and available land use/nonpoint source computer models to determine more precisely which areas are critical to water quality and thus should be protected through acquisition.

Lower limits (from one to ten acres) on the size of parcels that can be solicited may exclude environmentally sensitive lands from acquisition. It is recommended that these limits be relaxed and eventually dropped as the number of larger parcels diminishes and small adjacent parcels become available.

A local land trust for the Catskill/Delaware watershed region would be beneficial to furthering the goals of the land acquisition program. Currently, the methods for managing the lands acquired by the City, both through outright purchase and conservation easement, are complex and require oversight from the

City and State. A land trust can move more quickly than the Land Acquisition Program to acquire or accept gifts of land, and it may be attractive to those landowners who simply do not wish to sell to New York City.

The MOA Watershed Protection and Partnership Programs, most of which are administered by the Catskill Watershed Corporation (CWC), are intended to maintain and improve water quality while preserving the character and economic vitality of watershed communities. Land use plans are prerequisite to participation in some of the programs, in particular the new sewage treatment infrastructure program, the sewer extension program, and the phosphorus offset pilot program. Although each of these programs is in the initial stages, areas for improvement are apparent. First, the planning efforts under way are fragmented. That is, none of the efforts are truly comprehensive (i.e., housing, transportation, public utilities, environmental conditions, and other relevant factors are considered). This fragmentation is compounded by the numerous small constituencies responsible for plan development. Second, planning efforts do not require enough citizen participation to foster success. Finally, strategies for successfully implementing the plans have not been described.

The committee recommends that the CWC review local and county plans for the watershed area to make sure they are compatible. To coordinate planning efforts currently applied under the MOA, one alternative would be to create comprehensive plans for those watershed towns that do not have them, and to review and update existing plans to be in accord with the newly created plans. At the next level, counties should attempt to integrate the plans for towns and villages within their jurisdictions, and the CWC should review the county plans for compatibility and workability.

Expanding the mission of the CWC would allow it to play a greater role in local planning efforts of the smaller communities in the Catskill/Delaware region. The CWC could act as a forum for citizen involvement, provide technical assistance to support local implementation and enforcement, and provide a comprehensive framework for local watershed plans.

TOTAL MAXIMUM DAILY LOAD PROGRAM

As required by the City's filtration avoidance determination, TMDLs of the priority pollutant phosphorus have been calculated for all of the New York City water supply reservoirs in two phases (I and II). Several Croton reservoirs and the Cannonsville Reservoir consistently exceed their TMDL. According to land use model results, reductions in phosphorus loading from both point and nonpoint sources will be required to meet Phase II TMDLs in these reservoirs.

The Phase II TMDL calculations are a significant improvement over Phase I calculations because more accurate land use models were used and a more appropriate guidance value for total phosphorus in the reservoirs was chosen. Phase III

TMDLs are expected to be fully time-variable and spatially variable, and models to accomplish this are under development.

The 15-μg/L total phosphorus guidance value is appropriate for the Phase II TMDL process. The Phase I goal of 20 μg/L was not adequately conservative for a drinking water supply, as it is based on ecological and aesthetic considerations. Conservatism in the choice of a phosphorus standard is necessary because data for some of the New York City reservoirs (e.g., Cannonsville) show that algal productivities, estimated from average and maximal summertime algal biomass as chlorophyll *a*, are in excess of recommended values for drinking water systems.

NYC DEP should place a high priority on implementing all necessary nonpoint source control measures to reach Phase II TMDLs. The most frequently cited criticism of TMDL programs across the country is the failure of states to implement pollution control measures following TMDL calculations. Currently, it is unclear what specific measures are being taken to reduce phosphorus loading to the Cannonsville Reservoir and several Croton reservoirs other than WWTP upgrades.

Phase III of the TMDL program should focus more on public health protection by developing models that link phosphorus to DBP precursors and other relevant parameters. New York City is currently constructing input–output models for total phosphorus, measurements of dissolved and particulate phosphorus loadings and concentrations, and assessment of individual phosphorus retention coefficients in the Cannonsville Reservoir. Deterministic models are needed to link phosphorus and dissolved organic matter derived from in-reservoir and terrestrial sources to DBP formation potential, algae, chlorophyll, and taste and odor.

PHOSPHORUS OFFSET PILOT PROGRAM

The MOA includes a five-year phosphorus offset pilot program that allows for the construction of up to six new WWTPs in pollution-sensitive basins. The purpose of this program is to allow for continued growth in these watersheds, while preventing a net increase in phosphorus loading. The offset program is similar to watershed-based pollutant trading programs for water and emissions trading programs for air, in which a discharger of pollution is allowed to increase its discharge if another party will concomitantly reduce its discharge. The offset ratio indicates the amount of pollutant reduction that must occur to balance the new pollutant discharge, either 2:1 or 3:1 in the New York City program. A variety of offset mechanisms can be used to reduce pollutant loadings from other point and nonpoint sources. Recommendations for program improvement include the following.

Baseline, minimum requirements for phosphorus reduction must be in place and operating effectively before additional reductions can be defined

as surplus. This refers to such activities as all planned upgrades to WWTPs that will reduce phosphorus loadings and to phosphorus reduction mechanisms that are required as part of a Stormwater Pollution Prevention Plan, among others. Criteria are needed to determine whether baseline requirements are in place and operational and to further define surplus reductions that can be used as offsets.

New York City should reevaluate the 3:1 and 2:1 ratios and develop a technical basis for the ratios that reflects the unique conditions associated with specific proposed offset mechanisms. In addition to providing a safety margin for variability in the performance of best management practices (BMPs) and error in data collection, the offset ratio should reflect conditions present in the New York City watershed such as the spatial and temporal variability of offset mechanisms, the relative locations of the offset mechanism and the WWTP, and the different forms of phosphorus produced in the effluents of WWTPs versus the offset mechanisms. Because the ratios do not explicitly take these issues into consideration, they may not afford sufficient protection.

There is no evidence that the phosphorus offset pilot program will result in a net reduction in phosphorus loadings to the water supply reservoirs, because the offset ratios were not established with this intent. If a net reduction in phosphorus loading becomes a goal of the program, the offset ratio should be made more conservative.

ANTIDEGRADATION

Antidegradation is a federal regulation related to the CWA stating that waters must not be allowed to degrade in quality. It deals primarily with "assimilative capacity"—the amount of additional pollution that a waterbody can receive without exceeding its water quality criteria. States must develop an antidegradation policy that distinguishes three levels of water quality: Tier 1 (the lowest level of quality), Tier 2 (fishable and swimmable waters), and Tier 3 (outstanding natural resources). Tier 1 waters cannot be allowed to degrade any further, and their existing uses must be maintained. In Tier 2 waters, water quality is only allowed to degrade to the level of fishable/swimmable and only if important social or economic reasons for the lowering of water quality are given. Tier 3 waters, which have exceptional recreational and ecological significance, cannot be allowed to degrade at all. In New York State, antidegradation is implemented via three state regulations: stream reclassification, the State Pollutant Discharge Elimination System (SPDES) permitting program, and SEQR.

The SPDES Permit Program and the SEQR process are adequate tools for implementing antidegradation, but they could be improved with some revisions. For the most part, these regulations take into account the reasonable alternatives to a proposed action, the social and economic benefits of a proposed action, and the significance of potential environmental impacts, all of which are required under federal regulations. However, additional guidance on the types of economic and social benefits that should be part of environmental impact statements is needed. In addition, explicit consideration of a receiving water's assimilative capacity should be required as part of environmental impact statements. Because SEQR is the primary avenue for regulating new nonpoint sources that will adversely affect water quality in New York State, this requirement for addressing assimilative capacity is critical.

New York State should better define what a significant lowering of water quality is in a Tier 2 water. That is, it should set quantitative criteria for altering the assimilative capacity of a body of water. Other state antidegradation programs suggest that a 5 percent to 25 percent change in a water's assimilative capacity is significant.

ADDITIONAL TREATMENT OPTIONS

In addition to various watershed management activities, the MOA requires a series of studies on coagulation/filtration of the Catskill/Delaware system and a study on alternative disinfectants that might be used in lieu of filtration. Simultaneously carrying out watershed management and planning for additional treatment, such as coagulation/filtration and alternative disinfection, is a relatively new policy concept that has been dubbed the dual-track approach. This approach was designed to ensure no time is lost if other treatment processes are later determined to be necessary. Preliminary results from these studies led to the following recommendations regarding future additional treatment for the Catskill/Delaware supply.

The dual-track approach allows New York City to focus most of its resources on watershed management at this time. Watershed protection reduces source water pollutant concentrations that would be treated by future additional treatment. Thus, watershed management can reduce the cost of filtration. For this reason, future decisions to construct treatment processes beyond disinfection should in no way deter New York City from pursuing watershed management.

The results of the filtration pilot plant study show the New York City water supply can be effectively treated via coagulation/filtration. Treated water from direct filtration had a low turbidity and low total trihalomethane and haloacetic acid formation potential. A removal rate of at least 3 log for *Cryptosporidium* oocysts is expected for the entire treatment train (corresponding to a 3-log reduction in risk from potential waterborne *Cryptosporidium*). This level of treatment effectiveness depends upon high source water quality. Finished water quality would be expected to be even better with improvements in source water quality.

Additional studies to assess the potential of ozone as a treatment technique are needed. Ozone has the potential to increase levels of biodegradable organic carbon in finished water and foster regrowth of heterotrophic plate count organisms and possibly coliforms, although distribution system disinfectant residuals may counter this phenomenon. Thus, ozone should not be adopted as the only treatment other than a distribution residual unless the City undertakes an investigation to satisfy itself that bacterial regrowth can be kept in check.

NONPOINT SOURCE POLLUTION CONTROL PROGRAMS

Nonpoint source pollution is, by definition, widely dispersed in the environment and is associated with a variety of human activities that produce pollutants such as nutrients, toxic substances, sediment, and microorganisms. Activities that generate nonpoint source pollution cause changes in vegetative cover, soils, or flow paths that reduce the ability of the land to naturally remove pollutants in stormwater. Three programs within the MOA that address nonpoint source pollution are reviewed in this report: (1) the Watershed Agricultural Program, (2) the Watershed Forestry Program, and (3) Stormwater Pollution Prevention Plans. For each, implementation and long-term maintenance of BMPs will be critical to program success.

Watershed Agricultural Program

The Watershed Agricultural Program (WAP) is a voluntary program intended to standardize and improve environmental practices among farmers within the watershed. Unlike most other activities, agriculture is exempted from the Watershed Rules and Regulations in the MOA, including requirements for setback distances, discharge permits, and rules regarding pesticide application. The WAP is intended to "substitute" for such regulations, and its continuance is required by EPA's filtration avoidance determination for New York City.

The goal of the WAP is to design and implement Whole Farms Plans—comprehensive strategies for controlling pollution at individual farms in the watershed. Because dairy farms are the most common farm type in the Catskill/Delaware watershed, phosphorus and pathogens are the pollutants of concern, and common BMPs include barnyard stormwater management, improved manure storage, and the separation of calves from cows. Currently, program success is directly related to the numbers of practices installed and the numbers of farms participating.

A scientifically based phosphorus load reduction goal that will achieve the desired phosphorus concentration in the water supply reservoirs is needed for the WAP. The purpose of having this load reduction goal is to be able to apportion reductions among different phosphorus outputs from individual subwatersheds and farms. Simulation models that take into account in-reservoir generation of phosphorus, phosphorus cycling, and different forms of phosphorus can be used for this effort.

Although the WAP has met its required milestones (including over 95

percent farmer participation), these metrics do not indicate the net effect of the WAP on water quality. To accomplish this critical task, additional monitoring is needed at various scales. First, the use of demonstration farms for evaluating whole farm plans should be expanded beyond the two sites now being studied. Second, monitoring to determine the effectiveness of BMPs in reducing pathogen and phosphorus concentrations in nearby streams is urgently needed.

Lands enrolled in the newly established USDA Conservation Reserve Enhancement Program should be prioritized based on (1) frequency of flooding, (2) vegetation type, and (3) whether the landowner will voluntarily exclude livestock from riparian zones. If prioritization is not possible, rental and cost-share incentives should be increased to retire frequently flooded farmland into riparian forest buffers and to exclude livestock from streams.

Watershed Forestry Program

Modeled after the WAP, the voluntary Watershed Forestry Program (WFP) was created in 1996 to improve the economic viability of forestland ownership and the forest products industry in ways compatible with water quality protection and sustainable forest management. Although the WFP is new, progress toward program objectives has been rapid and substantial, mainly in areas such as program administration, education of potential program participants and foresters on the use of appropriate BMPs, and development (by independent scientists) of monitoring techniques to assess BMP effectiveness. To date, 78 landowners and a combined area of 21,000 acres are enrolled in the program.

Water quality monitoring should be implemented on the model forests and in conjunction with NYC DEP and New York State Department of Environmental Conservation (NYS DEC) monitoring networks on other tributaries. The WFP is encouraged to merge landowner, management plan, and field assessment data with water quality data to evaluate program performance.

The committee recommends that New York State consider tax policy changes that would promote sustained management of private forestland including those forests on relatively small parcels. A major impediment to the sustainability of forestry in the Catskill/Delaware region is the imbalance between taxes and expected revenues, such that short-term financial pressures discourage long-term investments in forest conservation. Because the benefits of the WFP are likely to extend beyond the local political boundaries, impacts on town revenues should be evaluated and mitigation payments to local governments by the State should be made as an integral part of the program.

Stormwater Pollution Prevention Plans

Most types of new, large-scale (greater than five acres) development in the New York City watershed region are required to submit a Stormwater Pollution

Prevention Plan (SPPP) for controlling the quantity and quality of stormwater runoff generated by new impervious cover. SPPPs specify BMPs that will prevent erosion and sedimentation during construction and prevent any increase in the rate of pollutant loading via stormwater after construction. They must include a quantitative analysis demonstrating that stormwater quality from post-construction conditions will be better than or equal to that of pre-construction conditions.

Prior to the MOA, fewer activities required the drafting of an SPPP, and the regulatory oversight was spread among multiple agencies. SPPPs have spawned confusion among engineers, developers, and local and state agencies about exactly how the plans should be interpreted and implemented, as most of these organizations have had no prior experience with stormwater quality control and/or stormwater BMP design. The report includes a variety of specific suggestions to improve the quality of SPPPs and promote their successful implementation.

For removal of both phosphorus and bacteria from stormwater runoff, current structural BMPs advocated by SPPP guidance material are only moderately effective. In almost all cases, they cannot reduce post-development loading to pre-development levels. At present, the only stormwater retrofit BMPs whose performance is quantifiable are ponds, wetlands, sand filters, and swales. To counteract these deficiencies in structural urban stormwater BMPs, SPPPs should also encourage use of nonstructural BMPs that will limit the amount and impact of impervious surfaces.

New York City should embrace a performance-based approach to stormwater management rather than the permit-based approach embodied by the current SPPPs. The SPPPs should rely less on quantitative, theoretical calculations and more on performance monitoring and strict requirements for BMP size and treatment efficiency. Guidance material for such a new approach should include information on performance monitoring of stormwater BMPs for a variety of pollutants.

SETBACKS AND BUFFER ZONES

One of the most prevalent features of the MOA Watershed Rules and Regulations is the use of setback distances for separating waters from potentially polluting activities. For activities such as storage of hazardous waste and petroleum products, siting of septic systems and landfills, and creation of new impervious surfaces, 25–1,000 ft of land must separate the activity from nearby reservoirs, reservoir stems, controlled lakes, wetlands, and watercourses.

The committee assessed the adequacy of these setback distances in three ways. First, it compared them with similar regulations found elsewhere. Second, information on pollutant removal in *buffer zones* was used to predict the effectiveness of setbacks. (Buffer zones are natural or managed riparian areas that protect waterbodies from adjacent nonpoint sources of pollution by retaining or transforming pollutants and producing a more favorable environment for aquatic ecosystems.) Finally, a predictive model analysis of pollutant movement through a setback was used to simulate the functioning of setbacks. All three of these evaluations are necessarily indirect because of the absence of actual monitoring data on pollutant concentrations in stormwater emanating from different land uses in the Catskill/Delaware watershed.

The setbacks prescribed in the MOA are not equivalent to buffer zones. Setback descriptions do not discuss the characteristics of setback land known to influence pollutant removal in buffer zones, such as slope, hydraulic conductivity, soil moisture, vegetation or surface roughness, and flow rates. When delineating setback areas, New York City should consider setting a slope threshold above which land cannot be included in setback considerations. Areas with a slope greater than 15 percent generally do not function as effective buffer zones.

Active management of setbacks is necessary to achieve the pollutant removal efficiencies ascribed to buffer zones. To improve the functioning of setbacks, BMPs that will convert channelized flow to sheet flow should be installed upslope from the setback. Setbacks should have appropriate vegetation to retain nutrients, sediment, and other pollutants. If setbacks are managed as buffers, periodic harvesting of vegetation may be necessary to remove accumulated pollutants and renew assimilative capacity.

Based on an analysis of travel times, existing literature, and a model analysis, the following setbacks are judged most likely to be inadequate: (1) the 100- and 500-ft setbacks for hazardous wastes, (2) the 100- and 500-ft setbacks for petroleum underground storage tanks, (3) the 100- and 500-ft setbacks for heating oil, (4) the 250- and 1,000-ft setbacks for landfills, and (5) the 100-ft setbacks for septic systems, WWTPs, and impervious surfaces because of microbial pathogens.

Because of their variable pollutant removal abilities, buffer zones should not be relied upon to provide the sole nonpoint source pollution control and are instead best used when integrated with appropriate source controls on pollutant releases. Although riparian buffers can ameliorate nonpoint source pollution in some circumstances, they are most effective when used as part of an overall pollution control or conservation plan.

WASTEWATER TREATMENT

Treatment and disposal of wastewater in the Catskill/Delaware watershed is a major factor in determining the quality of New York City drinking water. This is because almost all wastewater from the region is discharged either directly into streams tributary to the water supply reservoirs or into the subsurface where it can eventually reach the reservoirs. Individual septic systems (OSTDS) and centralized WWTPs are the major sources of wastewater in the watershed. The committee (1) used a qualitative approach to determine whether the watershed can sustain new WWTPs and OSTDS without declines in water quality, (2) evaluated the effluent standards for WWTPs mandated by the MOA and technologies for WWTPs and OSTDS, and (3) considered whether the rules governing the locations of new WWTPs and OSTDS are adequate.

Upgrades to WWTPs mandated by the MOA and the use of best available control technology for OSTDS should be effective in reducing effluent loadings of phosphorus, total suspended solids, coliforms, viruses, *Giardia* cysts, and *Cryptosporidium* oocysts. Although declines in water quality related to WWTPs and OSTDS will occur with population growth, in most cases from 40 to 100 years will pass before contaminant contributions from these treatment systems reach present levels. To arrive at these conclusions, best available control technology for OSTDS was assumed to be aerobic treatment units with essentially a zero failure rate.

Current technologies being used for new and replacement OSTDS in the Catskill/Delaware watershed are not adequate; they do not represent best available control technologies. Implementation of aerobic treatment systems for OSTDS, including a significant enforcement effort, could substantially reduce effluent concentrations of *Giardia*, *Cryptosporidium*, fecal coliforms, and viruses in all Catskill/Delaware watersheds. Therefore, aerobic treatment units should be mandated for new or replacement OSTDS, and enforcement efforts should include annual inspections. This recommendation is especially important for the Kensico watershed, because of its critical location in the water supply and because OSTDS serves a large percentage of the population.

To ensure that the Continuous Backwash Upflow Dual Sand Filtration units being used at WWTPs represent the best available control technology, these units should be subjected to rigorous long-term monitoring to verify that equivalency with microfiltration is maintained. Monitoring of particle counts and turbidity should be conducted to determine the effectiveness of the filtration process and backwashing, and it should detect operational problems that may occur, such as clogging of the filter.

The current 60-day value for siting WWTPs does not appear to be supported by available knowledge. A scientific rationale for such a zone of protection should be developed. Limited research suggests 60 days may be inadequate for significant inactivation of *Cryptosporidium* oocysts and *Giardia* cysts, which are known to be resistant to disinfection, particularly at low temperatures. New York City should undertake further studies of pathogen fate and transport that will refine this value.

SUMMARY

The New York City MOA, which outlines an ambitious program of watershed management, is a unique document in the history of water resource management. The program it advances is a prototype of the utmost importance to all water supply managers. The MOA provides for an extraordinary financial and legal commitment from New York City to (1) prevent existing and potential contaminants from reaching the source reservoirs, (2) monitor a broad range of water quality and drinking water parameters, (3) conduct new research on water-borne health risks and water quality, and (4) promote the economic and social well-being of the Catskill/Delaware watershed communities in an environmentally sustainable manner. The committee is encouraged by the evidence of progress achieved to date in realizing the goals of the MOA. Although its limitations are noted in this report and it is not a guarantee of permanent filtration avoidance, the MOA is a template for watershed management that, if diligently implemented, will maintain and improve source water quality.

1

THE PROBLEM

Nine million people in New York City and nearby areas enjoy access to abundant, clean, and inexpensive drinking water from that city's expansive and internationally admired water supply system. The Croton River system east of the Hudson River, originally constructed in the 1840s, drains 300 square miles and today provides about 10 percent to 12 percent of New York City's water. The other 90 percent is drawn from sources west of the Hudson River in the Catskill Mountains and the headwaters of the Delaware River. Water from this Catskill/Delaware system is collected from nearly 1,600 square miles of watershed land. Altogether, the New York City system provides an average 1.3 billion gallons of drinking water per day (Hazen and Sawyer, 1997).

New York City's drinking water is not filtered.[1] Like Boston, Portland, Oregon, San Francisco, and a few other cities, New York City has relied on the natural purity of its hinterland sources, along with chlorine disinfection, to provide high-quality water without filtration. During the 1990s, that *modus operandi* has been challenged by national public health concerns about disinfection byproducts and microbial pathogens, among other threats. Federal regulations pursuant to the Safe Drinking Water Act[2] (SDWA) now mandate filtration for public surface water supplies. However, this requirement may be waived by the U.S. Environmental Protection Agency (EPA) and state health agencies if a water supplier demonstrates it will "maintain a watershed control program, which minimizes the potential for contamination by *Giardia* cysts and viruses in the source

---

[1] The Croton System of the New York City water supply is under order from the EPA to be filtered.

[2] Safe Drinking Water Act, P.L. 93-523, as amended. 42 USCA, Secs. 300f et seq.

water" (40 CFR 141.71(b)(2)). Although EPA advocates watershed management at many levels and for a variety of activities (EPA, 1996), avoidance of filtration under the SDWA is the only example of a regulation that *requires* watershed management.

Like a handful of other cities with unfiltered water supply systems, New York City has spent considerable energy developing a watershed management program that would allow the Catskill/Delaware supply to remain unfiltered. After years of negotiation, a Memorandum of Agreement (MOA) was signed on January 21, 1997, between the City, communities in the Catskill/Delaware watershed, EPA, New York State, and certain environmental organizations. The MOA launched a massive program of watershed management and collateral payments to watershed interests in exchange for an EPA waiver from filtration effective until April 2002. Under the MOA, the City is implementing watershed management on a scale unprecedented in the United States.

Drinking water supply and other water resource issues have long relied upon the watershed as a basic management unit, and much of the science needed to support watershed-based management exists (NRC, 1999). However, because watershed boundaries rarely coincide with regional or political boundaries, the implementation of watershed management requires much more than scientific and technological progress to be effective. Social, economic, and political considerations that stem from a broad range of stakeholders with disparate interests are a necessary part of watershed management. EPA recognized this fact when it defined watershed management as "a holistic, integrated problem-solving strategy used to restore and maintain the physical, chemical, and biological integrity of aquatic ecosystems, protect human health, and provide sustainable economic growth" (EPA, 1993). As stated by the NRC (1999), "Watershed management is both institutionally and scientifically complex, and thus inherently difficult to implement."

New York City's watershed management strategy fits this characterization of the watershed approach to resource management. The MOA represents a new and complex combination of programs, policies, and management practices that was developed with input from a large number of diverse interest groups. Scientific, technological, social, and economic issues are dealt with throughout the MOA, encompassed by the overarching goal of maintaining the purity of the drinking water supply in the absence of filtration.

This report evaluates the scientific underpinnings of New York City's watershed management strategy and presents conclusions and recommendations that will contribute to its successful implementation. Because of the increasingly important role of watershed management in providing safe drinking water, many of the report's conclusions and recommendations extend beyond New York City and apply generally to surface water supplies elsewhere.

EVOLUTION OF THE NEW YORK CITY
WATERSHED MEMORANDUM OF AGREEMENT

In 1986, Congress amended the SDWA, tightening drinking water standards and generating new provisions for drinking water supplies. To implement this legislation, in 1989 EPA issued the Surface Water Treatment Rule (SWTR), requiring filtration for communities relying on surface water sources. This rule had tremendous fiscal implications for New York City. New York City estimated construction costs for Catskill/Delaware filtration facilities to be as much as $6 billion with annual operating expenses estimated to be more than $300 million (NYC DEP, 1993a; Paden and Shen, 1995). Although an EPA-appointed expert panel disputed the City's cost estimates (Okun et al., 1993), arguing that filtration of the Catskill/Delaware water supply could be achieved for as little as half the City's estimate, the cost of filtration was nevertheless perceived to be excessive.

Quest for Filtration Avoidance

The SWTR provides that under certain conditions, public water supplies may obtain a waiver from filtration. These conditions include creating a watershed management program, meeting standards for turbidity and fecal and total coliforms, providing adequate disinfection, and avoiding any waterborne disease outbreak. Eager to take advantage of this provision, in 1991 New York City began to develop a watershed management program that would maintain the high quality of its drinking water supply. As the agency primarily responsible for delivering safe drinking water, the New York City Department of Environmental Protection (NYC DEP) began drafting a watershed management program that would satisfy the SWTR and allow the city to avoid filtering the Catskill/Delaware supply. Water from the Croton watershed was not thought to be of sufficient quality for that system to qualify for a waiver from filtration as well. The City is currently under an order from EPA to construct a $687 million filtration plant for the Croton portion of the system.

New Watershed Rules and Regulations

The first step in filtration avoidance for the Catskill/Delaware system was improvement of the New York City Watershed Rules and Regulations dating from 1953. In 1990, NYC DEP produced a first draft of revised Rules and Regulations that, among other things, called for maintenance of buffer zones around watercourses and reservoirs and restrictions on siting and construction of sewerage facilities (NYC DEP, 1990). The proposed regulatory changes would have restricted a variety of developments, including the construction of roads, parking lots, and storage facilities for hazardous substances and waste. In addition to the Rules and Regulations, New York City proposed to acquire watershed

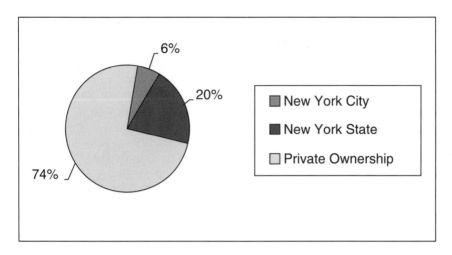

FIGURE 1-1 Land ownership in the Catskill/Delaware watershed.

land. New York City owns only about 6 percent of the Catskill/Delaware watershed, with another 20 percent being part of the New York State Catskill Forest Preserve (Figure 1-1). The other three-quarters of the watershed remain in private ownership and thus may yield contaminants from agriculture and other land use activities. Under the most extreme scenario, the City proposed to protect all developable land in the entire watershed by direct acquisition or conservation easements (NYC DEP, 1993b, p. 246). This scenario was estimated to require $2.7 billion to purchase fee title or easements on about 240,000 acres of vacant land (NYC DEP, 1993a; Pfeffer and Wagenet, 1999).

New York City's proposed actions aroused strong opposition from watershed residents who feared economic development would be stifled, property values would drop, and the local tax base would be eroded (Finnegan, 1997; Schneeweiss, 1997). The Coalition of Watershed Towns (CWT) was organized in 1991 to serve as a voice for 30 watershed towns west of the Hudson River. The CWT demanded the City compensate watershed communities for direct and indirect costs of its watershed protection program. The main goals of the CWT were to ensure the proposed regulations would not prevent reasonable community development and to limit the regulations to the minimum needed to protect water quality (Stave, 1998). The CWT pursued legal action as one means of meeting its objectives and took a lead role in opposing the proposed regulations (Finnegan, 1997).

Filtration Avoidance Threatened

The long-simmering animosity between New York City and watershed communities came to a head in late 1993 when the City (1) filed a state application for a water supply permit including plans to acquire 10,000 acres in the watershed and (2) submitted a long-term watershed protection and filtration avoidance program for the Catskill/Delaware System to EPA. Uncertainty over New York City's intent to use eminent domain to gain control of the land, and the perception that the City was shifting the costs of watershed protection to upstate communities, resulted in the weakening of relations between New York City and upstate communities. This culminated in the CWT and others filing suit to prevent the City from implementing its filtration avoidance plans. The CWT cited economic burdens on watershed residents resulting from restrictions placed on the use of privately owned land, and it claimed New York City would benefit almost exclusively from the regulations. The lawsuits led to an impasse in efforts to reach a compromise about a watershed management plan (Finnegan, 1997; Pfeffer and Wagenet, 1999).

In the face of this stalemate, EPA and others urged New York governor George E. Pataki to intervene by bringing the interested parties to the negotiating table. The initial April 1995 meeting included representatives from New York City, EPA, the CWT, Putnam and Westchester counties, and others. After 18 months of negotiations, the process yielded the New York City Watershed MOA, a mammoth document comprising nearly 1,000 pages of text and attachments. This unprecedented agreement, the legal equivalent of the Hoover Dam, was signed on January 21, 1997, by New York City, New York State, EPA, the CWT, environmental organizations (the Catskill Center for Conservation and Development, the Hudson Riverkeeper, the Trust for Public Land, the Open Space Institute, and the New York Public Interest Group), and some 70 upstate towns and villages. The agreement serves as a blueprint for the City's watershed management strategy for the Catskill/Delaware supply and will cost approximately $1.5 billion over ten years. At the same time, EPA granted New York City an interim filtration avoidance determination (FAD) for the Catskill/Delaware supply until April 2002.

Memorandum of Agreement

The MOA is a landmark agreement in watershed management that recognizes both the importance of preserving high-quality drinking water and the economic health and vitality of communities located within the watersheds. The MOA explicitly states that these goals are compatible and can be met through cooperation and partnership between New York City and the upstate watershed communities. For example, in the spirit of partnership and cooperation, New York City will not acquire land through eminent domain. Instead it will seek to gain fee simple title[3] or conservation easements to water quality sensitive lands through a willing buyer/willing seller process. Upstate towns affirmed their new relationship with New York City by dropping outstanding lawsuits brought by the CWT. Further, all parties to the MOA agreed not to file legal challenges to block any element of the agreement. Key provisions of the MOA are shown in Box 1-1; more detail is provided in Appendix A. The filtration avoidance determination, which was incorporated in its entirety into the MOA, is summarized briefly in Box 1-2.

Watershed Agricultural Program

One important group of stakeholders that was not directly involved in the MOA negotiations was the farming community in the Catskill/Delaware watershed. The proposed Rules and Regulations raised a great deal of concern among watershed farmers, particularly because of potential land use restrictions in the form of setback distances. Farmers estimated that as much as 25 percent of their tillable land would be taken out of production if agriculture was regulated under the MOA (Coombe, 1998). Having observed the transition from agriculture to residential development in the Croton watershed and subsequent declines in water quality, NYC DEP labeled agriculture a preferred land use in the Catskill/Delaware watershed (DelVecchio, 1997). Because agriculture is also a possible source of pollutants, NYC DEP and EPA made arrangements with the farming community to establish a Watershed Agricultural Program in lieu of including agriculture as a regulated activity under the MOA.

The $35.2 million Watershed Agricultural Program was established in 1994 to engage the voluntary participation of farmers and to develop Whole Farm Plans for participating farms. Whole Farm Plans are intended to reduce pollutant loadings by using innovative best management practices (BMPs). Efforts are coordinated via the Watershed Agricultural Council, which has representation from the farming community, NYC DEP, EPA, and state agencies. The program is being implemented in two phases. Phase I established Whole Farm Plans on ten demonstration farms, while Phase II extends to all interested farms within the Catskill/Delaware watershed. As a condition of its continued waiver from filtration, EPA insisted that milestones for the Watershed Agricultural Program be included in the City's filtration avoidance determination.

---

[3] Fee simple absolute is the totality of legal rights in a parcel of real property, including structures and improvements, vegetation, resources below the surface and the air space above it, subject to limitations imposed by federal, state, or local laws.

BOX 1-1
Key Provisions of the New York City Watershed
Memorandum of Agreement

Land Acquisition Program

This voluntary program allows New York City to acquire fee title or conservation easements to vacant water quality sensitive watershed lands on a "willing buyer/willing seller" basis. All titles and conservation easements are held in perpetuity. The City has committed more than $250 million for land acquisition. Areas closer to reservoir intakes and lower in the watershed are given higher priority for acquisition.

Watershed Rules and Regulations

The Watershed Rules and Regulations control a wide variety of pollution sources such as wastewater treatment plants, on-site sewage treatment and disposal systems (septic systems), stormwater runoff, and storage of hazardous materials. The regulations contain minimum treatment requirements for some technologies that control these sources of pollution and specify effluent standards that some of these treatment technologies must meet. In addition, the regulations restrict a variety of activities from occurring in close proximity to watercourses, reservoirs, reservoir stems, controlled lakes, and wetlands. The siting of wastewater treatment plants, septic systems, and storage facilities for hazardous materials, petroleum, and salt and the construction of impervious surfaces are all restricted within "setback" distances from the major waterbodies. Detailed descriptions of the regulations are provided in Appendix A.

Watershed Protection and Partnership Programs

These programs are intended to preserve the economic and social character of watershed communities while maintaining and enhancing water quality. The Catskill Watershed Corporation (CWC), a not-for-profit corporation, was formed to manage these programs for the Catskill/Delaware watershed region. The MOA calls for about $240 million to be allocated for these efforts, which include infrastructure improvements, development, conservation, and education. The MOA also provides funds for Westchester and Putnam Counties to develop a Croton Plan. The objective of this plan is to develop a comprehensive approach to identify significant sources of pollution, to recommend measures for improving water quality, and to protect the character of watershed communities east of the Hudson River. The MOA sets aside approximately $68 million for implementation of projects developed pursuant to the Croton Plans.

BOX 1-2
The New York City Filtration Avoidance Determination

The current filtration avoidance determination for New York City is one of several determinations that EPA has made for the City's Catskill/ Delaware water supply during the 1990s. Each new determination builds upon the tasks and accomplishments of the previous agreement. The City's current FAD expires in April 2002, at which time EPA and the NYS Department of Health will determine whether NYC DEP has complied with all elements of the FAD and whether another determination will be issued.

The FAD lays out hundreds of specific tasks, or deliverables, that NYC DEP must accomplish (EPA, 1997a). Many of these tasks correspond to activities mandated by the Watershed Rules and Regulations, such as upgrading the existing wastewater treatment plants. The general task categories are listed below:

- Compliance with objective criteria of the SWTR
- Design of a filtration facility for the Catskill/Delaware system
- Land acquisition
- Data gathering and Geographic Information System (GIS) development
- Multi-tiered water quality modeling
- Maintenance of the Watershed Agricultural Program
- Kensico Reservoir modeling and remediation efforts
- Nonpoint source pollution control
- Whole community planning
- Repair, replacement, and upgrade of septic systems
- Upgrade of wastewater treatment plants
- Non-typical activities
- Activities reports
- Distribution system activities
- Active disease surveillance
- Administration

KEY ISSUES IN NEW YORK CITY AND OTHER WATER SUPPLIES

The central focus of this study is an evaluation of New York City's strategy for maintaining its high-quality drinking water through active management of its water supply watersheds. As described below, the scope of the study is broad, the issues can be addressed at several levels, and the answers involve both science and value judgments. Although the committee was asked to address several specific scientific and technical issues regarding the MOA, these issues must be considered within the changing land use patterns and economic growth of the watershed region. The committee's findings and recommendations should enable the City, the State, and the watershed communities to make changes in the way management practices are being implemented in order to increase their effectiveness.

During the course of the study, the committee identified three particularly noteworthy issues that will shape the future of watershed management in water supplies across the country, both filtered and unfiltered: (1) priority setting in watershed management, (2) watershed management and economic development, and (3) filtration and other treatment options. The report revisits these issues in Chapter 12 and provides suggestions for their resolution in New York City. However, the applicability of these recommendations goes beyond New York City to other major metropolitan surface water supplies, including those that are currently unfiltered (e.g., Boston, Massachusetts; part of San Francisco, California; Seattle, Washington; Portland, Oregon; Portland, Maine; and Greenville, South Carolina).

Setting Priorities in a Watershed Management Strategy

Like most human activities, the consumption of drinking water involves certain inherent risks. Pathogenic microorganisms and chemicals that adversely affect human health are sometimes found in drinking water and pose a risk that is directly related to their concentration. In addition, poor water quality in drinking water reservoirs can pose a direct risk to the health of fish and other ecological receptors. In light of current scientific understanding of the relationships between water quality and human and ecological health, a central question for unfiltered supplies is whether a strategy based on watershed management without filtration can achieve a water supply with an acceptable level of risk to public health and the environment.

For this reason, it is often an overarching (though sometimes unstated) goal of watershed management to achieve the greatest overall risk reduction through allocation of resources among various watershed management components. In order for this to be achieved, prioritization must take place on multiple levels within a watershed management strategy. First, watershed management must be directed at the most polluting activities within a watershed. Currently, there are limited tools available for assessing the relative impacts of different sources of pollution and for mitigating their impacts. Confronted with this reality, water resource managers must be cautious when weighing the importance of different land uses and activities on overall reservoir water quality.

In addition to targeting the most polluting activities, watershed management must also consider the relative impact of various pollutants. Like many other eastern water supplies, New York City is focusing heavily on phosphorus as the pollutant of concern because of its role in eutrophication and the resultant formation of disinfection byproducts. This focus may direct resources away from potentially more significant pollutants, such as microbial pathogens. In addition, future regulations may direct more attention to other pollutants.

The difficulties inherent in guarding against multiple drinking water pollutants have long been recognized by EPA and were a primary motivation for amending the SDWA. In the future, water supplies will be expected to protect against microbial pathogens, toxic organic and inorganic compounds, and disinfection byproducts in the face of increasingly strict standards for each. This challenge is especially acute for water supplies that rely on chlorine disinfection as their sole treatment process. Chapter 12 discusses the issue of prioritization during watershed management and provides suggested new program directions for New York City that may further decrease overall risk.

Watershed Management in Relation to Economic Development

Watershed management will affect, for better or worse, the future economic growth in the region. This was recognized during the MOA negotiations, with the result that numerous programs were created to help maintain the watershed economy. The strategy taken by the MOA of compensating watershed residents for the imposition of watershed regulations by investing in their economy poses several significant questions. Will this type of monetary investment in the watershed region lead to development that adversely affects water quality? Can such a strategy reduce the impact of future population growth on water quality? Should this type of investment be used in regions with high population growth rates? These questions are difficult to answer unless the contribution of different land uses to overall pollution can be quantified. Nevertheless, the report describes how these issues could affect future watershed management in New York City and other communities considering similar strategies.

Filtration, Other Treatment Options, and Watershed Management

EPA views filtration and watershed management as two forms of "treatment" that are mutually supportive. In fact, EPA strongly supports a multiple-barrier approach in which filtration, watershed management, and other treatment processes are used to arrive at the cleanest possible drinking water (EPA, 1997b). However, most interested stakeholders in New York City perceive a distinct tradeoff between filtration and watershed management, primarily due to budgetary constraints (Cronin and Kennedy, 1997, p. 208; Goldstein, 1997; Marx and Goldstein, 1999). There is concern in New York City and other places that the adoption of a strategy that incorporates both filtration and watershed management would inevitably lead to a loss of public commitment to the high level of environmental conservation reflected in the present strategy of watershed management and chlorine disinfection (Goldstein, 1997; Howe, 1999). This perceived conflict, and its implications for future watershed management in New York City and other unfiltered supplies, is revisited in Chapter 12. Boxes 1-3, 1-4, and 1-5 describe how this conflict has developed in three other unfiltered water supplies: Boston, Seattle, and Greenville, South Carolina.

NATIONAL RESEARCH COUNCIL STUDY

The January 1997 MOA included several changes that were introduced by the New York City Office of the Comptroller. The Comptroller was not a party to the negotiation process that produced the MOA, but his signature of approval was necessary for the agreement to be finalized, partly because of the many lawsuits that were being terminated as a result of the MOA. The Comptroller objected to a number of provisions in the public draft released in late 1996, including (1) no apparent limit of the number of new wastewater treatment plants (WWTPs) in the watersheds, (2) the transfer of primacy from EPA to the New York State Department of Health within five years, and (3) the transfer of millions of dollars to upstate communities with "inadequate" oversight that might be used to further pollute the watershed (New York City Office of the Comptroller, 1997).

Although not all of the Comptroller's objections were alleviated in the second draft of the MOA, several suggested changes were made. EPA agreed to retain primacy for ten years rather than five, and greater oversight was provided for spending of funds in the upstate communities. The Comptroller requested funding for an independent scientific review of the MOA, resulting in this National Research Council (NRC) study. The completion of the study was planned to coincide with a midterm EPA review of the MOA, also a condition imposed by the Comptroller.

Study Statement of Task and Report Organization

The MOA is the focal point of this NRC study. Various management practices, programs, and policies embodied in the MOA were reviewed in response to specific requests from the study sponsor, including prescribed setback distances, the pilot phosphorus offset program, and antidegradation policy. During its two-year tenure, the committee identified other aspects of the watershed management strategy that also warranted consideration, and it has included these in the report for consistency and completeness. The committee's work plan addressed issues in the following areas:

- Evolving Safe Drinking Water Act
- Enhanced monitoring program and Geographic Information System (GIS)
- Public health protection

BOX 1-3
Boston Water Supply

Metropolitan Boston consists of over 100 cities and towns with a population of about 3.2 million in 1990. Situated on a deepwater harbor and surrounded by brackish estuaries, Boston has historically turned to inland sources for fresh water. Starting in the 1840s, following the model of New York City, Boston developed a series of reservoirs and aqueducts to draw water from its hinterlands. The main reservoirs in the regional water supply include the Wachusett Reservoir, 40 miles west of Boston, and Quabbin Reservoir, 70 miles west of the city. Today, Quabbin water flows to Wachusett through a 30-mile tunnel, and then into the metropolitan distribution system. One of the world's largest water supply reservoirs, Quabbin now provides about three-quarters of the Boston metropolitan water supply (Platt and Morrill, 1997). The system is jointly managed by the Metropolitan District Commission (MDC) and the Massachusetts Water Resources Authority (MWRA).

Like New York City, the metropolitan Boston system has long provided high-quality water without filtration. In both systems, land use change in privately owned portions of source watersheds threatens to reduce water purity in the future. Quabbin, whose watershed is nearly 80 percent in public ownership, has received a filtration avoidance determination (FAD) from EPA. But all Quabbin water must pass through Wachusett Reservoir, whose watershed is still predominantly (75 percent) in private ownership and which lies on the edge of the developing urban fringe of Worcester, MA. EPA has not awarded a FAD for Wachusett and has sued the MWRA and the state (in March 1998) to require a filtration plant to be constructed downstream of Wachusett. However, under a 1992 consent agreement with the state, with which EPA previously concurred, MWRA is pursuing watershed management for Wachusett while simultaneously designing the filtration plant (the dual-track approach also used by New York City). MWRA hopes to demonstrate that its watershed management program will meet EPA's requirements and justify a FAD for the Wachusett Reservoir (MDC/MWRA, 1991). A final decision is set for October 1999. Meanwhile, the MWRA is committed to installing ozonation in place of total reliance on chlorination. No *Cryptosporidium* oocysts have been detected in the Boston water supply (Eileen Simonson, WSCAC, personal communication, July 1, 1999).

Key elements of the Watershed Protection Plan for Wachusett include (1) septic system upgrades, (2) sewage treatment expansion,

continued

BOX 1-3 Continued

(3) nonpoint source pollution control, (4) land acquisition, (5) replacement of underground storage tanks, and (6) bird harassment to reduce fecal coliform levels. In 1992, the state legislature adopted the Watershed Protection Act (Mass. General Law 1992, Ch. 36), which mandates setbacks for new development along reservoirs and their tributaries (Hutchinson, 1993). In addition, since 1988, MWRA has succeeded in reducing per capita water demand by at least 16 percent through leak detection and repair, free retrofit of domestic plumbing devices, higher water rates, and public education. The MWRA demand management experience has been cited by Worldwatch Institute as an internationally significant success in sustainable water resource management (Postel, 1992).

BOX 1-4
Seattle Water Supply

Seattle has two surface water supplies: the Tolt Reservoir (one-third of the supply) and the Cedar Reservoir (two-thirds of the supply). All of the Cedar watershed is owned by the city and is undeveloped. The Tolt watershed is 75 percent city-owned, with the rest being owned by the USDA Forest Service. In both watersheds some logging has occurred, but it has recently ceased in the Tolt watershed. The combined systems serve 1.3 million customers.

Seattle received waivers from filtration for both water supplies in 1990, which are still in effect. The city has violated the Surface Water Treatment Rule once in the Cedar watershed. In 1992, fecal coliform standards were exceeded, which led to an automatic requirement for filtration. Seattle argued that this requirement is too strict because fecal coliforms are not themselves pathogenic, but are only indicators. EPA agreed that ozonation could be substituted for filtration. As a result, the Cedar watershed is having an ozonation plant built, and the Tolt is having both a filtration and an ozonation plant built. The projected costs for design and construction are $70 million.

Seattle has been preparing for filtration of the Tolt since 1989 for three main reasons:

1. Changing regulations meant that the city would probably not be able to meet the proposed disinfection byproducts standards. The source

continued

BOX 1-4 Continued

water is high in dissolved organic carbon from lignin and tannin, and Seattle did not think it would be able to comply with the proposed halo-acetic acid standard.

2. Seattle wanted the system to be reliable, even in the event that a water quality fluctuation upstream in the reservoir should occur. In particular, the Tolt system is susceptible to high levels of turbidity. Although turbidity in the main Tolt reservoir typically ranges from about 0.5 to 4.0 NTU, spot tests have found concentrations as high as 15 NTU during such conditions as low reservoir water levels, high winds, or heavy rains. During these times, the supply must be taken out of service.

3. The utility wanted to be able to draw the Tolt Reservoir down to obtain an additional increment of supply (approximately 9 mgd). Currently, the Tolt supply experiences fluctuating water quality caused by changing weather patterns. The water utility would like to be able to use more of this supply regardless of the weather conditions, which, to Seattle, means having a treatment process downstream that can control for variations associated with weather.

At full capacity, the 120 mgd filtration plant will be able to treat source water of up to 15 NTU of turbidity. It should provide 5 logs removal of *Cryptosporidium* as well as provide distributed water with less than 64 µg/L total trihalomethanes and less than 48 µg/L haloacetic acids when free chlorine is used for disinfectant residual maintenance.

Source: Adapted from Kelly et al. (1998).

BOX 1-5
Greenville, SC, Water Supply

Drinking water for 250,000 people in Greenville, South Carolina, is supplied by three watersheds: Table Rock, North Saluda, and Lake Keowee. The Greenville Water System (GWS) manages these sources and owns all of the 28,000 acres within the Table Rock and North Saluda watersheds. Table Rock and North Saluda are unfiltered supplies that have a combined capacity of about 15 billion gallons and a total safe yield of about 60 mgd. The quality of water from these two sources has been labeled as "two of the very best, if not the very best, in the United States"

continued

BOX 1-5 Continued

(Okun, 1992). In 1993, the watersheds were placed in a conservation easement with The Nature Conservancy.

On December 17, 1991, the South Carolina Department of Health and Environmental Control informed GWS that filtration must be provided for both the Table Rock and North Saluda sources. The reasons for this determination included failure to comply with total coliform standards between April and September 1991, a lack of redundant systems at the disinfection plants, and lack of a watershed control program. In defending the quality of its drinking water, GWS argued that its watershed was wholly owned by the water supplier and that enforcement activities prevented virtually any impacts on the watershed. In addition, GWS outlined a watershed protection and control program targeted at the only known sources of pollution: wildlife, runoff from two state highways, and acid rain.

In 1992, 1993, and 1994, GWS tests revealed additional violations of the Total Coliform Rule (although tests conducted by the South Carolina Department of Health and Environmental Control did not detect coliforms). GWS traced the source of the coliforms to biofilms within the pipes of the distribution system. This prompted EPA Region 4 to insist that plans for filtration of the supplies be accelerated. A detailed schedule for construction of the filtration plants was outlined, including penalties for any violations of the Administrative Order. GWS argued that flushing of the distribution system pipes would be the best method for combating the sources of coliform bacteria and that filtration, because it is upstream of the distribution system, would do nothing to reduce risk from these bacteria. Although GWS charged EPA with several factual errors, it decided not to contest EPA's order in the interest of time and money (GWS, 1995).

The project, consisting of coagulant addition, flocculation, dissolved air flotation, and anthracite mono-medium filtration, is well under way, and filtration should be ready for operation by the end of 1999, in accordance with the Administrative Order. The plant is expected to cost $75 million plus $4 million in annual operating costs. GWS predicts that water rates may be raised by as much as 13 percent to support the project (Thompson, 1993). Meanwhile, according to GWS, flushing of the distribution system has so far proven effective in eliminating detection of total coliform bacteria in the system (Gladfelter, 1995).

- Comprehensive planning efforts
- Total Maximum Daily Load program
- Pilot phosphorus offset program
- Antidegradation
- Use of the dual-track approach
- Buffer zones and setback distances
- New sewage treatment plants and septic systems

The first five chapters of this report provide background information needed to interpret and assimilate the conclusions and recommendations found in the final seven chapters. Chapter 2 describes the creation of the New York City water supply system and its current configuration and treatment processes, using chronological tables, maps, and system schematics. The chapter closes by summarizing salient biophysical and social characteristics of the upstate watershed region. Chapter 3 outlines the relevant federal, state, and local environmental laws controlling watershed management in New York City and the global public concerns that led to the creation of these laws. Chapter 4 discusses the essential components of an ideal watershed management strategy for source water protection. Although all components are rarely realized in most communities, they are described in detail for comparison to the MOA. Chapter 5 summarizes present-day environmental conditions within the Catskill/Delaware watershed region. Priority pollutants and potential sources of pollution are described. In addition, the current health of the watershed and the water supply are considered by evaluating water quality and compliance monitoring data. This chapter represents a transition between the report's introductory material and its evaluation of various scientific aspects of the MOA. The content of the final seven analytical chapters, organized by topic, is described below. For those readers who turn directly to Chapter 6, the Executive Summary provides a highly abridged version of the background information found in Chapters 1–5.

Implications of the Safe Drinking Water Act Amendments

The SWTR pursuant to the SDWA provides the regulatory framework that the MOA must satisfy to maintain New York City's waiver from filtration. Changes in the SWTR that may affect the City's waiver are likely to be promulgated within the next two years. Chapter 5 compares current New York City compliance data with projected future regulations in order to assess the City's ability to comply with projected regulatory changes. Special consideration is given to the Enhanced SWTR and the Disinfectants/Disinfection By-Products Rule.

Enhanced Monitoring Program and GIS

NYC DEP maintains a sophisticated monitoring system for measuring changes in water quality throughout the watershed region. Already the subject of external reviews, the adequacy of the enhanced monitoring program is considered in Chapter 6. The chapter builds upon previous studies by making specific suggestions for improving sampling techniques, the timing and numbers of samples taken, and the parameters measured. The different monitoring efforts are framed in terms of the goals they attempt to achieve.

To complement its enhanced monitoring program, NYC DEP is currently organizing its monitoring data for use in a geographical information and modeling system (GIS). Chapter 6 provides recommendations for improving the quality, usefulness, and accessibility of the GIS. Its relationship to other programs within the MOA is explored.

Public Health Protection and Microbial Risk Assessment

Because the end result of watershed management and water supply protection should be the protection of public health among consumers of the drinking water supply, this issue was considered by the committee from multiple viewpoints. First, active disease surveillance being conducted in New York City is evaluated. Chapter 6 considers whether there is adequate data collection on public health outcomes associated with drinking water quality, and it provides recommendations for linking observed gastrointestinal illness to potential drinking water sources. In addition, a quantitative microbial risk assessment is conducted using pathogen monitoring data collected at the source waters of the Catskill/Delaware water supply system. This information reveals potential disease rates that cannot otherwise be ascertained from active disease surveillance. It also highlights data analyses that could be used by New York City to define a baseline level of acceptable risk.

Comprehensive Planning

In order for the counties and municipalities in the watershed to receive certain funds under the Watershed Partnership and Protection Programs, they must develop a cooperative, comprehensive land use plan for water quality protection and growth management. Chapter 7 considers the multiple comprehensive planning efforts that have commenced in both the Catskill/Delaware and Croton watershed regions. These efforts are evaluated for their potential ability to promote environmentally sound economic development within the watersheds.

Total Maximum Daily Load Program

The Total Maximum Daily Load (TMDL) program stems from the Clean Water Act, which requires impaired waters nationwide to be assessed for point and nonpoint source pollutant loading. Although not all are classified as impaired, the 19 water supply reservoirs in the New York City watersheds have been the subject of TMDL calculations for the priority pollutant phosphorus. Chapter 8 reviews the methods used for Phases I and II of this program and provides suggestions for improvement during Phase III. This chapter considers whether phosphorus is the appropriate target of the TMDL calculation and whether existing guidance values for phosphorus should be used.

Phosphorus Offset Pilot Program

One of the most innovative programs found in the MOA is the phosphorus offset pilot program that allows for construction of new wastewater treatment plants (WWTPs) in polluted watersheds. The construction of new WWTPs, or the expansion of existing WWTPs, is allowed if an offset in phosphorus loading can be identified from another source within the same basin. Either a 2:1 or a 3:1 offset ratio must be achieved. This program, and its potential for reducing overall pollutant loading in the watersheds, is evaluated in depth in Chapter 8. The few documented cases of effluent trading are compared to New York City's program to assess potential reliability and enforcement. The chapter provides recommendations on how to improve the program if it is to be implemented at full scale.

Antidegradation

Federal regulations require states to create and implement an antidegradation policy that will prevent further deterioration of water quality in all waterbodies. Although not encompassed by the MOA, antidegradation could be used to enhance the protection currently afforded to the New York City water supply reservoirs. Chapter 8 compares the New York State antidegradation policy to similar policies of other states, focusing on its specific provisions and implementation. The adequacy of New York's antidegradation policy is assessed by evaluating two state regulations used to carry out antidegradation: the State Environmental Quality Review Act and the State Pollutant Discharge Elimination System (SPDES) permitting program.

Dual-Track Approach

As part of its filtration avoidance determination, New York City must simultaneously plan for the construction of a filtration plant while implementing the watershed protection program. Chapter 8 considers the merits of this "dual-track" approach and compares it to the multiple-barrier approach strongly endorsed by EPA and the American Water Works Association. Pilot filtration studies and studies on alternate disinfectants are both reviewed.

Buffer Zones and Setback Distances

One of the committee's most challenging tasks was to assess the capability of setback distances for improving water quality in the water supply reservoirs. Setback distances ranging from 25 feet to 1,000 feet are evaluated in Chapter 10 for their ability to remove such pollutants as phosphorus, sediment, pesticides, and landfill leachate. Several suggestions are given for managing setback areas to achieve the greatest pollutant removal possible. In addition, the constitutionality of requiring setbacks, and the use of setbacks in other states, is presented for comparison with the MOA.

Setback distances and buffer zones represent only one specific type of BMP used to control nonpoint source pollution. Thus, Chapter 9 considers the wide variety of BMPs used throughout the watershed region and their implementation. The discussion is organized around three important programs designed to limit nonpoint sources of pollution: the Watershed Agricultural Program, the Watershed Forestry Program, and Stormwater Pollution Prevention Plans.

New Sewage Treatment Plants and Septic Systems

Because of its potential for introducing pollutants into the watershed, wastewater is the focus of multiple programs under the MOA. The MOA imposes technology and effluent standards and siting requirements for new WWTPs as well as siting requirements for septic systems, also known as on-site sewage treatment and disposal systems (OSTDS). Chapter 11 analyzes the relative impact of new WWTPs and OSTDS on overall water quality under current and future conditions. In addition, it considers the suitability of rules that govern the locations of these wastewater treatment systems. The chapter provides recommendations regarding technology requirements for WWTPs and OSTDS, and it provides an evaluation of the adequacy of effluent standards for WWTPs.

In making its assessments of the MOA, the committee attempted to consider long-, mid-, and short-term impacts to the water supply system. Important long-term considerations include (1) the discovery of scientific information relating various physical, chemical, and biological components of water quality to human health, (2) evolving federal public policies and regulatory frameworks, (3) the changing demand for water from these watersheds over time, (4) the value of preserving natural ecosystems within the watersheds for water quality-related and other reasons, and (5) the pressures of economic change within the watersheds. Midterm events that can substantially affect the water supply system and the watershed management strategy include hydrologic conditions such as sustained drought and fluctuations in wastewater discharge caused by the large seasonal population of the Catskill/Delaware watershed region. Finally, short-term impacts that must be taken into account include severe storm events, possible terrorist activities, and accidents such as chemical spills. These considerations define special management challenges that may be materially affected by the strategy chosen to protect the drinking water supply.

Transferability of Report Contents

Although no other community and its water supply is quite like New York City, information within this report applies to other communities that use surface water sources for drinking water supplies. Watershed management for protecting surface water supplies (source water protection) is valuable as a first barrier, regardless of the treatment technologies that are subsequently applied. Chapter 4 provides guidance to help communities implement various elements of a source water protection program, a "how to" for source water protection that has not been presented elsewhere. Most watershed management publications focus on regulatory requirements or on case studies of how communities have gone about developing source water protection programs. In this report, more general advice and guidance are given for communities embarking on similar programs.

Many of the ideas presented in this report have relevance for a wide range of conditions that various water supplies might face, including pristine supplies that are wholly owned and protected by the surface water supplier, largely uncontrolled watersheds, and transbasin watersheds that are far removed from the water supplier's jurisdiction. The key lessons and areas of information within the report that are particularly transferable are as follows:

- Chapter 4 presents a generic framework for watershed management for source water protection, discussing several necessary components. Two components that are often overlooked are (1) goal setting to provide the foundation for a sound program and (2) effectiveness monitoring to measure the success of programs. Examples of full-scale implementation mechanisms are also presented.
- A detailed description of potential drinking water quality constituents of concern in the New York City watersheds is found in Chapter 5. These constituents are common pollutants in other surface water supplies as well.
- Chapter 6 discusses the elements of a monitoring program to support source water protection, including means to check program effectiveness.
- Innovative policies to support source water protection efforts across the country are explored and reviewed in Chapter 8, such as effluent trading and the TMDL concept.
- Nonpoint source practices for source water protection, including recommended programs for septic system improvements to protect drinking water supplies, are discussed extensively in Chapters 9–11.

REFERENCES

Coombe, R. 1998. Watershed Agricultural Council. Presentation given at the third NRC Committee Meeting, May 13–16, 1998, Oliverea, NY.

Cronin, J., and R. F. Kennedy. 1997. The Riverkeepers. New York, NY: Scribner.

DelVecchio, J. R. 1997. The Agricultural Program to Protect the Drinking Water of New York City. In Proceedings of the 5th National Watershed Conference. Reno, NV.

Environmental Protection Agency (EPA). 1993. The Watershed Protection Approach. EPA 840-S-93-001. Washington, DC: EPA.

EPA. 1996. Watershed Approach Framework. Washington, DC: EPA.

EPA. 1997a. New York City Filtration Avoidance Determination. New York, NY: EPA.

EPA. 1997b. Notice of Data Availability ESTWR. Washington, DC: Office of Water, EPA.

Finnegan, M. C. 1997. New York City's watershed agreement: A lesson in sharing responsibility. Pace Environmental Law Review 14:577–644.

Gladfelter, Melinda. 1995. Water System Ordered to Monitor for Illnesses. The Greenville News.

Greenville Water Supply (GWS). 1995. Watershed Protection Program. Greenville, SC: Greenville Water System.

Goldstein, E. 1997. Natural Resources Defense Council. Presentation given at the first NRC Committee Meeting, September 25–26, 1997, New York, NY.

Hazen and Sawyer/Camp Dresser and McKee. 1997. The New York City Water Supply System. New York, NY: Hazen and Sawyer/Camp Dresser and McKee.

Howe, P. J. 1999. Judge Orders Trial in Water Quality Dispute. The Boston Globe. Friday, May 7, 1999.

Hutchinson, D. P. 1993. A Setback for the Rivers of Massachusetts? An Application of Regulatory Takings Doctrine to the Watershed Protection Act and the Massachusetts Rivers Protection Act. Boston University Law Review 73:237–270.

Kelly, E. S., S. Haskins, and P. D. Reiter. 1998. Implementing a DBO project. Journal of the American Water Works Association 90(6):34–46.

Marx, R., and E. A. Goldstein. 1999. Under Attack: New York's Kensico and West Branch Reservoirs Confront Intensified Development. New York, NY: Natural Resources Defense Council.

Metropolitan District Commission/Massachusetts Water Resources Authority. 1991. Watershed Protection Plan: Wachusett Reservoir Watershed. Boston, MA: MDC/MWRA.

National Research Council (NRC). 1999. New Strategies for America's Watersheds. Washington, D.C.: National Academy Press.

New York City Department of Environmental Protection (NYC DEP). 1990. Discussion Draft of Proposed Regulations for the Protection from Contamination, Degradation and Pollution of the New York City Water Supply and Its Sources. Corona, NY: NYC DEP.

NYC DEP. 1993a. Draft Generic Environmental Impact Statement for the Draft Watershed Regulations for the Protection From Contamination, Degradation, and Pollution of the New York City Water Supply and Its Sources. Corona, NY: NYC DEP.

NYC DEP. 1993b. Watershed Protection through Whole Community Planning: A Charter for Watershed Partnership. Ithaca: New York State Water Resources Institute, Center for the Environment, Cornell University.

New York City Office of the Comptroller. 1997. Avoiding Disaster: The Need to Make the Upstate-Downstate Watershed Protection Partnership Work. New York, NY: New York City Office of the Comptroller.

Okun, D. 1992. Remarks on the Greenville Watersheds. In Properties of the Table Rock and Point Set Reservoirs: Their Future. Greenville, SC: Greenville Watershed Study Committee.

Okun, D. A., G. F. Craun, J. K. Edzwald, J. R. Gilbert, E. Pannetier, and J. B. Rose. 1993. Report of the Expert Panel on New York City's Water Supply. Washington, DC: EPA.

Paden, C., and A. Shen. 1995. New York City water under pressure. Inside DEP 1(1):1–8.

Pfeffer, M. J., and L. P. Wagenet. 1999. Planning for Environmental Responsibility and Equity: A Critical Appraisal of Rural/Urban Relations in the New York City Watershed. Pp. 179–205 in Lapping, M. B., and O. Furuseth (eds.), Contested Countryside: The Rural Urban Fringe of North America.

Platt, R., and V. Morrill. 1997. Sustainable water supply management in the United States: Experience in metropolitan Boston, New York, and Denver. Pp. 292–307 In Shrubsole, D., and B. Mitchell (eds.) Practicing Sustainable Water Management: Canadian and International Experiences. Cambridge, Ontario: Canadian Water Resources Assn.

Postel, S. 1992. Last Oasis: Facing Water Scarcity. New York and London: W. W. Norton.

Schneeweiss, J. 1997. Watershed protection strategies: A case study of the New York City Watershed in light of the 1996 Amendments to the Safe Drinking Water Act. Villanova Environmental Law Journal. 9:77–119.

Stave, K. A. 1998. Water, Land, and People: The Social Ecology of Conflict over New York City's Watershed Protection Efforts in the Catskill Mountain Region, NY. Ph.D. Dissertation, Yale University.

Thompson, S. 1993. Panel Considers 13% Water Hike. Greenville Piedmont. October 11, 1993.

2

The New York City
Water Supply System

New York City between the 1840s and the 1960s developed the largest and, some would argue, the best urban water supply system in the world in terms of quality, reliability, and innovative management. The 1997 Memorandum of Agreement (MOA) reflects a new era of creative management in response to the dual realities posed by (1) the need to comply with the federal Safe Drinking Water Act (SDWA) and (2) the unavailability of new sources to augment or replace existing supplies. This chapter sketches the historical evolution of the New York City water system in relation to socioeconomic growth of the city during the nineteenth and early twentieth centuries. It then describes the basic physical elements of the system as it exists today, together with the biophysical geography of the Catskill/Delaware headwaters from which 90 percent of the City's water is derived. This chapter thus provides the background and context essential to the detailed examination of the MOA that follows in later chapters.

BRIEF HISTORY OF THE NEW YORK CITY WATER SUPPLY

At the dawn of the nineteenth century, American cities were few in number, small in size, and coastal in location. Infrastructure inherited from the colonial period was primitive even by the standards of a half century later: streets were crooked and unpaved, public buildings were spartan, street lighting was rare, waste collection was practically unknown, and water supplies were totally inadequate. With immigration, industrialization, and the growth of an urban middle class, the population of towns began to swell in the early decades of the century. New York City in particular grew from 60,000 to 200,000 between 1800 and 1830. This rapid growth in population was accompanied by a succession of epidemics, vaguely understood to be related to impure water, as well as frequent outbreaks of fires that could scarcely be contained with available water supplies.

New York City was surrounded by tidal, brackish water with no immediate access to fresh water streams. Residents depended initially upon wells or rainwater cisterns for their water needs. The water table aquifer on which it depended was easily contaminated with surface wastes. Wells close to the shore became brackish because of saltwater intrusion. And with limited surface recharge of local groundwater, the reliable yield of springs and wells was insufficient. Rainwater cisterns added little to the general supply. Adding further to water demand was the patenting of the flush toilet in 1819, which greatly increased water consumption per capita as waterborne sewerage gradually replaced on-site privies and night soil collection (Weidner, 1974, p. 55).

During the early nineteenth century, the provision of urban water supply was regarded as a private rather than a public function (Blake, 1956). New York, Boston, Baltimore, and several small towns relied initially on enfranchised private companies in preference to assuming the burden directly. An exception was Philadelphia, where recurrent outbreaks of yellow fever at the turn of the nineteenth century prompted a more aggressive municipal response. In 1801, Philadelphia constructed at public expense a pumping plant on the Schuylkill River powered by two steam engines. This project was designed and promoted by the noted engineer Benjamin Latrobe. It marked both a technological and an institutional breakthrough, namely in the use of the steam engine to pump water and the use of public taxation to establish a municipal water supply (Blake, 1956). In the case of Boston, the Jamaica Pond Aqueduct Company was chartered in 1796 to bring water to that city through a series of hollow log pipes.

In New York City, the Manhattan Water Company was chartered in 1799 with an exclusive franchise to supply the city with water. It constructed a reservoir in lower Manhattan to supply 400 families from local groundwater.

> But this water proved both scarce and bad; the company, neglecting the ostensible purpose of its organization, soon turned its attention almost exclusively to banking affairs and thus lost the confidence of the community, and it was not long before the new works were voted a failure. (Booth, 1860)

In 1811, a plan for the future expansion of New York was prepared by a special commission established by the state legislature. The "Commissioners' Plan" projected future streets marching miles into the countryside of upper Manhattan as far as "155th Street." The plan was an accurate forecast of the spatial growth of the city. The opening of the Erie Canal in 1825, which connected the Hudson River with the Great Lakes, established the City's economic preeminence in the nation and contributed to its rapid population growth and prosperity. Local water sources were hopelessly inadequate to serve this rapid rate of growth in terms of quantity, quality, and pressure. Various schemes were debated fruitlessly, with many in favor of a nearby diversion from the Bronx River just outside the city limits. This source, however, could not meet the prospective needs of the expanding city for long (Blake, 1956).

Damming the Croton River

The solution of the New York City water crisis ultimately required a synthesis of innovation in technology, in public administration, and in civic responsibility previously unknown in urban history. These factors, in conjunction with growing public desperation and fear, contributed to a municipal achievement that still today is the envy of other cities worldwide. In particular, public action to establish a water supply could no longer be debated or delayed after the city was wracked by fires in 1828 and 1835 and by cholera in 1832. The City retained an engineer, Colonel DeWitt Clinton, Jr., to study the water crisis and propose a solution. He predicted that Manhattan would reach a population of 1 million by 1890 (which proved to be late by 12 years). To meet the crisis, he proposed tapping the Croton River 40 miles north of the city to obtain a reliable supply of 20 million gallons per day (mgd) of pure upland water, a project of stunning simplicity in concept but daunting in terms of cost and engineering challenge. The elevation of a Croton River reservoir at 200 ft above sea level would permit the water to flow by gravity through an aqueduct to be constructed with enough "head" to serve the needs of taller buildings and fire fighting in the city (Weidner, 1974, pp. 28–31). To reach far beyond the city limits would ensure (at the time) that water would be relatively pure and that private landowners would be powerless to object to the diversion of local stream water to the city. The project also attracted the interest of civic leaders and politicians in both city and state government. (The federal government had no role in the project whatsoever.)

The Croton River project required the construction of storage and conveyance facilities unprecedented since the Roman Empire. With the total cost estimated at several million dollars, the project was considered too large and too important for private enterprise. Accordingly, the City of New York, under authority from the state legislature, undertook to plan and execute the Croton River project directly. A water commission was quickly appointed, financing was approved by the City's voters in 1835, and construction began in 1837 (Blake, 1956).

The project involved five major structural elements: (1) a masonry dam 50 ft high and 270 ft long impounding a reservoir with a surface area of 440 acres and a storage capacity of 600 million gallons, (2) a 40-mile covered masonry aqueduct with a cross section of seven by eight ft, (3) a 1,450-ft-long "high bridge" to convey the aqueduct across the Harlem River into Manhattan, (4) a 35-acre receiving reservoir located within the future site of Central Park, and (5) a four-acre, masonry-walled distributing reservoir located on the present site of the New York Public Library at Fifth Avenue and 42nd Street. The first Croton River

Croton Dam. Source: The Hudson (Lossing, 1866. © 1866 by H.B. Nims & Co.).

water arrived in Manhattan on July 4, 1842, an event celebrated with church bells, cannon, and a five-mile-long parade. The event was both a technological and an institutional threshold: the City of New York had arrived of age (Weidner, 1974, pp. 45–46; Platt, 1996, p. 187). Within a decade, the City could celebrate another milestone—the opening of Central Park—which contained the key receiving reservoir for the Croton system. A map of the current Croton water supply watershed is given in Figure 2-1.

Evolving Public Health Laws

During this time, two important public health laws relating to drinking water safety were being developed in New York State that would make continued expansion of the New York City water supply possible. First, the New York Metropolitan Health Act was adopted by the New York State legislature in 1866 as the first major American public health law. It was directly inspired by the findings of English sanitary reformer James Chadwick, whose 1842 "Report of

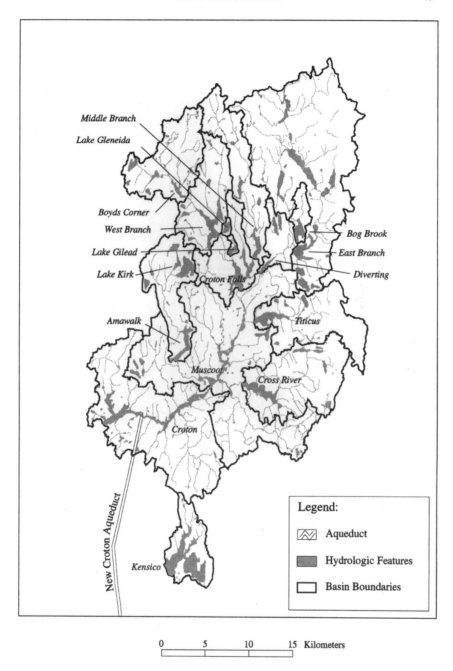

FIGURE 2-1 The Croton water supply system. Courtesy of the NYC DEP.

the Poor Law Commissioners Concerning the Sanitary Condition of the Labouring Population of Great Britain" prompted a parallel investigation in New York City by John Griscom in 1845. These reports began to convince the informed public that the incidence of infectious disease such as cholera and typhoid was closely related to the purity and abundance of the water supply.

The second important piece of legislation was the New York State Public Health Law of 1905. This statute allowed the city to regulate land use in the upstate watershed region to protect City drinking water (Article 11). The law also gave New York City the authority to acquire land through eminent domain, and it authorized the State Department of Health to promulgate rules and regulations to protect the City's drinking water (Nolan, 1993, p. 534).

Expanding the Water Supply

As the population of New York City crossed the one million threshold in the 1870s, the City began to address the need to expand the Croton system. A new Croton aqueduct was opened in 1892, and a massive new Croton River Dam that entirely submerged the old dam was completed in 1905. With the construction of several smaller dams by 1911, the present-day Croton River system was in place, providing a potential maximum supply of 336 mgd (which exceeds the entire supply available to metropolitan Boston today).

But even this expanded Croton system would be insufficient to keep pace with the City's tremendous growth. In 1898, Greater New York with a population of 3.5 million was formed through the consolidation of Manhattan, the Bronx, Queens, Brooklyn, and Staten Island. New York water engineers in the early decades of the twentieth century began to look further afield, to sources in the Catskill Mountains beyond the Hudson River. Between 1907 and 1929, the City acquired water rights and constructed the Schoharie and Ashokan reservoirs in the Catskill Mountains. A new 92-mile Catskill aqueduct conveyed water from Ashokan to the City, crossing under the Hudson River by means of an "inverted siphon" 3,000 ft long and 1,100 ft below sea level (Weidner, 1974, p. 161). The Catskill system also included the construction of the Kensico and Hillview reservoirs just north of the City, City Tunnel No. 1, and the terminal Silver Lake Reservoir on Staten Island.

This feat was repeated in the 1940s when the City reached out to the headwaters of the Delaware River over 100 miles away. Unlike the Catskill reservoirs, which tapped streams entirely within New York State, the Delaware is an interstate river basin. The proposed diversion of substantial quantities of water from the headwaters to New York City aroused opposition from the downstream states of New Jersey and Pennsylvania, where many communities draw on the Delaware River for their own water supplies. Under the U.S. Constitution, disputes between states may be taken directly to the U.S. Supreme Court. Ultimately, New York City's Delaware River diversion was authorized up to a maxi-

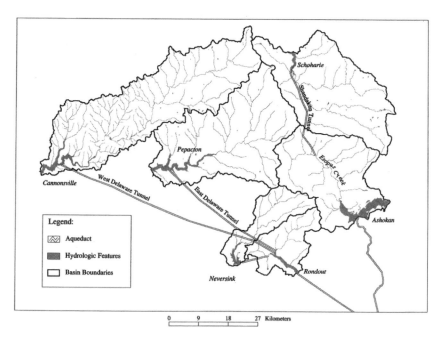

FIGURE 2-2 The Catskill/Delaware water supply system. Courtesy of the NYC DEP.

mum limit by the Court in decisions in 1931 and 1953. The 105-mile Delaware River Aqueduct was the world's longest continuous aqueduct tunnel (Weidner, 1974, p. 300). It meets the Catskill Aqueduct at Kensico Reservoir east of the Hudson River, which it crosses via a deep inverted siphon. Today, the combined systems (Croton, Catskill, and Delaware) are capable of supplying New York City with about 1.3 billion gallons per day, of which approximately 40 percent is derived from the Catskill system, 50 percent from the Delaware System, and 10 percent from the Croton System. A map of the Catskill/Delaware watershed is presented in Figure 2-2. Figure 2-3 shows both the Croton and Catskill/Delaware systems. A chronology of important water supply events is found in Table 2-1.

Both New York City and metropolitan Boston followed roughly similar strategies of water supply development between the 1840s and the 1960s, impounding hinterland sources of water to be conducted via gravity flow to the user region. But the two systems differ markedly in terms of their institutional organization. New York City itself established and continues today to operate its water supply system, even in upstate New York. By contrast, Boston conveyed its system in 1895 to a newly created regional entity, the Metropolitan Water District, which in turn was absorbed into a state agency—the Metropolitan District Commission, established in 1919. The latter completed the 400-billion-

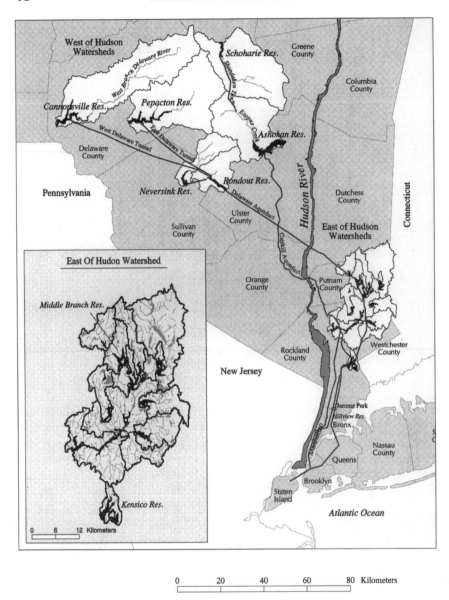

FIGURE 2-3 New York City water supply. Courtesy of the NYC DEP.

TABLE 2-1 Chronology of the New York City Population, Water Supply, and
Related Events

NYC Pop.	Year	Event
60,000	1800	
	1811	Commissioners Plan for New York City—future growth envisioned
	1819	Albert Giblin patents the silent valveless wastewater preventor (flush toilet) in Great Britain
	1825	Erie Canal opened
200,000	1830	
	1832	Cholera epidemic strikes New York City and other cities
	1834	NYC Board of Water Commissioners established by state legislature
	1835	Great Fire burns much of New York City
	1837	Croton River Dam and 41-mile aqueduct begun
	1842	First water reaches New York City from Croton River—celebrations
	1858	Central Park opened—including reservoir
	1866	New York State Metropolitan Health Act
1 million	1878	
	1893	Completion of New Croton River aqueduct
3.5 million	1898	Consolidation of "Greater New York City"—five boroughs
	1905	State Legislature gives New York City power to regulate upstate watershed land
4.6 million	1911	New Croton River system completed—10 percent of present New York City system
	1925	Interstate compact re Delaware River allocation—only ratified by New York
	1927	Catskill Mountain system completed—40 percent of present New York City system
	1931	*New Jersey v. New York* (283 U. S. 336) New York City is authorized to withdraw up to 440 mgd from Delaware River headwaters
7.5 million	1936	Delaware River system begun
	1953	*New Jersey v. New York* (347 U. S. 995)—NYC authorized to divert up to 800 mgd from the Delaware headwaters subject to maintaining minimum downstream flows. River Master appointed. New York City Watershed Regulations published
7.7 million	1961	Delaware River Basin Interstate Compact adopted (four states and U. S.)—Delaware River Basin Commission established
	1964	Delaware River system completed—50 percent of present New York City system
	1964-65	Northeast Drought
7.9 million	1970	
7.3 million	1990	
	1997	Watershed Management Agreement signed

gallon Quabbin Reservoir, 70 miles west of Boston, during the 1940s. In 1985, responsibility for distributing water to metropolitan Boston was reassigned to a new regional agency, the Massachusetts Water Resources Authority. Thus, while New York City has retained control over its system, and has encountered much anti-City hostility in its source regions, the metropolitan Boston system has long been administered by agencies serving a region containing half the state's population and therefore has gained broader political support in its efforts to protect its water supplies. As both systems follow parallel tracks in seeking to avoid filtration under the SDWA in the 1990s, this contrast in administrative structure may prove to be significant. In particular, the metropolitan Boston system has employed new state laws to promote watershed management in the source areas, while New York City has had to engage in lengthy and delicate negotiations with upstate interests to achieve the set of commitments that underlie the landmark MOA.

It is worth noting that the New York/Boston model of interbasin diversion from distant, upland sources influenced many other American and foreign cities in their own quests for pure water. Sometimes long-distance diversions are not required, as with some Great Lakes cities that enjoy ample water at their doorsteps. Western cities such as Los Angeles, San Francisco, Denver, and Phoenix have aggressively pursued scarce water wherever it could be found, developing whatever technical and institutional measures that were necessary to procure it. In some cases, such as the Owens Valley project of Los Angeles, the legal means were questionable and the technical execution was flawed, as demonstrated in the collapse of the St. Francis Dam in 1905 (Reisner, 1993). But the basic premise behind these western water wars was the same as that which motivated the early New York and Boston projects: employ modern technology and political power to ensure an abundant and inexpensive supply of pure water to the urban populace (Platt, 1996, p. 188).

Future Demands on the New York City Water Supply System

Average daily demand served by the New York City water supply system declined from 1,547 mgd in 1990 to 1,449 mgd in 1995 (Hazen and Sawyer/ Camp Dresser & McKee, 1997, p. 11). Although the latter figure is still higher than the system's estimated safe yield of 1,290 mgd, clearly the City's demand management program is paying off. Continued expansion of metering and the installation of water-saving plumbing devices to date have caused the average daily demand to continue to fall since 1995 (Warne, 1999a). Box 2-1 discusses the history of New York City's demand management activities.

Water demand management, which has proven to be effective in both the New York and Boston water systems, is nevertheless vulnerable to being counteracted by increasing numbers of water users. Future increases in users, and therefore in demand on the system, could arise from population growth within the

BOX 2-1
Demand Management in New York City

By the late 1980s, the New York City water supply system was called upon to meet an average demand of 1,400–1,450 mgd for the City proper plus an additional 120 mgd provided to eligible suburban communities. With a dependable safe yield from the total New York City water system estimated at about 1,300 mgd, the system was technically deficient. But further augmentation through development of new sources was infeasible. New out-of-basin transfers from upstate rivers including the Delaware are likely to be blocked by legal action. New York City has a seldom-used pumping station on the Hudson River upstream from the salt front, but it would face environmental challenges if the station were to be used on more than an emergency basis. Groundwater underlying Long Island theoretically could supply additional water, but this would be opposed by communities already relying on that source (Platt and Morrill, 1997).

Until the 1980s, water usage in New York City remained almost completely unmetered, which of course discouraged water conservation by households and other users. This nonsustainable situation was addressed in a Universal Water Metering program announced by the mayor in 1986. More than 600,000 meters have been installed at a cost of $350 million. When metering is complete, New York City will be able to monitor the use of water and employ pricing as a strategy to limit both waste and increases in demand.

In 1991, the City launched a pilot water conservation program to stem rising demand. The program offered free leak detection and installation of water-saving plumbing devices such as low-flow showerheads and faucets, aerators, toilet tank displacement bags, and low-flow toilets. These services were provided to 10,000 homes with 1–3 families citywide (Nechamen et al., 1995). Starting in 1993, a larger scale water conservation program conducted leak detection for some 8,000 homes with 1–3 families and 80,000 apartments. The City provided an expanded range of water-saving showerheads and toilet devices, new outreach and public education, and energy conservation in cooperation with the electrical utility Consolidated Edison. The City has developed an audit of leakage as a basis for estimated long-term benefits from subsidized water conservation measures. By 1995, in-city average demand had dropped to about 1,300 mgd. Much greater savings are anticipated from continuation of leak detection and plumbing retrofit efforts. It is anticipated that one-third of the City's residential toilets will be replaced with 1.6-gallon-per-flush units by 1998.

New York's Environmental Conservation Law (ECL), Sec 15-0105, has long declared that "the Waters of the State be conserved." A newer law (ECL, Sec. 15-0314) mandates the use of water-saving plumbing fixtures in all new construction or replacements of existing fixtures, including limits of 3 gallons per minute for sink faucets and showerheads and 1.6 gallons per flush for toilets. The State expects that this requirement will save over 500 mgd statewide (Nechamen et al., 1995).

present service area (New York City and certain suburbs in Westchester County) as well as from expanding the service area to include new communities. (For purposes of this discussion, it is assumed that water use per capita will continue to be steady or will decline in the future.)

The likelihood of a significant increase in population in the present service area is slight. During the 1970s, New York City lost 900,000 inhabitants, declining from 7.9 million in 1970 to 7.0 million in 1980. With the improvement in the City's economy and the influx of new migrants from abroad, the City's population in 1990 stood at 7.3 million. Although further small increases may occur, New York City is not expected to regain its former population of nearly 8 million. Since the water system served that level of usership prior to demand management measures and through most climatic perturbations, there is no reason to foresee any problem with meeting the expected demand of the City itself. Nor is it likely that the existing suburban user communities, which in 1995 averaged 123 mgd in demand, will significantly increase in population or water usage. These are mostly older communities that are largely built out. Although some intensification of residential development may occur through the replacement of single family homes with higher-density development, the net effect on the City's water system should not be significant.

A more likely source of future increase in demand on the New York City system would arise from the expansion of its service area to include communities and population currently served by other sources. This could occur in the event that present sources become contaminated or become insufficient to meet the needs of areas within potential reach of the City's distribution system (with installation of necessary connectors). Three possible areas of future shortfalls may be envisioned: (1) Long Island, especially Nassau County, (2) portions of Westchester or Putnam counties within reach of the city's aqueducts, and (3) northern New Jersey. The committee has not conducted any research on the likelihood of water failures in any of these regions, but it is aware of perennial concerns about possible contamination of the deeper aquifers that are the sole sources of water for Long Island. Moreover, all three of the identified regions may be expected to experience further population growth over the next two decades which, even without contamination of present sources, may pose the need for at least supplementary augmentation from the New York City system.

To the extent that the City's system is serving a smaller population now than it did three decades ago, it may be viewed by surrounding jurisdictions as a logical source of future water supply. Ironically, this perception may be enhanced by the City's success in lowering average per capita demand through its conservation measures (see Box 2-1).

The addition of further communities to the New York City system would be a political decision, very likely involving many stakeholders at the federal, state, and local levels. The outcome of a request by a water-deficient jurisdiction cannot be predicted. However, if public health is threatened, emergency connections may well be established that could remain in place indefinitely, leading to de facto permanent dependence on the New York City system, at least in times of drought. The committee is not aware of how the issue of future connections to the City system may have been addressed in legislation or other policy statements to date. However, if the issue remains unaddressed, the relevant parties should take whatever steps may be necessary to protect existing user sources against contamination and against rising demand in order to minimize the potential for future demands on the New York City system.

DESCRIPTION OF THE NEW YORK CITY
WATER SUPPLY SYSTEM

The water supply system for New York City is under the jurisdiction of the New York City Department of Environmental Protection (NYC DEP). Drinking water delivered to the city from upstate watersheds is impounded in the Croton, Catskill, and Delaware systems (see Figures 2-1, 2-2, and 2-3). The three watersheds contain 19 reservoirs and three controlled lakes with a total available storage capacity of about 558 billion gallons. During periods of normal rainfall, the total of the average yield for the three systems is estimated at 2,400 mgd (Hazen and Sawyer/Camp Dresser & McKee, 1997). The three water systems are interconnected at multiple locations to enhance flexibility and allow the exchange of water between systems. During periods of severe drought, Hudson River water can be used to augment the water supply by 100 mgd.

Water is delivered by gravity from the reservoirs of the Croton, Catskill, and Delaware watersheds to the City via large aqueducts and two balancing reservoirs. Three tunnels, two of which are deep bedrock tunnels, and two distribution reservoirs are then used to distribute drinking water to consumers. A third deep bedrock tunnel (Tunnel No. 3) has been under construction since 1970 and will supplement the two deep tunnels currently used for Catskill and Delaware water. The flow profile for the Catskill/Delaware system illustrating the changes in elevation as water passes from the Catskill Mountains to the City is shown in Figure 2-4.

A small part of the southeastern section of Queens is supplied by wells in addition to the main city distribution system. These wells, formerly under the operation of the Jamaica Water Supply Company, have been operated by the City since 1987 and supply between 17 and 24 mgd, less than half of the total average daily demand in the service area. A more detailed overview of the entire system is presented below. Table 2-2 lists the hydrological characteristics of each watershed.

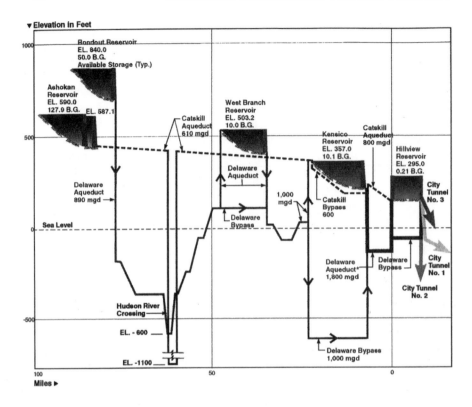

FIGURE 2-4 Flow profile for the Catskill/Delaware system. Source: Reprinted, with permission, from Hazen and Sawyer/Camp Dresser & McKee (1997). © 1997 by Hazen and Sawyer/Camp Dresser and McKee.

TABLE 2-2 Hydrological Characteristics of the Three New York City Watersheds

Parameter	Croton	Catskill	Delaware
Average percent of total water supply	10	40	50
Safe yield[a] (mgd)	240	470	580
Available storage[b] (10^9 gallons)	86.6	150.6	320.4
Total storage[c] (10^9 gallons)	94.6	147.5	325.9
Drainage basin (sq. miles)	375	571	1,010
Permanent population	132,000	36,000	45,000
Population per square mile	352	63	45

[a]Safe yield refers to the maximum amount of water that can be safely drawn from a watershed during the worst drought in the period of record.

[b]Available storage is the amount that can be withdrawn from a reservoir through its outlet structure and aqueduct.

[c]Total storage is the estimated volume between the crest of the spillway and the lowest elevation of the outlet of the reservoir.

Source: Reprinted, with permission, from Hazen and Sawyer/Camp Dresser & McKee (1997). © 1997 by Hazen and Sawyer/Camp Dresser and McKee.

New York City Drinking Water Reservoirs

Croton System

The Croton system normally provides approximately 10 percent to 12 percent of the City's daily water supply and can provide up to 25 percent during drought conditions. The system consists of 12 reservoirs and three controlled lakes on the Croton River, its three branches, and three other tributaries (Figure 2-1). Water from upstream reservoirs flows through natural streams to downstream reservoirs rather than through aqueducts or tunnels. Water from the terminal New Croton Reservoir is then conveyed through the New Croton aqueduct to the Jerome Park Reservoir in the Bronx.

The Croton watershed is largely within New York, but it includes a small portion of Connecticut. During 1990, permanent and seasonal population of this area was about 132,000, translating into 350 persons per square mile. In 1991, the watershed also contained about 3,000 head of livestock (mostly nondairy). In addition to these population pressures, the quality of the Croton water supply (particularly color and odor) has diminished over the years. Since 1992, NYC DEP has been under an enforceable agreement with the New York State Department of Health to install filtration for the Croton system, although the plant will not be completed until March 1, 2007.

A limited amount of water can be transferred between the Croton system and the Delaware system via two hydraulic pumping stations and the West Branch reservoir (see Figure 2-5 and the discussion below). In addition, water can be transferred by gravity from the Catskill system into the New Croton aqueduct. These operational possibilities and other factors allow the entire Croton system service area within the city to be supplied by the Catskill/Delaware system if necessary.

Catskill System

The Catskill system was constructed in multiple stages. The Ashokan Reservoir, Catskill aqueduct, Kensico Reservoir, Hillview Reservoir, City Tunnel No. 1, and terminal Silver Lake Reservoir in Staten Island were completed in 1917. Schoharie Reservoir and the Shandaken tunnel were completed in 1927. Seven years later, Tunnel No. 2 from Hillview Reservoir to Brooklyn was completed. Finally, in 1971, Silver Lake Reservoir was replaced with covered tanks.

The Catskill watershed is a sparsely populated area in the central and eastern portions of the Catskill mountains (Figure 2-2). The main tributaries that drain the watershed are the Esopus and Schoharie creeks. The Esopus Creek, a tributary of the Hudson River, is impounded by Ashokan Reservoir, which is divided by a dike into the East and West basins. Schoharie Reservoir, the northernmost basin in the entire system, is fed by Schoharie Creek and other tributaries. Most of the water from the Ashokan and Schoharie watersheds is stored in Ashokan Reservoir, with the balance being stored in Schoharie Reservoir. Water from Schoharie Reservoir is withdrawn through the 18-mile-long Shandaken tunnel to Esopus Creek, where it then flows 15 miles to the Ashokan Reservoir.

Delaware System

The Delaware system was the last of the three systems to be constructed and is furthest from New York City (Figure 2-2). Because of the large number and size of its reservoirs, the Delaware system normally provides 50 percent to 55 percent of the City's daily water supply. Cannonsville Reservoir, Pepacton Reservoir, and Neversink Reservoir collect water from tributaries of the Delaware River. Water from each reservoir travels through individual rock tunnels (West Delaware, East Delaware, and Neversink tunnels) to Rondout Reservoir, where the Delaware aqueduct begins. Rondout Reservoir drains Rondout Creek, a tributary of the Hudson River; however, most of Rondout's water is from the three upstream reservoirs on the branches of the Delaware River.

From Rondout, the Delaware aqueduct is connected to the West Branch Reservoir in the Croton system. Under normal operating conditions, water from the West Branch watershed is transferred to the Delaware system rather than flowing into the Croton system (and this is expected to continue after completion

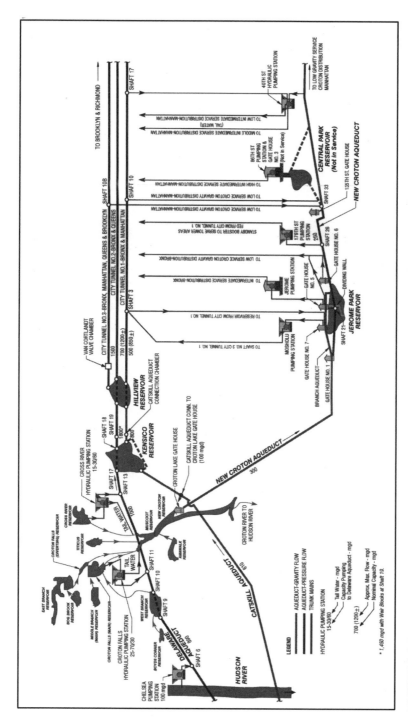

FIGURE 2-5 Reservoirs, aqueducts, tunnels, and pumping stations east of the Hudson River. Source: Reprinted, with permission, from Hazen and Sawyer/Camp Dresser & McKee (1997). © 1997 by Hazen and Sawyer/Camp Dresser and McKee.

of the Croton filtration plant). The Delaware aqueduct continues from West Branch to the Kensico Reservoir. The Delaware bypass at West Branch allows water to travel directly from Rondout to Kensico.

Although the Delaware watershed has a lower human population density than the other two watersheds (only 45 persons per square mile), it contains 35,000 head of livestock, most of which are dairy cattle. These agricultural operations can have a significant effect on the water quality of the Delaware supply.

Kensico and Hillview reservoirs, both of which are open, serve as balancing and distribution reservoirs, respectively, for the Catskill and Delaware systems. Either system can bypass the Kensico Reservoir and connect directly into Hillview. Under normal operations, water from both systems is discharged into Kensico Reservoir, thus mixing the two supplies. Water is then conveyed to Hillview Reservoir either through the Delaware aqueduct (which usually bypasses Hillview) or the Catskill aqueduct. Because of the mixing at Kensico, the Kensico and Hillview reservoirs, the sections of the Catskill and Delaware aqueducts between these two reservoirs, and the three City tunnels leaving Hillview Reservoir are referred to collectively as the Catskill/Delaware System.

Chelsea Pumping Station

In times of drought, Hudson River water can be pumped into Shaft 6 of the Delaware Aqueduct from an emergency pumping station at Chelsea. The station was originally constructed in the early 1950s to augment the City water supply prior to completion of the Delaware aqueduct. Although it was dismantled in 1957, it was reconstructed following several drought periods and has subsequently been put into service on three occasions (droughts of 1965–66, 1985, and 1989) (Warne, 1999a). The station is permitted for 100 mgd. Water delivered from the Hudson River is chlorinated prior to entering the Delaware aqueduct, and equipment is in place for coagulation if deemed necessary.

Brooklyn/Queens Aquifer

During the early part of this century, groundwater from the aquifer beneath Brooklyn and Queens provided a significant portion of the City's water supply. However, low water levels and saltwater intrusion have caused the abandonment of all wells except for some that service southeastern Queens. The City is currently investigating the use of renewed groundwater pumping to provide supplemental sources of drinking water.

Water Supply Infrastructure

Aqueducts and Tunnels

The principal reservoirs, aqueducts, tunnels, and pumping stations east of the Hudson River are shown in Figures 2-5 and 2-6. Under normal operating conditions, the Catskill and Delaware aqueducts convey water from the Catskill and Delaware watersheds to Kensico Reservoir, where significant mixing occurs. Monitoring data from NYC DEP have shown that the water quality at the two Kensico effluent locations (DEL 18 and CATLEFF) is similar, except for slightly higher coliform concentrations at DEL 18 (NYC DEP, 1998). Water from both of these effluent locations is conveyed to Hillview Reservoir, where it enters Tunnels No. 1 and 2 for conveyance to the city. Trunk mains convey water from the shafts on Tunnels No. 1 and 2 to the Catskill/Delaware system service area.

Additional flexibility in water delivery between Hillview and the City will be provided by Tunnel No. 3, which is being constructed at a cost of approximately $6 billion. Construction is expected to be complete in 2020. Stage 1 of Tunnel No. 3, completed at a cost of approximately $1 billion, went on-line in July 1998. This 13-mile segment runs from Hillview through the Bronx into Manhattan, under the East River and Roosevelt Island and into Queens. In the future, Tunnel No. 3 also will be connected with Kensico Reservoir directly, allowing water to be delivered from Kensico to the City without passing through Hillview. Operation of City Tunnel No. 3 will allow inspection of Tunnels No. 1 and 2 for the first time since they were put in service in 1917 and 1938, respectively.

Distribution System

The New York City water distribution system consists of a grid network of water mains ranging in size from 6 to 84 inches in diameter. It contains approximately 6,000 miles of pipe, 87,000 mainline valves, and 98,000 fire hydrants. Water pressure is regulated within a range of 35–60 pounds per square inch (psi) at street level; generally, 40 psi is sufficient to supply water to the top of a five- or six-story building. About 95 percent of the total city consumption is normally delivered by gravity.

The distribution system in each borough is divided into three or more zones in accordance with pressure requirements that are determined by local topography. The ground elevation in the City varies from a few feet above sea level, along the waterfront, to 403 ft at Todt Hill in Staten Island. The storage facilities at Hillview and Jerome Park reservoirs and Silver Lake tanks handle the hourly fluctuations in demand for water throughout the City as well as any sudden increase in draft that might arise from fire or other emergencies. With the exception of some communities in the outlying areas of the City, which may experience

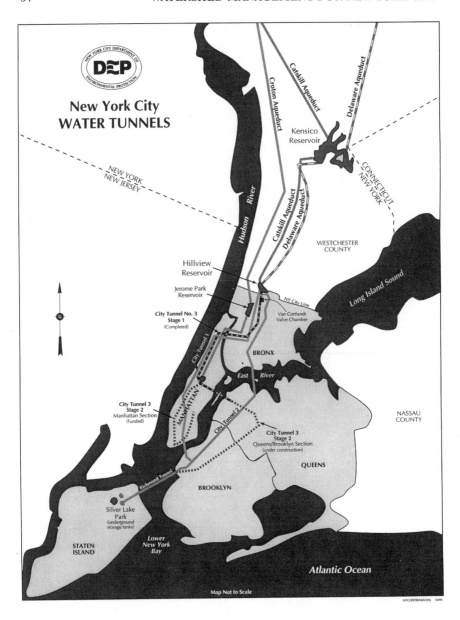

FIGURE 2-6 New York City water tunnels. Courtesy of the NYC DEP.

low-pressure service in peak hours during summer months, the water distribution system generally provides excellent service. Engineering consultants to the New York City Water Board have judged the distribution system to be in adequate condition, based on criteria such as watermain breaks, pressure tests, flow tests, and leak detection (Warne, 1999b).

Existing Treatment

Table 2-3 shows average water quality conditions in the Catskill/Delaware reservoirs from 1992 to 1996. Because of their high quality, waters from the Croton and Catskill/Delaware systems have historically only been treated with chlorine disinfection, beginning in 1910 for the Croton System and at startup of the Catskill and Delaware systems. Although chlorination has been shown to be effective for killing or inactivating bacteria, viruses, and *Giardia*, the level of chlorination used by New York City (less than or equal to 2 mg/L free chlorine) is not effective for inactivation of *Cryptosporidium* oocysts (Liyanage et al., 1997).

Chlorination takes place at multiple locations in the Catskill/Delaware and Croton systems. Croton water is chlorinated at the Croton Lake Gate House to achieve a level of disinfection sufficient to satisfy the Surface Water Treatment Rule (SWTR) in the New Croton aqueduct. Additional chlorine is added at the Jerome Park Reservoir to maintain a residual within the distribution system. Similarly, the Catskill/Delaware system is chlorinated twice prior to the distribution of water into the City. Chlorine is initially added to both the Catskill and Delaware aqueducts as the water leaves Kensico. These chlorine levels are used to determine compliance with the SWTR. Additional chlorine is added to the Catskill aqueduct prior to Hillview and at the Hillview downtake chambers to maintain residual levels in the distribution system.

In addition to chlorination, water treatment facilities can also provide alum to increase settling in the reservoirs during periods of high turbidity. Alum is used most frequently for the Catskill system, where it is added to the Catskill aqueduct prior to the Kensico Reservoir. Coagulated solids then settle within Kensico. There is also a facility for adding alum to the Delaware system between West Branch and Kensico. However, this facility has not been used at any time during the last 10 years.

In the 1960s, the City began adding fluoride at the Kensico and Dunwoodie reservoirs for dental health reasons. Sodium hydroxide addition takes place at Hillview Reservoir to assist corrosion control and to neutralize acidity arising from fluoride addition and intermittent alum applications. More recently, orthophosphate facilities have been installed at the Hillview downtake chambers and the Jerome Park Reservoir gate houses. Like sodium hydroxide, orthophosphate also accomplishes corrosion control for the distribution system. Finally, copper sulfate treatment is used on an irregular basis to control algae. Copper sulfate can

TABLE 2-3 Average Water Quality Conditions in the Catskill/Delaware
Reservoirs

Reservoir	pH[a]	Alkalinity[a] (mg/L)	Total hardness[a] (mg/L CaCO$_3$)	Turbidity[b] (NTU)	Total organic carbon[a] (mg/L)	Ammonia[a] (mg/L)
Ashokan W	6.9	9.68	15.92	4.54	1.7	0.02
Ashokan E	7.0	9.87	16.05	2.09	1.8	0.02
Cannonsville	7.3	14.21	24.77	4.09	2.1	0.021
Neversink	6.4	2.16	8.92	1.58	1.7	0.016
Pepacton	7.2	10.10	18.99	1.69	1.6	0.01
Rondout	7.0	7.94	16.16	1.20	1.7	0.01
Schoharie	7.0	12.81	19.82	8.98	2.2	0.02
Kensico[c]	6.9	10.40	18.79	1.05	1.7	0.016
West Branch	7.0	13.50	25.44	1.12	2.0	0.02

[a]1986–1998: Whole lake data. Courtesy of the NYC DEP.
[b]1993–1996: TMDL Reports (NYC DEP, 1999a–h).
[c]Kensico Reservoir data best represent the overall quality of the raw water supply prior to treatment.

be added to the system at a large number of locations, both upstream and down-stream of the New Croton Reservoir and upstream of Kensico Reservoir. Figure 2-7 illustrates the principal chemical feed locations in both the Croton and Catskill/Delaware systems.

CATSKILL/DELAWARE WATERSHED

The Catskill and Delaware watersheds from which New York City draws the majority of its drinking water encompass an area collectively referred to as the Catskill Mountains. This region, though sparsely populated, has supported a wide variety of enterprises, from agriculture to heavy industry. This section describes the physical attributes of the Catskills, including those that account for New York City's historically excellent water quality. It then describes the evolving land uses within the region that have influenced the quality of New York City drinking water.

Biophysical Setting of the Catskills

As Alf Evers notes in his widely acclaimed 1993 book *The Catskills: From Wilderness to Woodstock,* the precise regional boundaries of the Catskills are difficult to define. Perhaps the most colorful description can be attributed to an old man who lived in the shadow of Plattekill Mountain. When asked by Evers

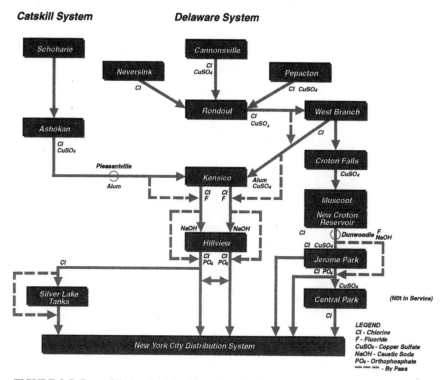

FIGURE 2-7 Potential chemical feed locations for the New York City water supply (not all chemicals and feed locations are used on a regular basis). Source: Reprinted, with permission, from Hazen and Sawyer/Camp Dresser & McKee (1997). © 1997 by Hazen and Sawyer/Camp Dresser and McKee.

just where the Catskills began, he replied, "You keep on going until you get to where there's two stones to every dirt. Then b'Jesus you're there." Native Americans evoked the misty blue appearance of the Catskills in their name, "Onteora," meaning "mountains of the sky." Both are apt descriptions of site, soil, and hydrologic characteristics of the Catskills.

Geology and Soils

The Catskills are an uplifted, maturely dissected portion of the Allegheny Plateau (Isachsen et al., 1991; Titus, 1993). Murdoch and Stoddard (1992) describes the Catskills as an eroded peneplain at the northern end of the Appalachian Plateau, with flat-lying sandstone, shale, and conglomerate bedrock deposited as part of a Devonian Age delta. A wedge-shaped formation, the Catskill

Escarpment rises steeply (about 1,640 ft) within 5–6 miles of the Hudson River. The highest elevations (3,000–4,200 ft) occur in the Esopus and Schoharie Creek watersheds. From the "High Peaks" area, the plateau tapers to the west into the Delaware Valley and the Finger Lakes region.

Catskill Mountains. Source: The Hudson (Lossing, 1866. © 1866 by H.B. Nims & Co.).

During the fall and winter when the deciduous trees are bare, the horizontally bedded sedimentary rock layers are clearly discernible along the Catskill Escarpment when viewed from the Hudson River and from many other vantage points. While hiking through virtually any part of the Catskills, the casual observer will note the horizontal layers and seams in rock outcrops or large boulders. The hydraulic properties and generally horizontal orientation of bedrock largely limit vertical water movement to bedrock fractures. Therefore, most subsurface water flow occurs through the soil mantle.

Most of the Catskills region is blanketed with a thin veneer of glacial till. Rock outcrops and boulder fields are common at higher elevations. Soils on mountaintops, ridgelines, and steep slopes are shallow (~0.7–4.3 ft) and stony, usually classified as stony sandy loams. The combination of high permeabilities and steep gradients produces rapid rates of lateral subsurface flow and droughty conditions during the growing season (USDA SCS, 1979). In saddles (rounded ridges between two peaks), along lower slopes, and in valley bottoms and floodplains, soils have a larger proportion of fine-textured material (silts and clays) as well as extensive deposits of alluvial sands and gravels. Local landforms and bedrock topography strongly influence the location, spatial extent, and heterogeneity of soils. For example, sand and gravel deposits are prevalent along the steeply incised valleys and relatively narrow floodplains of the Woodland Creek watershed (tributary to the Esopus Creek in the High Peaks region). By contrast, deeper silt loams form broad, flat floodplains along the Beaver Kill, another major tributary entering the Esopus Creek from the east.

Climate and Streamflow

At an approximate latitude and longitude of 42° N and 74–75° W, respectively, the Catskills region is affected by continental and maritime air masses. This results in frontal storms from the west and north, coastal storms from the south (along with an occasional Nor'Easter), and local thunderstorms (Murdoch and Stoddard, 1992; Stoddard and Murdoch, 1991; USDA SCS, 1979). Average annual precipitation is about 47 inches. The average air temperature is –4°C (24°F) in January and 22°C (71°F) in July. Orographic effects[1] and differences in land slope and aspect can strongly influence precipitation, air temperature, wind velocity, and relative humidity. Evidence of these microclimatic differences can be found in the species composition and growth form of vegetation, in snow accumulation and melt rates, and in other readily observed ecosystem characteristics.

---

[1]Orographic effects refer to increases in precipitation with increasing elevation as air masses flow up and over mountains. This is common in the Catskill/Delaware region as continental air masses move east from the Ohio Valley or maritime air masses move north up the Hudson River valley.

Although highly variable in water equivalent (the total amount of water), spatial extent, and duration, a snowpack typically forms in late November and persists through midwinter thaws until melt occurs in late March and early April. As evidenced by the January 1996 flood, rain-on-snow and snowmelt tend to be the largest and most destructive flow events in the Catskills region. Although precipitation is relatively uniform in temporal distribution, streamflow can vary markedly between dormant and growing seasons, as shown in Figure 2-8(A). In addition to snowmelt events described above, large rain events that occur between leaf-fall and the beginning of snow accumulation can produce very high flows. This occurs when air temperatures and evapotranspiration rates decrease, causing a simultaneous and rapid increase in soil water content, in the extent of saturated source areas, and in streamflow. Finally, large, high-intensity convective storms can produce rapid streamflow response during the growing season or spring (leaf-out) and fall (leaf-fall) transition periods. However, stormflow from large summer thunderstorms is generally less significant than streamflow from snowmelt or dormant season rains. All of these sources contribute to the estimated 2,000 miles (from the U.S. Geological Survey) of watercourses and intermittent streams found in the Catskill/Delaware watershed.

In summary, precipitation is relatively uniform in quantity and timing throughout the year, but total annual streamflow volume is not. The streamflow regimen often is dominated by a few large events. Therefore, vegetative cover, land use, and resulting pathway(s) and rate of flow play an important role in determining how climatic and hydrologic characteristics affect reservoir water quality.

Reservoir Dynamics

Water supply reservoirs have many characteristics that are similar to those of natural lakes. However, there are important differences in spatial and temporal dynamics that influence reservoir productivity[2] and water quality (Wetzel, 1990). The drainage basins of reservoirs are consistently much larger relative to the reservoir surface area than is the case for most natural lakes. Thus, reservoirs receive runoff water and associated pollutant loadings from watershed areas that are many (10–200) times larger than those of most natural lakes. In addition, source water for reservoirs is transported mainly via high-order streams, which results in high energy for erosion, large sediment load carrying capacities, and extensive loading of dissolved and particulate contaminants into reservoir water.

---

[2]Productivity is defined as the rate of formation of organic matter over some defined period of time, usually a year (never just the "growing season"). Net productivity is the production of new organic matter by photosynthetic organisms, or the acquisition and storage of organic matter by nonphotosynthetic heterotrophic organisms, less losses from respiration and egestion, divided by the time interval.

(A)

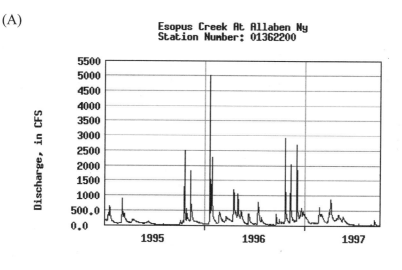

Esopus Creek At Allaben Ny
Station Number: 01362200

Legend: ——— Discharge, in CFS
         ——— Estimated Discharge, in CFS

(B)

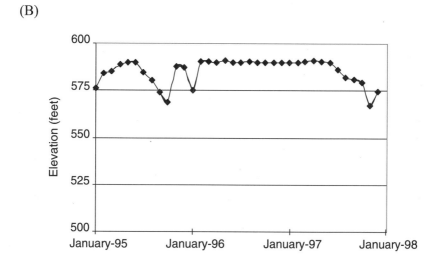

FIGURE 2-8 (A) Streamflow for the Esopus Creek in the Ashokan Watershed, 1995–1998. Source: http:/waterdata.usgs.gov. (B) Water levels in the West Basin of Ashokan Reservoir. Courtesy of the NYC DEP. (A) and (B) show how stream flow volatility is dampened in the reservoir because of the reservoir's large size, selective withdrawals, and the distribution of water between the West and East basins. The crest of the spillway between the West and East basins is 587 ft.

Because reservoir inflows are primarily channelized and often are not intercepted by energy-dispersive buffer zones, runoff inputs are larger, are more directly coupled to precipitation events, and extend much farther into the reservoir per se than is the case for most natural lakes. All of these properties result in high, but irregularly pulsed, pollutant loading to water supply reservoirs.

Within the reservoir itself, the irregular dynamics of inflow and rapid, variable flushing rates can markedly alter environmental conditions for biota. A reservoir can be viewed as a very dynamic lake in which a significant portion of its volume possesses characteristics of a river (Wetzel, 1990). Often the riverine portion of a reservoir functions like a large, turbid river in which turbulence, sediment instability and high turbidity, reduced light availability, and other characteristics preclude extensive photosynthesis, despite high nutrient availability. This reduction in photosynthesis is only partially ameliorated by turbulent, intermittent recirculation of algae into the photic, or light-penetrating, zone. In more lacustrine (lakelike) regions of reservoirs, greater light penetration is possible, the depth of the photic zone increases, and primary productivity increases. Nutrient limitations, so characteristic of natural lakes of low to moderate productivity, can then occur to varying degrees as losses of nutrients exceed loading renewal rates.

Internal loading of nutrients from sediments or deep-water areas of storage, normally low in natural lakes, can be high in reservoirs. Much of the internal nutrient loading in reservoirs is associated with the irregular inflow and withdrawal dynamics, which can disrupt thermal and dissolved oxygen stratification patterns that suppress sedimentary nutrient release and redistribution in more physically stable, natural lakes.

The average depth of the eight primary reservoirs of the Catskill/Delaware system is quite large (approximately 50 ft; Table 2-4). Despite the relatively large storage capacities of the reservoirs, water residence times are variable, between 0.06 to 0.71 years. Residence times of less than a third of a year tend to be relatively unstable and force organisms into rapid growth cycles (Ford, 1990). Pepacton Reservoir's residence time approaches residence times of natural lakes, which can contribute to greater stability, particularly in relation to displacement of nutrients to sediment storage sites.

Withdrawals constitute a removal of maximum storage capacity of approximately 0.0025 percent per day. Although most withdrawals are from surface waters, water can be withdrawn from varying depths within all Catskill/Delaware reservoirs except Schoharie Reservoir. The chosen depth of withdrawal depends on turbidity, dissolved oxygen levels, color, and other water quality parameters (Warne, 1999a).

Wetlands

Wetlands are relatively flat land areas that are partially or wholly flooded throughout the year. Soils in these areas are usually deprived of dissolved oxygen

TABLE 2-4 Storage Capacities, Drainage Areas, and Residence Times of the Catskill/Delaware Reservoirs

Drainage Basin	Total Storage Capacity[a]		Drainage Basin Area[a]		Mean Depth[b]	Residence Time[c]	Average Supply
	10^6 m^3	10^9 gal.	km^2	mi^2	(ft)	(yr)	(10^6 gpd)
Catskill System							
Schoharie	74.2	19.6	813.3	314.0	44.3	0.10	
Ashokan—E[d]	305.4	80.7	63.5	24.5	41.3	0.31	
Ashokan—W	178.7	47.2	602.2	232.5		0.20	
Total	558.3	147.5	1,479.0	571.0			501
Delaware System							
Cannonsville	366.1	96.7	1,165.5	450.0	53.5	0.43	
Pepacton	543.9	143.7	963.5	372.0	69.5	0.71	
Neversink	134.2	35.5	240.9	93.0	62.0	0.42	
Rondout	189.4	50.0	246.1	95.0	66.3	0.14	
Total	1,233.6	325.9	2,616.0	1,010.0			846
East-of-Hudson							
West Branch[e]	38.2	10.1	52.8	20.4	25.9	0.38	
Kensico	115.8	30.6	34.4	13.3	41.0	0.06	

[a]Hazen and Sawyer/Camp Dresser & McKee, 1997.
[b]Table 2.5 In NYC DEP (1993a).
[c]From 1999 TMDL reports (NYC DEP, 1999a–h); average of 1992–1996 data. It should be noted that the TMDL reports use "annual residence times" rather than "mean monthly residence time." These different methods for calculating residence time generate different numbers, with the TMDL method generating smaller values.
[d]East Basin area calculated by subtracting the West Basin area from the total Ashokan drainage basin area (257 sq mi).
[e]West Branch area is often given as 42.9 sq mi. This value includes areas of Boyd Corners, Lake Gleneida, and Berret's Pond.

and consequently only support the growth of specialized aquatic plants. Much like riparian zones, wetlands often connect upland areas to adjacent waterbodies. Thus, they are critical in the protection of water quality and aquatic habitats and in flood and erosion control. Examples of wetlands include marshes, swamps, ponds, wet meadows, and seasonally flooded floodplains (Tiner, 1997).

Wetlands comprise a small (yet important) proportion of the Catskill/ Delaware watershed (Tiner, 1997). They may have organic, mineral, or mixed substrates, and they tend to occur in riparian areas throughout the region. Figure 2-9 shows the general distribution of wetlands and deepwater habitats of greater than ten acres in the Catskill/Delaware watershed. The dominant wetland types

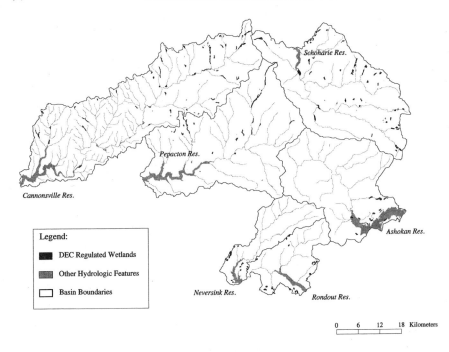

FIGURE 2-9 General distribution of wetlands and deepwater habitats of greater than 12.4 acres in the Catskill/Delaware watershed. Courtesy of the NYC DEP.

in the Catskill/Delaware region are (1) *emergent* (commonly called marshes and vegetated with grasses, sedges, cattails, or common reed), (2) *scrub-shrub* (dominated by woody vegetation less than 20 ft tall, such as speckled alder or willows), and (3) *forested* (comprised of tree species greater than 20 ft tall such as red maple, eastern hemlock, and associated woody shrubs). Table 2-5 summarizes the types and corresponding areas of wetlands in the Catskill/Delaware region. Detailed discussion of community types, classification methods, and wetland functions can be found in Cowardin et al. (1979), Mitsch and Gosselink (1993), and NRC (1995).

    In the Catskill/Delaware watershed, wetlands occur in three landscape positions. First, a small number of relatively large wetlands exist in riparian/floodplain areas, the transition zone between upland and aquatic ecosystems. In these critical landscapes, wetlands reduce flow velocity, trap sediment, assimilate and transform nutrients, and provide valuable and unique habitats. Second, many small wetlands occur in depressional areas in alluvial soils of broad valleys, mainly, if not entirely, in the Cannonsville, Pepacton, and Schoharie watersheds.

TABLE 2-5 Wetland Types and Areas in the Catskill, Delaware, and Croton Watersheds of the New York City Water Supply System (table includes wetlands of less than 10 acres)

Wetland Type	Catskill acres (% of total wetland area)		Delaware acres (% of total wetland area)		Croton acres (% of total wetland area)	
Emergent	838	(22%)	1,607	(19%)	750	(5%)
Scrub-Shrub	615	(16%)	769	(9%)	754	(5%)
Shrub/Emergent	169	(4%)	322	(4%)	812	(5%)
Deciduous Forested	894	(23%)	478	(6%)	11,036	(70%)
Evergreen/Mixed Forested	337	(9%)	416	(5%)	158	(1%)
Pond	827	(21%)	1,505	(18%)	2,152	(14%)
Reservoir/Lake/ River shallows and shores	192	(5%)	3,190	(39%)	147	(<1%)
Totals	3,872	(100%)	8,287	(100%)	15,809	(100%)
Watershed area (acres)	365,440		648,320		247,680	
Wetland area (%)	1.0		1.3		6.4	

Source: Tiner (1997).

Finally, wetlands are found in isolated depressions high in the watershed(s). First order perennial streams sometimes originate from these sites, most of which are above the 2,000-ft elevation line and so receive de facto protection from the Catskill State Park and other regulations.

The MOA specifically identifies wetlands for water quality protection, similar to that afforded to reservoirs, major tributaries, and streams. Unlike the U.S. Army Corps of Engineers, which has no minimum acreage requirement for determining its jurisdiction, the MOA protects only those wetlands greater than 12.4 acres in size. This arbitrarily established size limit is not expected to substantially affect water quality in the reservoirs because wetlands smaller than 12.4 acres in the Catskills do not constitute the more important riparian and headwater wetlands discussed above. Rather, they are isolated depressions on agricultural land in the Cannonsville, Pepacton, and Schoharie watersheds (the second type described above) that have an indirect effect on reservoir inflow and quality during the dormant season and have limited or insignificant effects during the growing season. Although depressional wetlands comprise a small fraction of the total watershed area, they can be important nutrient sinks and sediment traps in the subwatersheds of minor tributaries. In addition, they provide critical habitat for a range of plant and animal species. They are most valuable ecologically and economically if kept intact.

Vegetation

The Catskills landscape is predominantly forested, although the total area of forest varies considerably across the region. The Catskill watershed, which includes the Ashokan and Schoharie reservoirs, also includes the core area of the 335,000-acre Catskill Forest Preserve. The majority of the uplands support mixed-species stands of deciduous trees (red and white oak, beech, yellow, paper, and gray birch, red and sugar maple, and common understory associates). Many stands, especially those on well-drained south aspects, have a dense understory of mountain laurel. Conifers such as eastern hemlock and eastern white pine are more prevalent in valley bottoms and along stream channels at higher elevations. The summits of the highest peaks—Wittenberg, Cornell, and Slide—have remnant stands of balsam fir and red spruce along with small areas of alpine grasses and heath shrubs.

The age and condition of the Catskills forest reflect several centuries of land use and economic change, the most prominent of which are outlined in the following section. Although the forests of the Catskills have been, at various times and for various reasons, repeatedly exploited and ignored, their protective function remains largely intact. Natural regeneration and planted stands have restored the protective litter layer on the forest floor, have increased soil organic matter, and have encouraged colonization by organisms ranging from microbes to small mammals and the proliferation of roots throughout the shallow profile. As a result, wherever forests are present, subsurface (rather than overland) flow has been restored to the watershed. In light of the summary statistics for forest and vacant land, this bodes well for the protection and maintenance of water quality, especially because forests occur on steep slopes and along countless intermittent and ephemeral streams throughout the watershed.

Land Use

Table 2-6 and Figure 2-10 show land use and land cover categories for the Catskill/Delaware watershed. The most significant agricultural land use is centered in valley bottoms of the Delaware watershed. Dairy farming became established in the 1800s when railroads linked the region to lucrative markets in the Hudson Valley and New York City. Post-World War II expansion of dairy farming in western New York and the Midwest intensified competition and led to widespread farm abandonment (evidenced in the vacant land category, abandoned fields not yet classified as forest, in Table 2-6). Residential and commercial development accounts for a relatively small proportion of watershed land use. However, because this development tends to be concentrated in stream valleys, it can have a disproportionately large influence on water quality.

Without active intervention and management, intensive agricultural use (e.g., tilled fields, barnyards, and feedlots) and impervious surfaces associated with

TABLE 2-6 Land Use in the Catskill/Delaware Watershed by Reservoir Basin (in acres and as percent of total)

Basin	Agri-culture	Low-Density Resi-dential	High-Density Resi-dential	Commercial/ Industrial/ Government	Vacant Land	Forests	Other Open Space	Total Acres
Ashokan	54	6,945	6,945	555	2,617	142,975	512	153,873
	<1%	5%	<1%	<1%	2%	93%	<1%	
Schoharie	5,975	37,521	748	2,819	25,737	123,760	3,793	200,353
	3%	19%	<1%	1%	13%	62%	2%	
Cannonsville	30,523	61,072	751	3,316	37,057	148,256	4,121	285,096
	11%	21%	<1%	1%	13%	52%	1%	
Neversink	136	2,376	16	47	1,487	52,955	661	57,678
	<1%	4%	<1%	<1%	3%	92%	<1%	
Pepacton	6,960	43,332	622	1,516	23,259	150,957	3,980	230,624
	3%	19%	<1%	<1%	10%	65%	2%	
Rondout	1,152	5,157	239	125	2,253	48,614	320	57,862
	2%	9%	<1%	<1%	4%	84%	1%	
Totals	44,800	156,403	2,590	8,378	92,410	667,517	13,387	985,486
	5%	16%	<1%	<1%	9%	68%	1%	

Note: The low density residential, high density residential, and commercial/industrial/government categories do not correspond with the NYC DEP "urban" category used in TMDL calculations and the pathogen monitoring studies. NYC DEP's "urban" category is a measure of impervious surfaces, major roadways, commercial, industrial, and high- and medium-density residential areas as derived from LANDSAT ™ scenes that have a 28.5-m resolution.
Source: NYC DEP, 1993b.

residential and commercial development inevitably generate overland flow. If these areas are connected with the stream channel network, larger quantities of lower quality water reach downstream areas more rapidly than does subsurface flow through native forest. Thus, although nonforest land uses may only comprise single-digit percentages of reservoir watersheds (Table 2-6), they are the principal challenge for mitigation and management.

Land Use History

A chronology of natural resource use in the Catskill Mountains is presented in Table 2-7. A historical review of changes in local economic activities and settlement patterns helps to put into perspective contemporary social and ecological conditions and the cumulative effects they have had on forests, soils, and water resources.

For thousands of years before Hendrick Hudson sailed up the "Great River of Mountaynes" in 1609, Native Americans made limited use of the Catskills. They

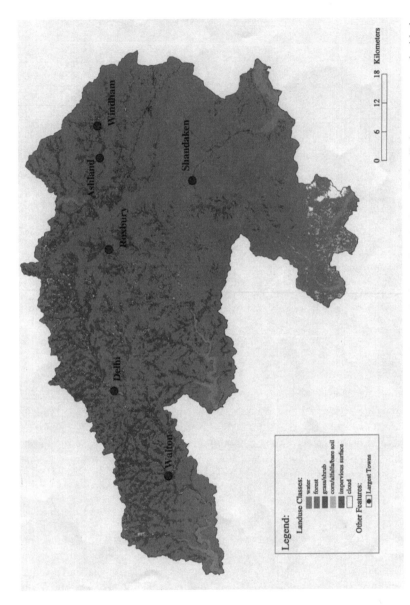

FIGURE 2-10 Land cover of the Catskill/Delaware watershed. Note that impervious surfaces ("urban") do not correspond with the residential, commercial, and industrial land use in Table 2-6. Actual impervious cover is a lower percentage of land than the categories used in Table 2-6.

TABLE 2-7 A Chronology of Natural Resource Use in the Catskill Mountain Region, New York

??? ~ 1630	Native American hunting, gathering, and war parties made extensive use of mountain slopes and summits. Farming and seasonal settlements were clustered in and along floodplains of major streams and along the Hudson River.
1609	On a voyage of discovery sponsored by the Dutch East India Company, Hendrick Hudson sails the Half Moon up the "Great River of Mountaynes" [later the North or Mauritius River then, finally, the Hudson River] in search of the Northwest Passage. He claims the region for the Netherlands eleven years before the Mayflower reaches Plymouth.
1610	Active trade with Native Americans, primarily for beaver pelts, begins during Hudson's second voyage.
1624	Dutch settlement proceeds slowly in proximity of the Hudson River, and later the Esopus and Rondout Creeks. In addition to New Amsterdam [now New York City], fur trading posts are established: Wiltwyck [now Kingston] and Fort Nassau [now Albany].
1664	September 8th, Governor Petrus Stuyvesant surrenders New Netherland to a fleet of five English warships. The colony is named New York in honor of the King's brother.
1665	Disease and armed conflicts with white settlers lead to the extirpation of Native Americans from the region.
1700 ~ 1800	Land speculation, namely the 1.5 million acre Hardenbergh Patent, slows the colonization of the Catskills. Not until 1754 is the patent, held by Johannis Hardenbergh and seven partners, divided among their 52 heirs.
~1750–pres.	Agriculture begins in the region. Early subsistence farms evolve into livestock, truck crop, and dairy operations.
1765	Early roads reach the interior. The Plank Road up the Esopus Creek valley [now NYS Route 28] first appears on a map.
1777	October 13th, British troops burn Kingston, the first capital of the State of New York.
1800–1870	Leather tanneries are established near major streams and virgin hemlock (*Tsuga canadensis*) forests throughout the region. When in peak production raw hides are imported from as far away as South America. Except for small, inaccessible areas, the late successional hemlock forest was destroyed to provide tanbark. Watercourses are fouled with tannery wastes and by deforestation. The native trout fishery is decimated.
1820 ~ 1930	The dense second-growth deciduous forest is repeatedly cut for barrel hoops, mass-produced furniture parts (largely for mail-order catalog companies), and fuelwood (for steamboats and locomotives).
1820 ~ 1930	Urban dwellers enamored of the writings of Washington Irving (*Rip Van Winkle,* 1729), James Fenimore Cooper (*The Leatherstocking Tales,* 1823–1841), and John Burroughs (1837–1921) as well as paintings of the Hudson River School, flock to early mountain houses and boarding establishments.
1830 ~ 1895	Bluestone is quarried throughout the region. Most is hauled to Island Dock on the Rondout Creek near Kingston, then by barge to New York City where it is used for curbs, sidewalks, buildings, and terraces.

continued

TABLE 2-7 Continued

1870–1954	The Ulster & Delaware Railroad extends the New York Central Railroad lines along the Hudson River to the interior of the Catskill region. Improved transportation benefits tourism and agriculture.
1885	The Catskill Forest Preserve is established as "Wild forest, forever" by amendment to the New York State Constitution. Most of the lands are mountaintop and sidehill parcels of second- or third-growth forest, that after being abandoned by owners, revert to state and local governments in lieu of delinquent taxes.
1905	New York State Public Health Law authorizes New York City to secure lands, by eminent domain if necessary, to build reservoirs and infrastructure needed to expand the water supply.
1907–1965	Beginning with the Ashokan Reservoir (Esopus Creek) and ending with the Cannonsville Reservoir, the New York City system encompasses 985,486 acres of watershed land.
1950s	Alpine skiing increases in popularity with the construction of several major ski areas (Belleayre, Highmount, Hunter Mountain, Ski Windham, and others). Deer hunting, trout fishing, hiking, backpacking, camping, and recreational activities increase in popularity. Fall foliage attracts visitors from far and wide.
1960s–present	Forest products industry is reestablished as forest on private lands reach sawtimber size classes. Other forest owners expand the production of maple syrup and sugar.
1970s–present	Improvements in the local and regional highway system and lifestyle changes increase the demand for vacation and second homes.

Adapted from Evers (1993), Stave (1998), and Wilstach (1933).

hunted in deciduous forests along the lower slopes, grew corn, beans, and squash in small floodplain fields, and occasionally moved through the interior hemlock and upper-elevation spruce-fir forests in hunting or war parties. Colonial settlement of the Hudson River Valley by the Dutch and later the British along the lower Esopus, Rondout, and Schoharie creeks loosely encircled the Catskills. Beaver flourished in the numerous headwater streams, and this led to a burgeoning fur trade between Indians and Dutch and English traders. This was the first phase of intensive natural resource use in the region. Trade and other interaction occurred for several decades until fundamental misunderstandings over land transactions led to war and to extirpation of Indians from the region. Indians believed they were giving permission for friendly people to settle amongst them; whites believed they were buying the land in accordance with European norms and traditions (Evers, 1993).

Through an almost unbelievable sequence of events, in 1708 a Kingston merchant, Johannes Hardenbergh, was granted 2 million acres of land, encompassing virtually the entire region (Evers, 1993). This enormous land patent had

Workers peeling and piling bark showing the tools needed—barking axes, peeling irons, and a "bob sled" with bark racks to secure the bark for travel to the tannery. Engraving c. 1840s. Collection of the Zadock Pratt Museum. Source: Reprinted, with permission of the Zudock Pratt Museum, from Bare Trees *by Patrick Millen (Black Dome Press: 1995).*

the effect of forestalling extensive pioneer settlement until the mid-1700s. It also established a precedent for contentious "landlord–tenant" interaction that lingers to this day. The Catskills have been transformed by multiple enterprises, including leather tanning, "cut-and-run" logging, stone quarrying, and subsistence and commercial farming until tourism, the establishment of the Forest Preserve, and finally the New York City water supply system led to the current landscape and community structure (Evers, 1993).

The long and volatile chronology summarized in Table 2-7 reminds us that more changes are, no doubt, in store. The forest continues to mature, farming seems poised to shift from dairy to other crops, and economic and societal changes induce people to seek rural areas as places to live, work, and recreate.

Population Trends in the Watershed Region

One of the reasons the Catskill/Delaware region was tapped for water early in the twentieth century was that it was sparsely populated. There are 40 towns west of the Hudson River with some land in the Catskill/Delaware watershed. An examination of the populations of these communities in the late nineteenth and early twentieth centuries reveals little population growth (Curve A, Figure 2-11). In fact, the town populations actually declined in the initial decades of this century. Although a growth spurt was experienced between 1930 and 1980, since 1980 there has been very little population growth in the region.

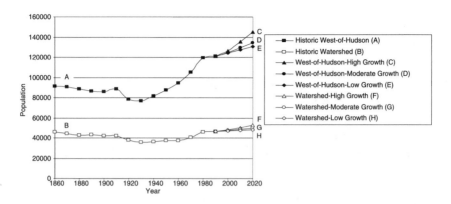

FIGURE 2-11   Historical (1860–1996) and predicted future (1990–2020) population trends for the Catskill/Delaware watershed region and 40 West-of-Hudson communities.

This analysis, however, can be misleading because less than half the population of the West-of-Hudson watershed towns actually resides within the watershed boundary. In order to determine actual watershed population figures, the 1990 percentages of the town population residing in the watershed (NYC DEP, 1993b) were used to estimate permanent watershed populations since 1860, with results shown as Curve B in Figure 2-11. These estimates demonstrate that the actual Catskill/Delaware watershed population hardly changed between 1860 and 1990. In fact, the 1990 watershed population exceeded the estimated 1860 population by just 235 persons.

Based on these historic trends, future permanent population trends into the year 2020 were estimated assuming three different rates of growth. Average annual growth rates were determined for the periods 1900–1990 (long-term or moderate growth), 1940–1990 (high growth), and 1990–1996 (recent or low growth) using U.S. Census population figures. The corresponding average annual percentage growth rates for the total population of the 40 towns are given in Table 2-8. For the estimated population residing within the watershed boundaries, the three corresponding annual percentage growth rates are somewhat lower, as expected. Population projections based on these alternative growth rates are given in Figure 2-11, Curves C–H. It is clear from the trend lines that the estimated permanent watershed population will change little between now and 2020. If the 1990–1996 growth trend were to continue (Curve H), the permanent population in 2020 would be greater than the 1996 population by 2,500 persons. Figure 2-11 also indicates that growth in the 40 West-of-Hudson communities is likely to come in those areas lying outside the watershed.

The Croton watershed region has experienced significantly more population growth than the Catskill/Delaware watershed. The associated increase in resi-

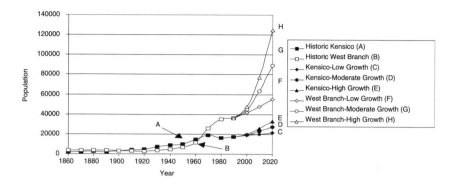

FIGURE 2-12 Historical (1860–1996) and predicted future (1990–2020) permanent population trends for the Kensico and West Branch watershed regions.

dential and commercial construction, the loss of forests, and the increased waste-water demands have seriously degraded the water quality of the Croton water-shed reservoirs. Twelve of the 13 reservoirs (including West Branch) are classified as eutrophic. Although the Croton system provides only 10 percent of the City's drinking water during nondrought conditions, almost all of the City's water passes through the Kensico and/or West Branch reservoirs, which reside within the Croton watershed boundaries. Population growth projections for the Croton watershed are likely to reflect conditions within the Kensico and West Branch watersheds that could have a significant impact on water from the Catskill/Delaware watershed.

The historical and future permanent populations residing in these two basins were analyzed by the same method used for the population residing in the Catskill/Delaware watershed. U.S. Census data were used to develop Figure 2-12, which shows that the population living in the Kensico basin has leveled off since about 1970 (Curve A). By contrast, the population residing in the West Branch water-shed has grown at higher rates and more continuously than that of the Kensico watershed (Curve B).

Based on historic population trends for each basin (data not shown), three different growth rates were calculated for each, as shown in Table 2-8. These growth rates were applied to estimate future permanent population in the Kensico and West Branch watersheds (Figure 2-12, Curves C–H). Population growth in the Kensico watershed is likely to be relatively moderate into the next century, even under the high growth assumption (2.52 percent per annum), which would raise the population residing in the Kensico basin by about 14,000 between 1996 and 2020. Based on the high growth rate characteristic of 1940–1990 (4.9 percent per year), the West Branch basin population would grow by 83,146 persons,

TABLE 2-8 Annual Percentage Growth Rates for Different Regions in
the New York City Watershed

Area	1990–1996 Low growth	1900–1990 Moderate growth	1940–1990 High growth
40 West-of-Hudson Towns	0.25	0.38	0.79
Catskill/Delaware Watershed	0.11	0.22	0.49
Kensico Watershed	0.69	1.68	2.52
West Branch Watershed	1.38	3.45	4.90

Adapted from U.S. Census data.

or 200 percent by 2020. These projected population trends underscore the importance of (1) monitoring reservoir water quality, (2) reducing present pollutant loadings via best management practices within the watershed, and (3) preventing future pollutant loading to the Kensico and West Branch reservoirs by restricting certain activities. These will be formidable tasks in the face of future population pressures.

Seasonal residence in the West-of-Hudson region is a factor that was not taken into account in this analysis, partly because estimating this population is difficult. The 1993 environment impact statement (EIS) for the New York City watershed regulations projected that in the absence of regulations, the seasonal population would grow at a rapid rate (NYC DEP, 1993b). It concluded that if the watershed regulations then proposed were adopted, all population growth in the area would be minimal. (The EIS provided no information on how this conclusion was reached.) Unfortunately the analysis in the EIS contains errors, and the EIS relied on a method of estimating seasonal population that yields unreliable results.[3]

In order to independently assess the role of seasonal population, the committee turned to census data on housing, which support some concern about in-

---

[3] To estimate the seasonal population in 2010, the EIS used a census count of vacant houses in 1980 and 1990. The EIS estimated the seasonal population by multiplying the number of vacant houses by the average family size in each township. Using this estimation method for 1980 and 1990, the EIS projected that seasonal population would double between 1990 and 2010 in the West-of-Hudson watershed areas. This procedure is likely to yield exaggerated estimates of seasonal population growth because it does not differentiate between seasonal housing and other vacant housing. For example, areas like Delaware County had the largest number of unoccupied houses and also experienced net population loss during the 1980s. Much of this housing is likely to be vacant, not seasonal housing. In addition, the EIS reported that in 1990 about one-third of all housing units in the West-of-Hudson watershed area were vacant. This appears to be an error resulting in an overestimate of seasonal change in housing. Our review of census figures indicates that about 19 percent of all housing units in the area were vacant in 1990.

creases in seasonal population in watershed counties west of the Hudson during the 1980s. Between 1980 and 1990, permanent population growth throughout the watershed counties stagnated while the housing stock increased 5 percent, or 0.5 percent annually (which would yield a 10 percent increase in the housing stock if projected to year 2020). This growth in housing in excess of permanent population growth might indicate seasonal home development.

One can also evaluate new housing construction rates and approximate corresponding increases in housing stock. Between 1990 and 1996, the new housing construction rate west of the Hudson dropped by more than half (based on Census Bureau reports of building permits issued), and the rate of growth of the permanent population slightly *exceeded* that of new construction (0.25 percent compared with 0.20 percent per annum), suggesting little seasonal home development (Department of Commerce, 1990, 1992, 1994, 1996). If the 1990–1996 trend in new construction is projected to 2020, a modest 4 percent increase in the housing stock can be expected. Recent anecdotal accounts of second-home development in some watershed areas west of the Hudson, however, indicate some renewed interest in the area as a seasonal destination (Hall, 1998). In any case, the construction of seasonal homes will result in some growth of part-time residents in the watershed, and attention should be given to the distinctive watershed impacts of such residents.

REFERENCES

Blake, N. F. 1956. Water for the Cities: A History of the Urban Water Supply Problem in the United States. Syracuse, NY: Syracuse University Press.

Booth, M. L. 1860. History of the City of New York. New York, NY: W. R. C. Clark & Meeker.

Cowardin, L. M., V. Carter, F. C. Golet, and E. T. LaRoe. 1979. Classification of wetlands and deepwater habitats in the United States. FWS/OBS-79/31. Washington, DC: U.S. Fish and Wildlife Service.

Department of Commerce. 1990, 1992, 1994, 1996. County and City Data Book. Washington, D.C.: U.S. Dept. of Commerce, Bureau of the Census.

Evers, A. 1993. The Catskills: From Wilderness to Woodstock. Woodstock, NY:The Overlook Press. (fourth printing, revised and updated from the 1972 edition, Doubleday, Garden City, NY).

Ford, D. E. 1990. Reservoir transport processes. Pp. 15–41 In Thornton, K. W., B. L. Kimmel, and F. E. Payne (eds.) Reservoir Limnology: Ecological Perspectives. New York, NY: J. Wiley & Sons, N.Y.

Hall, T. 1998. For sales of second homes: No vacation. New York Times, November 1, 1998.

Hazen and Sawyer/Camp Dresser and McKee. 1997. The New York City Water Supply System. New York, NY: Hazen and Sawyer/Camp Dresser and McKee.

Hazen and Sawyer/Camp Dresser and McKee. 1998. Task 4 Phase II. Pilot Testing Report EPA Submittal. Corona, NY: NYC DEP.

Isachsen, Y. W., E. Landing, J. M. Lauber, L. V. Rickard, and W. B. Rogers, eds. 1991. Geology of New York: A Simplified Account. Educational Leaflet No. 28. Albany, NY: New York State Museum/ Geological Survey.

Liyanage, L. R. J., G. R. Finch, and M. Belosevic. 1997. Effect of aqueous chlorine and oxychlorine compounds on *Cryptosporidium parvum* oocysts. Environmental Science & Technology 31(7): 1992.

Lossing, B. 1866. The Hudson: From the Wilderness to the Sea. Troy, NY: H.B. Nims & Co.

Millen, P. E. 1995. Bare Trees. Hensonville, NY: Black Dome Press Corp.

Mitsch, W. J., and J. Gosselink. 1993. Wetlands. 2nd Edition. New York, NY: Van Norstrand Reinhold.

Murdoch, P. S., and J. L. Stoddard. 1992. The role of nitrate in acidification of streams in the Catskill Mountains of New York. Water Resources Research 28(10):2707–2720.

National Research Council (NRC). 1995. Wetlands: Characteristics and Boundaries. Washington, DC: National Academy Press.

Nechamen, W. S., S. Pacenka, and W. Liebold. 1995. Assessment of New York City Residential Water Conservation Potential. New York, NY: NYC DEP.

New York City Department of Environmental Protection (NYC DEP). 1993a. Implications of Phosphorus Loading for Water Quality in NYC Reservoirs.

NYC DEP. 1993b. Final Generic Environmental Impact Statement for the Proposed Watershed Regulations for the Protection from Contamination, Degradation, and Pollution of the New York City Water Supply and its Sources. November 1993. Corona, NY: NYC DEP.

NYC DEP. 1998. Kensico Watershed Study Annual Research Report: April 1997-March 1998. Valhalla, NY: NYC DEP.

NYC DEP. 1999a. Proposed Phase II Phosphorus TMDL Calculations for Ashokan Reservoir. Valhalla, NY: NYC DEP.

NYC DEP. 1999b. Proposed Phase II Phosphorus TMDL Calculations for Cannonsville Reservoir. Valhalla, NY: NYC DEP.

NYC DEP. 1999c. Proposed Phase II Phosphorus TMDL Calculations for Neversink Reservoir. Valhalla, NY: NYC DEP.

NYC DEP. 1999d. Proposed Phase II Phosphorus TMDL Calculations for Pepacton Reservoir. Valhalla, NY: NYC DEP.

NYC DEP. 1999e. Proposed Phase II Phosphorus TMDL Calculations for Rondout Reservoir. Valhalla, NY: NYC DEP.

NYC DEP. 1999f. Proposed Phase II Phosphorus TMDL Calculations for Schoharie Reservoir. Valhalla, NY: NYC DEP.

NYC DEP. 1999g. Proposed Phase II Phosphorus TMDL Calculations for Kensico Reservoir. Valhalla, NY: NYC DEP.

NYC DEP. 1999h. Proposed Phase II Phosphorus TMDL Calculations for West Branch Reservoir. Valhalla, NY: NYC DEP.

Nolan, J. R. 1993. The erosion of home rule through the emergence of state interests in land use control. Pace Environmental Law Review 10(2):497–562.

Platt, R. H. 1996. Land Use and Society: Geography, Law, and Public Policy. Washington, D.C.: Island Press.

Platt, R., and V. Morrill. 1997. Sustainable water supply management in the United States: Experience in Metropolitan Boston, New York, and Denver. Pp. 292–307 In Shrubsole, D., and B. Mitchell (eds.) Practicing Sustainable Water Management: Canadian and International Experiences. Cambridge, Ontario: Canadian Water Resources.

Reisner, M. 1993. The American West and its disappearing water. Pp. 97–99 In Cadillac Desert. New York, NY: Penguin Books.

Stave, K. A. 1998. Water, Land, and People: The Social Ecology of Conflict over New York City's Watershed Protection Efforts in the Catskill Mountain Region, NY., Ph.D. Dissertation, Yale University, Department of Forestry and Environmental Studies, New Haven, CT.

Stoddard, J. L., and P. S. Murdoch. 1991. Catskill Mountains. Pp. 237–271 In Charles, D. F. (ed.) Acidic Deposition and Aquatic Ecosystems: Regional Case Studies. New York, NY: Springer-Verlag.

Tiner, R. W. 1997. Wetlands in the Watersheds of the New York City Water Supply System: Results of the National Wetlands Inventory. Prepared for the New York City Department of Environmental Protection, Valhalla, NY. Hadley, MA.:U.S. Fish and Wildlife Service, Ecological Services, Northeast Region.

Titus, R. 1993. The Catskills: A Geologic Guide. Fleischmanns, NY: Purple Mountain Press.

USDA Soil Conservation Service (SCS). 1979. Soil Survey of Ulster County. New York, NY: USDS SCS.

Warne, D. 1999a. NYC DEP. Memorandum to the National Research Council dated April 1999.

Warne, D. 1999b. NYC DEP. Memorandum to the National Research Council dated May 1999.

Weidner, C. H. 1974. Water for a City. New Brunswick, NJ: Rutgers University Press.

Wetzel, R. G. 1990. Reservoir ecosystems: Conclusions and speculations. Pp. 227–238 In Thornton, K. W., B. L. Kimmel, and F. E. Payne (eds.) Reservoir Limnology: Ecological Perspectives. New York, NY: John Wiley and Sons.

Wilstach, P. 1933. Hudson River Landings. Port Washington, NY: Ira J. Friedman, Inc.

3

Evolution of Key Environmental Laws, Regulations, and Policies

The Watershed Rules and Regulations of the New York City Memorandum of Agreement (MOA) are among the most comprehensive and detailed regulations regarding watershed activities found in this country. However, they cannot be considered in isolation given the large number of federal, state, and local statutes and regulations with which New York City must comply. This chapter briefly describes the initial impetus for the creation of environmental regulations relating to drinking water, it outlines certain federal regulations that pertain to the New York City drinking water supply, and it describes how the MOA attempts to fill gaps between federal, state, and local environmental regulations.

GLOBAL PUBLIC CONCERNS ABOUT DRINKING WATER SAFETY

Although environmental regulations have not traditionally focused on the watershed as a management unit, this approach has become more common in the last ten years (NRC, 1999). Watershed approaches to water resource management require the integration of traditional water quality concerns (such as the safety of drinking water) with more general ecological and aesthetic concerns (such as the health of aquatic ecosystems). For watershed protection programs to succeed, a multitude of interests, stakeholders, and priorities must be considered, creating daunting and relatively new challenges for water supply managers and environmental regulators. Some stakeholders are most concerned with water-borne diseases, others with the chronic effects of chemicals used in water treatment, and others with ecological and aesthetic considerations.

In addition to multiple stakeholder concerns, water supply managers and environmental regulators must also contend with increased public awareness regarding water safety. The emergence of the media has had a powerful effect on public perceptions of environmental quality, including water quality. A 1993 nationwide study of consumer attitudes about water quality conducted by the American Water Works Association (AWWA, 1993) found that about 40 percent of all persons interviewed nationwide had "seen or heard something [in the media] that made them worry about their local water quality," although fewer (26 percent) were aware of a local threat or incident. Although there have been few documented cases of contaminated water adversely affecting public health, the public's perception of water safety varies greatly. This variability is evident in Box 3-1, which describes the results of a survey conducted in New York City in which residents were questioned about their drinking water. Consumer confidence in the water supply is a complex issue that must be taken into consideration in making management decisions.

The most prominent public concerns regarding drinking water quality and water supply systems are discussed below, followed by a description of the relevant environmental laws and regulations that have been developed to address these concerns. These regulations are not only relevant to public health and safety and to aesthetic improvements in water supply reservoirs, but they are also economically prudent for society over the long term.

Waterborne Infectious Disease

As with most other environmental issues, the greatest public concern regarding drinking water is that it can potentially endanger human health and safety (Freudenberg and Steinsapir, 1992; Szasz, 1994). Waterborne infectious diseases are the most recognized type of danger because numerous microbial agents transmitted by ingestion of contaminated water have the potential to cause acute or chronic illness.

Disease outcomes associated with waterborne infections include mild to life-threatening gastroenteritis, hepatitis, conjunctivitis, respiratory infections, and generalized infections. Most disease-causing microorganisms (pathogens) originate in the enteric tracts of humans or animals and enter water sources via fecal contamination from human or animal sources. However, there are some indigenous aquatic microorganisms that are also capable of causing disease under certain circumstances. Although most waterborne pathogens must infect and reproduce within a human host in order to be considered virulent, there are waterborne microorganisms that affect humans through the production of toxins. The outcomes of exposure to pathogenic organisms can be highly variable. Some waterborne pathogens will cause disease in healthy exposed persons, while others pose little or no threat to healthy adults but can pose a threat to children, the elderly, or anyone with a weakened immune system.

The connection between drinking water quality and human health has been recognized since ancient times, as evidenced by the development of technologies

BOX 3-1
Public Perception of Drinking Water Safety
in New York City

This box discusses a recent study on public perception of drinking water quality in New York City and compares the results to trends observed nationwide. Public perception of drinking water safety is an important indicator of the success of public utilities, although it is not as objective as water quality monitoring data. Water suppliers and the public often have very different impressions of drinking water safety. For example, a nationwide study of multiple stakeholders concluded, "Most public officials, water utility managers, and environmentalists believe the greatest threat to the quality of drinking water today is problems with water quality at the source. But the public is just as likely to perceive that problems with [water] treatment methods pose the greatest threat" (AWWA, 1993). In addition, two-thirds of American adults believe that they receive very little or no information about water quality, while 151 water utility managers nationwide overestimated the amount of information the public perceives it is getting about the quality of its drinking water (AWWA, 1993).

Survey of Water Consumers. Because New York City's water is unfiltered, it has garnered increased attention compared to other communities. In a recent consumer survey of 1,560 randomly selected City residents (Pfeffer and Stycos, 1996), respondents were asked whether they agreed or disagreed with the following statement: "New York City has one of the safest and cleanest water supplies in the world." 54 percent of respondents agreed, while 46 percent disagreed. This indicator of confidence in the water supply reveals clear differences between City residents' perceptions of the overall quality of the water supply and the favorable compliance record of the NYC DEP (discussed in detail in Chapter 5).

Reasons for Lacking Confidence in Tap Water. To help determine why a substantial proportion of New York City's population lacks confidence in the water supply, survey respondents were asked more specific questions about their perception of tap water quality, their satisfaction with federal drinking water standards, whether they drink bottled water and why, whether they have had problems with tap water in the past, and other issues. Results of the survey are given in Table 3-1.

More than 69 percent of the New Yorkers sampled rate tap water quality as good or excellent (1087 out of 1560). This proportion compares quite favorably with results from a nationwide study in which about 58 percent of central city residents rated their tap water as either good or excellent (AWWA, 1993). As shown in Table 3-1, half (50.3 percent) of those who lack confidence in the water supply (i.e., disagree that New York City's water is safe and clean) rate the City's overall tap water quality as poor, while 87.1 percent of those who are confident in the water supply rate the water as good or excellent.

Differences in confidence in the water supply are also related to trust in the NYC DEP. Sixty (60) percent of those confident in the water supply give NYC DEP an overall job rating of good or excellent, while just one-third of those survey respondents who lack confidence in the water supply gave NYC DEP such high marks. Those who lack confidence are also much more likely to think that federal water quality standards are "not strict enough," whereas those confident in the system are more likely to think the guidelines are "about right."

Differences in levels of confidence in the water supply are also reflected in water consumption. Those lacking confidence are twice as likely to drink bottled water only. Only 31 percent of all New Yorkers drink tap water exclusively (482 out of 1560), a rate that is significantly lower than the rates in other parts of the country (AWWA, 1993). About 30 percent of New Yorkers say they drink bottled water for the taste; however, more than half report that they drink it for health reasons (data not shown).

The strongest predictor of confidence in the water supply were past problems with tap water taste, color, and clarity. Those New Yorkers who experienced problems with the taste of tap water were almost twice as likely to lack confidence in New York City water. Finally, differences in confidence in the water supply relate, to a lesser degree, to one's environmental awareness, income, education, sex, race, and place of residence. In general, the observed public perceptions of drinking water quality in New York City are typical of nationwide patterns reported by the AWWA (1993).

TABLE 3-1 Selected Factors Related to Public Confidence in New York City's Water Supply

	Total number responding	NYC's Water is Safe and Clean	
		Percent Agree	Percent Disagree
All respondents	1,560	54.0	46.0
Perception of overall tap water quality:			
Poor	465	12.3	50.3
Good	853	62.0	46.0
Excellent	234	25.1	3.1
Trust in DEP-provided information:			
No Trust	183	9.0	14.9
Some Trust	1,167	75.6	73.9
Complete Trust	197	14.4	10.6

continued

Box 3-1 Continued

Table 3-1 Continued

	Total number responding	NYC's Water is Safe and Clean	
		Percent Agree	Percent Disagree
Overall DEP job rating:			
Poor	153	4.9	15.6
Fair	608	31.7	47.6
Good	661	50.9	32.4
Excellent	90	9.5	1.4
Federal drinking water quality standards:			
Too strict	35	2.0	2.5
About right	601	50.4	24.5
Not strict enough	869	43.3	70.3
Water Consumption:			
Tap only	483	31.9	21.3
Bottled only	280	10.7	26.5
Tap and bottled	760	48.2	49.4
Problems with Tap Water			
Taste	723	36.8	63.2
Color/clarity	940	45.3	54.7
Were Aware of			
NYC DEP	1,452	54.0	46.0
NYC watershed	1,047	56.9	43.1
MOA	614	61.4	38.6
Cryptosporidium	309	52.8	47.2
Total Annual Household Income			
<$35,000	633	49.0	51.0
$35-70,000	485	52.8	47.2
>$70,000	225	68.4	31.6
Education			
Less than high school	151	53.6	46.4
High school	404	49.8	50.2
Post secondary	982	55.6	44.4
Borough of Residence			
Brooklyn	487	50.5	49.5
Bronx	487	54.2	45.8
Manhattan	308	56.5	43.5
Queens	411	54.7	45.3
Staten Island	80	61.3	38.8

Source: Pfeffer and Stycos (1996). Data were collected in New York City's five boroughs in Spring 1996 via telephone interviews of individuals living in households selected by means of random digit dialing. Individuals answering the call were asked for the person in their household age 18 or older who last had a birthday. That person was then interviewed, for an average time of 20 minutes. The interviews covered a wide range of topics including the subject's evaluation of world, national, and local environments, specific environmental concerns (especially those related to water quality), attitudes toward environmental problems, knowledge of environmental law, population, and environmental behavior, and their sociodemographic characteristics.

to treat contaminated water. Baker (1949) reports that descriptions of water treatment are found in Sanskrit medical lore and Egyptian inscriptions dating back to the fifteenth century B.C. Boiling water and filtration through porous vessels or through sand and gravel have been used to improve water quality for thousands of years. The writings of Hippocrates (460–354 B.C.) discuss the relationship between water quality and health and recommend boiling rain water and straining it through a cloth bag (NRC, 1977).

In the nineteenth and twentieth centuries, installation of filtration and disinfection processes in community water supplies has been associated with decreased morbidity and mortality from infectious diseases in the United States and Europe. Filtration was first used to remove particles from drinking water in New York State during the 1870s. Later experiments demonstrated that filtration had a direct impact on observed disease rates of typhoid fever. Disinfection of drinking water supplies, developed around the turn of the century, has virtually eliminated many waterborne microbial diseases in developed countries. Chlorine was and continues to be the most popular disinfectant because of its relatively low cost and high potency. Like other disinfectants, chlorine inactivates bacteria, viruses, and other microbes via nonspecific oxidation of the organism. Table 3-2 lists landmark advances in the discovery and control of waterborne infectious disease.

Common Waterborne Pathogens

The most commonly recognized waterborne pathogens consist of several groups of enteric and aquatic bacteria, enteric viruses, and enteric protozoa (Table 3-3). Data collected on waterborne disease outbreaks from 1920 to the present indicate that there has been a shift in the microorganisms responsible for waterborne disease. Recognized outbreaks during the first half of the twentieth century were caused by bacterial agents—primarily *Salmonella typhi* and *shigella* sp. Since the 1970s, recognized waterborne disease outbreaks have been caused predominantly by enteric protozoa such as *Giardia* or *Cryptosporidium* (when an etiologic agent is identified) or viral agents (Craun, 1986). This shift may be due to the greater resistance of protozoa to chlorination, which is shown in Table 3-4.

Table 3-3 lists only those organisms documented to have caused waterborne disease. Other pathogens that can be transmitted via ingestion of water include adenoviruses, *Helicobacter pylori,* atypical (nontuberculosis) mycobacteria, and Microsporidia (*Enterocytozoon* and *Septata*).

Treatment Options for Controlling Waterborne Pathogens

Most waterborne pathogens are removed or inactivated by conventional water treatment processes such as coagulation and sedimentation, filtration, and disinfection. Enteric viruses and bacteria are particularly susceptible to disinfection

TABLE 3-2 Advances in the Recognition and Control of Waterborne Disease

Year	Event
1829	First well-documented water filter built by James Simpson for the Chelsea Water Company of London.
1849	An estimated 110,000 people die from cholera in the UK.
1854	John Snow removes the handle from the Broad Street pump in an effort to stop the transmission of cholera in London.
1872–1874	First water filtration plants in the U.S. built in Poughkeepsie, NY, and Hudson, NY.
1884	Robert Koch identifies *Vibrio cholera* as the causal agent of cholera and describes the germ theory of disease.
1887	Experiments on water filtration conducted in Lawrence, MA. This leads to the first rapid sand filter in 1893 and an observed 79 percent decrease in typhoid fever mortality over the next 5 years.
1892	Rienecke observes that increases in the bacterial content of drinking water in Hamburg, Germany, corresponded to increases in infant mortality and reports a 50 percent decline in infant mortality from diarrheal disease in the year after Hamburg started to filter the public water supply.
1893	Chlorination used to treat sewage effluent in Brewster, NY, to protect New York City drinking water.
1897	Chlorination of drinking water in Maidstone, Kent, UK, after an outbreak of typhoid fever.
1902	First continuous chlorination of a water supply in Belgium.
1904	10 percent of U.S. urban population receives filtered water.
1907	46 U.S. cities using filtration to treat drinking water.
1908	First continuous, large-scale use of chlorination for an urban water supply in the U.S. in Jersey City, NJ.
1914	36 percent of U.S. urban population receives filtered water. Allan Hazen writes enthusiastically about the benefits of water chlorination.
1920	Earliest data on occurrence and causes of waterborne disease outbreaks in the U.S. is collected.
1930	27 percent of community water supplies in the U.S. have disinfection facilities.
1920–1935	Typhoid fever is the most commonly recognized waterborne disease in the U.S.
1936–1961	Shigellosis is the most commonly recognized waterborne disease in the U.S.
1965	Outbreak (16,000 cases) of waterborne salmonellosis in Riverside, CA. First documented waterborne outbreak of giardiasis in the U.S. occurs at Aspen, CO.
1971–1980	Giardiasis becomes the most commonly recognized waterborne disease.
1975	First recognized outbreak of waterborne disease caused by toxigenic *E. coli* in Crater Lake National Park, OR.
1984	First recorded waterborne outbreak of cryptosporidiosis occurs in Texas.
1989	First recorded waterborne outbreak of *E. coli* O157:H7 occurs in Missouri (243 cases, 4 deaths).
1993	Largest recorded waterborne disease outbreak in U.S. history caused by *Cryptosporidium* in Milwaukee, WI (estimated 400,000 cases).

Sources: Craun (1986), Hunter (1997), ILSI (1993), Longmate (1966), NRC (1977), Sedgwick and MacNutt (1910).

TABLE 3-3 Illnesses Caused by Microbial Agents Acquired by Ingestion of Water

Agent	Source	Incubation Period	Clinical Syndrome	Duration
Viruses:				
Astrovirus	human feces[1]	1-4 days	Acute gastroenteritis	2-3 days; occasionally 1-14 days
Enteroviruses (polioviruses, coxsackieviruses, echoviruses)	human feces	3-14 days (usually 5–10 days)	Febrile illness, respiratory illness, meningitis, herpangina, pleurodynia, conjunctivitis, myocardiopathy, diarrhea, paralytic disease, encephalitis, ataxia	Variable
Hepatitis A	human feces	15-50 days (usually 25-30 days)	Fever, malaise, jaundice, abdominal pain, anorexia, nausea	1-2 weeks to several months
Hepatitis E[2]	human feces	15-65 days (usually 35-40 days)	Fever, malaise, jaundice, abdominal pain, anorexia, nausea	1-2 weeks to several months
Norwalk-like viruses	human feces	1-2 days	Acute gastroenteritis with predominant nausea and vomiting	1-3 days
Group A rotavirus	human feces[1]	1-3 days	Acute gastroenteritis with predominant nausea and vomiting	5-7 days
Group B rotavirus[2]	human feces[1]	2-3 days	Acute gastroenteritis	
Bacteria:				
Aeromonas hydrophila	fresh water		Watery diarrhea	Average 42 days
Campylobacter jejuni	human and animal feces	3-5 days (1-7 days)	Acute gastroenteritis, possible bloody and mucoid feces	1-4 days occasionally >10 days

continued

TABLE 3-3 Continued

Agent	Source	Incubation Period	Clinical Syndrome	Duration
Enterohemorrhagic E. coli O157:H7	human and cattle feces	3-5 days	Watery, then grossly bloody diarrhea, vomiting, possible hemolytic uremic syndrome	1-12 days Average 7-10 days
Enteroinvasive E. coli	human feces	2-3 days	Possible dysentery with fever	1-2 weeks
Enteropathogenic E. coli	human feces	2-6 days	Watery to profuse watery diarrhea	1-3 weeks
Enterotoxigenic E. coli	human feces?	12-72 hours	Watery to profuse watery diarrhea	3-5 days
Plesiomonas shigelloides	fresh surface water, fish, crustaceans, animals	1-2 days	Bloody and mucoid diarrhea, abdominal pain, nausea, vomiting	11 days average
Salmonellae	human and animal feces	8-48 hours	Loose, watery, occasionally bloody diarrhea	3-5 days
Salmonella typhi[2]	human feces and urine	7-28 days (average 14 days)	Fever, malaise, headache, cough, nausea, vomiting, abdominal pain	Weeks to months
Shigellae	human feces	1-7 days	Possible dysentery with fever	4-7 days
Vibrio cholera O1[2]	human feces	9-72 hours	Profuse, watery diarrhea, vomiting, rapid dehydration	3-4 days
Vibrio cholera non-O1[2]	human feces	1-5 days	Watery diarrhea	3-4 days
Yersinia enterocolitica	animal feces and urine	2-7 days	Abdominal pain, mucoid, occasionally bloody diarrhea, fever	1-21 days average 9 days

Protozoa:

Balantidium coli[2]	human and animal feces	Unknown	Abdominal pain, occasional mucoid or bloody diarrhea	Unknown
Cryptosporidium parvum	human and animal feces	1-2 weeks	Profuse, watery diarrhea	4-21 days
Entamoeba histolytica[2]	human feces	2-4 weeks	Abdominal pain, occasional mucoid or bloody diarrhea	Weeks to months
Cyclospora cayetenensis[2]	human feces	1 week average	Watery diarrhea, profound fatigue, anorexia, weight loss, bloating, abdominal cramps, nausea	Weeks if untreated
Giardia lamblia	human and animal feces	5-25 days	Abdominal pain, bloating, flatulence, loose, pale, greasy stools	1-2 weeks to months and years

Algae:

Cyanobacteria (*Anabaena* spp., *Aphanizomenon* spp., *Microcystis* spp.)[2]	Algal blooms in water	A few hours	Toxin poisoning (blistering of mouth, gastroenteritis, pneumonia)	Variable

Helminths:

Dracunculus medinensis[2] (Guinea worm)	larvae	8-14 months (usually 12 months)	Blister, localized arthritis of joints adjacent to site of infection	Months

[1]Animal strains of these viruses are not believed to be pathogenic for humans.
[2]Waterborne infections in the U.S. are rare or undocumented.

TABLE 3-4 Chlorine Disinfection Requirements for 99 Percent Inactivation of Waterborne Pathogens

Organism	Temperature (°C)	pH	Chlorine (mg/L)	CT Value[a]
Escherichia coli	5	6.5	0.02–0.10	0.02
	23	7.0	0.10	0.014
Giardia lamblia	5	7.0	2.0–8.0	25.5–44.8
	25	7.0	1.5	<15
Cryptosporidium parvum	25	7.0	80	7,200

[a]CT = concentration of chlorine (mg/L) times the contact time (minutes)
Adapted from Sterling (1990).

by chlorine and ozone. Enteric protozoa, on the other hand, are relatively resistant to chlorine disinfection but are typically removed by the physical processes of coagulation and flocculation followed by filtration. Unfortunately, these treatment processes do not guarantee complete removal of microbial pathogens from drinking water, as evidenced by outbreaks of waterborne disease associated with water supplies that have conventional treatment. These disease outbreaks have generally resulted from (1) source contamination and the breakdown of one or more of the treatment barriers, (2) contamination of the distribution system, or (3) the use of untreated water. Box 3-2 describes how the failure of a water filtration plant and other factors contributed to a massive outbreak of the pathogenic protozoan *Cryptosporidium* in Milwaukee in 1993. Although treatment has reduced the incidence of waterborne disease in the United States since the nineteenth century, waterborne pathogens continue to pose a significant threat to public health in this country. Numerous outbreaks of waterborne disease are reported each year to the Centers for Disease Control and Prevention and represent only a fraction of the true burden of waterborne disease. Evidence of endemic waterborne disease comes from recent epidemiologic studies (described in Chapters 4 and 6 of this report).

Chemical-Related Health Effects

Although microbes cause the bulk of the waterborne disease in this country, chemicals also cause problems. For example, according to Craun and McCabe (1973), during the period 1946–1970, four chemical poisonings occurred in public supplies and eight occurred in private supplies (individual wells). Between 1961 and 1970, there were two and seven documented outbreaks in public and private supplies, respectively. In addition to acute poisonings, many chemicals cause chronic problems. For example, exposure of the human body to arsenic leads to both skin and lung cancer and nervous system toxicity. Of the known chronic problems, cancer is thought by many to be the most serious.

BOX 3-2
Outbreak of Cryptosporidiosis in Milwaukee

In 1993, the largest documented outbreak of waterborne disease in United States history, affecting an estimated 403,000 people, occurred in Milwaukee, WI. This outbreak is noteworthy for several reasons: (1) it involved a newly recognized waterborne pathogen, *Cryptosporidium*, (2) *Cryptosporidium* oocysts were detected in historical samples of ice so that past waterborne exposure could be estimated, (3) drinking water met standards for total coliforms and average turbidity, (4) the Milwaukee public water supply used standard water treatment processes of coagulation/flocculation, filtration, and disinfection; and (5) the frequency of testing patients for *Cryptosporidium* in this community was inadequate to detect this outbreak (MacKenzie et al., 1994).

The outbreak was first recognized around April 5, 1993, after reports of numerous cases of gastroenteritis and high rates of absenteeism in schools and hospital employees. Initially many cases were misdiagnosed as "intestinal flu" on the basis of the clinical symptoms and were not further investigated. Because laboratory testing for *Cryptosporidium* was not a routine procedure, recognition of *Cryptosporidium* as the causative agent of this outbreak was delayed. From March 1 through April 16, a total of 2,300 stool specimens were submitted to the 14 clinical laboratories in the Milwaukee area for routine examination of enteric pathogens. Twelve of these laboratories tested for *Cryptosporidium* only at the request of the physician, and by April 6, only 42 stool specimens had been examined for *Cryptosporidium* (29 percent were positive). On April 7, two laboratories identified *Cryptosporidium* oocysts in the stools of 7 adults in the Milwaukee area and, at the request of public health officials, the other 12 laboratories began to test all stool specimens for *Cryptosporidium*. From April 8 through April 16, *Cryptosporidium* oocysts were detected in 331 of 1,009 specimens (33 percent). Between March 1 and May 30, 739 *Cryptosporidium* infections were detected by the 14 laboratories.

Persons with laboratory-confirmed cases of *Cryptosporidium* infection with illness between March 1 and May 15 (N = 285) were compared with 201 persons who had experienced watery diarrhea during the same time period (as identified from telephone surveys in the Milwaukee area). The epidemiologic features and dates of illness were similar in both groups and suggested that many of the watery diarrhea cases were also caused by *Cryptosporidium*.

The total extent of the outbreak was estimated on the basis of a random digit telephone survey of 840 households in the greater Milwaukee area. Of the 1,663 respondents, 26 percent reported watery diarrhea in the period from March 1 though April 28. By extrapolating this rate to the total population of the greater Milwaukee area (1.6 million people) and

continued

BOX 3-2 Continued

subtracting a background rate of diarrhea of 0.5 percent per month (16,000 cases), it was estimated that 403,000 cases of *Cryptosporidium* were associated with this outbreak.

An early survey of nursing home residents, a relatively immobile population, indicated that *Cryptosporidium* infection was significantly higher during the first week of April among nursing homes supplied by water from the southern treatment plant. The household telephone survey provided additional information on the geographical distribution of the cases. The risk of watery diarrhea was 2.7 times higher among residents of the Milwaukee Water Works service area than it was among residents outside the service area. Within the service area, the highest attack rate (52 percent) was among residents served by the southern water treatment plant, and the lowest attack rate (26 percent) was among residents served by the northern water treatment plant. Residents that lived in the middle of the service area and could be exposed to water from either or both treatment plants reported an infection rate of 33 percent.

The Milwaukee Water Utility supplies water to approximately 800,000 people (Fox and Lytle, 1996). Two treatment plants treat Lake Michigan source water by conventional treatment processes including coagulation, sedimentation, filtration, and disinfection with chloramines. Waterborne transmission was suspected by April 7, when public health officials issued a boil water advisory and then closed the southern water treatment plant on April 9. The southern plant had observed highly variable treated water turbidity since around March 21, with peaks of 1.7 NTU on March 28 and March 30 and 2.7 NTU on April 5. Consumer complaints about poor quality drinking water were also reported to the Milwaukee Water Works during this period (Milwaukee Journal, 1993). At all times during this period, samples of treated water were negative for coliforms and met the Wisconsin regulations for turbidity. Investigation of the water treatment plant determined no evidence of an obvious mechanical breakdown in the flocculation and filtration system. However, difficulty in determining the appropriate dose of coagulant to aggregate particulates, failure to continuously monitor the turbidity from each filter bed, and recycling of filter backwash water were cited as possible factors contributing to this outbreak. In particular, the southern plant had switched to using a poly-aluminum chloride coagulant in August 1992 in order to have a higher finished water pH for corrosion control, to reduce sludge volume, and to improve coagulation effectiveness for cold raw water conditions (Fox and Lytle, 1996). When challenged with heavily contaminated raw water, the plant had little experience in adjusting the dosage of the polyaluminum chloride to optimize the chemical coagulation conditions. Jar-test data and consultation with the chemical supplier were used to guide adjustments to the coagulant dosage in response to fluctuations in turbidity. However, the short residence time for the water in the plant and the rapidly changing influent water quality made dosage optimization difficult, and on April 2 the operators switched back to alum as the primary coagulant.

Methods to detect *Cryptosporidium* oocysts in water are laborious and relatively inefficient. In this outbreak investigation, samples of ice made on March 25 and April 9 were melted and filtered to concentrate *Cryptosporidium* oocysts that were later detected by an immunofluorescent technique. Estimates of oocyst concentrations in these samples ranged from 2.6 to 13.2 oocysts per 100 L and from 0.7 to 6.7 oocysts per 100 L for March 25 and April 9, respectively. However, the epidemiologic data suggested that these may have been gross underestimates of contamination (possibly because of the effect of freezing and thawing on the oocysts and/or poor recovery from the filters). This is based on the fact that visitors to the Milwaukee Water Works service area who only consumed very small amounts ($\leq$ 240 mL) of water developed laboratory-confirmed cryptosporidiosis.

The source of the *Cryptosporidium* oocysts and the timing of the water contamination are still unknown. The number of cases with onset of illness before March 23 (when the filtered turbidity increases were noted) indicates that oocysts must have entered the water supply before the turbidity rise. Speculation about the effect of heavy rains and runoff from cattle farms and slaughterhouses into nearby rivers and Lake Michigan has yet to be confirmed. Several water treatment plants along Lake Michigan reported turbidity problems during March and April 1993 (Fox and Lytle, 1996).

Several public health recommendations came from this investigation. The adequacy of current microbiological water standards and the turbidity standard to protect the public from waterborne transmission of enteric protozoa was questioned. Continuous monitoring of treated water for turbidity and tightening the turbidity standard to $\leq$0.1 NTU were recommended. Particle counting was also suggested as a tool for monitoring treatment performance (Fox and Lytle, 1996). Changes in water treatment procedures related to filter maintenance and backwashing were implemented, and laboratory facilities for *Cryptosporidium* monitoring of raw and finished waters were established. MacKenzie et al. (1994) advocated the routine examination for *Cryptosporidium* oocysts in stools, although the infection is self-limited in the immunocompetent host and no effective treatment is available. Furthermore, the study's authors advised making cryptosporidiosis a reportable disease to improve the recognition of *Cryptosporidium* outbreaks in the United States. The AWWA (1993) observed that media coverage during the outbreak "may have influenced the public's agenda of concerns and made water quality a more salient issue."

The potential for chemicals in drinking water to cause cancer plays a major role in determining maximum contaminant levels (MCLs) and maximum contaminant level goals (MCLGs) set by the Environmental Protection Agency (EPA). MCLGs are nonenforceable standards, specified for all chemicals, that correspond to contaminant concentrations at which no known or anticipated adverse health effects will occur. For carcinogens and pathogens, MCLGs have currently been set at zero in order to provide complete public health protection. For some contaminants, EPA has developed MCLs, which are enforceable concentrations of chemicals that must not be exceeded in a drinking water supply. By law, a contaminant's MCL must be set as close to its MCLG as practicable.

MCLs represent minimum requirements for defining "safe" drinking water. Because our understanding of the health effects of contaminants will improve, today's standards are likely to differ from future standards. As a consequence, many forward-thinking water utilities strive to produce drinking water of better quality than the minimum safety requirements represented by MCLs.

Water utilities must monitor the concentrations of a large number of regulated contaminants. Most of these contaminants, such as pesticides, gasoline components, and cleaning solvents, are usually present at levels below their detection limit. However, one class of carcinogenic chemicals—disinfection byproducts (DBPs)—is commonly found at detectable levels in distributed drinking water.

Disinfection Byproducts

In the early 1970s, J. Rook discovered that free chlorine, a common drinking water disinfectant, reacts with nontoxic natural organic matter (derived mainly from decaying vegetation and associated fulvic and humic acids) to form trihalogen-substituted single carbon organic compounds (DBPs). Initially, only four DBPs were identified—the trihalomethanes (THMs) chloroform (trichloromethane), bromodichloromethane, dibromochloromethane, and bromoform (tribromomethane). Early animal testing indicated that chloroform was a suspected human carcinogen via ingestion. In spite of the lack of information on the health effects of bromodichloromethane, dibromochloromethane, and bromoform, EPA decided in 1979 to regulate all four compounds as a group because (1) they were all formed by the same mechanism, (2) all of the compounds were expected to show adverse health effects once testing was performed, and (3) the techniques for controlling their concentration were the same.

Intense study of DBPs over the last 25 years has added greatly to our knowledge of this issue. Among the more important findings has been the discovery that many more byproducts are formed when free chlorine reacts with natural organic matter than the four THMs originally identified. As of 1998, nearly 30 different organic byproducts have been identified, and even more are likely to exist. A comparison of identified compounds in chlorinated drinking waters to the total amount of halogen substitution, as measured by the group parameter total organic halogen, indicates that only about a third (on average) of the byproducts formed have been identified (Singer et al., 1995).

Another important outcome of DBP research has been the discovery of another group of nine compounds formed in concentrations near or in excess of THMs. These are the mono-, di-, and trihalogen-substituted acetic acids, or haloacetic acids (HAAs). As it has done with THMs, EPA is regulating HAAs as a group because the control techniques for each compound in the group are similar.

Evidence is mounting to show that the health effects of individual compounds within the DBP groups are quite variable. Box 3-3 contains a summary of the current known health effects of DBPs formed subsequent to chlorination of drinking water.

Treatment Options for Controlling DBPs

Water utilities faced with the challenge of controlling DBPs have two basic choices: (1) they can change to a disinfectant that is a weaker halogenating agent than free chlorine (such as chloramines) or to a non-halogen-containing disinfectant (such as ozone), or (2) they can control the concentrations of the natural organic matter precursors with which free chlorine reacts. The use of alternate disinfectants can pose considerable problems. Ozone is difficult to maintain in a water supply distribution system because of its rapid degradation. In addition, it can create ozone byproducts (e.g., bromate) and biodegradable organic matter, the latter of which can promote bacterial regrowth in the distribution system. Chloramines are weaker oxidants than chlorine; they are of lower effectiveness against viruses and *Giardia,* and they are virtually useless against *Cryptosporidium.* In addition, chloramines also produce DBPs, notably the dihalogen-substituted acetic acids. Chlorine dioxide, another possible disinfectant, can lead to the accumulation of chlorite and chlorate in water supplies, both of which pose health risks. Water supplies relying on alternate disinfectants are attempting to overcome these obstacles by using multiple, sequential chemical applications, such as ozone followed by chloramines.

At the present time, the most common long-term approach to preventing the formation of DBPs is precursor control. Lowering concentrations of humic and fulvic acids in stormwater and lowering concentrations of algae in water supply reservoirs are important first steps. Once in the source water, the only way to ensure removal of DBP precursors is by treatment prior to chlorination, for example by the addition of coagulants to promote settling of natural organic matter in turbid waters or by coagulation/filtration.

BOX 3-3
Summary of the Health Effects of DBPs
Derived from Chlorination

Cancer

Bladder, colon, and rectal cancer may be associated with drinking chlorinated water (EPA, 1998a, p. 69394). Bladder cancer has received considerable attention because of the abundance of high-quality studies compared to other types of cancer. According to EPA, the number of potential bladder cancer cases that could be associated with exposure to DBPs in chlorinated surface water is estimated to be in an upperbound range of 1,100–9,300 per year (EPA, 1998b). However, given past epidemiological studies, there are insufficient data to conclusively demonstrate a causal association between exposure to DBPs in chlorinated surface water and cancer. Nonetheless, EPA believes the overall weight of evidence from available epidemiological and toxicological studies on DBPs and chlorinated surface water supports a hazard concern and has therefore decided to regulate these compounds as a prudent public health measure.

Chloroform. Since 1996, EPA has advocated a nonlinear approach for estimating the additional lifetime carcinogenic risk associated with exposure to chloroform via drinking water. However, after considerable debate, the current MCLG for chloroform currently stands at zero (EPA, 1998c). EPA believes that the MCL of 0.080 mg/L for THMs is appropriate for the protection of the public from chloroform. The current MCL is thought to not only provide protection against chloroform, but also to provide protection against several other potentially hazardous DBPs, such as bromodichloromethane and bromoform.

Dichloroacetic Acid. Since 1994, EPA has maintained that dichloroacetic acid (DCA) is a probable human carcinogen (i.e., a Group B2 carcinogen). Because the data needed to determine a dose-response relationship for DCA are inadequate, the MCLG for this compound remains at zero to assure maximum public health protection (EPA, 1998d).

Adverse Reproductive and Developmental Effects

A recent study has demonstrated that consumption of tap water containing high concentrations of THMs, particularly bromodichloromethane, is associated with an increased risk of early-term miscarriage (Waller et al., 1998). Although this association did not constitute proof that exposure to THMs causes early term miscarriages, miscarriage has added to the list of possible adverse effects of DBPs on human health. Examining the relationship between DBPs and adverse reproductive and developmental effects is the primary activity of EPA's epidemiology and toxicology research program. In addition to conducting follow-up studies to the 1998 Waller et al. report, EPA is working with the California Department of Health Services to improve estimates of exposure to DBPs in a study population by developing a DBP exposure database. EPA is also collaborating with the Centers for Disease Control and Prevention (CDC) and the National Toxicology Program to screen individual DBPs for reproductive and developmental effects. In addition to miscarriage, the effect of bromodichloromethane on male reproduction is being investigated. Data gathered from these efforts will be used to tighten regulatory goals for both THMs and HAAs.

Ecological and Aesthetics Considerations

Although adverse health impacts are of primary importance, additional concerns about the quality of a drinking water supply include taste, odor, color, and turbidity. In addition, eutrophication of water supply reservoirs caused by nutrient enrichment can negatively affect both human and ecological receptors in multiple ways. All these problems are directly or indirectly related to biological growth within water supply reservoirs and to how that growth responds to changing environmental conditions.

Drinking Water Aesthetics

The public expects its drinking water to be clear and free of taste and odors. Pure water is a neutral medium and alone cannot produce either odor or taste sensations. The compounds responsible for water taste and odor include most organic and some inorganic chemicals, although some substances (e.g., certain inorganic salts) produce taste without odor. Nearly all the natural compounds that generate offensive tastes and odors in water are the result of dissolved organic substances released by actively growing microorganisms or released during decomposition of algae and higher vegetation. Although a number of assays exist to estimate the relative taste and flavor quality of water, most are highly subjective (e.g., American Public Health Association, 1995). (It should be noted that drinking water held for long periods within the distribution system can also acquire tastes and odors from metals and from organic compounds released by microorganisms within the distribution system.)

The quality of water is affected by both color and turbidity. The observed color of water is the result of light being scattered upward after it has passed through the water and has undergone selective absorption en route. In pure water, molecular scattering of light is a function of frequency, making the high-frequency color blue the dominant observed color. Beyond these physical factors, most of the true color of natural waters results from dissolved organic compounds such as humic and fulvic acids, which are derived from the partial decomposition of structural tissues of higher plants. Waters containing large amounts of dissolved organic compounds have a light yellow or brown organic tint. Under certain conditions of high acidity, natural metallic ions (iron and manganese) can result in a variety of water colors.

Because of its importance in determining consumer satisfaction, color in drinking water is routinely measured. Color has two decisive characteristics—color intensity (brightness) and light intensity (lightness)—that make discriminating between colors extremely subjective. For this reason, a number of color scales have been devised to empirically compare the true color of lake water, after filtration to remove suspended particles (cf. Wetzel and Likens, 1991; American Public Health Association, 1995). Consumers generally do not complain about drinking water with a color indicator of less than 15 Standard Color Units (SCU) (NYC DEP, 1999).

Turbidity in natural waters is caused by suspended and colloidal particulate matter such as clay, silt, and other finely divided inorganic and organic (both living and dead) matter. These particles scatter light of specific wavelengths and can be responsible for the observed color of water supply reservoirs. For example, colloidal $CaCO_3$, common to hard-water lakes and some of the New York City reservoirs, scatters light in the greens and blues and gives these waters a very characteristic blue-green color. Although some turbidity is generated by phytoplankton in water supply reservoirs, most turbidity problems emanate from watershed erosion during heavy precipitation events. Although low to moderate levels of turbidity are not generally harmful to human health, excessive levels (above 5 NTU) are considered aesthetically displeasing and will result in consumer complaints.

Eutrophication

Eutrophication is a general term that refers to an array of conditions associated with increased growth and productivity of organisms in aquatic ecosystems. Eutrophication occurs when elevated supplies of macronutrients, particularly phosphorus and nitrogen, are delivered to surface waters. These nutrients promote the growth of algae, photosynthetic and heterotrophic bacteria, and higher aquatic plants. The increased production of algae and their associated organic matter can negatively alter conditions within the reservoirs in many ways. Turbidity may be affected because of the presence of algal material and algal byproducts. The increase in total organic carbon derived from algal biomass can lead to formation of DBPs in the water distribution system. Algae can produce potentially toxic compounds, some of which may create taste and odor problems. And heterotrophic bacteria introduced by natural or human sources can consume the increased dissolved organic matter, thereby depleting dissolved oxygen levels within the reservoirs and destroying fish habitat.

In the past several decades, society has recognized the cost effectiveness of reducing pathogen, chemical pollutant, and nutrient loadings to recipient lakes and reservoirs for maintenance of high water quality. Relatively simple control measures within drainage basins can be implemented to minimize pollutant loadings, and such measures are often much more economical than treating degraded water supplies.

FEDERAL ENVIRONMENTAL REGULATIONS

One of the most important mechanisms used to ensure the delivery of safe drinking water is enforcement of environmental regulations. Water supplies must comply with a plethora of environmental regulations stemming from the Clean Water Act (CWA, 33 USCA, Section 1151 et seq., 1972) and the 1974 Safe Drinking Water Act (SDWA), laws passed in response to degraded water quality in major bodies of water and drinking water across the country. These laws target microbial pathogens and associated waterborne disease, chemical contaminants of drinking water, and aesthetic and ecological considerations related to water quality. The CWA focuses jointly on human and aquatic ecosystem health by establishing a water quality standard of "fishable and swimmable" that is applied to all bodies of water, including sources of drinking water. The SDWA focuses on human health by setting drinking water standards, a process that began as early as 1925. Most of the federal laws written since 1970 have been amended several times to incorporate new science and technology. For example, amendments to the SDWA require regulation and monitoring of new biological and chemical contaminants. Table 3-5 shows how federal regulations regarding water quality have evolved during the 20th century.

Over the last ten years, there has been a growing interest in taking a watershed approach to evaluating water quality. This is particularly true within EPA, which has developed a watershed framework (EPA, 1996; NRC, 1999). In general, there are no federal laws mandating watershed management for source water. However, rules developed from the 1996 SDWA amendments require states to assess watershed conditions and create watershed control programs for surface water supplies that are not filtered. A number of bills have been proposed in Congress that would mandate the watershed approach on a more widespread basis, but none has yet made it into law.

TABLE 3-5 Federal Regulations Timeline

1893	**Interstate Quarantine Act**
1899	**Rivers and Harbors Act**
1912	First drinking water-related regulation (McDermott, 1973).
1914	First federal drinking water standards, binding only to interstate carriers (Borchardt and Walton, 1971).
1925	First inclusion of chemicals in drinking water standards (U.S. Public Health Service, 1925).
1946	Hexavalent chromium added to list of regulated drinking water contaminants.
1948	**Federal Water Pollution Control Act**
1962	Last drinking water standards issued by the U.S. Public Health Service under the Interstate Quarantine Act; 28 regulated contaminants; microbial standards binding only on about 700 interstate carrier systems; chemical standards nonbinding.
1972	**Federal Water Pollution Control Act Amendments** (P.L. 92-500)
1974	First publications on formation of trihalomethanes following chlorination.
1974	**Safe Drinking Water Act** (P.L. 93-523). Specified drinking water regulations that apply to all community water supplies.
1975	Interim Primary Drinking Water Regulations, based on the 1962 U.S. Public Health Service drinking water standards.
1977	**Clean Water Act** (P.L. 95-217). This law amends the Federal Water Pollution Control Act.
1979	**Reauthorization of the SDWA** (P.L. 96-63).
1979	Trihalomethane final rule.
1983	First use of the term "best available technology" (BAT).
1986	Fluoride final rule.
1986	**SDWA amendments** (P.L. 99-339). 83 specified contaminants must be regulated. 25 additional contaminants are to be added every three years.
1987	Volatile organic contaminants, final rule.
1988	**Lead Contamination Control Act** (P.L. 100-572)
1989	Surface Water Treatment Rule, final rule.
1989	Total Coliform Rule, final rule.
1991	Lead and Copper Rule, final rule.
1996	Information Collection Rule, final rule.
1996	**SDWA amendments** (P.L. 104-182) Eliminated requirement for 25 new regulated contaminants every three years. Requires development of Candidate Contaminant List.
1998	Drinking Water Candidate Contaminant List, final.
1998	Consumer Confidence Reports, final rule.
1998	Stage 1 Disinfectants/Disinfection By-Products Rule, final rule.
1998	Interim Enhanced Surface Water Treatment Rule, final rule.

Note: Bolded entries indicate federal statutes.

Safe Drinking Water Act

In the early 1970s, several scientific factors came together that prompted Congress to draft new legislation on drinking water quality. First, a community water supply study showed that 41 percent of 969 water supplies did not meet current drinking water standards set by the U.S. Public Health Service (McCabe

et al., 1970). Further evidence of drinking water contaminated by organic compounds was found in New Orleans (Harris and Brecher, 1974). Meanwhile,
ongoing research in the Netherlands and at an EPA research laboratory confirmed
that chlorine could combine with harmless natural organic compounds in water to
produce chloroform and other THMs (Bellar and Lichtenberg, 1974; Rook, 1974).
During the same time period, the National Cancer Institute announced that
chloroform was a suspected human carcinogen (National Cancer Institute, 1976).
These combined factors led Congress, after four years of effort, to pass the
SDWA in 1974.

The SDWA contains multiple provisions for assessing, controlling, and preventing biological and chemical contamination of drinking water supplies, both
surface water and groundwater. In addition to other requirements, drinking water
suppliers must abide by a variety of rules that target specific contaminants and
water supplies. The most recent rules reflect EPA's desire to balance the risks of
microbial pathogens with risks from chemical contaminants, most notably DBPs.

Surface Water Treatment Rule

The Surface Water Treatment Rule (SWTR, 40 CFR Part 141), promulgated
by EPA on June 29, 1989, requires that all surface water systems treat their water
by filtration unless it can be proven to be unnecessary. The SWTR describes
improved criteria for filtration and disinfection treatment processes to control for
Giardia and viruses. In doing so, two important metrics of success for treatment
operations are introduced: (1) log removal of microbial pathogens, which refers
to a decrease in an organism's concentration by a factor of ten, and (2) CT, the
product of disinfectant concentration (C) and the contact time (T), as the control
parameter for disinfection.

For those water supply systems that were not filtering at the time of the
SWTR promulgation, the rule requires such systems to either begin filtration or
develop a monitoring program demonstrating that filtration is unnecessary. Continued avoidance of filtration is a possibility, provided that a water supply system
satisfies a variety of conditions outlined below. These conditions (and other
important criteria) are documented in a Filtration Avoidance Determination
(FAD) that is issued by EPA to the water supply system.

Source Water Quality. Avoidance of filtration is intended to be applicable
only to those waters that have historically excellent water quality, as measured by
a variety of parameters. Prior to disinfection, the fecal coliform concentration in
the source water must be less than 20 colony forming units (CFU)/100 mL in at
least 90 percent of the samples taken, or the total coliform density must be less
than 100 CFU/100 mL in at least 90 percent of the samples taken. These data
must be based on monitoring results during the previous six months. For utilities

serving a population of over 25,000, a minimum of five coliform samples must be taken each week.

Turbidity of the source water prior to disinfection cannot exceed 5 NTU, based on continuous sampling or grab samples collected every four hours that the system is in operation. A system occasionally may exceed the 5-NTU limit and still avoid filtration as long as (1) the state determines that each event occurred because of unusual or unpredictable circumstances and (2) not more than two such events have occurred in the past 12 months, or not more than five such events in the past 120 months. An event is defined as a series of consecutive days in which at least one turbidity measurement each day exceeds 5 NTU. If the system is unable to comply with these criteria, it is required to install filtration.

Disinfection Criteria. Disinfection of unfiltered water must achieve 99.9 percent (3-log) inactivation of *Giardia* cysts and 99.99 percent (4-log) inactivation of viruses. The effectiveness of the disinfection process must be demonstrated every day by meeting minimum CT values specified in the SWTR guidance manual (EPA, 1990). Filtration must be installed if the system fails to meet this requirement on more than one day a month, during two or more months within a consecutive 12-month period.

In order to prevent regrowth of bacteria in the distribution system, the SWTR requires that a "residual" concentration of disinfectant exist in all finished waters. The disinfectant residual at the entry point of the distribution system water cannot be less than 0.2 mg/L for more than 4 hours. If the residual at the entry point falls below 0.2 mg/L for any length of time, the water utility must notify the state, regardless of the type of disinfectant used. Continuous monitoring for disinfectant residual is required of systems serving more than 3,300 persons.

Deeper within the distribution system, the disinfectant residual cannot be undetectable in more than 5 percent of the samples in a month for any two consecutive months that the system serves water to the public. A system may measure for heterotrophic plate count (HPC) in lieu of disinfectant residual in the distribution system. A sampling site with an HPC level of less than 500 CFU/mL is considered to have a "detectable" residual for compliance purposes. Finally, systems providing disinfection as the only treatment must provide redundant disinfection equipment. This includes auxiliary power, automatic start-up, and an alarm or an automatic shutoff of water delivery to the distribution system when the disinfectant residual level at the entry point drops below 0.2 mg/L.

Site-Specific Criteria. Four other extremely important criteria have been established for unfiltered water supply systems. First, an effective watershed control program must be established and maintained. Second, an annual on-site inspection conducted by the state (or a third party approved by the state) is required. This inspection must consider the effectiveness of the watershed control program, the condition and protection of the source intake(s), the physical condition of operating equipment, the adequacy of operating procedures, the existence of monitoring records, and areas for improvement. Third, an unfiltered system must demonstrate an absence of waterborne disease outbreaks. If the system has been identified as the source of a waterborne disease, then it must be modified sufficiently to prevent further outbreaks. Finally, the water supply system must comply with the total coliform and total trihalomethane MCLs.

Total Coliform Rule

The purpose of the Total Coliform Rule (TCR) is to prevent waterborne microbial disease by requiring water suppliers to test drinking water for potentially harmful microorganisms. The total coliform group of bacteria is used to indicate the presence or absence of pathogenic organisms. Measurements of total coliform are often supplemented by the more specific bacterial indicators, fecal coliforms and *Escherichia coli.*

For water systems analyzing at least 40 water samples per month, no more than five percent of the samples may be positive for total coliforms, while for systems analyzing fewer than 40 samples per month, no more than one sample per month may be positive. In addition, if repeated testing of a water sample demonstrates sequential positive tests for total or fecal coliforms, the water supply is considered out of compliance with the TCR (Pontius, 1990). Total coliform measurements are generally made at the same frequency and locations as measurements for disinfectant residual (EPA, 1989).

Lead and Copper Rule

The Lead and Copper Rule (LCR) requires increased evaluation of treatment processes that control corrosion, with the goal of optimizing these processes. Several chemical additives are commonly used in water supply systems to prevent the chemical and biological corrosion of lead and copper pipes. For example, orthophosphate and sodium hydroxide are common additives that increase pH and neutralize corrosion-causing acid compounds. The LCR mandates enhanced sampling of distribution system water to determine the extent of corrosion and the efficacy of treatment processes.

Information Collection Rule

EPA's attempt to balance microbial and chemical risks of drinking water is most apparent in three recent rules that were the product of negotiated rulemaking in the early 1990s: the Information Collection Rule (ICR), the Disinfectants/Disinfection By-Products (D/DBP) Rule, and the Enhanced Surface Water Treatment Rule (ESWTR). The ICR mandates water supply systems to collect water quality data that will be used to form a national database of important parameters such as microbial pathogens and DBPs. These data, which will be available in late 1999, will guide EPA in the development of future drinking water regulations.

Rules following the 1996 SDWA Amendments

The 1996 amendments to the SDWA specified that EPA must promulgate two new rules by November 1998, both of which would substantially impact filtered and unfiltered water supplies (EPA, 1997a,b). The D/DBP Rule is intended to better control concentrations of disinfectants and disinfection byproducts, while the ESWTR is targeted at controlling the pathogenic protozoan *Cryptosporidium*. The D/DBP Rule will be implemented in two stages (Stage 1 D/DBP and Stage 2 D/DBP); the ESWTR has been broken up into three parts: the Interim ESWTR (IESWTR), the Long-term 1 ESWTR (LT1ESWTR), and the Long-term 2 ESWTR (LT2ESWTR). Target dates for the proposal, promulgation, and enforcement of these rules are shown in Table 3-6.

D/DBP Rule. The D/DBP Rule specifies updated MCLs for total trihalomethanes (TTHMs) and new MCLs and MCLGs for the sum of five haloacetic acids (HAA5) and the inorganic DBPs bromate and chlorite. Bromate has recently become an issue in water supply systems that use ozone as a primary disinfectant. Ozonation can result in the production of bromate in waters that contain bromide. Chlorite in drinking water is also a relatively recent discovery. It can be derived from the breakdown of sodium hypochlorite ($NaOCl$), or from the primary disinfectant chlorine dioxide.

The Stage 1 MCLs are 0.080 mg/L for TTHM, 0.060 mg/L for HAA5, 1.0 mg/L for chlorite, and 0.010 mg/L for bromate. The rule also specifies new maximum residual disinfectant levels (MRDLs) for chlorine, chloramine, and chlorine dioxide. The upper bound on disinfectant residuals was formerly determined by the taste and odor of the finished water. The MRDL for chlorine, the

TABLE 3-6 Anticipated Regulatory Schedule for the Disinfectants/Disinfection By-Products Rule and the Enhanced Surface Water Treatment Rule

	Proposed	Final	Effective
Stage 1 D/DBP	1994	Dec. 1998	Dec. 2001
Interim ESWTR	1994	Dec. 1998	Dec. 2001
LT1ESWTR	Nov. 1999	Nov. 2000	Nov. 2003
Stage 2 Reg. Neg.[a]	Begin 1999	End 2000	
Stage 2 D/DBP	Nov. 2000	May 2002	May 2005[b]
LT2ESWTR	Nov. 2000	May 2002	May 2005[b]

[a]Negotiated Regulation Process.
[b]Date uncertain.

most commonly used primary disinfectant, will be 4 mg/L. Recommendations are given for best available technologies to achieve these new MCLs, MCLGs, MRDLs, and maximum residual disinfectant level goals (MRDLGs).

Although negotiations for the Stage 2 rule are ongoing, "placeholder" MCLs have been discussed for TTHMs and HAA5, and they are stricter than the Stage 1 MCLs. These placeholder MCLs, which represent starting points for the negotiation process, are 0.040 mg/L for TTHMs and 0.030 mg/L for HAA5.

Enhanced Surface Water Treatment Rule. The IESWTR requires a tightening of turbidity MCL requirements for filtered systems from 0.5 NTU to 0.3 NTU. The new turbidity MCL is being required of filtered systems to enhance treatment performance and associated health benefits. (Turbidity happens to be an excellent indicator of filtration performance.) Similar requirements for unfiltered systems, which currently must meet a turbidity requirement of 5 NTU, are not anticipated until the LT2ESWTR. Box 3-4 explains why the turbidity requirements for filtered systems and unfiltered systems differ by more than an order of magnitude.

Although the final ESWTR will not be promulgated soon, several additional issues are likely to be incorporated. Microbial benchmarking will be required of systems that have problems with DBPs.[1] A system can be forced to benchmark its disinfection process if either its TTHM or HAA5 levels are 80 percent of the MCL as an annual average. EPA is currently advocating a multiple-barrier approach to pathogen removal that includes source water protection, filtration, and disinfection. To demonstrate that this approach is being used, the ESWTR will require drinking water systems to show that both filtering and disinfection are reducing concentrations of *Cryptosporidium*, an organism that is currently not regulated as part of the SDWA. Unfiltered supplies will specifically be asked to amend watershed control programs to control for *Cryptosporidium*.

Drinking Water Candidate Contaminant List. As part of the 1996 amendments to the SDWA, EPA must decide, on a regular basis, whether to regulate new and emerging drinking water contaminants. In order to gather information on the occurrence of new contaminants to assist in these judgments, EPA has issued a Drinking Water Candidate Contaminant List, published on March 2, 1998 (EPA, 1998e). This list contains 50 chemical and 10 microbial contaminants/contaminant groups for possible future regulation. Water utilities will be collecting occurrence data on these contaminants over the next few years.

Source Water Assessment Program. Although much of the SDWA consists of drinking water standards for particular contaminants, some elements of

---

[1] This process involves calculating daily levels of *Giardia* inactivation and plotting them over a year as a "benchmark" of how well the disinfection system is working.

Box 3-4
Turbidity Requirements under the ESWTR

Under the ESWTR, filtered systems will be required to meet a turbidity MCL of 0.3 NTU while unfiltered systems must meet a turbidity MCL of 5 NTU. The reason for the considerable discrepancy has to do with the impact of turbidity on the primary treatment process. In filtered systems, turbidity is mainly a cause for concern because it indicates ineffective filtration. Lowering the turbidity requirement for filtered systems is seen as a regulatory mechanism to improve the performance of filtration. In unfiltered systems, turbidity is most likely to disrupt the disinfection process. Particles responsible for turbidity have been shown to physically shield pathogens from disinfectants (Symons and Hoff, 1976). Five (5) NTU is thought to be sufficient to prevent turbidity from impairing disinfection. This has been supported by anecdotal evidence from managers of unfiltered systems, who argue that they have not experienced a documented waterborne disease outbreak in complying with the turbidity MCL of 5 NTU.

In the committee's opinion, however, the turbidity standard of 5 NTU for unfiltered systems has not been adequately justified. This is because other factors may increase the relative risk of turbidity in unfiltered systems compared to filtered systems. The particles present in unfiltered source water, as measured by turbidity, are much more dangerous than the particles in filtered water that have escaped filtration. That is, unfiltered water turbidity may contain pathogenic microorganisms, while filtered water turbidity is likely to consist almost entirely of small floc. This difference between turbidity in source waters and turbidity in filtered waters was borne out in Colorado several years ago. At that time, the filtered water turbidity standard was 1.0 NTU. Several water suppliers with relatively clear mountain streams as sources thought that if the source water turbidity was below 1.0 NTU, the resulting drinking water would also meet the turbidity standard. On an occasion when the source water turbidity was below 1.0 NTU, the operators stopped adding coagulant, which prevented filtration from occurring efficiently. This practice resulted in several outbreaks of giardiasis, indicating that sources with turbidities of less than 1.0 NTU are not equivalent to filtered water with a turbidity of 1.0 NTU (Hopkins et al., 1983).

the law specify regulations for watershed management. In particular, the 1996 SDWA amendments established a new state program, the source water assessment program (SWAP), to provide a strong basis for developing, implementing, and improving source water protection. State source water assessment programs were submitted to EPA for review by February 6, 1999, and implementation of the programs should begin immediately after their approval. A major goal of the SWAP is to delineate the boundaries of watersheds that supply drinking water systems, using all reasonably available hydrogeologic information. Once these areas are defined, the SWAP should determine the susceptibility of public drinking water systems in these areas to regulated contaminants (and unregulated contaminants as specified by the state) (Pontius, 1997).

Enforcement Authority

EPA is required by law to allow states to assume responsibility for implementing and enforcing the SDWA. In order to obtain this responsibility, known as primacy, states must pass, implement, and enforce laws that EPA must review and approve. If state policies are not approved, enforcement responsibilities for the SDWA remain with EPA.

Like most states, New York has primacy for the SDWA, which is delegated to the New York State Department of Health (NYS DOH). However, given the unfiltered status of the Catskill/Delaware water supply, EPA was asked to retain SDWA primacy specifically for this system until May 15, 2007. For all other public water supplies in New York State, NYS DOH is the primary enforcement agency.

Clean Water Act

The CWA is the act under which EPA regulates the quality of the nation's surface waters, including wetlands. In addition to strengthening the nation's water quality standards system, this legislation regulates discharges into bodies of water, it encourages the use of the best available technology for pollution control, and in the past it provided billions of dollars for construction of wastewater treatment plants (WWTPs). Thus, while the SDWA targets the quality of drinking water at or near the point of use, the CWA focuses more on the quality of source waters and on discharges into those waters.

The CWA has two primary objectives: (1) to regulate the discharge of pollutants into the nation's waters and (2) to achieve water quality levels that are fishable and swimmable. The act and accompanying regulations from EPA provide a comprehensive framework of standards, technical tools, and financial assistance to address the many causes of poor water quality, including municipal and industrial wastewater discharges, polluted runoff from urban and rural areas, hydrologic modification, and habitat destruction.

Water Quality Standards

The CWA requires states to establish specific water quality standards appropriate for their waters and to develop pollution control programs to meet these standards. First, the states must designate their waters by a use classification such as drinking water, aquatic life, recreational, or industrial, among others. Second, they must designate narrative and numeric water quality criteria for specific chemicals that correspond to the particular use classifications. Both these measures constitute water quality standards (i.e., use classification + water quality criteria = water quality standard). Every two years, the states must report to EPA the overall health of their waters and whether the waters are meeting water quality standards. The water quality parameters that are most important for a drinking water supply include certain microbes, turbidity, nutrients, total organic carbon, and toxic compounds, examples of which are listed in Table 3-7.

Water quality standards across the states vary considerably for specific parameters. Obviously, for waters with different use classifications, there are variable water quality criteria. For example, in New York, a Class A, AA, B, or C water must not exceed 190 µg/L dissolved arsenic, while a Class D water must not exceed 360 µg/L dissolved arsenic. For those waters that have been classified as a drinking water source, the water quality criteria that must be met are sometimes MCLs. For those parameters that do not have an MCL, there can be considerable variability among the states regarding the allowable concentrations of these pollutants. Table 3-7 compares seven states' water quality criteria for surface waters that are a source of drinking water. A comprehensive list of New York's water quality standards is found in Appendix B.

For the common source water parameters listed in Table 3-7, there is some chemical-specific variation among the seven states. New York State standards are similar to those of Virginia and North Carolina, eastern states with a significant number of surface water supplies. All three states have narrative standards for nitrogen and phosphorus, although New York has interpreted its narrative standard for phosphorus into a guidance value of 20 µg/L.

It should be kept in mind that not all the water quality standards for the parameters listed in Table 3-7 are based on drinking water uses. For example, the New York standard for phosphorus is based on aesthetics and contact recreation, while the chlorine and ammonia standards listed for all states are based on fish survival and propagation. Interestingly, the western states (except for Washington) have slightly less stringent water quality criteria than the eastern states, making water treatment more important to those states and deemphasizing the role of source water protection.

Total Maximum Daily Loads

Many waters are not currently meeting their designated beneficial uses established by states to satisfy Section 303 of the CWA. For those waters that are

TABLE 3-7 State Water Quality Standards for Surface Waters Used as a Source of Drinking Water (all values in mg/L unless specified otherwise)

Parameter	New York[a]	Arizona[b]	Colorado[c]	North Carolina[d]	Utah[e]	Virginia[f]	Washington[g]
Fecal Coliform	monthly geo. mean: < 200/ 100 mL	1,000 per 4,000 mL	mean: 2,000/ 100 mL	monthly geo. mean: < 50–200/ 100 mL	max: 2,000/ 100 mL	monthly geo. mean: < 200/ 100 mL	mean: 100/ 100 mL
Phosphorus	Narrative[h]			Narrative[i]		Narrative[j]	
Nitrogen	Narrative[h]			Narrative[i]		Narrative[j]	
Chlorine[k]	0.005			0.017	0.025	0.011	
Chloroform[l]	0.007						
Ammonia[k]	2		0.5			0.08–2.5	
DO[m]	> 5	6.0	3.0	> 5	5.5	> 5	8.0
pH	6.5–8.5	5–9	5–9	6–9	6.5–9	6–9	6.5–8.5
Mercury[n]	0.002	0.002		1.2×10^{-5}		5×10^{-4}	

[a]NYS DEC (1994).
[b]Arizona Department of Environmental Quality (1996).
[c]Colorado Department of Public Health and the Environment (1996).
[d]North Carolina Department of Environment and Natural Resources (1998).
[e]Utah Department of Environmental Quality (1994).
[f]Virginia Department of Environmental Quality (1988).
[g]Washington Department of Ecology (1993).
[h]None in amounts that will result in the growth of algae, weeds, and slime that will impair the waters for their best uses, currently interpreted by the NYS DEC to be 20 μg/L.
[i]For those waters classified as nutrient-sensitive waters, no increase over background levels.
[j]Nonspecific narrative for nutrients.
[k]Chlorine and ammonia standards based survival of aquatic species.
[l]Chloroform standard in New York State based on carcinogenicity.
[m]Dissolved oxygen standards become stricter for waters supporting trout habitats.
[n]The MCL for mercury is 0.002 mg/L.

out of compliance, the states must develop Total Maximum Daily Loads (TMDLs). A TMDL is a written, quantitative assessment of water quality problems and contributing pollutant sources. It specifies the amount of a pollutant or other stressor that needs to be reduced to meet water quality standards, allocates pollution control responsibilities among pollution sources (both point and nonpoint) in a watershed, and provides a basis for taking actions needed to restore a body of water. NYC DEP has developed phosphorus TMDLs for all of the drinking water reservoirs in the Croton and Catskill/Delaware watersheds.

Current regulations require states to submit lists of impaired waters every two years and to target those waters for which TMDLs will be developed during the next two years. Not all states have successfully met these strict requirements, subjecting the TMDL program to considerable controversy. The most debated issue is whether EPA can enforce the implementation of nonpoint source pollution control measures to reach TMDLs. Opponents argue that Section 303d is not applicable to nonpoint source pollution because such language is not specifically found within the CWA (Larson, 1999; Water Report, 1999). Supporters argue that the TMDL program should regulate nonpoint source pollution because voluntary implementation of nonpoint source pollution control measures has met with limited success.

Nonpoint Source Pollution

As suggested above, there is currently no regulatory mechanism by which EPA can control nonpoint source pollution, a position that has recently been upheld in federal courts for some sections of the CWA (Larson, 1999). Unlike point source pollution, nonpoint source pollution is not derived from a single discharge point, such as industrial and sewage treatment plants, but rather from diffuse sources. It can originate from agricultural and construction activities, on-site sewage treatment and disposal systems, and atmospheric deposition, among others. Nonpoint source pollution enters nearby water from dry and wet atmospheric deposition and from runoff that has collected nutrients, toxic chemicals, sediment, and microorganisms from the land.

Section 319 of the CWA encourages states to implement programs for nonpoint source pollution control by offering grant assistance for development and implementation. Future attempts by EPA to regulate nonpoint sources of pollution are likely to stem from section 303(e) of the CWA, which might be used as a "framework for implementing TMDLs, especially the nonpoint source load allocations" (EPA, 1997c). EPA is suggesting revisions of state nonpoint management programs under Section 319 and may take even stronger steps, such as reduction of federal grant dollars to states that fail to carry out nonpoint source implementation measures (EPA, 1997c).

NPDES Permits

All discharges to waters of the nation (point sources) are required to meet performance standards and effluent quality standards. This is accomplished through the CWA's National Pollutant Discharge Elimination System (NPDES) program. The NPDES program requires EPA (or the states with delegated programs) to issue enforceable permits for point source discharges such as industrial process wastewater, municipal WWTPs, or stormwater discharges from urban areas. Permits also regulate industrial point sources and concentrated animal feeding operations that discharge into other wastewater collection systems or that discharge directly into receiving waters. More than 200,000 sources are regulated by NPDES permits nationwide. In New York, the state issues the permits, which are referred to as SPDES (State Pollutant Discharge Elimination System) permits. There are 41 sewage treatment plants and seven industrial treatment plants in the Catskill/Delaware watershed and 85 WWTPs in the Croton watershed that operate under SPDES permits (Marx and Goldstein, 1993).

When the NPDES program was created, most effluent limits were specified for *technology-based* parameters that measure the performance of a WWTP, such as biological oxygen demand (BOD) and total suspended solids (TSS). If a WWTP achieved the effluent concentrations typical of secondary treatment (e.g., 30 mg/L for BOD), then it was meeting effluent standards. Over time, the impact of point source discharges on the water quality of receiving bodies has become important, and there has been a corresponding evolution away from technology-based effluent limits toward *water quality-based* effluent limits (WQBEL). Common types of regulated pollutants for a WWTP have been expanded to include fecal coliform, oil and grease, organics (such as pesticides, solvents, polychlorinated biphenyls, and dioxins), metals, and nutrients (such as nitrogen and phosphorus). As a result, chemicals that affect water quality but may give no indication of WWTP performance, such as nutrients, metals, and ammonia, have been added to NPDES permits.

Besides setting effluent limits, NPDES permits outline both standard and site-specific compliance monitoring and reporting requirements for the discharger. Finally, they detail enforcement actions that will be taken when and if regulated facilities fail to comply with the provisions of their permits. EPA uses a variety of techniques to monitor compliance, including on-site inspections and review of data submitted by permittees.

EPA has developed specific NPDES programs designed to combat particular types of effluent discharge that are exacerbated as the result of wet weather conditions and associated runoff and high flows. In particular, the NPDES Stormwater Program establishes a two-phased approach to addressing stormwater discharges. Phase I, which is currently being implemented, requires permits for separate stormwater systems serving medium- and large-sized communities (those with over 100,000 inhabitants) and for stormwater discharges associated with industrial and construction activity involving at least five acres. There are currently three SPDES-permitted stormwater discharges associated with construction activities in the Catskill/Delaware watershed. Phase II, which is currently under development, will address remaining stormwater discharges. The goal of this program is to encompass as many activities and areas as possible, including urban areas with populations under 100,000, smaller construction sites, and retail, commercial, and residential activities.

Wetlands

Section 404 of the CWA establishes a program to regulate the discharge of dredged and fill material into the waters of the United States, which have been defined to encompass wetlands. Activities that are regulated under this program include water resource projects such as dams, infrastructure development such as highways, fill for development such as housing subdivisions, and conversion of wetlands to uplands for farming and forestry. The guiding principle of the program is that if a practical alternative that is less damaging to the aquatic environment exists or if the nation's waters would be significantly degraded, a proposed discharge of dredged or fill material into a wetland cannot be permitted. The U.S. Army Corps of Engineers administers the program within a policy framework established by EPA. Individual permits are usually required for activities with potentially significant impacts. As a means of expediting program administration, general permits governing particular categories of activities such as minor road crossings and utility line backfill are issued on a nationwide, regional, or state basis.

Antidegradation

Antidegradation is a federal regulation related to the CWA stating that waters must not be allowed to degrade in quality such that their use classification and water quality criteria are violated (CFR Part 131, 1983). The concept of antidegradation deals primarily with "assimilative capacity"—the amount of additional pollution that a body of water can receive without exceeding its water quality criteria or use classification. Antidegradation policies are meant to be developed by each state and must contain language similar to that promulgated by EPA. Although not always achieved, one goal of the EPA policy is to describe how the state will allocate the assimilative capacity of its waters (R. Shippen, EPA, personal communication, 1998).

After a state has established use classifications and water quality criteria, it must develop an antidegradation policy that distinguishes three levels of water quality: Tier 1 (the lowest level of quality), Tier 2 (fishable and swimmable waters - High Quality Waters), and Tier 3 (Outstanding Natural Resources). Tier 1 waters cannot be allowed to degrade any further, and all existing uses of all waters must be maintained. This is meant to guarantee a baseline level of protection for all waters, especially those that have no additional assimilative capacity. Tier 2 waters have a water quality greater than or equal to fishable/swimmable quality (i.e., they have some assimilative capacity). Water quality in these waters is only allowed to degrade to the level of fishable/swimmable (never lower) and only if certain significant criteria are met: (1) there must be important social or economic reasons for the lowering of water quality, (2) the public must be informed prior to any activity, and (3) the highest level of statutory and regulatory requirements for point sources, and the use of best management practices for nonpoint sources, must be achieved. Tier 3 waters, which have exceptional recreational and ecological significance (and consequently considerable assimilative capacity), cannot be allowed to degrade at all. These waters are generally found in national parks or other areas designated as pristine.

As one might expect, Tier 2 waters are the most controversial because a lowering of water quality may in some cases be allowed, using up some of the waters' assimilative capacity. EPA requires that any activities that will result in a lowering of Tier 2 water quality be reviewed to determine whether the activity will violate state antidegradation policy. How this review will be conducted, and which activities will trigger a review, is left to the states to decide.

Enforcement Authority

As with the SDWA, the CWA calls for state enforcement of the regulations issued by EPA. Having been granted primacy by EPA, the New York State Department of Environmental Conservation (NYS DEC) must ensure New York's compliance with the CWA. NYS DEC determines the use classifications and water quality criteria for all state waters, it oversees the TMDL program, it operates the SPDES permitting program for WWTPs, and it is responsible for developing and implementing the state antidegradation policy. Guidelines and policies for implementing the CWA have been written into New York State law (Article 17 of the Environmental Conservation Law).

LOCAL ENVIRONMENTAL REGULATIONS AFFECTING NEW YORK CITY

The SDWA and the CWA, while comprehensive, are not expected to provide the sole protection of surface and groundwater supplies used as sources of drinking water. Rather, they are meant to serve as minimum requirements with which the states must comply in creating their own environmental programs. All states have written environmental laws to interpret the federal statutes and have created additional requirements as well. These state laws, regulations, and policies fill gaps between the federal regulations and (in some cases) increase the level of protection afforded to bodies of water.

Gaps between the federal regulations often stem from the various, sometimes-unrelated targets of these laws. For example, the SDWA lists specific quantitative standards for many known pollutants in drinking water, while the CWA mandates similar standards for source waters. Although both acts have watershed-based components, neither act requires or outlines a truly comprehensive watershed management strategy. The New York City MOA is an example of a state/local effort to design a watershed management strategy that complements and enhances federal environmental regulations.

New York City Watershed Memorandum of Agreement

The most extensive legal document governing activities in the New York City water supply watersheds is the 1997 New York City Watershed MOA. As described in Chapter 1, many sections of the MOA satisfy conditions of the City's filtration avoidance determination and are thus necessary for New York City to comply with the SWTR.

Contents of the Memorandum of Agreement

Land Acquisition Program. Because only 26 percent of Catskill/Delaware watershed is owned by New York City or New York State, an important element of the MOA is a land acquisition program that aims to increase the percentage of publicly owned land. This voluntary program allows New York City to acquire fee titles or conservation easements to vacant water quality-sensitive watershed lands on a "willing buyer/willing seller" basis. All titles and conservation easements are held in perpetuity. New York City has committed more than $250 million for land acquisition, and the MOA requires that New York City solicit participation of the owners of more than 350,000 acres by the year 2007. As discussed and analyzed in more detail in Chapter 7, areas closer to reservoir intakes and the distribution system are given higher priority for acquisition. The MOA also created a $17.5 million land acquisition program for the Croton watershed, but different criteria are used to prioritize the subwatershed areas, and there are no solicitation goals.

Watershed Rules and Regulations. Because most of the watershed is in private ownership, the Watershed Rules and Regulations were created to control activities within the watershed that may increase pollution. Both point and nonpoint pollution sources are targeted, including WWTPs, on-site sewage treatment and disposal systems (OSTDS), stormwater runoff, and storage of hazardous materials. The regulations contain minimum treatment requirements for technologies that control these sources of pollution and specify effluent standards that some of these treatment technologies must meet.

In addition, the regulations restrict a variety of activities from occurring in close proximity to watercourses, reservoirs, reservoir stems, controlled lakes, and wetlands. For example, the siting of WWTPs, OSTDS, storage facilities for hazardous materials, petroleum, and salt, and the construction of impervious surfaces are all restricted within "setback" distances from the major bodies of water. There are some exemptions to these restrictions that were designed to promote responsible growth in existing areas while protecting water quality via increased regulation.

Watershed Protection and Partnership Programs. The purpose of the Watershed Protection and Partnership Programs is to preserve the economic and social character of the Catskill/Delaware watershed communities while maintaining and enhancing water quality. The Catskill Watershed Corporation (CWC), a not-for-profit corporation, was formed to manage some of these programs for the Catskill/Delaware watershed region. The MOA calls for about $240 million to be allocated for these efforts, which include infrastructure improvements for stormwater and wastewater, development, conservation, and education (see Chapter 7).

The MOA also provides funds (approximately $68 million) for Westchester and Putnam counties to develop and implement a Croton Plan. The objective of this plan is to develop a comprehensive approach to identify significant sources of pollution, to recommend measures for improving water quality, and to protect the character of watershed communities east of the Hudson River.

In addition to the Watershed Protection and Partnership Programs, the MOA established the Watershed Protection and Partnership Council to provide broad oversight of New York City watershed management. The 27-member Council consists of representatives from New York City and State government agencies, watershed counties, environmental groups, the CWC, EPA, and the Watershed Agricultural Council. Its three main functions are (1) to serve as "a forum for the exchange of views, concerns, ideas, information, and [non-binding] recommendations relating to Watershed protection and environmentally responsible economic development," (2) to "periodically review and address efforts undertaken by governments and private parties to protect the Watershed," and (3) to "solicit input from governmental agencies, private organizations, or persons with an interest in the Watershed and the New York City drinking water supply" (MOA, Article IV). The Executive Committee of the Council, consisting of 16 members, has more specific functions.

Regulatory Authorities Involved in MOA Implementation

EPA, the New York State Department of Environmental Conservation (NYS DEC), and the New York State Department of Health (NYS DOH) are the regulatory agencies responsible for enforcing the provisions of the MOA. In particular, EPA Region 2 and NYS DOH oversee the New York City filtration avoidance determination, while NYS DEC oversees the TMDL and the SPDES permitting programs. The efforts of these three agencies are focused almost exclusively on the New York City Department of Environmental Protection (NYC DEP), which implements these programs in the New York City watersheds. NYC DEP monitors the major bodies of water, inspects WWTPs, and runs a variety of nonpoint source control programs. Because of its predominant role in the watershed, NYC DEP is the primary focus of this report's reviews, critiques, and recommendations.

Other government agencies also play a role during implementation of the MOA. The New York State Department of State, Division of Local Government, provides training and technical assistance to local governments and community organizations to help them take advantage of the Watershed Protection and Partnership Programs. A local office has been established in the Catskills to increase access to watershed governmental officials and citizens. New York City operates a Department of Health (NYC DOH) that is responsible for safeguarding the health of city-dwellers. As part of its many responsibilities under the filtration avoidance determination, NYC DOH monitors waterborne disease occurrence in the City by a variety of methods discussed in Chapter 6. NYC DOH and NYC DEP work together to determine the prevalence of waterborne disease and its possible sources in the watershed, the water supply system, or the distribution system. Finally, county health departments are involved in ensuring the success of the MOA. Along with NYS DOH and NYC DEP, county health departments oversee the construction and maintenance of new sewerage and septic systems. This includes new residential septic systems and commercial, institutional, and multifamily systems up to a maximum flow of 10,000 gallons per day.

Other State and Local Laws

Several state and local laws complement the regulations found in the MOA, the CWA, and the SDWA. The most important of these is the New York State Environmental Quality Review Act of 1975 (SEQR, Article 8 of the Environmental Conservation Law). Similar to the National Environmental Policy Act of 1975, SEQR requires that an environmental impact statement (EIS) be written for all major ministerial actions. For example, an EIS was prepared in 1993 following the drafting of the New York City Watershed Rules and Regulations (NYC DEP, 1993).

Because of their size and scope, many of the actions required by and allowed under the MOA will require the preparation of an EIS. This process enables some public involvement in decision-making, but it also increases the time required for planning and implementation of new projects. Few states have a strong environmental review statute, and SEQR represents an extremely progressive environmental policy on the part of New York. Its provisions are described in greater detail in Box 3-5.

BOX 3-5
State Environmental Quality Review (SEQR) Act

The purpose of the New York State Environmental Quality Review Act (as regulated in 6 NYCRR Part 617) is "to incorporate the consideration of environmental factors into planning, review, and decision-making processes of state, regional, and local government agencies at the earliest possible time." SEQR requires that all agencies determine whether the actions they undertake, fund, or approve of will have a significant impact on the environment. If so, an environmental impact statement (EIS) must be prepared that systematically considers significant adverse environmental impacts, alternatives, and mitigation. The EIS must also describe the potential social and economic benefits associated with the proposed project.

SEQR spells out a number of Type I actions that are likely to require the preparation of an EIS. These include changes in any zoning district affecting 25 acres or more, nonresidential projects physically altering 10 or more acres of land, and the acquisition, sale, lease, annexation, or transfer of 100 acres or more by a local agency. Examples with particular relevance to water quality include the construction of a WWTP or a large residential housing unit. Type II actions, which are likely to not require the preparation of an EIS, include repaving existing highways, maintaining existing landscaping, constructing a single-family home, adding a carport or swimming pool, and public or private forestry best management practices on less than 10 acres of land.

The process for developing an EIS is similar to that of the National Environmental Policy Act process. First, the proposed action is classified and an environmental assessment form is completed. The lead agency (usually NYS DEC) then decides if the action includes the potential for at least one significant adverse environmental impact, which will cause "substantial adverse change in existing air quality, ground or surface water quality or quantity, traffic and noise levels; a substantial increase in solid waste production, or a substantial increase in the potential for erosion, flooding, leaking, or drainage problems." Significant impacts also include removal of vegetation, interference with wildlife migration, creation of a conflict with a community's current plans or goals, creation of human health hazard, and substantial change in land use or intensity of land use. If there is no potentially significant impact, than a negative declaration is published and filed. For projects that may negatively affect the environment, the applicant must complete a draft EIS.

The draft EIS fully describes the proposed action, including its relevant social and economic benefits. The environmental setting is described,

continued

BOX 3-5 Continued

and all potential environmental impacts are analyzed. In justifying the necessity for the action, the draft EIS must describe all reasonable alternatives to the action, including the "no-action" alternative. Finally, ways to reduce adverse environmental impacts must be explored. The draft EIS is a public document that must be available for a 30-day public review and comment period. After receiving and appropriately responding to public comments, the draft EIS is finalized. The project is then approved or denied by all relevant government agencies, ending the SEQR process.

Other important state laws pertaining to watershed management can be found elsewhere in New York's Environmental Conservation Law (ECL). With regard to wetlands, a permit is required for any construction, dredging, or dumping in a freshwater wetland (ECL §24-0701). Fishing, hunting, and trapping are allowed without permit in tidal wetlands, as is farming in freshwater wetlands.

Finally, local government laws can play a role in watershed management, primarily via zoning provisions within the counties, towns, and villages of the watershed region. The Croton watershed contains portions of three counties— Westchester, Putnam, and Dutchess counties—while the Catskill/Delaware watershed wholly or partially includes Delaware, Schoharie, Greene, Sullivan, and Ulster counties. The role of local government regulations in dictating activities within the watershed region is described more extensively in Chapter 7.

REFERENCES

American Public Health Association. 1995. Standard Methods for the Examination of Water and Wastewater. 19th Edition. Washington, DC: American Public Health Association.

American Water Works Association (AWWA). 1993. Consumer Attitude Survey on Water Quality Issues. Denver, CO: American Water Works Association.

Arizona Department of Environmental Quality. 1996. R18-11-1 Appendix A Numeric Water Quality Standards. April, 1996.

Baker, M. N. 1949. The Quest for Pure Water. New York: American Water Works Association.

Bellar, T. A., and J. J. Lichtenberg. 1974. Determining volatile organics at microgram-per-litre levels by gas chromatography. Journal of the American Water Works Association 66(12):739–744.

Borchardt, J. A., and G. Walton. 1971. Water quality. In: Water Quality and Treatment, American Water Works Association. 3rd ed. New York, NY: McGraw-Hill.

Colorado Department of Public Health and the Environment. 1996. Basic Standards and Methodologies for Surface Waters, 3.1.0 (5CCR 1002-8). March, 1996.

Craun, G. F. 1986. Statistics of waterborne outbreaks in the U.S. (1920-1980) In Craun, G. F. (ed.) Waterborne Diseases in the United States. Boca Raton, FL: CRC Press.

Craun, G. F., and L. J. McCabe. 1973. Review of the Causes of Waterborne-Disease Outbreaks. Journal of the American Water Works Association 65(1):74–84.

Environmental Protection Agency (EPA). 1989. National Primary Drinking Water Regulations; Total Coliforms (Including Fecal Coliforms and *E. coli*) Final Rule. Federal Register 54(124):27544–27568.

EPA. 1990. Guidance manual for Compliance with the Filtration and Disinfection Requirements for Public Water Systems Using Surface Water Sources. Washington, DC: EPA.

EPA. 1996. Watershed Framework. Washington, DC: EPA.

EPA. 1997a. National Primary Drinking Water Regulations: Disinfectants and Disinfection Byproducts; Notice of Data Availability, Proposed Rule. Federal Register 62(212):59388–59484.

EPA. 1997b. National Primary Drinking Water Regulations: Enhanced Surface Water Treatment Rule; Notice of Data Availability, Proposed Rule. Federal Register 62(212):59388–59484.

EPA. 1997c. Draft TMDL Policy on Pace and Implementation. Office of Water. Washington, DC: Office of Water, EPA.

EPA. 1998a. Final D/DBP Rule. Federal Register 63(241):68389–69476.

EPA. 1998b. Proposed D/DBP Rule. Federal Register 63(61):15673–15692.

EPA. 1998c. EPA Rationale for Chloroform Zero MCLG. Water Policy Report 7(25):4–6.

EPA. 1998d. Dichloroacetic acid: Carcinogenicity Identification Characterization Summary. Washington, DC: National Center for Environmental Assessment, Office of Research and Development, EPA.

EPA. 1998e. Contaminant Candidate List Rule. Federal Register 63(40):15063–15068.

Fox, K. R., and D. A. Lytle. 1996. Milwaukee's cryptosporidiosis outbreak: investigation and recommendations. Journal of the American Waterworks Association 88(9):87–94.

Freudenberg, N., and C. Steinsapir. 1992. Not in Our Backyards: The Grassroots Environmental Movement. Pp. 27-38 in Dunlap, R. E., and A. G. Mertig (eds.), American Environmentalism: The U.S. Environmental Movement, 1970–1990. Philadelphia, PA: Taylor and Francis.

Harris, R. H., and E. M. Brecher. 1974. Is the water safe to drink? Consumer Reports June:436–443.

Hopkins, R. S. , Gaspard, B., Eisnach, L., and R. J. Karlin. 1983. Waterborne Disease in Colorado–Report on Two Years Surveillance and Eleven Waterborne Outbreaks, Final Report, EPA Contract 68-03-2927.

Hunter, P. 1997. Waterborne Disease: Epidemiology and Ecology. Chichester: John Wiley.

International Life Sciences Institute (ILSI). 1993. U.S. Perspective on Balancing Chemical and Microbial Risks of Disinfection. In Safety of Water Disinfection: Balancing Chemical and Microbial Risks, G. F. Craun, editor. Washington, DC: ILSI Press.

Larson, R. L. 1999. Court Rules the CWA Does Not Regulate Nonpoint Source. Water Environment and Technology 11(1): 72.

Longmate, N. 1966. King Cholera: The Biography of a Disease. London: Hamish Hamilton.

MacKenzie, W. R., N. J. Hoxie, M. E. Proctor, M. S. Gradus, K. A. Blair, D. E. Peterson, J. J. Kazmierczak, D. G. Addiss, K. R. Fox, J. B. Rose, and J. P. David. 1994. Massive Waterborne Outbreak of *Cryptosporidium* Infection Associated with a Filtered Public Water Supply, Milwaukee, Wisconsin, March and April 1993. New England Journal of Medicine 331(3):161–167.

Marx, R., and E. Goldstein. 1993. A Guide to New York City's Reservoirs and Their Watersheds. New York, NY: National Resource Defense Council.

McCabe, L. J., J. M. Symons, R. D. Lee, and G. G. Robeck. 1970. Survey of community water supply systems. Journal of the American Water Works Association 62(11):670–687.

McDermott, J. H. 1973. Federal Drinking Water Standards—Past, Present, and Future. Jour. Envir. Engr. Div., ASCE, EE4:99:469.

Milwaukee Journal. 1993. Watering down a disaster. Milwaukee, WI: Milwaukee Journal.

National Cancer Institute. 1976. Report on Carcinogenesis Bioassay of Chloroform. Technical Information Service No. PB 264018/AS. Bethesda, MD: National Cancer Institute.

National Research Council (NRC). 1977. Drinking Water and Health. Washington, DC: National Academy Press.

NRC. 1999. New Strategies for America's Watersheds. Washington, DC: National Academy Press.

New York City Department of Environment and Protection (NYC DEP). 1993. Final Generic Environmental Impact Statement for the Proposed Watershed Regulations for the Protection from Contamination, Degradation, and Pollution of the New York City Water Supply and its Sources. November 1993. Corona, NY: NYC DEP.

NYC DEP. 1999. Development of a water quality guidance value for Phase II TMDLs in the NYC Reservoirs. Valhalla, NY: NYC DEP.

New York State Department of Environmental Conservation (NYS DEC). Water Quality Regulations: Surface Water and Groundwater Classifications and Standards. Effective January 9, 1994. Albany, NY: NYS DEC.

North Carolina Department of Environment and Natural Resources. 1998. Classifications and Water Quality Standards. 15A NCAC 2B .0200. Raleigh, NC: NCDENR.

Pfeffer, M. J., and J. M. Stycos. 1996. Public opinion survey on environment and water quality in the New York City Watershed. Ithaca, NY: Cornell University. Unpublished Data.

Pontius, F. W. 1990. New regulations for total coliforms. Journal of the American Water Works Association 82(8):16, 20–22.

Pontius, F. W. 1997. Overview of SDWA source water protection programs. Journal of the American Water Works Association 89(11):22–24, 123. See also USEPA, State Source Water Assessment and Protection Programs, Final Guidance, Office of Water, EPA 816-R-97-009, Washington, DC (Aug. 1997).

Rook, J. J. 1974. Formation of haloforms during chlorination of natural water. Water Treatment and Examination 23(2):234–243.

Schwartz, J., R. Levin, and K. Hodge. 1997. Drinking water turbidity and pediatric hospital use for gastrointestinal illness in Philadelphia. Epidemiology 8(6):615–620.

Sedgwick, W. T., and J. S. MacNutt. 1910. On the Mills-Reincke phenomenon and Hazen's theorem concerning the decrease in mortality from diseases other than typhoid fever following the purification of public water supplies. Jour. Inf. Dis. 7(4):489–564.

Singer, P. C., A. Obolensky, and A. Greiner. 1995. DBPs in chlorinated North Carolina drinking waters. Journal of the American Water Works Association 87(10):83–92.

Sterling, C. R. 1990. Waterborne crytosporidiosis. In Dubey, J. P., C. A. Speer, R. Fayer (eds.) Cryptosporidiosis of Man and Animals. Boca Raton, FL: CRC Press Inc.

Symons, J. M., and J. C. Hoff. 1976. Rationale for turbidity maximum contaminant level. Paper 2A-4a In American Water Works Association Technology Conference Proceedings, Water Quality Technology Conference, December 8-9, 1975, Atlanta, GA., American Water Works Association, Denver, CO.

Szasz, A. 1994. EcoPopulism: Toxic Waste and the Movement for Environmental Justice. Minneapolis: University of Minnesota Press.

U.S. Public Health Service. 1925. Report of the Advisory Committee on Official Water Standards. Public Health Reports 40:693 (April 10).

Utah Department of Environmental Quality. 1994. Utah R317-2-14. Numeric Criteria. February, 1994.

Virginia Department of Environmental Quality. 1988. Water Quality Standards. July.

Waller, K., S. H. Swan, G. DeLorenze, and B. Hopkins. 1998. Trihalomethanes in drinking water and spontaneous abortion. Epidemiology 9(2):134–140.

Washington Department of Ecology. 1993. Washington WAC 173-201A-030, 040, and 050. General Water Use and Criteria Classes. Toxic Substance Criteria from 40 CFR 131.36 (b)(1), Column D1, July, 1993.

Water Report. 1999. Clean Water Act Reauthorization. Key House Member Says CWA Should Not Regulate Nonpoint Pollution. Water Policy Report March 31:16.

Wetzel, R. G., and G. E. Likens. 1991. Limnological Analyses. 2nd Edition. New York, NY: Springer-Verlag.

4

Watershed Management for
Source Water Protection

Watershed management can be defined as any program or collection of strategies that positively influence activities and land characteristics within a drainage basin. Watershed management is necessary for many reasons, including meeting requirements of the Clean Water Act (CWA) (to maintain the physical, chemical, and biological integrity of the nation's waters), the Safe Drinking Water Act (SDWA) (to protect drinking water supplies), and state programs, or meeting local community needs. The purpose of this chapter is to provide an overview of watershed management for the protection of drinking water supplies—that is, *source water protection*—as a basis for review and critique of the watershed strategies included within the Memorandum of Agreement (MOA).

Source water protection programs are essentially watershed management programs with the specific goal of protecting drinking water supplies. In recent years, it has been more widely recognized that source water protection is a critical first step in the multiple-barrier approach to providing safe drinking water advocated by the Environmental Protection Agency (EPA) and the American Water Works Association (AWWA, 1997; EPA, 1997a). Examples of barriers include selecting the highest-quality source water, practicing watershed management, using the best available treatment technologies, maintaining a clean distribution system, practicing thorough monitoring and accurate data analyses, having well-trained operators, and maintaining operating equipment. Although the term "multiple barriers" is a recent one, the concept is old, as evidenced by the following quotation:

"This all tends to bring into prominence the necessity of using barriers of protection to the best possible advantage. The first of these is the reservoir system through which all raw waters should flow, even if the source is of sufficiently good quality to be by-passed to the works direct. The effect of promoting a continuous flow throughout storage systems in which water is sedimented results in an enormous reduction of bacterial and organic components. Next, an efficient filter barrier is interposed and, lastly, a chemical process which should be afforded time for effective action, either in contact tanks or in the mains before the draw-off. Although, on the one hand, one counsels the use of chemicals as a finishing process, and as a supplementary final barrier, it is necessary with such effective agents as the chloramines to voice a warning as regards their position in the scheme of water purification. The aim of application should be as a corrective to counter an emergency rather than the be-all and end-all of water purification, otherwise there is a definite risk of allowing our filtration, sedimentation, and storage barriers to fall to a less important position and to depend upon chloramine entirely." (Harold, 1934)

As perceptive as this 65-year-old quotation is, it omits a critical part of the equation—the active management of reservoir watersheds such that reservoir water quality is enhanced (watershed management). Today, it is well understood that water utilities should both control contaminant discharges into watersheds and prevent any unwise land use in watersheds that would adversely influence downstream water quality.

Source water protection programs include actions, policies, and practices to protect and enhance sources of drinking water (AWWA, 1997; EPA, 1997b). Figure 4-1 depicts the components of a watershed management framework for source water protection advocated by the committee. Several components or steps are necessary: (1) establishment of goals and objectives, (2) an inventory of the watershed and assessment of possible contaminant sources, (3) development and (4) implementation of protection strategies, and (5) monitoring and evaluation of program effectiveness. Stakeholder involvement in each of these components is key to ensuring that watershed management programs are effective. The remainder of this report analyzes sections of the New York City watershed management strategy that correspond to these components. For example, Chapter 6 focuses on the New York City Department of Environmental Protection's (NYC DEP) enhanced monitoring system and data analysis (effectiveness monitoring and evaluation). It also evaluates several important tools used in site and contaminant assessment. Chapters 7, 9, 10, and 11 discuss different protection strategies and their implementation, from structural best management practices (BMPs) used for stormwater management to nonstructural activities such as land acquisition.

Watershed management can be implemented at varying scales, from small subbasins (e.g., first-order streams or small lakes) to very large and complex systems (e.g., higher-order streams, large lakes, or estuaries). Previous experi-

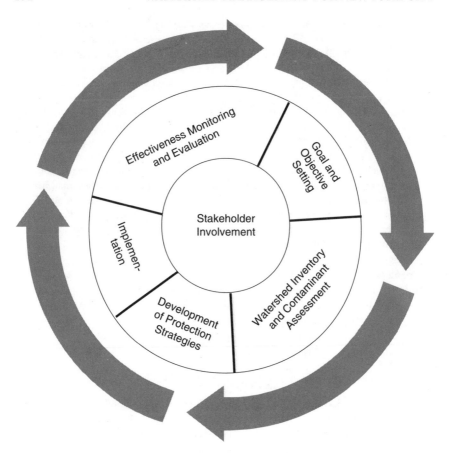

FIGURE 4-1 Components of source water protection. Adapted from EPA (1994) and Clements et al. (1996).

ence has found that watershed management is most effective when it is conducted at the scale of medium-sized tributary and subtributary watersheds rather than at larger or smaller scales (NRC, 1999).

The watershed management process is dynamic and iterative, and the steps in Figure 4-1 are repeated as conditions within the watershed evolve and new information becomes available. It should be kept in mind that this chapter describes an ideal water management strategy that might be used for source water protection. It is rare to find a community in which all these components are successfully carried out, although many of the elements can be found in the watershed management program of the Tennessee Valley Authority (Box 4-1).

BOX 4-1
Watershed Management in the Tennessee Valley Authority

The Tennessee Valley Authority (TVA), an independent federal agency established in 1933, is responsible for flood control, energy production, economic development, and natural resource conservation in the Tennessee Valley. TVA serves a geographic area of 80,000 sq mi that includes 125 counties in seven states. The Tennessee River is 652 miles long and is the fifth largest river system in the United States. The necessary components of watershed management for source water protection, as described in Figure 4-1, are aptly demonstrated by several activities of the TVA.

Goal and Objective Setting: The TVA has recently placed a greater emphasis on water quality and has adopted a goal of making the Tennessee river system the cleanest and most productive commercial river system in the United States by the year 2000. To this end, it adopted the Clean Water Initiative in 1993 and, using a watershed approach, it has divided the region into seven subbasins, each with its own self-directing interdisciplinary River Action Team.

Watershed Inventory and Contaminant Assessment: To pinpoint water-quality problems, the River Action Teams are using rapid biological assessment methods and conventional physical and chemical methods in conjunction with aerial photographs and Geographic Information System (GIS) maps. Data and input from other agencies and individuals are being used to characterize the region and improve the health of the watershed.

Protection Strategies: A variety of structural and nonstructural protection strategies are being used to improve water quality in the region, from creation of wetlands for removal of phosphorus in wastewater to installation of pump-out stations on lakes. The strategies that are used differ, depending on the type of pollutant present, the value of the water resource (evaluated by six indicators: human health, ecological integrity of the resource, human use of the resource, ecological integrity of the downstream resource unit, downstream human use, and economic sustainability), and the likelihood of success (as determined by assessing local political, economic, and regulatory realities).

Implementation: The River Action Teams ensure implementation by partnering with federal and state agencies and individual landowners affected by the choice of protection strategies. Successful projects are

continued

BOX 4-1 Continued

used by the River Action Teams as examples to demonstrate the benefits of the watershed approach.

Effectiveness Monitoring and Evaluation: There is an extensive biological, chemical, and physical monitoring system to track water quality. Results are fed to the River Action Teams, which use the information to further develop and refine protection strategies.

Stakeholder Involvement: The TVA's holistic, multiobjective watershed approach to the Tennessee River and its tributaries relies on the input of multiple stakeholders. Projects undertaken to improve water quality are joint actions of individuals, businesses, local and state agencies, and other federal agencies, along with TVA. Public education and open disclosure of the environmental problems are also part of the solution. For example, RiverPulse, a colorful, easy-to-read document, reports annually on the conditions for swimming and fishing and on overall ecological health of the waters in the Tennessee Valley, and it discusses the progress of some of the projects under way in the watershed.

GOAL AND OBJECTIVE SETTING

Goals provide general direction for source water protection programs by broadly stating the intent of the management plan. A primary goal of all municipal water suppliers is to provide an adequate supply of high-quality water, as defined by its taste, odor, color, clarity, and concentration of contaminants. Beyond this general goal, specific goals can be tailored to a watershed's physical characteristics, existing water quality concerns, contaminant sources, and regulatory constraints (AWWARF, 1991). Detailed numeric objectives often complement general goals by providing quantifiable and measurable direction for source water protection programs.

Source Water Protection Goals

Source water protection goals generally reflect the specific needs and conditions of the watershed and the entity implementing the program. Beyond providing high-quality source water, such programs may strive to (1) reduce or limit sources of contamination, (2) minimize the risk of hazardous chemicals entering the water supply, (3) mitigate the effects of natural disasters, (4) provide flexibility within water system operations, (5) minimize treatment costs, and (6) comply with regulatory requirements. Realistic goals must recognize the need for balance and compromise among competing and often conflicting demands for various uses within the watershed (AWWARF, 1991), such as protection of aquatic life, recreation, water supply, agriculture, forestry, and urban development. This need for compromise is particularly pressing in water supply watersheds that are not substantially or wholly owned and managed by the water supplier. In addition, the limitations of regulatory authorities that implement and enforce program goals must be acknowledged. Goals must have the necessary supporting legal and regulatory authority to ensure effective implementation.

One example of a goal is "to protect the water quality and supply reliability by seeking to balance the watershed uses such as the rights of private property owners and public recreational activities with the protection and management of natural resources" (Santa Clara Valley Water District, 1995). The Santa Clara Valley Water District goal recognizes the need for public support and for cooperation from other stakeholders to ensure a successful source water protection program. The Upper South Platte Watershed Protection Association (1998) in Colorado has established a similar goal that reflects a desire to balance activities and effectively engage stakeholders. The Salt Lake City Watershed Management Plan is more strongly focused on maintenance of water quality over other factors: "[Watershed] management emphasis prioritizes water quality first and multiple use of the watershed second. The Wasatch Canyons are protected to maintain a healthy ecological balance with stable environmental conditions, healthy streams and riparian areas, and minimal sources of pollution" (Salt Lake City Department of Public Utilities, 1998).

Specific Water Quality Objectives

In addition to basic goals, water agencies may also establish specific numeric or narrative objectives for their drinking water sources. As required under the CWA, most states or EPA regions have established use designations and associated water quality criteria for the protection of drinking water supplies. To date, water quality criteria to protect drinking water supplies have been limited and somewhat inconsistent from one state to another (Table 3-7). Based on a recent survey of western states, existing criteria generally focus on nitrate, metals, and a few other inorganic constituents; some organic and radiological constituents; and fecal coliform bacteria (Paulson and Vlier, 1997). Other constituents of concern in drinking water supplies, such as organic carbon and specific pathogens, have not yet been widely addressed by water quality criteria.

Maximum contaminant levels (MCLs) have been specified by the SDWA for application to treated water supplies and are not directly applicable as water quality objectives for source waters prior to treatment. Natural waters, even pristine supplies, may not be able to meet MCLs for many constituents. MCLs can, however, serve as a basis for the derivation of water quality objectives for source water supplies. This is done by accounting for the effects of transport, storage, and treatment, a challenging task because treatment levels can vary and water quality transformations are not always fully understood. Hypothetical numeric objectives that might be set for source water are given in Table 4-1.

WATERSHED INVENTORY AND CONTAMINANT ASSESSMENT

An inventory of the watershed and an assessment of potential contaminants help to define the boundary conditions and form the basis for source water protection programs. A watershed inventory begins with a delineation of the source water protection area and an evaluation of existing water quality conditions. A complete inventory includes information on natural characteristics, land uses, activities, and other factors that can affect water quality. Land uses and ownership are particularly important because they help determine the need for regulatory authority to support the source water protection plan. The potential suscep-

TABLE 4-1 Example Numeric Objectives for a Surface Water Supply Reservoir

Constituent	Protection Goals	Example Numerical Objectives
Total phosphorus	Limit algal production, eutrophication, and associated problems	0.01 to 0.03 mg/L
Chlorophyll *a*	Limit algal sources of organic carbon; avoid taste and odor problems	0.005 to 0.15 mg/L
Dissolved organic carbon	Limit disinfection byproduct precursors; avoid higher levels of treatment required by the Enhanced Surface Water Treatment Rule for waters with various levels of organic carbon	<2 to <8 mg/L
Fecal coliform	Avoid gastrointestinal and systemic illnesses caused by pathogenic organisms that may be indicated by fecal coliform	1/100 mL to 100/100 mL
Turbidity	Provide clear water; limit interference with disinfection process; reduce levels of particulate-phase pollutants	<5 NTU
Others (metals, synthetic organics)	Meet treated water goals, as defined by MCLs	10% to ≥ 100% of MCLs

Note: Numerical objectives in this table are presented as examples only and are not endorsed by the committee for global application to reservoirs. The committee recommends that actual numeric objectives be developed on a site-specific basis to account for site-specific conditions (e.g., treatment processes, natural conditions, and uses).

tibility of water supplies to natural and human sources of contamination within the source protection area is then evaluated in a contaminant assessment.

Source Water Protection Area Delineation

For surface water systems, the source water protection area generally includes the watershed area upstream of a water supplier's intake. It is delineated by the boundaries of drainage basins that supply streams, lakes, and reservoirs that serve as source water. Basins can also be divided into smaller subbasins that drain to tributary systems. In areas with transbasin diversions, the entire source water protection area may include watersheds that are geographically far removed from the point of use.

For groundwater systems, the source water protection area, also known as the wellhead protection area, is defined as the zone of recharge around a well. The wellhead protection area can be delineated using one of several methods, including the following: an arbitrary radius around a well (e.g., 2–3 miles), a calculated fixed radius that is determined as a function of hydraulic gradients, analytical modeling, or hydrogeologic mapping (Colorado State Department of Public Health and Environment, 1998; EPA, 1989). In delineating the source water protection area, zones and pathways through which contaminants could migrate and reach surface or groundwater systems must be considered.

Existing Water Quality Conditions

Existing water quality conditions can be assessed by comparing all available data on physical, chemical, and biological parameters to water quality objectives for source protection. In some states, the water quality objectives to which data should be compared are encompassed by a water's use classification and water quality criteria. Published assessments of water quality conditions (e.g., CWA Section 305(b) reports and Section 303(d) listings) can serve as readily available summaries of water quality conditions relative to use classifications and water quality criteria. For source water protection, some constituents are of particular concern because of their potential impact on water supply treatment and finished water quality. Some of these constituents are listed in Table 4-1.

In addition to specific constituents of concern, other chemical, physical, and biological measures can also provide strong indications of watershed ecosystem "health" (Meyer, 1997). For example, the relative condition of biological communities, particularly benthic macroinvertebrate communities, can serve as a screening tool to identify potential water quality problems that could be overlooked in chemical-specific monitoring. It should be noted that macroinvertebrate monitoring is most useful for meeting aquatic health goals and is somewhat limited as an indicator of water quality for drinking water sources.

In a preliminary assessment, water quality data can be compared with source water quality objectives to determine where objectives are being met and to identify areas for improvement.

Background Conditions

Natural watershed characteristics define conditions in the absence of human impacts. These can serve as a baseline against which changes in water quality associated with human development are monitored. An inventory of natural conditions within a watershed includes information on hydrology, topography, soils, vegetative cover, erosion potential, wetland and riparian areas, wildlife, and disturbance potential. Hydrologic information is needed for all surface waters (e.g., major rivers, tributaries, lakes, and reservoirs) and groundwater (water level contours, deep aquifers) in order to delineate basin and subbasin boundaries. Erosion potential can be determined by considering data on slope, soils, and land cover (Universal Soil Loss Equation; see Wischmeier and Smith, 1978). Disturbance potential refers to forest fires, hurricanes, and insect and disease outbreaks, among other things.

Land Cover and Uses

Source water quality is directly and profoundly influenced by land cover and land uses, both natural and human. For this reason, quantifying land cover and land uses is a necessary step in watershed management. Information on land ownership, land jurisdictions, and water rights is also needed to help determine the potential to manage land uses and mitigate impacts.

Important categories of land cover to measure include forests, open spaces, bodies of water, agricultural cover (e.g., pastures, row crops), and impervious surfaces. Activities within watersheds can affect water quality by producing contaminants from discrete point sources or from diffuse nonpoint sources. Those deserving special consideration in a watershed inventory include industrial and municipal wastewater treatment plants, individual sewage disposal systems, permitted stormwater discharges, agricultural activities, forestry, mineral extraction, and the generation, storage, and disposal of hazardous materials. Recreational activities, both within watersheds and on bodies of water, can also affect water quality.

Contaminant Assessment

Existing and potential contaminant sources identified during the watershed inventory should be assessed for their future impact on source water quality. Factors such as location of the contaminant source, relative toxicity and mobility of the contaminant, and size and quantity of the contaminant source must be considered. In addition, existing levels of the contaminant measured in the watershed and their relationship to water quality goals can help determine the relative significance of various contaminant sources. Although work in this area is limited to date, both qualitative and quantitative contaminant assessment can help prioritize water quality problems and better direct limited resources.

Tools

Geographic Information Systems (GIS) and water quality models can be applied independently or jointly to support source water protection efforts. In combination, these tools are particularly powerful, with GIS as a tool to streamline data input and facilitate effective presentation of modeling results. As technology develops, specific linkages for these tools will continue to evolve.

Geographic Information Systems

Watershed inventories require the integration of many types of data and spatial location of that data to effectively evaluate potential impacts on source water supplies. This is best facilitated by a GIS, a computerized system for the storage, display, and manipulation of geographic data. It has been widely applied in source water protection and watershed planning initiatives across the nation (EPA, 1996).

A GIS can be built to house multiple "layers" of information to support a watershed inventory. Data layers, such as topography, water features, roads, and vegetation type, can be stored in two forms. Linear features such as roads or streams are often stored in vector form as a series of compass directions or azimuths[1] and distances tracing the path of the landscape feature. Other attributes such as topography, soil type, land use, or vegetation type may be represented with vectors enclosing a polygon(s) or as matrices of position and attribute data in *primary* layers. *Secondary* layers are formed with attribute data (as real numbers) pertaining to a primary layer. For example, a primary layer of data on soil type can be used to generate secondary layers of data on soil thickness, infiltration capacity, and permeability. GIS layers also may be comprised of data on such points as buildings, wells, septic systems, or other features of limited or discrete size.

One of the powerful features of GIS is that point, line, and area data can be combined to map and model interrelationships, calculate areas, or create new layers. For example, a slope layer can be created by calculating the change in elevation between adjacent grid cells. The slope layer can then be used to estimate flow paths. These *derivative* layers are invaluable for watershed modeling and management. It should be noted that the quality and value of derived layers is dependent on the available grid size used in the GIS.

---

[1]Azimuth is an angle in degrees clockwise from the north.

Information stored in a GIS can be linked to water quality databases and can be used to display water quality conditions at various locations, as part of the inventory. GIS can also be applied, in conjunction with water quality modeling, to help trace a problem identified downstream back to its source in the watershed. For example, GIS land cover data are used in New York City in conjunction with a terrestrial runoff model to determine the relative importance of phosphorus sources. In assessing the susceptibility of source water supplies to contaminants, a GIS can be used to locate areas that have the greatest potential for adverse impacts, quantify relative impacts for various subbasins, and determine where to concentrate management efforts.

Modeling

Water quality models can be an important tool to support source water assessment and protection initiatives, particularly in areas where data are limited, or to project future conditions. Water quality modeling objectives can range from predicting impacts of future development scenarios, to identifying key impact areas, to looking at the effectiveness of various control strategies. A wide array of water quality models has been developed (EPA, 1992) that can be roughly divided into two general categories: (1) loading models, which address pollutant loads from various point and nonpoint sources within a watershed (e.g., P8, Watershed Management Model, AGNPS), and (2) receiving water models, which address the impacts of various loading conditions on water quality in rivers (e.g., QUAL2E), lakes (e.g., eutrophication models), and groundwater (e.g., Modflow). Recent research and development efforts have emphasized the linkage of GIS and parametric methods to build spatially distributed models.

DEVELOPMENT OF PROTECTION STRATEGIES

Comprehensive source water protection includes multiple structural and nonstructural protection strategies to remove or reduce contaminant sources of surface and groundwater supplies. Effective protection strategies focus on high-priority watershed activities and contaminants and target the ones with the greatest potential to affect water sources. The challenge is to select the appropriate combination of practices to prevent or treat point and nonpoint pollution sources in the watershed. The potential structural and nonstructural practices that are often considered in watershed management are listed in Table 4-2.

Structural Practices

Structural practices are defined as those that treat or reduce pollution discharges from an existing source. Examples include upgrades to wastewater treatment plants (WWTPs) to reduce point source pollution and BMPs to reduce

TABLE 4-2 Structural and Nonstructural Practices to Be Considered in a Watershed Management Strategy

Structural	Nonstructural
Wastewater Treatment Upgrades	Watershed Land Acquisition
Stormwater Practices for New Development	Conservation Easements
Stormwater Retrofit Practices	Zoning and Development Ordinances
On-Site Sewage Treatment and Disposal System Upgrades or Rehabilitation	Riparian Buffer Ordinances
	Hazard Exclusions or Setbacks
Riparian Buffer Systems	Wetland Conservation
Agricultural Best Management Practices	OSTDS Siting Requirements
Forestry Best Management Practices	OSTDS Inspection and Cleanout
In-stream Treatment (aeration/liming)	Stewardship Incentives
Ecosystem Restoration	Economic Incentives
Erosion and Sediment Control	Incentives to Reduce Impervious Cover
Correction of Sanitary Sewer Overflows	Stormwater Pollution Prevention
Correction of Illegal or Illicit Connections	Public Education and Outreach
Land Reclamation	

Note: The structural and nonstructural classifications, as defined by the committee, refer to practices that control present pollution sources or prevent the creation of future pollution sources, respectively.

nonpoint source pollution. BMPs can mitigate nonpoint source impacts from agriculture, forestry, mining, construction, and urban runoff. Nonpoint source practices for agriculture generally address erosion control and sediment transport in disturbed areas, including the use of buffer zones, and management of significant activities such as grazing and irrigation, with special emphasis on controlling nutrients, pesticides, and herbicides (EPA, 1993). Forestry BMPs address timber harvest, road construction and maintenance, revegetation of disturbed areas, management of chemicals and fire hazard, and maintenance of streamside buffer zones (EPA, 1993). Urban stormwater BMPs include several conventional practices that have been widely applied: vegetated swales and filter strips, stormwater infiltration and percolation, extended detention (dry) basins, retention ponds (wet), and constructed wetlands (WEF and ASCE, 1998). More detail on the design and performance of BMPs can be found in the literature (Brown and Schueler, 1997; EPA, 1993; Schueler et al., 1992; WEF and ASCE, 1998) and in Chapters 8–11.

Nonstructural Practices

Nonstructural controls prevent or reduce potential pollution discharges from future sources. Many of the nonstructural controls involve institutional or policy tools used during the local land use planning process that governs how and where new developments can occur. Public education can also be an important nonstructural tool to change specific behaviors that create pollution discharges (e.g., lawn care, pet waste cleanup, and pesticide application).

Nonstructural practices (also known as source control measures) are often the most effective tools in addressing potential water quality problems, both in terms of cost and reliability. Depending on the ownership of, jurisdiction over, and authority for the land within the watershed, nonstructural strategies may include land use controls, land acquisition, and limits on activities and on the presence of hazardous materials. In addition, special management practices can be applied to limit the impacts of agriculture, forestry, and mining activities within key watershed areas. Some of the nonstructural practices found in the New York City MOA are evaluated in Chapter 7.

IMPLEMENTATION

Several steps are necessary to ensure that the combinations of structural and nonstructural practices selected to protect the watershed are effectively implemented on the ground and maintained over the long term. The seven most frequently used elements to support implementation, often administered at the local or regional level, include performance criteria, review and inspection, maintenance, compliance enforcement, education and training, funding and incentives, and tracking.

1. *Performance Criteria.* Specific minimum engineering criteria, standards, or specifications are needed for each practice to ensure it is properly applied and can achieve its target performance for pollutant treatment or prevention. Using on-site sewage treatment and disposal systems (OSTDS) as an example, this may involve specific soil tests to determine whether subsurface disposal is feasible at a site, tank and field sizing criteria, setback distances, reserve fields, and technology requirements. Performance criteria are typically included in local, regional, or state manuals.

2. *Review and Inspection.* Ensuring that the watershed protection practices are designed properly and installed correctly occurs in two stages. In the first stage, development plans are scrutinized in a local review process to ensure that they meet applicable performance criteria and are appropriate for the site. In the second stage, field inspections are conducted at the site to ensure that the practices have been installed or constructed according to the plan. In the case of OSTDS, a developer might submit a plan for on-site wastewater disposal for local review, followed by an inspection at the site during construction to confirm that the system is installed properly and works as designed.

3. *Maintenance.* Structural and nonstructural practices must be continuously maintained in order to ensure effectiveness over time. Maintenance tasks for structural practices include routine pollutant cleanouts and rehabilitation of the practices as they approach the end of their design life. With OSTDS, routine maintenance entails periodic cleanouts of the septic tank and repairs to the distri-

bution field. OSTDS rehabilitation occurs as the system approaches the end of its design life and may include switching to a reserve field and/or tank replacement. For nonstructural practices such as a conservation easement, maintenance can involve routine inspections and vegetative management measures. Each watershed practice requires a legally binding maintenance agreement that identifies the party responsible for future maintenance and clearly outlines maintenance tasks, costs, and schedules.

4. *Compliance Enforcement.* An enforcement authority is usually needed to ensure that all owners or developers are in compliance with watershed regulations or criteria. Enforcement may be needed in the event an owner avoids watershed requirements, fails to properly implement plans, or does not adequately maintain a practice. Periodic site inspections are needed to detect watershed violations, and a range of enforcement tools can be used to induce compliance (correction notices, fines, stop-work orders, and even criminal penalties). In the case of OSTDS, compliance issues could include failure to install the system, improper installation or location, or failure to perform routine cleanouts or periodic rehabilitation.

5. *Training and Education Outreach.* Effective implementation of structural practices often requires a community investment in intensive training on how to design and apply watershed practices, since some consultants, engineers, and planners may be unfamiliar with current design practice. For example, contractors and engineers may require training on new OSTDS testing or technology requirements, and outreach may also be needed for OSTDS owners to improve operation and maintenance. Education and technical assistance are also critical in the implementation of nonstructural practices that involve private stewardship, pollution prevention, and land use management.

6. *Funding and Incentives.* The costs of applying and maintaining watershed practices can be significant for landowners or developers. Consequently, to achieve more widespread implementation, it may be desirable to provide innovative financing systems and/or economic incentives. This is particularly important for nonstructural protection strategies, such as conservation easements and other stewardship of private lands. Innovative financing systems such as stormwater or septic system "utilities" are often useful in providing long-term funding for the maintenance of structural watershed protection tools.

7. *Watershed Tracking.* Over time, implementation results in the application of practices at hundreds of development sites, farms, properties, and other locations within a watershed. In many cases, dozens of local, state, and federal agencies, utilities, land trusts, and watershed organizations are involved in various aspects of implementation. Consequently, it is important to track the location, management status, and maintenance record of all practices to gauge the cumulative progress toward implementation in a coordinated fashion, namely with a multipurpose database that can be linked to a GIS.

EFFECTIVENESS MONITORING AND EVALUATION

Water quality monitoring is integral to measuring the success of any source water protection program. Monitoring data can be used to refine programs as necessary, by assessing the effectiveness of specific controls and identifying needed adjustments. In addition, monitoring is essential to earlier stages of watershed management, such as the watershed inventory and contaminant assessment steps. This section broadly discusses the multiple purposes of monitoring for source water protection, including water quality monitoring, monitoring of health outcomes among the consumers of drinking water, and monitoring of social and economic parameters to determine the success of watershed management. Subsequent data analysis and program evaluation are highlighted because these are frequently the weakest steps in a watershed management strategy.

Physical, Chemical, and Biological Monitoring of Water Supplies

Monitoring for source water protection is implemented to address four general objectives: compliance with environmental regulations, systems operations, performance of BMPs, and modeling activities. *Compliance monitoring* evaluates specific physical, chemical, and biological parameters for compliance with local, state, and federal regulations, including the SDWA. The methodologies used for monitoring most parameters are rigorously specified and regulated. An example of compliance monitoring might be daily sampling of turbidity, pathogens, and disinfection byproducts at the water intake. *Operational monitoring* is conducted on a broad set of parameters needed to effectively assess the ongoing and successional quality of water and reservoir dynamics and to determine the sources of pollution that influence water quality. This type of sampling is typically conducted at a broad spatial scale (e.g., in the reservoirs and upstream tributaries), and it is directed at specific water quality issues such as eutrophication. *Performance monitoring* is needed to evaluate the effectiveness of watershed management practices and policies and to isolate design factors that influence the variability of pollutant removal. Performance monitoring often involves intensive sampling of flow and pollutant mass as it passes through a particular management practice (such as a stormwater pond, OSTDS, or riparian buffer). Finally, monitoring data can be used *to support modeling* of projected changes in water quality under different conditions that are due to land use change or watershed management actions. Modeling-support monitoring involves both intensive and extensive sampling of a reservoir and its watershed to define parameter values, set initial conditions, and physically characterize the watershed. These four types of monitoring are not mutually exclusive, and often data collected for one activity can be used in other contexts. Physical, chemical, and biological parameters to target in monitoring for source water protection are described below.

Pathogenic Microorganisms of Direct Potential Risk to Human Health

Viruses, bacteria, and protozoans originating from animal and human sources are often found in drinking water supplies prior to treatment, albeit at low concentrations. As noted in Chapter 3, the protozoans *Cryptosporidium* and *Giardia* are especially troublesome because of their resistance to disinfection with chlorine. Pathogen monitoring programs must be capable of measuring dilute concentrations of these organisms, determining their viability, and assessing the impact of treatment processes on their survivability in drinking water.

Eutrophication

As described in Chapter 3, eutrophication refers to an increase in the rate of organic matter production in surface waters as a result of nutrient (e.g., phosphorus) addition and subsequent microbial growth. This increased organic matter can cause reductions in dissolved oxygen, alter taste and create odors in drinking water, and it can cause destruction of fish and aquatic plant habitat. Comprehensive monitoring that encompasses a wide variety of eutrophication parameters (including phosphorus, chlorophyll *a*, organic carbon compounds, and dissolved oxygen among others) is critical to evaluating the success of source water protection programs.

Disinfection Byproducts

Natural organic compounds in a drinking water supply can react with disinfection agents (e.g., chlorine) and form potentially carcinogenic chemical compounds (disinfection byproducts or DBPs) in water distribution systems. An effective monitoring program must be able to determine the sources and quantities of dissolved organic compounds and their potential for forming DBPs, while taking into account seasonal variations and the operational flexibility of the water supply system.

Inorganic Turbidity

High turbidity levels are indicative of sediment transport generated by heavy or abnormal precipitation events in the drainage basins of source waters. Erosion either from land runoff or stream banks and the associated sediment transport can be detrimental to water supply systems and aquatic habitats for many reasons. Turbidity has been shown to interfere with the disinfection process (Symons and Hoff, 1976). In addition, sediments can introduce particulate-phase pollutants into water supply reservoirs. Assessing and understanding these effects are important goals of an effective monitoring program.

Toxic Substances

The presence of toxic chemicals in a drinking water supply is a more site-specific problem than the four pollutant categories discussed above. Those chemicals known to have adverse impacts on drinking water quality include metals and metalloids, synthetic organic chemicals, volatile organic chemicals, and pesticides. Persistent bioaccumulating toxic chemicals, such as polychlorinated biphenyls (PCBs) and mercury, are particularly troublesome and are a potential concern for the long-term safety of water supplies. The concentrations of PCBs can be substantial in the Great Lakes area as well as some reaches of the Hudson and Housatonic rivers. Monitoring of toxic chemicals should take into account these site-specific considerations.

Monitoring of Public Health

In addition to monitoring water quality, a comprehensive watershed management strategy includes monitoring of public health to confirm that no waterborne disease outbreaks or unacceptable levels of endemic illness are associated with the water supply. In some communities, this type of monitoring is mandatory for compliance purposes. There are several approaches for disease surveillance that can be implemented.

Disease Surveillance

Public health surveillance activities typically consist of reporting cases of specific infections to local and national public health agencies, such as county health departments or the Centers for Disease Control and Prevention (CDC). Information on disease rates can serve several purposes as part of a watershed management program. Its main role is to provide baseline data on disease trends over time in a target population. Baseline disease rates can then be compared with new information to determine whether specific disease rates are generally increasing or declining, to elucidate seasonal trends in specific disease rates, and to delineate high-risk populations or geographic areas.

Passive surveillance refers to the voluntary reporting of cases of "notifiable" diseases to public health authorities by a health care provider or laboratory. The list of reportable diseases varies from state to state. Waterborne infections that are usually reported include salmonellosis, shigellosis, hepatitis A virus infection, typhoid fever, cholera, *E. coli* O157:H7 (48 states, including New York), cryptosporidiosis (44 states, including New York), and giardiasis (44 states, including New York) (www.cste.org). Passive surveillance systems are often very insensitive—especially for mild conditions where an ill person may not seek medical care—and thus they detect only a small fraction of the true incidence of cases.

Active surveillance is a system where the source of case information (health care provider or laboratory) is called on a regular basis to determine if any cases of a specific condition have been observed. Although this process is considerably more sensitive than passive surveillance, the sensitivity of both active and passive disease surveillance is limited at multiple steps in the process.

Finally, *enhanced* surveillance systems are those where "special additional efforts are made to encourage disease reporting" (Frost et al., 1996). Examples include surveillance for gastrointestinal disease in sentinel nursing homes, monitoring absenteeism in schools or among hospital employees, monitoring Health Maintenance Organization nurse hotline calls about gastrointestinal illness, and monitoring sales of antidiarrheal medications. These activities require far more intensity of effort and resources than passive surveillance but can provide real-time monitoring of the health of the target population.

Depending on the time for development of symptoms and on when a case is diagnosed relative to exposure, some active or enhanced surveillance systems may be able to alert health authorities about immediate, sharp increases in disease rates and enable detection of disease outbreaks. However, in most instances, surveillance for disease is too slow to detect waterborne disease outbreaks.

Epidemiological Studies of Waterborne Disease

Surveillance systems collect information on illness rates in the community but cannot determine risk associated with drinking water because illness reported to state surveillance systems may be associated with contaminated food or other common transmission routes. Determining whether an observed pattern of disease is associated with drinking water requires epidemiologic studies specifically designed to link health outcomes to specific exposures.

Most epidemiologic investigations of drinking water and health have been conducted following an outbreak of waterborne disease. Outbreak investigations have provided valuable information on risk factors and etiologic agents associated with waterborne disease. At least 740 recognized waterborne outbreaks occurred in the United States between 1971 and 1994, and enteric protozoa were the most frequently identified cause of waterborne outbreaks (20 percent) and illnesses requiring hospitalization (78 percent) (Craun et al., 1998).

Epidemiologic studies are also designed to examine *endemic* (baseline level) waterborne disease or other health risks that may be associated with low levels of microbial or chemical contaminants in drinking water. In situations where baseline levels of waterborne disease may be low, the study population must be large enough to detect any difference between the disease patterns in an "exposed" and an "unexposed" population. Although most epidemiologic study designs are too expensive, time-consuming, and long-term to be conducted on a regular basis, a well designed and conducted study can provide the ultimate test of the safety of a water supply.

Several epidemiologic study designs have been applied to the study of endemic waterborne disease (Box 4-2). *Ecologic* studies examine patterns of illness or infection (collected from surveillance systems) and concurrent data on water quality (such as turbidity) to determine if any correlation can be observed over space and time. These descriptive studies are relatively inexpensive and easy to perform because they usually take advantage of existing data on health outcomes and water quality. However, they are limited because they examine aggregate data from groups of populations and cannot take into account individual risk factors (such as contact with child daycare centers or overseas travel) or individual water exposure (such as use of bottled water or length of residence in the study community).

In *case-control* studies, the exposure histories of individuals with the disease of interest ("cases") are compared to the exposures of individuals without the disease ("controls"). Individual study subjects are queried about their residence history, water consumption habits, and risk factors for gastrointestinal disease. The analysis of these results allows the association between exposure and a single health outcome to be evaluated while controlling for individual risk factors. Case-control studies cannot prove that exposure caused the adverse health outcome because they do not provide evidence that the exposure preceded the disease. However, they are useful for examining risk factors for specific health outcomes and require fewer participants than cohort studies (described below).

Cohort studies also collect information on individual exposure, risk factors, and health outcomes. The illness rates in a group of people who are exposed to a water supply of interest are compared to the illness rates in a group of people exposed to a different water supply (such as water receiving a different type of treatment, bottled water, or water receiving additional in-home treatment). Because this design identifies the study population and measures exposure before the development of disease, it can be used to determine the temporal relationship between exposure and disease. These studies are typically the most expensive and time-consuming, especially if a long follow-up period is used.

For all of these epidemiologic approaches, accurately measuring actual exposure to microbial pathogens or chemical contaminants in drinking water and choosing appropriate health outcomes from the wide array of possible water-associated health effects is extremely challenging. Approaches and considerations for exposure assessment and outcome measurement are reviewed in detail elsewhere (NRC, 1998).

Monitoring of Social and Economic Parameters

Although it is not explicitly addressed in most programs, monitoring of social and economic factors can play an important role in measuring the success of a watershed management strategy. The chosen parameters must be tailored to the goals of the specific watershed management plan. Social metrics of interest

BOX 4-2
Examples of Epidemiological Studies of Endemic
Waterborne Enteric Disease in Industrialized Countries

Ecologic Studies

A variety of ecologic studies have been conducted to demonstrate a correlation between health outcomes and water quality parameters. Batik et al. (1980) attempted to find a relationship between endemic rates of hepatitis A infection reported in 75 counties and municipal source water quality and/or level of water treatment for all water supplies in the country. However, no statistically significant associations were observed. A longitudinal study of French alpine villages that used untreated groundwater for their drinking water supplies observed a weak relationship between rates of acute gastrointestinal disease and the presence of fecal streptococci indicator bacteria in the public water system over a 15-month study period (Zmirou et al., 1987). Illness data were collected through active surveillance by physicians, pharmacists, and schoolteachers, while weekly water samples collected from frequently used taps in the distribution system of each village were analyzed for several bacterial indicator organisms. Schwartz et al. (1997) attempted to link turbidity in the Philadelphia water supply with hospital emergency room visits for gastrointestinal symptoms. However, a number of methodological problems with this investigation make the study results questionable.

The role of filtration in affecting disease rates has been investigated in two contrasting studies. The first study examined *Cryptosporidium* antibodies in 86 blood samples collected from children as part of the lead-testing program in Massachusetts (Griffiths, 1999). Samples from children living in towns with an unprotected, filtered surface water supply were more likely to have *Cryptosporidium* antibodies (74 percent) and higher antibody levels (mean optical density = 0.250) than samples from children living in towns with an unfiltered, protected surface water supply (40 percent seropositive and mean optical density = 0.138). The authors concluded that there is significantly more exposure to *Cryptosporidium* for children supplied by an unprotected, filtered surface water than for children served by an unfiltered, protected supply and that increased watershed protection and more stringent filtration methods are need to reduce waterborne exposure to *Cryptosporidium*.

The second study evaluated cryptosporidiosis among AIDS patients in Los Angeles County by comparing prevalence in two communities with different types of water treatment (Sorvillo et al., 1994). One community used standard flocculation and sand filtration; the other community had

continued

BOX 4-2 Continued

only water clarification and chlorine disinfection (which was later modified, in December 1986, to include pre-ozonation, coagulation, flocculation, high-rate filtration through anthracite media, and chlorination). The
water sources for both communities included surface waters from which
oocysts had been recovered. From 1983 through 1986, AIDS patients in
both communities had a similar prevalence of cryptosporidiosis (6.2 and
4.2 percent), although the rate in the community with filtered water was
slightly higher. During the 4-year period after filtration was installed in the
second community (1986-1990), cryptosporidiosis rates declined in both
communities to 3.3 percent and 3.4 percent. The authors concluded that
municipal drinking water was not an important risk factor for AIDS patients
in Los Angeles County. However, the authors noted that the ecologic
nature of the study did not allow examination of the quantity and sources
of water consumed by individuals, and there was no information on the
levels of contamination in the different catchment areas for the two communities.

Case-Control Studies

Two case-control studies have examined the relation between water
supply and endemic giardiasis (Chute et al., 1987; Dennis et al., 1993).
In these studies, cases of giardiasis were identified from a clinic or from a
state registry, and controls (individuals without giardiasis), matched for
age and sex, were recruited from the same clinic, from acquaintances of
the case, or by random digit dialing. Cases and controls were either
interviewed by telephone or they filled out a mail survey about potential
risk factors such as source of drinking water, child daycare utilization,
animal contacts, foreign travel, camping, and swimming in a natural body
of fresh water. Both studies found that giardiasis was significantly associated with the use of a shallow dug well or surface water as the household water source.

Cohort Studies

Cohort studies in the form of randomized intervention trials were conducted in Canada to examine the risk of gastrointestinal illness associated with the consumption of conventionally treated municipal drinking

water that met current microbiological standards (Payment et al., 1991). The first study used 606 households, 299 of which were supplied with reverse-osmosis filters that provided additional in-home water treatment. Gastrointestinal symptoms were recorded in family health diaries. Water samples from the surface water source, treatment plant, distribution system, and study households were analyzed for several indicator bacteria and culturable viruses. Over a 15-month period, a 35 percent higher rate of gastrointestinal symptoms was observed in the 307 study households drinking municipal tap water without in-home treatment compared to the 299 study households supplied with reverse-osmosis filters. Symptomatology and serologic evidence suggested that much of this increased illness may be due to low levels of enteric viruses in the municipal water supply that originated from a river contaminated by human sewage.

Using a similar design, a second intervention study was conducted in the same community in Montreal with 1,400 families randomly allocated to four groups of 350. One group consumed conventionally treated municipal tap water that met current North American drinking water standards. The second group consumed tap water from a continuously purged tap. The third group consumed tap water that was bottled at the treatment plant, and the fourth group consumed purified bottled water (tap water treated by reverse osmosis or spring water). The health of the families was monitored for a 16-month period. The groups consuming tap water and continuously purged tap water experienced 14 percent and 19 percent more illness, respectively, than did the families consuming purified bottled water. Greater illness rates were observed in children 2–4 years of age. The authors concluded that 14 percent to 40 percent of the gastrointestinal illnesses reported were attributable to tap water meeting current water standards and that contamination in the water distribution system was partly responsible for these illnesses (Payment, 1997; Payment et al., 1997). Congress has recently mandated that the CDC and EPA provide a national estimate of waterborne disease occurrence by August 2001. To address this issue, these agencies will be conducting two epidemiologic studies (household intervention studies) similar to those conducted by Payment and others. One study will be conducted in a city receiving drinking water from a surface water source and the other in a city receiving drinking water from a groundwater source.

may include (1) population growth in the watershed, (2) awareness levels of watershed residents regarding watershed management, (3) compliance with inspection and maintenance schedules for BMPs or OSTDS, and (4) rates of residential pesticide application in the watershed. Information about such social parameters can be gained by conducting surveys of watershed residents on a regular basis. Economic parameters may include (1) employment rates and opportunities for watershed residents, (2) types of new development in the watershed (as tracked by building permits), and (3) acreage of land acquired during watershed management. Depending on the scope of the watershed management program, such factors as these are likely to change during program implementation and merit monitoring because of their importance to overall program success.

Evaluation of Monitoring Data and Information

Monitoring, by itself, has limited management value in source water protection efforts unless it is integrated within a larger framework of watershed evaluation. In many watersheds around the country, significant resources are being expended on comprehensive monitoring programs. Often, however, data are not being thoroughly evaluated to draw scientifically based conclusions about the effectiveness of watershed management plans. In other words, current implementation of watershed management plans may be data-rich and information-poor.

The evaluation of monitoring data is needed for three of the four general categories of monitoring outlined above: compliance, operation, and performance. Modeling is itself a tool that can be applied to evaluate data for a variety of objectives. Effective monitoring programs include provisions to collect the data needed to fully support anticipated modeling efforts.

Compliance Evaluations

Determining compliance with federal, state, and local environmental regulations is the most frequent use of monitoring data. Generally, compliance evaluations consist of comparing water quality monitoring data and information to specific numeric objectives and determining the frequency and magnitude of any exceedances. Compliance can be measured for source water quality, finished water quality, and human health effects. It can also include an assessment of implementation rates (e.g., number of structural stormwater practices constructed in compliance with design specifications, or number of construction sites inspected and found to comply with erosion control ordinances) versus objectives for planned watershed practices.

Operational Evaluations

Monitoring data can be evaluated to assess the condition of water sources and treated supply operations. Trend analysis—one means of evaluating improvements in or degradation of water quality over time—can be applied to reservoirs, upstream tributaries, and treated water. Water quality at points throughout a watershed can also be evaluated to identify the relative significance of various land uses or activities within the watershed and related impacts on water quality. Evaluating the relationships among water quality variables can also be useful in identifying cause-and-effect linkages. These kinds of evaluation can help define problem areas within the watershed and focus future management efforts.

Often, simple graphical presentations of data can provide visual overviews for the ongoing tracking of operations. Such graphical data displays are also useful in communicating to the public.

Performance Evaluations

One of the more valuable, yet often overlooked, applications of water quality monitoring data is to evaluate performance. On a broad scale, the overall performance of a watershed plan can be measured by looking at key indicator variables in selected locations over time (see discussion below on risk analysis). On a smaller scale, data can be used to evaluate the effectiveness of specific structural or nonstructural practices. Information about the effectiveness of practices being implemented throughout the watershed is needed to determine the actual value of these practices in benefiting water quality. This information can then be used to support future decision-making. A particularly valuable activity is to evaluate the effectiveness of newer technologies or approaches on a small, pilot scale before widespread implementation.

Formal risk analysis provides an example of how monitoring data might be evaluated to assess the adequacy of a water supply from a public health point of view. Risk is defined as the possibility of suffering harm from a hazard, and risk analysis provides the tools by which the magnitude and likelihood of such consequences are evaluated. If available, pathogen or toxic chemical monitoring data can be used to estimate the risk of "infection" given a certain daily consumption of drinking water. Trends in the risk estimate over time can be used to evaluate the watershed management program over time. These and other types of evaluations of monitoring data are suggested throughout this report.

STAKEHOLDER INVOLVEMENT

Successful watershed planning requires careful attention to the nature of public participation (NRC, 1999). Furthermore, watershed management must get both the "right participation" and the "participation right" (Stern and Fineberg, 1996). That is, the relevant stakeholders must be represented throughout the planning process, and their viewpoints and concerns must be adequately understood in a timely fashion. A stakeholder is defined as an agency, group, organization, or person who has an interest in a process, has decision-making responsibility or authority over that process, or is affected by the outcome of that process.

Involvement of relevant stakeholders is complicated by the almost certain lack of correspondence between political jurisdictions and watershed boundaries. This raises the question of how a community of interest within a watershed context is defined. When a watershed covers a large area, geographically dispersed and socially diverse groups must be brought together to solve a common problem. In some cases, an institutional arrangement to facilitate such community formation may not be available.

The mix of stakeholders may differ depending on specific watershed problems, and the community of interest must be defined on a case-by-case basis. This is a daunting task, given the usually contingent and fortuitous nature of community formation. However, watershed plans can only be effectively implemented if such community definition and formation take place. There are several examples of major federally funded river system projects that have completed numerous scientific studies but have resulted in little change in the way land is managed because of the lack of stakeholder involvement. Particularly complex and context-specific issues like nonpoint source pollution, aquatic and terrestrial habitat preservation, and protection of wetlands are usually resolved only at the local level involving all the relevant stakeholders in planning and decision making (NRC, 1999).

Inclusion of the complete range of stakeholders in the watershed management process does not assure that all interests will be served. There are always tradeoffs and conflicts between competing economic, social, and environmental interests. Methods of conflict resolution can serve as useful tools in helping stakeholders to achieve an acceptable balance among tradeoffs. However, the successful resolution of conflicts requires that stakeholders develop a shared knowledge base regarding the watershed and that they work toward common goals. For this reason, expanding environmental awareness and improving stakeholder education are critical to the success of a watershed management program.

Education may broadly be thought of to include not only formal school curricula, but also media coverage and government dissemination of information. At a minimum, meaningful citizen participation is based on wide and open access to information (Popovic, 1993). Of particular importance are spatially distributed databases (such as GIS) and user-friendly models that can be used to understand both the present structure of the watershed and future configurations. Providing maps of watersheds and subbasins is a necessary step in disseminating information to stakeholders. Stakeholder education helps to create a more complete understanding of the consequences of local actions. For example, public support for certain protection strategies requires an understanding of the "downstream" environmental consequences of local actions. This knowledge may force individuals to recognize public goods that conflict with their particular interests.

Because it requires recognition of a full range of key interests and tradeoffs, the building of trust, and identification of effective and equitable solutions, involving stakeholders in watershed management is an ongoing and lengthy process. There are a variety of tools to encourage stakeholder involvement, but none are completely effective. Typical methods include public hearings, citizen advisory committees and task forces, workshops, alternative dispute resolution, citizens' juries, citizens' panels, and public opinion surveys. The best approach given the current state of knowledge is to open various channels for involvement and to create and maintain an openness to a wide range of methods for integrating stakeholder inputs.

CONCLUSION

This chapter has defined the elements of a watershed management strategy for source water protection. As is apparent in Figure 4-1, watershed management is iterative or cyclical, designed to respond to natural, legal, social, political, and economic changes. Watershed management is a continuous process, with progress measured in discrete steps. Implementation of protection strategies is eventually undertaken, and monitoring is used to assess plan success. Watershed management must be adaptive and flexible in order to incorporate new information as it becomes available. Data inputs that allow assumptions and strategies to be refined keep the watershed management process going.

Ideally, a watershed management plan integrates complete knowledge of the watershed with known methods for water quality protection into a holistic strategy. In some watersheds, the main problems have been identified, and prevention and mitigation methods are available. In most other instances, however, information is limited. Watershed and water resource managers must use best professional judgment to implement protection strategies, monitor their effectiveness, and then make adjustments accordingly. The result of this approach is that watershed management, instead of being a single, comprehensive plan, is instead a series of smaller, individually oriented, plans. Although this expedites the implementation of management practices, it can also work at cross-purposes, since the solution to one problem can create new problems or exacerbate an already existing problem.

Experience has shown that although many communities have used watershed plans as a key tool to protect their local water resources, their success is highly variable (Schueler, 1995). (See Box 4-1 for a description of a successful, comprehensive watershed management strategy.) In general, local watershed plans are developed based on mapping, monitoring, and modeling studies and are adopted through a management process that involves key stakeholders (EPA,

1994). Ideally, the planning process results in the implementation of a future land use pattern and of appropriate management practices that are capable of meeting the water resource objectives developed for the watershed. In reality, however, many local watershed plans have not realized this promise (Schueler, 1995). The remainder of the report focuses on the watershed management strategy used for the New York City watershed, highlighting important successes of the program and making recommendations for improvement.

REFERENCES

American Water Works Association Research Foundation (AWWARF). 1991. Effective Watershed Management for Surface Water Supplies. Arlington, VA: AWWARF.

American Water Works Association (AWWA). 1997. Source Water Protection Statement of Principles. AWWA MainStream.

Batik, O., G. F. Craun, R. W. Tuthill, and D. F. Kraemer. 1980. An epidemiologic study of the relationship between hepatitis A and water supply characteristics and treatment. Am. J. Public Health 70:167–168.

Brown, W., and T. Schueler. 1997. National Pollutant Removal Performance Database for Stormwater BMPs. Center for Watershed Protection, prepared for Chesapeake Research Consortium, Inc.

Chute, C. G., R. P. Smith, and J. A. Baron. 1987. Risk factors for endemic giardiasis. Am. J. Public Health 77(5):585–587.

Clements, J. T., C. S. Creager, A. R. Beach, J. B. Butcher, M. D. Marcus, and T. R. Schueler. 1996. Framework for a Watershed Management Program, Project 93-IRM-4. Water Environment Research Foundation.

Colorado Department of Public Health and Environment. 1998. Colorado Source Water Assessment and Protection—Draft Program, Colorado Water Quality Control Division.

Craun, G. F., S. A. Hubbs, F. Frost, R. L. Calderon, and S. H. Via. 1998. Waterborne outbreaks of cryptosporidiosis. Journal of the American Waterworks Association 90(9):81–91.

Dennis, D. T., R. P. Smith, J. J. Welch, C. G. Chute, B. Anderson, J. L. Herndon, and C. F. vonReyn. 1993. Endemic giardiasis in New Hampshire: a case-control study of environmental risks. J. Infect. Dis. 167(6):1391–1395.

Environmental Protection Agency (EPA). 1989. Wellhead Protection Programs—Tools for Local Governments. EPA 440/6-89-002. Washington, DC: EPA.

EPA. 1992. Compendium of Watershed-Scale Models for TMDL Development. EPA 841-R-92-002, Office of Water. Washington, DC: EPA.

EPA. 1993. Guidance Specifying Management Measures for Sources of Nonpoint Pollution in Coastal Waters. EPA 840-B-92-002. Office of Water. Washington, DC: EPA.

EPA. 1994. The Watershed Protection Approach: Statewide Basin Management. Office of Water. WH-55. Washington, DC: EPA.

EPA. 1996. Better Assessment Science Integrating Point and Nonpoint Sources—BASINS Version 1.0 User's Manual. EPA-823-R-96-001. Washington, DC: EPA.

EPA. 1997a. Notice of Data Availability for the Enhanced Surface Water Treatment Rule. Washington, DC: Office of Water, EPA.

EPA. 1997b. State Source Water Assessment and Protection Programs Guidance—Draft Guidance. EPA 816-R-97-007. Office of Water. Washington, DC: EPA.

Frost, F. J., G. F. Craun, and R. L. Calderon. 1996. Waterborne disease surveillance. Journal of the American Waterworks Association 88(9):66–75.

Griffiths, J. K. 1999. Serological testing for cryptosporidiosis. Presentation at U.S. EPA Office of Ground Water and Drinking Water Stage 2 Microbial/Disinfection Byproducts Health Effects Workshop. February 10–12. Washington, DC.

Harold, C. H. H. 1934. Res. Chem. Bact. Biol. Exam. London Waters 29.

Meyer, J. K. 1997. Stream health: Incorporating the human dimension to advance stream ecology. Journal of the North American Benthological Society 16(2):439–447.

National Research Council (NRC). 1998. Issues in Potable Reuse: The Viability of Augmenting Drinking Water Supplies With Reclaimed Water. Washington, D.C.: National Academy Press.

NRC. 1999. New Strategies for America's Watersheds. Washington, DC: National Academy Press.

Paulson, C., and J. Vlier. 1997. Implementation of Source Protection Plans—Where is the Authority?, American Water Works Association Water Resources Conference—Water Resources Management: Preparing for the 21st Century. August 1997.

Payment P., L. Richardson, J. Siemiatycki, R. Dewar, M. Edwardes, and E. Franco. 1991. A randomized trial to evaluate the risk of gastrointestinal disease due to consumption of drinking water meeting current microbiological standards. Am. J. Public Health 81(6):703–708.

Payment P., J. Siemiatycki, L. Richardson, G. Renaud, E. Franco, and M. Prevost. 1997. A prospective epidemiological study of gastrointestinal health effects due to the consumption of drinking water. Int. J. Env. Health Res. 7:5–31.

Payment, P. 1997. Epidemiology of endemic gastrointestinal and respiratory diseases: incidence, fraction attributable to tap water and costs to society. Water Sci. Tech. 35:7–10

Popovic, N. A. F. 1993. The right to participate in decisions that affect the environment. Pace Environmental Law Review 19(2):683–709.

Salt Lake City Department of Public Utilities. 1998. Salt Lake City Watershed Management Plan. Final draft. Salt Lake City Department of Public Utilities.

Santa Clara Valley Water District. 1995. Comprehensive Reservoir Watershed Management—Phase 1 Project Report.

Schueler, T. 1995. Crafting better urban watershed protection plans. Watershed Protection Techniques 2(2):329–337.

Schueler, T. R., P. A. Kumble, and M. A. Heraty. 1992. A Current Assessment of Urban Best Management Practices–Techniques for Reducing Non-point Source Pollution in the Coastal Zone. Anacostia Restoration Team, prepared for U.S. EPA Office of Wetlands, Oceans, and Watersheds.

Schwartz, J., R. Levin, and K. Hodge. 1997. Drinking water turbidity and pediatric hospital use for gastrointestinal illness in Philadelphia. Epidemiology 8(6):615–620.

Sorvillo, F., L. E. Lieb, B. Nahlen, J. Miller, L. Mascola and L. R. Ash. 1994. Municipal drinking water and cryptosporidiosis among persons with AIDS in Los Angeles County. Epidemiol. Infect. 113:313–320.

Symons, J. M., and J. C. Hoff. 1976. Rationale for turbidity maximum contaminant level. Paper 2A-4a In American Water Works Association Technology Conference Proceedings, Water Quality Technology Conference, December 8–9, 1975, Atlanta, GA, American Water Works Association, Denver, CO.

Upper South Platte Watershed Protection Association. 1998. Memorandum of Understanding.

Water Environment Federation and American Society of Civil Engineers (WEF and ASCE). 1998. Urban Runoff Quality Management. WEF Manual of Practice No. 23 and ASCE Manual and Report on Engineering Practice No. 87.

Wischmeier, W. H., and D. D. Smith. 1978. Predicting rainfall-erosion losses. USDA Agriculture Handbook No. 537. Washington, DC: USDA.

Zmirou, D., J. P. Ferley, J. F. Collin, M. Charrel and J. Berlin. 1987. A follow-up study of gastrointestinal diseases related to bacteriologically substandard drinking water. Am. J. Public Health 77(5):18–20.

5

Sources of Pollution in the New York City Watersheds

Like most areas of the United States, the source waters of the New York City supply are affected, to varying degrees, by a range of pollutants. This chapter describes water quality constituents of primary concern in the New York City drinking water supply—microbial pathogens, nitrogen, phosphorus, organic carbon compounds, sediment, and toxic compounds—as well as their ecological and operational significance. Discrete, point sources (sewage treatment plants and other sources with discharge permits) and diffuse, nonpoint sources (namely on-site sewage treatment and disposal systems or OSTDS, agriculture, residential and commercial development, forestry, and atmospheric deposition) are discussed in detail. The chapter concludes by considering (1) the relative proportion of point and nonpoint source pollution in major watersheds of the New York City systems, (2) reservoir water quality and eutrophication, and (3) compliance with the Safe Drinking Water Act (SDWA) and forthcoming amendments.

POLLUTANTS

Microbial Pathogens

Surface water supplies can be affected by pathogenic microorganisms (bacteria, viruses, and protozoa) originating from various sources. Many bacterial pathogens (e.g., *Salmonella, Shigella, Vibrio*) have long been known to be potentially waterborne (Black et al., 1978; Geldreich, 1990; West, 1989), and they may emanate from both human (wastewater) and nonhuman sources. However, they are at least as sensitive to disinfection with chlorine as are coliforms (Butterfield et al., 1943; Wattie and Butterfield, 1944). Therefore, in disinfected systems such as the New York City supply, the presence of bacterial pathogens is not a significant threat compared to other microbial pathogens.

There are many human enteric viruses that may be present in surface water supplies, such as the rotaviruses, coxsackieviruses, and echoviruses (Clarke and Chang, 1959; Cooper, 1974; Gerba and Rose, 1990; Melnick et al., 1978). Human wastes (including wastewater discharges) are the most frequently documented sources of human enteric virus in surface waters (Gerba and Rose, 1990). Viruses are more sensitive to disinfection than are cysts of *Giardia* (Hoff and Akin, 1986), and thus the amount of inactivation provided to viruses during treatment will be at least as great as that provided to *Giardia*. Thus, like bacterial pathogens, the presence of viruses in water supplies that are disinfected is not of great concern.

Sources of Giardia

As noted in earlier chapters, two protozoa have received increasing attention in the United States over the past several decades—*Giardia lamblia* and *Cryptosporidium parvum*. These organisms form stages, known as cysts and oocysts, respectively, that are resistant to disinfection with chlorine. There is a large body of evidence demonstrating the occurrence of *G. lamblia* in human wastewater as well as in animal wastes. Some animals are believed to serve as reservoirs for human pathogenic strains, with much attention being given to beavers and other aquatic animals. However, evidence that animal-derived cysts have actually resulted in human outbreaks is not conclusive (Erlandsen and Bemrick, 1988). Nevertheless, it is clear that aquatic animals, domestic dogs and cats, and cattle may serve as sources of measurable cysts in surface waters.

Monitoring studies of wildlife in the New York City watershed found that 6.9 percent of those animals tested were infected with *Giardia* cysts (NYC DEP, 1998a). This correlates well with infection rates of cattle in the watershed. Research conducted by Cornell University measured a 7 percent *Giardia* infection rate among previously uninfected cattle (NYS WRI, 1997). This study also demonstrated that cattle previously infected with *Giardia* are more likely to develop infections of other protozoan pathogens.

Sources of Cryptosporidium

Knowledge of the life history of *Cryptosporidium* in water is somewhat less developed than that of *Giardia*. First, there are serious methodological problems with many commonly used environmental detection methods for oocysts (Clancy et al., 1994). Second, there are recently developing lines of evidence suggesting the possibility of different subspecies/strains of *C. parvum* with different preferential hosts (Carraway et al., 1997). Hence, prior evidence may undergo reinterpretation if the taxonomy of *C. parvum* is reconsidered.

The literature indicates that domestic animals, livestock, and wildlife may serve as reservoirs of *C. parvum*, with infected calves and lambs excreting large amounts of oocysts (Smith, 1992). In the New York City watershed, 0.66 percent of tested wildlife (6,261 total specimens) and approximately 2 percent of cattle tested were found to be infected with *C. parvum* (NYC DEP, 1998a; NYS WRI, 1997). There have been studies of the cross-infectivity between humans and other animal species (O'Donoghue, 1995). Human *C. parvum* isolates have been found to infect calves, lambs, goats, pigs, dogs, cats, mice, and chickens. Isolates from calves have been found to infect humans. There is also epidemiological evidence of transmission from cats or pigs to humans (Rose, 1997). Thus, both nonpoint, animal-derived sources as well as human wastewater must be considered as sources of waterborne *Cryptosporidium* in the New York City watershed.

For all microbial pathogens, determining the relative importance of wildlife, domestic animals, and humans as sources is a challenge. To elucidate the contribution of wildlife, field research of the kind demonstrated in Box 5-1 is needed in the Catskill region to document the link between wildlife populations and water quality. In the interim, management practices that encourage increases in wildlife populations (e.g., supplemental feeding of deer, posting to prevent hunting, trapping bans on beaver, and some habitat-enhancement techniques), especially in riparian areas, are ill-advised. A wide variety of forest management strategies are available to enhance wildlife habitat for nongame species (e.g., songbirds), overstory tree growth, and forest health without increasing the size of the deer populations. Beaver populations may require active management in some parts of the watersheds.

Occurrence of Infection in Humans

There is evidence of the occurrence of both giardiasis and cryptosporidiosis in the New York City population. The most recent surveillance data (NYC DEP, 1999a) show rates of giardiasis and cryptosporidiosis of 25.7 and 2.8 per 100,000 persons per year (over the period 1994–1998), respectively. It should be noted that this includes illness from all causes, not just drinking water.

The magnitude of this disease burden can be placed in context with other active surveillance efforts conducted by the Centers for Disease Control and Prevention (CDC). In the Foodnet program, a number of infectious diseases are monitored at several target locations. Although the motivation for this monitoring is understanding foodborne illness, all illnesses from particular pathogens (regardless of source) are tracked. Summaries of incidence rates are indicated in Table 5-2 for cryptosporidiosis and for the total of all infectious diseases tracked. Interestingly, the reported New York City cryptosporidiosis rate is identical to the 1997 Foodnet average cryptosporidiosis rate among all reporting sites, and a number of sites (in Minnesota and California) have higher detected occurrences.

BOX 5-1
Investigating Wildlife as Sources of Pathogens

Wildlife populations in the Catskill region include the species, perhaps as many as 350 vertebrates, typically associated with northeastern forest ecosystems (De Graaf et al., 1992). Once extirpated from the region by marked hunting and trapping, white-tailed deer and beaver have rebounded during the last century. The current distribution and abundance of deer and beaver in the Catskills is not well documented.

A recent study (Fraser, 1999; Fraser et al., 1998) in the Saw Kill watershed on the east side of the Hudson River measured fecal coliform bacteria and other water quality constituents in 12 subwatersheds. Two reference subwatersheds had no domestic livestock or residences; the other ten had livestock (beef and dairy cattle, sheep, and horses) ranging in number from 25 to 1,474 and unknown numbers of wildlife and pets (Pinney and Barten, 1997, 1998). Except for the reference subwatersheds, there are scattered residences with septic systems that may confound the fecal coliform data. Table 5-1 summarizes watershed characteristics, animal numbers, and mean fecal coliform concentrations during the summer of 1996. Although fecal coliform concentrations in the reference subwatersheds are lower than most, if not all, of the concentrations in the treatment subwatersheds, the background contamination from wildlife populations is apparent.

TABLE 5-1 Summary of 1996 Water Quality and Livestock Data for 12 Subwatersheds of the Saw Kill, Tivoli Bays (Hudson River) National Estuarine Research Reserve, near Red Hook, New York (the italicized rows are the reference subwatersheds with no domestic livestock but unknown numbers or types of wildlife)

Subwatershed ID#	Area (km^2)	Total number of domestic livestock	Mean fecal coliform (CFU/100 mL)
W1	3.0	25	626
W2	4.2	91	797
W3	*1.9*	*0*	*594*
W4	5.9	38	597
W5	1.2	90	389
W6	2.1	122	1,657
W7	2.9	122	341
W8	12.1	54	450
W9	50.0	1,474	1,118
W10	17.0	690	142
W11	*1.5*	*0*	*167*
W12	4.0	600	453

Source: Reprinted, with permission, from Fraser (1999). © 1999 by Fraser.

TABLE 5-2 Disease Incidence Rates Reported by 1997 Foodnet Surveillance

| | Number of Cases per 100,000 per year | |
Locale	All illnesses[a]	Cryptosporidiosis
All sites	53.1	2.8
California	89.2	3.4
Connecticut	46.6	1.3
Georgia	44.9	No data
Minnesota	51.8	5.2
Oregon	42.9	0.8

[a]Includes illnesses from *Campylobacter, Cryptosporidium, Cyclospora, E. coli* O157:H7, *Listeria, Salmonella, Shigella, Vibrio,* and *Yersinia.*
Source: CDC (1998).

Nitrogen

Nitrogen has many forms and functions in watersheds and other environmental systems. As shown in Table 5-3, it is found in nature in all three environmental media (air, water, and soils) and in six of its eight possible oxidation states (+V to –III). Nitrogen can act as a pollutant in surface waters in four principal ways: (1) as a nutrient for photosynthetic activity in streams, lakes, and reservoirs (eutrophication), (2) by producing toxic effects on fish when present as ammonia (NH_3), (3) as an acid when present as nitric acid (HNO_3), and (4) by exerting an oxygen demand when in the form of organic N, ammonium (NH_4^+), or NH_3. It can also exert a significant chlorine demand and can therefore affect the disinfection process.

Sources of nitrogen in the Catskill and Delaware watersheds are diverse. Atmospheric deposition, both wet and dry, is substantial. It can include gaseous NO_x [the sum of nitrogen dioxide (NO_2), nitrous oxide (N_2O), and nitrogen oxide (NO)], HNO_3, and NH_3, as well as ammonium and nitrate particulates. Other nonpoint sources include septic tanks, agricultural runoff (as discussed below), urban stormwater, and groundwater. These can contain NH_4^+, NH_3, nitrate (NO_3^-), and organic nitrogen. Point sources of NH_4^+, NH_3, NO_3^-, and organic nitrogen include municipal wastewater treatment plants, industrial discharges, and stream or reservoir inflows.

Point sources of nitrogen are controlled by the limits that have been placed on total ammonia concentrations [$NH_4^+ + NH_3$] in wastewater discharges. Biological treatment to meet these limits reduces the possibility of fish toxicity related to NH_3 and also lowers the nitrogenous oxygen demand entering surface waters by oxidizing reduced nitrogen to NO_3^-.

Nonpoint sources of nitrogen are more difficult to control. Manure is a significant source of nitrogen in agricultural watersheds such as the Cannonsville

TABLE 5-3 Forms of Nitrogen

Compound	Media	Oxidation State
Elemental nitrogen	Air	0
Ammonia(um)	Soils, air, water	–III
Nitrate	Soils, water	V
Nitrite	Soils, water	III
Amino acids	Soils, water	–III
Nitric acid	Air, atmospheric precipitation	V
Nitrous acid	Air	III
Nitrogen oxides	Air, water	II, III
Dissolved organic N	Water	III
Nitrous oxide	Air	I

watershed. Although nitrogen is lost through ammonia volatilization and denitrification, there can be large amounts of organic and inorganic nitrogen in runoff from barnyards and manured fields. Additionally, both manured and fertilized fields tend to have high levels of nitrate leaching to groundwater. The Cannonsville Reservoir shows a tendency to nitrogen limitation on algal production during mid- to late summer (Effler and Bader, 1998), indicating that nitrogen loading, in some instances, is important.

Deposition of nitrogen currently exceeds the biological demand for nitrogen in the Catskills because vegetative growth rates have slowed as the forest has matured (Murdoch, 1999). Nitrogen in excess of demand is stored in the soil and/or leached into streams. Increases in nitrate concentrations in streams have been accompanied by measured decreases in pH in the East Branch of the Neversink River.

The effects of atmospheric deposition of nitrogen in the Catskill/Delaware watershed include impaired fish habitat in streams. In addition, the associated soil acidification can lead to slower growth rates for vegetation, less nutrient retention in the forest, and greater erosion and nutrient leaching to streams. Previous work has shown enhanced eutrophication in downstream reservoirs when nitrate concentrations increase in the presence of high phosphorus concentrations (Dodds et al., 1989; Morris and Lewis, 1988). Finally, temporary acidification of the reservoirs themselves is possible. This is most likely in the Neversink Reservoir, which is not sufficiently alkaline to buffer acid inputs from the Neversink River (Murdoch, 1999).

Phosphorus

The pollutant in the New York City water supply system that has received the most attention is phosphorus. Phosphorus has been identified as the dominant limiting macronutrient for algal and other plant growth in the New York City reservoirs (NYC DEP, 1999b). When phosphorus concentrations become elevated in these waters, enhanced growth of algae, photosynthetic and heterotrophic bacteria, and higher aquatic plants can occur. In turn, the increased production of organic matter by these organisms can alter conditions within the reservoirs and lead to eutrophication.

The most important negative consequences of eutrophication include (1) an increase in water turbidity caused by algal material and algal byproducts, (2) an increase in total organic carbon derived from algal biomass that can lead to formation of disinfection byproducts (DBPs), (3) algal production of potentially toxic compounds, some of which may create taste and odor problems, and (4) a decrease in dissolved oxygen levels within the reservoirs and the associated negative impacts on fish habitat. Many of these conditions are particularly difficult to overcome in water supplies that do not undergo extensive treatment. Thus, the suppression of phosphorus loading rates to the Catskill/Delaware reservoirs is a particularly important goal for the New York City watershed management strategy.

Total phosphorus is divided into soluble and particulate forms. Particulate phosphorus (PP), which must be mineralized or hydrolyzed prior to uptake by algae, is much less bioavailable than soluble phosphorus, and it often settles to reservoir sediments. Therefore, it is less likely to lead to increased algal growth and eutrophication. The fraction of particulate phosphorus that can be hydrolyzed and mineralized to release soluble reactive phosphorus back to the water column varies. Studies in the Cannonsville watershed found that 25 percent to 48 percent of tributary particulate phosphorus was bioavailable (Auer et al., 1998). Soluble forms of phosphorus include soluble organic phosphorus, total dissolved phosphorus, and soluble reactive phosphorus (SRP). Soluble reactive phosphorus is the most commonly measured dissolved form because of its important role in eutrophication.

Phosphorus sources in the Catskill, Delaware, and Croton watersheds are many and varied. As with most of the pollutants discussed in this chapter, phosphorus emanates from a variety of point and nonpoint sources such as wastewater treatment plants (WWTPs), OSTDS, agriculture, urban stormwater, and (in low concentration) forests. However, the dominant form of phosphorus created by each of these activities can be very different. Sewage treatment plants are the primary point source of phosphorus in the New York City water supply system because of the paucity of industrial waste discharges. Effluent from municipal sewage treatment plants consists mainly of soluble reactive phosphorus (essentially phosphate) plus some particulate phosphorus from plants that do not have efficient secondary sedimentation and/or filtration of the effluent. Similarly, OSTDS with poorly operating drain fields can result in leaching and discharge of soluble reactive phosphorus to streams.

Stormwater transports both particulate and dissolved phosphorus. In particular, runoff from agricultural lands treated with manure and from barnyard areas has high levels of dissolved phosphorus with average volume-weighted concentrations of 230 µg/L (Robillard and Walter, 1984). Model calculations for Cannonsville Reservoir in the early 1980s estimated that 77 percent of the dissolved phosphorus load to Cannonsville Reservoir came from either direct runoff (both surface and subsurface) or baseflow. Urban stormwater has been shown to contribute approximately equal loadings of particulate and dissolved phosphorus (Schueler, 1995).

Organic Carbon Compounds

Organic carbon compounds in water supply reservoirs are problematic because some can react with chlorine to form DBPs in the water distribution system. Before describing important sources of organic carbon in the Catskill/ Delaware watershed, it is necessary to define the many parameters used to measure organic carbon concentrations. The most generic classification is natural organic matter (NOM). This term has been used to differentiate between organic carbon compounds from natural versus human-synthesized sources. The most common measure of organic carbon compounds used in environmental engineering is total organic carbon (TOC). TOC includes all particulate and dissolved organic matter, ranging in size from simple dissolved molecules to particles several millimeters in diameter and larger, and it includes both natural organic matter as well as artificial human created compounds.

Organic carbon is often characterized by dividing it into dissolved and particulate forms. Particulate organic carbon matter (POC) is organic carbon in particulate form greater than 0.5 µm in diameter. Dissolved organic carbon (DOC) is analytically separated from POC by filtration through 0.5-µm glass fiber filters and includes both dissolved and colloidal organic carbon. Both TOC and DOC refer only to organic carbon and should not include inorganic compounds such as CO_2, HCO_3^-, and CO_3^{-2}. In addition, volatile organic compounds removed from samples during TOC analyses are not reflected in these measurements. DOC is the dissolved (and colloidal) portion of TOC and is the primary focus of the following discussion.

DOC can be a precursor to the formation of DBPs in chlorinated water supplies. Reservoirs, including those in the Catskill/Delaware system, have two general sources of such precursors: (1) allochthonous organic carbon (both dissolved and particulate) that flows into the reservoirs with surface and groundwater runoff and (2) autochthonous organic carbon that is created within the reservoirs as a result of microbial activity.

Allochthonous organic carbon is largely imported in dissolved and colloidal form as humic substances and is derived from partial microbial degradation of lignin-cellulose based carbon compounds of higher plants. These compounds are chemically recalcitrant to rapid biodegradation. (However, rates of biological degradation are increased markedly if the dissolved organic compounds are exposed to ultraviolet radiation [Wetzel et al., 1995]). Loading of allochthonous TOC to reservoirs is directly correlated with rainfall intensities (e.g., Jordan et al., 1985; Mickle and Wetzel, 1978; Wetzel and Otsuki, 1974). Autochthonous organic carbon is produced largely by algal and cyanobacterial photosynthesis in reservoirs with little or no development along the shoreline. As a result, autochthonous organic carbon has a definite seasonal cycle and is relatively readily biodegradable.

Several studies have attempted to compare the DBPs formed from allochthonous versus autochthonous carbon precursors (Briley et al., 1980; Hoehn et al., 1980; Hoehn et al., 1984). DBP yields from humic substances were found to fall within the ranges reported for algal biomass and extracellular products (Hoehn et al., 1980). Recent studies by the New York City Department of Environmental Protection (NYC DEP) have shown that allochthonous precursor carbon produces DBPs that are primarily dissolved (94 percent) and are primarily chloroform (98 percent) (Stepczuk et al., 1998). The same was found for autochthonous sources of precursor carbon (NYC DEP, 1997a). If the types of DBPs formed from these different sources are similar, then the most important factors affecting DBP formation are the overall quantities of autochthonous and allochthonous precursor carbon present in each reservoir and their relative rates of degradation. However, if one of these sources has a greater tendency to form DBPs than the other [which is possible for haloacetic acids (HAAs) and other DBPs that have yet to be studied], then successful precursor control will depend heavily on identifying the most significant source.

In all natural lakes, most (> 60–80 percent) of the DOC occurring at any one time in the reservoirs is composed of recalcitrant allochthonous DOC. Because of its complex structural chemistry, allochthonous DOC degrades more slowly than autochthonous carbon and dominates in-reservoir DOC. Allochthonous DOC has been shown to degrade at a rate of about 1 percent per day with a turnover time of about 80 days (Cummins et al., 1972; Wetzel and Manny, 1972). This is not to say that autochthonous sources are not, during certain seasons, in greater abundance than allochthonous sources. DOC production by phytoplankton certainly increases during the summer relative to allochthonous DOC production, but it is not dominant until hypereutrophic conditions are reached. Thus, on an annual basis, allochthonous DOC will dominate the pool of precursor carbon that is available to react with the disinfectants in the water supply system.

Determining which source of precursor carbon is more important to DBP production in the Catskill/Delaware watershed has recently become a priority for NYC DEP. In 1997, modeling of the West Branch of the Delaware River and the Cannonsville Reservoir suggested that autochthonous sources of DBP precursors are more important than allochthonous sources (NYC DEP, 1997a). Other data in the report, however, demonstrated that allochthonous precursors are present and should not be ignored. The report states that "net autochthonous production of precursors in the epiliminion, apparently driven by primary productivity of phytoplankton, was found to be a major source of precursor for the reservoir (Cannonsville), representing about two-thirds of the cumulative mass input over the April to mid-summer interval." This finding indicates that about a third of the precursor pool in the epilimnetic water was allochthonous. Although the volume of the epilimnion is less than half of the total, it is not a trivial quantity. It is also clear from the data presented that these ratios are highly dependent on seasonal variations. For example, the ratio of autochthonous precursors to allochthonous precursors was estimated to vary from a high of 2.5 in June 1995 to a low of 0.75 in November 1995. This work has focused on trihalomethane formation; information about HAA formation derived from DOC in the water supply is not yet available.

There are other important considerations. During dry years, the contribution of allochthonous precursors may be low because of limited streamflow. Also, because algae tend to accumulate at the surface of waterbodies, there can be significant variations in the ratio of autochthonous to allochthonous precursors throughout the depth of a reservoir. This stratification may have played a role in the observed results.

Management strategies for controlling allochthonous versus autochthonous precursor carbon are significantly different. If allochthonous sources are suspected, best management practices that reduce DOC loadings from rainfall and snowmelt runoff will be beneficial. In addition, if it is exposed to sunlight, the recalcitrant, allochthonous DOC within reservoirs will degrade faster (both photochemically and microbially) than if it is not exposed (Tranvik and Kokalj, 1998; Wetzel et al., 1995). Thus, management practices to reduce the pool of allochthonous DOC should strive to increase reservoir residence time and exposure to light in the epilimnion. Despite the more rapid decomposition of autochthonous DOC compared to allochthonous DOC, it is essential that algal growth be controlled, especially during those times in which autochthonous precursor carbon dominates. For the New York City reservoirs, this has usually been accomplished by reducing phosphorus loadings to the reservoirs from both point and nonpoint sources.

Sediment

Sediment decreases the clarity of water, thereby increasing turbidity and its undesirable effects, such as interference with chlorination. In addition, other pollutants (e.g., nutrients, metals, and pathogens) may be adsorbed to sediment particles, which can mask a fraction of the total pollutant load from detection. Wind-induced mixing can disperse sediment and associated pollutants throughout the reservoir. This may be especially pronounced when the reservoirs are thermally stratified and vertical mixing is largely limited to the epilimnion. Coarse-textured sediment (sands and larger silt particles) and bedload may settle to the bottom of a reservoir near the point of entry. However, clay particles may remain in suspension for days to weeks.

In the absence of anthropogenic disturbance, the Catskill and Delaware watersheds, like most forest ecosystems, have very low rates of sediment production (Patric, 1976; Patric et al., 1984). Forest vegetation occupies virtually all of the available growing space. Forest fires and insect and disease outbreaks severe enough to kill overstory trees are rare in the region and throughout the eastern deciduous forest. Therefore, the loss of the forest's protective influence occurs infrequently and over relatively limited areas.

Soils in the region have high hydraulic conductivity augmented by extensive, interlocking root systems, by an organic litter layer beneath the forest vegetation, and by actions of organisms ranging from microbes to small mammals. Hence, overland flow is exceedingly rare. Even when overland flow occurs, unless there is significant detachment of soil particles by raindrop splash and subsequent transport by overland flow, erosion and sediment production occur in minuscule amounts in areas away from the stream channel network (e.g., the upturned root mass of a fallen tree, the exposed soil below a rock outcrop). Hence, the primary source of sediment in forested watersheds (not subject to landslides and other forms of mass erosion) is erosion occurring within the stream channel. Under present conditions, most turbidity problems in the Catskill/Delaware supply are caused by inorganic particles and sediment derived from surface and channel erosion during heavy precipitation events, primarily in the Schoharie and Ashokan watersheds (Longabucco and Rafferty, 1998).

Further problems with sediment could arise if land use in the region changes. When changes in land use lead (1) to the removal of vegetation (or the conversion of forest vegetation to lawns), (2) to disruption or removal of the litter layer, (3) to disintegration or compaction of the soil surface, or (4) to concentration of stormwater from impervious surfaces, overland flow supplants subsurface flow as the dominant pathway. Reductions in interception and transpiration exacerbate the process by increasing stormwater volumes. Roads and storm drain systems frequently short-circuit natural pathways of flow to discharge into nearby streams or wetlands. Unless management practices are used to limit the force of raindrop splash and overland flow (by seeding and mulching exposed soil, dispersing overland flow into forested areas, terracing slopes, etc.) or to collect and clarify water (riparian buffers, settling ponds, created wetlands, catch basins), the downhill path of sediment and associated nonpoint source pollutants is inexorable.

Changes in the quantity and quality of flow affect the dynamic equilibrium of stream channels. Sediment deposition in low-velocity stream reaches limits cross-sectional area when larger flows occur. As variation in water level and velocity increases, channel scour and realignment become commonplace. Sedimentation of particulate matter from nonpoint source runoff can cause ecological impacts because of siltation and destruction of habitat for fish and macroinvertebrates.

Another potential source of turbidity that should be considered is biogenic turbidity generated by phytoplankton in the reservoirs. Colloidal $CaCO_3$, common to hard-water lakes and some of the New York City reservoirs, contributes to both reservoir turbidity and the characteristic blue-green color of the reservoirs. Colloidal carbonates are inorganic particulates that are largely induced by the photosynthetic activities of phytoplankton. Although thought not to be a significant problem at this time, increased algae growth within the reservoirs may eventually lead to increased turbidity, among other things.

Toxic Compounds

Because much of the Catskill/Delaware watershed region consists of undisturbed forest, it is far less likely than industrial regions to harbor hazardous wastes and toxic compounds that might pose a human or environmental health risk. However, there are a few specific types of hazardous compounds used on a regular basis in the watershed region, and there are also a small number of regulated hazardous waste sites. Because of the considerable uncertainty that surrounds the transport, fate, and toxicity of many of these compounds, additional information is needed to better assess and prevent exposure of humans and ecological receptors to these chemicals.

Pesticides are used on a regular basis in the Catskill/Delaware watershed, primarily on agricultural lands in the Cannonsville region. Residential and commercial use of pesticides also occurs throughout the Catskill, Delaware, West Branch, and Kensico watersheds. A comprehensive compilation of pesticide use in the watershed can be found in the 1993 environment impact statement for the watershed regulations (NYC DEP, 1993a). Some of the more prevalent compounds found in the watershed include alachlor, aldicarb, atrazine, carbaryl, carbofuran, chlorpyrifos, cyanazine, 2,4-D, and metolachlor, among others (NYC DEP, 1997b,c; Quentin, 1996). NYC DEP monitoring of pesticides has occurred primarily in the distribution system to comply with the Safe Drinking Water Act (SDWA), in aqueduct entry points (annually), and at Kensico stream sites. Regular and widespread monitoring of pesticides has not been conducted by NYC DEP or the Watershed Agricultural Program.

Sediment and surface water (mainly river water) samples have been found to have low levels of all the pesticides listed above (NYC DEP, 1997b; Quentin, 1996). In almost all cases, measured concentrations were below maximum contaminant levels (MCLs) for those compounds regulated under the SDWA. Concentrations of unregulated compounds were below 2 µg/L. The most frequently detected compound was 2,4-D, a weed-killing herbicide used throughout the watershed region for residential and commercial lawns (Quentin, 1996).

In addition to pesticides, other hazardous substances may be generated, stored, and disposed of in the Catskill/Delaware watershed. As of 1990, there were 35 petroleum storage facilities, 15 hazardous waste storage facilities, 1 hazardous waste generator, and no hazardous waste disposal facilities in the West-of Hudson region (NYC DEP, 1993a). The most significant of these are the Richardson Hill Road landfill (a Superfund site) and the Rotron-Olive site, both of which are currently inactive (NYC DEP, 1993a). The Richardson Hill Road landfill lies within 500 ft of a tributary to the Cannonsville Reservoir. Volatile organic compounds and polychlorinated biphenyls (PCBs) have been detected in soil at the site and in groundwater beneath the site. Contaminated groundwater has also been detected in the drinking water wells of residents living in close proximity to the site. Potential health risks from this site are being mitigated by treating contaminated groundwater at private wells, by excavating contaminated sediments at the site, and by collecting and treating landfill leachate (EPA, 1999). In addition, each town in the Catskill/Delaware watershed has a municipal land-fill that may or may not be closed. However, information about conditions at these sites is extremely limited.

General information about the types and amounts of hazardous substances generated at the other hazardous substance storage facilities in the Catskill/Delaware region (NYC DEP, 1993a) reveals that 44 percent of the 177,672 gallons of hazardous waste stored west of the Hudson River is acids and bases, 24 percent is inorganic compounds, 23 percent is metals or salts, 5 percent is volatile organic compounds, and 4 percent is unclassified. Those hazardous waste storage facilities that hold SPDES permits (and hence are regulated point sources) are discussed in the next section. It should be noted that none of New York State's top ten facilities for release of toxic compounds is located in either the West-of-Hudson or East-of-Hudson watershed regions (Toxics Release Inventory, 1996).

POINT SOURCES OF POLLUTION

Domestic Wastewater Treatment Plants

Domestic wastewater contains substantial concentrations of pathogenic microorganisms and must be discharged in an area that will ensure removal of pathogens before the effluent reaches groundwater (Veneman, 1996). Individual septic systems (OSTDS) and centralized WWTPs represent the two main strategies used to collect, treat, and dispose of domestic wastewater. Centralized sewage treatment systems collect wastewater from large numbers of residential and commercial facilities through pipes (sanitary sewers). The sewage travels through the pipes by gravity flow (which may be facilitated by occasional "lift stations") to a treatment works. Small-scale treatment works that serve perhaps a single subdivision are known as package plants; larger-scale systems serving entire communities are referred to as municipal treatment plants (Purdom, 1971). WWTPs are considered to be point sources of pollution because they discharge treated wastewater from discrete locations (effluent pipes) into a receiving water. In the Catskill/Delaware watershed, there are 39 WWTPs that discharge waste into adjacent streams and two that discharge to the subsurface. Table 5-4 lists important parameters for the WWTPs in the Catskill/Delaware watershed.

Regardless of size, the basic treatment methodology of a centralized sewage treatment system can be summarized in a few basic steps. The first step involves screening large debris from the wastewater and grinding or macerating the remaining sewage. This slurry is then allowed to settle and the resulting sludge is digested aerobically or anaerobically in large tanks. This process, termed primary treatment, removes about 40 percent of the 5-day biochemical oxygen demand (BOD_5) in the wastewater and 50–90 percent of the bacteria (Chanlett, 1979).

During secondary treatment, the liquid fraction of wastewater is subjected to degradation through contact with large numbers of microorganisms in an aerobic environment. Several technologies are used to maintain high levels of oxygen and maximize the contact between the sewage and the microorganisms. These include (1) trickling filters—rocks or plastic media covered with microorganisms, (2) activated sludge—a slurry of microorganisms subjected to intense mixing with atmospheric oxygen, and (3) rotating biological contactors—discs of large surface area colonized by microorganisms. The rotating action of the latter method alternately exposes the microorganisms to the atmosphere and wastewater. Secondary treatment removes approximately 85 percent to 90 percent of the BOD_5 and 90 percent to 95 percent of the bacteria from the wastewater (Chanlett, 1979).

Tertiary treatment involves removal of nutrients such as phosphorus and nitrogen, sand filtration, and microfiltration, or other techniques that remove an additional five percent of the BOD_5 (Chanlett, 1979). A few WWTPs in the Catskill/Delaware watershed currently use chemical precipitation in conjunction with tertiary sand filters to remove phosphorus, and more plants are expected to require chemical precipitation to comply with the MOA (D. Warne, NYC DEP, personal communication, 1999). Finally, most WWTPs chlorinate (and sometimes dechlorinate) their wastewater prior to discharge.

The entire treatment train (sequence) for one of the larger WWTPs in the region is shown in Figure 5-1. This treatment train, among the most sophisticated in the Catskill/Delaware watershed, includes all of the tertiary treatment upgrades mandated by the MOA, including microfiltration (see discussion below). Figure 5-2 shows a treatment train more typical of one of the smaller WWTPs in the watershed and does not reflect upgrades that will be installed as part of the MOA.

As part of the Watershed Rules and Regulations in the MOA, all sewage treatment plants in the watershed are being upgraded to meet new performance criteria and effluent standards, with upgrades scheduled for completion by 2002 (EPA, 1997). The most significant of these requirements for WWTPs that discharge to surface waterbodies are the following:

TABLE 5-4 Municipal Wastewater Treatment Plants in the Catskill/Delaware Watershed

Reservoir/Watershed	SPDES Number	SPDES Permitted Flow (gpd)	1998 Actual Flow[a] (gpd)	1998 Effluent [Phos.][b] (mg/L)	Months of Operation	Class of Receiving Stream[c]
Kensico						
None						
West Branch						
Clear Pool Camp	NY-0098621	20,000	3,858	0.63	Year-round	B
Ashokan						
Belleayre Mt. Ski Center	NY-0034169	15,000	2,230	1.61	Year-round	B(T)
		14,000	4,220	1.00		D
Camp Timberlake	NY-0240664	34,000	10,000	2.43	June–Oct. 1	B(T)
Chichester	NY-0233943	9,900	NA	NA	Year-round	subsurface
Mountainside Rest.	NY-0251241	3,076	650	3.59	Year-round	B
Onteora Schools	NY-0099856	27,000	9,320	1.88	Year-round	B
Pine Hill	NY-0026557	500,000	285,250	0.43	Year-round	B(T)
E.G.&G Rotron, Inc.	NY-0098281	12,750	0	NV	Year-round	B/B(T)
Schoharie						
Camp Loyaltown	NY-0104965	21,000	NV	0.13	June–Oct. 31	C
Colonel Chair Estates	NY-0101001	30,000	14,242	0.58	Year-round	C(TS)
Crystal Pond Twnhs.	NY-0223638	36,000	0	NA	Winter	C(TS)
Elka Park	NY-0092991	10,000	NV	0.77	May–Oct.	C(TS)
Forester Motor Lodge	NY-0100374	3,900	0	1.4	Year-round	C(TS)
Frog House Rest.	NY-0224731	1,788	985.5	17.48	Year-round	subsurface
Golden Acres Farm	NY-0069957	5,800	NV	1.48	July–October	C
		1,100		1.73		
		2,300		1.99		

Name	ID					
Grand Gorge	NY-0026565	500,000	114,875	0.06	Year-round	C
Harriman Lodge	NY-0100277	20,000	10,000	0.80	June–Sept.	C(TS)
Hunter Highlands	NY-0061131	80,000	8,414	0.09	Year-round	C
Latvian Church Cmp.	NY-0072192	7,000	NV	1.4	July–August	C(TS)
Liftside at Hunter Mt.	NY-0212288	81,000	29,600	0.24	Year-round	C(TS)
Mountain View Estates Home	NY-0241261	7,000	2,717	4.05	Year-round	C
Mountain View Estates, Inc.	NY-0212407	6,000	2,550	1.8	Year-round	C
Ron-De-Voo Rest.	NY-0124672	1,000	< 1,000	0.28	Year-round	NA
Snow Time	NY-0065315	120,000	3,800	2.33	Year-round	C
Tannersville	NY-0026573	800,000	334,714	0.10	Year-round	C(TS)
Thompson House Inc.	NY-0101168	4,775	2,200	5.33	May–Oct.	A(TS)
Whistletree Dev.	NY-0310821	12,450	NV	0.56	Year-round	C(TS)
Cannonsville						
Delaware-BOCES	NY-0097446	2,500	1,000	1.35	Year-round	C(T)
Delhi	NY-0020265	515,000	390,000	3.12	Year-round	C(T)
SEVA Institute	NY-0205800	6,600	0	NA	Spring–fall	C(T)
		1,200				
Allen Center	NY-0029645	20,000	10,477	7.59	Year-round	C(T)
Stamford[d]	NY-0021555	500,000	517,075	1.28	Year-round	C(T)
Village of Hobart	NY-0029254	160,000	40,000	3.20	Year-round	C(T)
Walton[d]	NY-0027154	1,170,000	1,387,125	0.52	Year-round	B(T)
Pepacton						
Camp NuBar	NY-0023787	12,500	6,000	1.42	June–August	C(TS)
Camp T'ai Chi	NY-0104957	7,500	0	0.25	July–Sept.	B
Margaretville	NY-0026531	400,000	268,000	1.46	Year-round	A(T)
Mountainside Farms, Inc.	NY-0084590	49,800	11,700	17.44	Year-round	GA & GW
Regis Hotel	NY-0100382	9,600	6,167	1.50	April–Nov.	B(TS)
Roxbury Run	NY-0099562	100,000	37,914	1.59	Year-round	C

continued

TABLE 5-4 Continued

Reservoir/Watershed	SPDES Number	SPDES Permitted Flow (gpd)	1998 Actual Flow[a] (gpd)	1998 Effluent [Phos.][b] (mg/L)	Months of Operation	Class of Receiving Stream[c]
Rondout						
Grahamsville	NY-0026549	180,000	96,666	0.3	Year-round	A(T)
Neversink						
None						

[a]Actual flow data are 1998 averages of monthly (30-day average) data from discharge monitoring reports to the NYS DEC. Only those months indicated in the months of operation column are included in the value.

[b]Average 1998 effluent phosphorus data are averages of monthly or biweekly data reported by NYC DEP. In most cases, these concentrations represent conditions prior to the installation of upgrades mandated by the MOA.

[c]See Appendix B for explanation.

[d]WWTP in exceedance of permitted flow on a regular basis.

NA = not applicable

NV = not available

Note: Bolded entries have received the WWTP upgrades mandated by the MOA. They are all city-owned facilities.

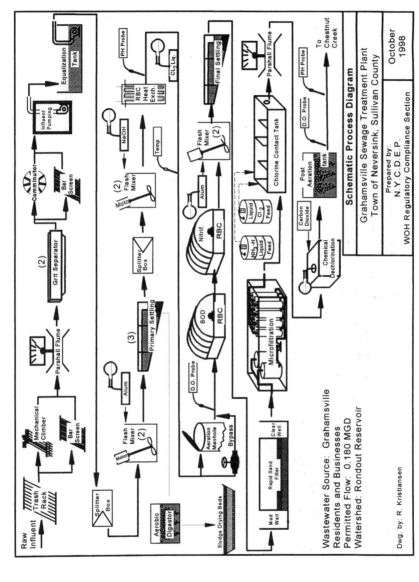

FIGURE 5-1 Primary, secondary, and tertiary treatment at the Grahamsville WWTP. Source: NYC DEP (1998b).

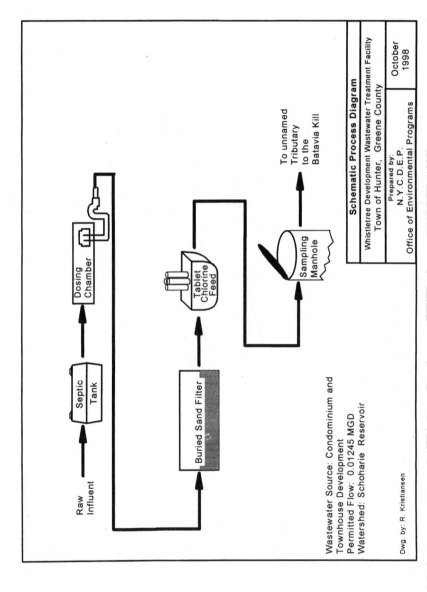

FIGURE 5-2 Sewage treatment at the Whistletree Development WWTP. Source: NYC DEP (1998b).

- phosphorus effluent standards: 1.0 mg/L for <50,000 gallons per day (gpd)
                                    0.5 mg/L for 50,000–500,000 gpd
                                    0.2 mg/L for >500,000 gpd
- 99.9 percent removal and/or inactivation of *Giardia* and enteric viruses
- upgrades of existing WWTPs to include sand filtration, disinfection, phosphorus removal, and microfiltration (or an approved alternative technology).

Some of these requirements have overlapping goals. For example, microfiltration can be used to achieve particulate phosphorus and microbial pathogen removal. It should be noted that five of the New York City-owned WWTPs are installing microfiltration (as illustrated in Figure 5-1) while most of the plants not owned by New York City will be using an alternative type of technology known as Continuous Backwash Upflow Dual Sand Filtration (EPA, 1998b). The equivalence of this technology to microfiltration and its ability to meet the required effluent standards are discussed in Chapter 11.

One significant benefit of WWTPs is the SPDES permit requirement for a trained operator (see Chapter 3). In addition, samples of the effluent are routinely analyzed to ensure that effluent quality standards are met. These factors are meant to ensure that the system functions as designed at all times and that corrective actions are applied in a timely manner.

In addition to treated wastewater, WWTPs produce partially digested sludge, or residuals. These residuals contain substantial amounts of nutrients and pathogens and some heavy metals. They are typically rendered free of detectable pathogens, or stabilized, by adding lime until the pH is too basic to support pathogenic microorganisms. These residuals must be properly managed to avoid creating a secondary source of contaminants. Though sludge treatment practices vary throughout the watershed, most solid waste generated by WWTPs is eventually buried in landfills within the watershed region or is moved outside the watershed boundaries (Warne, 1998a).

Although combined sewer overflows are not found in the Catskill/Delaware watershed, it is possible that stormwater can impact WWTP flows via inflow and infiltration (I/I) into sanitary sewer pipes and other infrastructure. These processes can greatly increase treatment plant flow, which may result in overflow conditions and a short-circuiting of the treatment processes. Much of the sewerage infrastructure in the Catskill/Delaware watershed predates the advent of sturdy and leak-proof pipe materials. At least 11 WWTPs west of the Hudson River have been identified as having problems with I/I (NYC DEP, 1993a). NYC DEP estimates that nearly every West-of-Hudson WWTP experiences I/I at a level that is approximately 25 percent of its average daily flow, with some plants receiving I/I that is equal to or greater than their permitted flow during large storm events (Warne, 1998b).

Other Point Sources

There are a small number of point sources other than municipal WWTPs in the Catskill/Delaware watershed, some of which are permitted (Table 5-5). Unlike sewage, the composition of industrial wastewater varies widely, depending on the type of industry. In addition, chemical concentrations may be extremely high in comparison to domestic wastewater. The industrial wastewater and cooling water discharges in the Catskill/Delaware region derive from dairy processing plants and hazardous waste treatment processes. There are also stormwater permits for industrial operations. As mentioned in Chapter 3, stormwater permits are often required for construction and industrial activities that affect five or more acres of land. For the purposes of this discussion, these permitted discharges are labeled point sources because they are regulated under Section 402 of the Clean Water Act (CWA).

Although few in number in comparison to municipal WWTPs, industrial point sources may still be responsible for pollutant loadings in the Catskill/ Delaware watershed because of high effluent concentrations derived from these sources. For example, the average phosphorus concentration of wastewater from the UltraDairy facility that is sprayed onto nearby fields was 25.7 mg/L in 1998. This effluent phosphorus concentration is considerably higher than that associated with any municipal WWTP during the same year. Thus, industrial point sources cannot be ignored when determining the overall contribution of point sources to pollutant loadings in the water supply reservoirs.

NONPOINT SOURCES OF POLLUTION

Over the last quarter century, water quality across the country has improved dramatically, primarily as a result of technologies that have greatly reduced pollution from point sources. Despite this progress, EPA estimates that 40 percent of the nation's waterbodies still do not meet CWA standards of fishable and swimmable quality (EPA, 1998a). In New York State, 7 percent of rivers and streams and 53 percent of lakes and reservoirs do not fully support their designated uses (NYS DEC, 1996).

The dominant threat to water quality today is nonpoint source pollution, or polluted runoff, which derives from multiple diffuse sources. Technologies for reducing nonpoint source pollution are, in some respects, more difficult to design, implement, and monitor than those for point source pollution. In addition, the regulatory strategies that have proven successful in reducing point source pollution are inadequate for combating nonpoint source pollution (EPA, 1998a). Little regulatory pressure exists specifically to deal with nonpoint source pollution. For these reasons, the states have been slow to develop, implement, and enforce effective strategies for reducing polluted runoff.

Nonpoint source pollution has been blamed for several significant national pollution problems such as coastal eutrophication, leaching of nitrate into ground-

TABLE 5-5 Other Point Sources in the Catskill/Delaware Watershed

Reservoir/Watershed	Type of Discharge	SPDES Number	SPDES Permitted Flow	Class of Receiving Stream[a]
Ashokan				
Shokan Post Office	Treated outflow from chemical spill	NY-0233889		A(T)
E.G.&G Rotron, Inc.	Treated outflow from solvent-contaminated site	NY-0098281		B(T)
Schoharie				
Agway Petro	Stormwater and tank test water			C
Falke's Quarry	Stormwater runoff	NY-0223506		C(T)
Hunter Synagogue Remediation	Treated outflow from chemical spill	NY-0241041		C(TS)
Town of Hunter Landfill	Stormwater runoff from landfill	NY-0103107		D
Cannonsville				
Kraft Inc. Dairy	Cooling water	NY-0008494	1,080,000	B(T)
Walton Town Garage	Discharge from an oil/water separator	NY-0249483	No limit	
Dairyvest at Fraser (Ultra Dairy/DMV)	Dairy processing waste	NY-0068292	200,000	GA (spray irrig.)
	Cooling water		720,000	C(T)
Mallincraft/Grahm Labs	Water with pharmaceutical and industrial chemicals			
Audio-Sears, Inc.	Water/air with acids, heavy metals, electroplating			
Pepacton				
Railway Laundry Dry Cleaning	Water/air with dry cleaning chemicals			
General				
Town/County DPW Buildings	Water via floor drains from garage areas			

[a] See Appendix B for an explanation.
Sources: Marx and Goldstein (1993) and Warne (1999a).

water, and siltation of riverways and waterbodies. Consistent with national trends showing increased awareness of nonpoint source pollution, New York City has acknowledged the role of polluted runoff in degrading water quality in the Catskill/Delaware watershed (NYC DEP, 1998c). Both the New York State Department of Environmental Conservation (NYS DEC) and NYC DEP have developed programs for combating nonpoint source pollution from a variety of sources.

The activities that produce nonpoint source pollution in the Catskill/Delaware watershed are similar to those across the nation. In general, the predominant land uses of an area will determine what the major types of nonpoint pollution are. Table 2-6 reveals that agriculture, forestry, and urban development are likely contributors to nonpoint source pollution in the Catskill/Delaware watershed. The following section discusses these activities as well as the contributions of OSTDS and atmospheric deposition.

Agriculture

Modern agriculture focuses large amounts of energy, materials, and management on relatively small portions of the landscape. The high productivity gained in this way also leads to the potential for nonpoint source pollution. In general, as agriculture is practiced in most parts of the United States today, row crops are treated with fertilizers and pesticides and have the potential for contaminating surface waters with nutrients, pesticides, and sediment. Animal-based agriculture typically involves land application of manures or manure mixed with bedding material, which may contaminate surface waters with nutrients, organic matter, and biological pathogens. Both row-crop agriculture and animal-based agriculture can supply nutrients to groundwater.

Loadings per unit area of nonpoint source pollutants to surface water and groundwater resources are greater for most agricultural land uses than they are for undisturbed areas. Certain nonpoint source pollutants such as pesticides are wholly human-made and will not occur at all in unmanaged areas unless they are transported there through atmospheric or hydrologic processes. Other types of nonpoint source pollution such as sediment, nutrients, and organic matter may be increased as a result of agricultural management but also occur in unmanaged areas.

Agricultural land uses in the Catskill/Delaware watershed are primarily dairies; thus, pollutants associated with animal manure and dairy barnyards are the main concern. Manure contains high concentrations of nutrients, particularly nitrogen and phosphorus. Manure is also a potential source of parasitic protozoa such as *Giardia* and *Cryptosporidium*, pathogenic viruses and bacteria, and organic chemicals used for promoting animal survival (such as antibiotics). The other major sources of nonpoint pollution from dairy operations are row crops, hayfields, and pastures that are commonly grown for silage. Nonpoint source pollution can result from the application of fertilizers, manures, and pesticides to these lands. There is also the threat of erosion and sediment transport from crops or pasturelands to nearby waterbodies. This may be exacerbated by agricultural activities such as plowing, harvesting, building of roads, and the movement of animals.

Agriculture comprises 4.5 percent of the total land use in the Catskill/Delaware watershed. The greatest impacts of agricultural runoff occur in the West Branch of the Delaware River and the Cannonsville Reservoir, where agriculture occupies 10 percent to 11 percent of the watershed. Runoff from barnyards, overland and shallow groundwater flow from manured fields, inundation of manured fields during floods, and general enrichment of groundwater phosphorus related to agriculture have been shown to contribute to eutrophication of the Cannonsville Reservoir (Brown et al., 1984; NYC DEP, 1997d; Robillard and Walter, 1984). Runoff from dairy barnyards was found to have an average total phosphorus concentration of 11.9 mg/L (Robillard and Walter, 1984). Mean phosphorus concentrations measured in surface water at a farm site (Robertson Farm) were as high as 3.8 mg/L compared to concentrations of less than 0.1 mg/L at a nearby control site (Shaw Road) (WAP, 1997). Finally, mean phosphorus concentrations during baseflow conditions on the West Branch of the Delaware River were estimated to be twice as high as mean phosphorus concentrations in groundwater from undisturbed areas (Brown et al., 1984).

Monitoring of streams above and below a dairy farm has also demonstrated that downstream water is enriched with *Cryptosporidium* oocysts in comparison to reference sites (NYC DEP, 1998a). Pesticides are also found in surface waterbodies of the Cannonsville watershed, but almost always at levels well below their regulated MCL.

Urban Stormwater

A second important class of activities that produce nonpoint source pollution are classified as "urban." This term encompasses a wide variety of commercial, residential, and industrial activities such as road building, construction of housing, and the creation of golf courses, among other things. There are discrepancies regarding the amount of land in the Catskill/Delaware watershed that is urban. According to calculations done for the Total Maximum Daily Load (TMDL) Program, the amount of land classified as urban is less than 1 percent of the land area west of the Hudson River, 3.8 percent of the West Branch watershed, and 15.1 percent of the Kensico watershed (NYC DEP, 1999c–j). This analysis is likely to underestimate the total percentage of land in urban uses because it includes only impervious surfaces. (In fact, the TMDL documents state that some residential land was classified as forests.) Other analyses (Table 2-6) indicate a much higher percentage of the Catskill/Delaware watershed (17 percent) is urban.

Urban development can have a profound influence on the local hydrologic cycle and water quality. The hydrologic changes begin during the clearing and grading that accompany construction. Trees that had intercepted rainfall are felled, and natural depressions that had temporarily ponded water are graded to a uniform slope. The leaf litter and organic layer on the soil surface are scraped off, eroded, or severely compacted. Having lost its natural storage capacity, a cleared and graded site can no longer prevent rainfall from being rapidly converted into runoff.

Roof tops, roads, parking lots, driveways, and other impervious surfaces prevent rainfall from soaking into the ground. Consequently, most rainfall is directly converted into stormwater runoff. This phenomenon is illustrated by the strong correlation between site imperviousness and the runoff coefficient (R_v), which expresses the fraction of rainfall volume converted to stormwater runoff (rather than infiltrating into the soil). For example, a one-acre parking lot produces on average about 15 times more annual runoff than does a one-acre meadow in good condition (Schueler, 1995). Extensive drainage networks (using curbs and gutters, enclosed storm sewers, and lined channels) are necessary to rapidly collect and convey this additional stormwater runoff.

Increased stormwater runoff can significantly alter stream geometry. Following urban development, stormflow frequency and magnitude increase dramatically, causing a greater number of bankfull and sub-bankfull flow events. When streambeds and banks are exposed to destabilizing flows for long periods of time, their cross-sectional area increases (either by channel widening, down cutting, or both). Under extremely high flows, streams may undergo severe streambank erosion and habitat degradation. Indeed, the presence of impervious surfaces has been linked to declines in nearby stream insect, freshwater mussel, and fish diversity (Maxted et al., 1994; May et al., 1997; Schueler, 1995).

During storm events, accumulated pollutants are quickly washed off of impervious areas and are rapidly delivered to downstream waters. Some common pollutants found in urban stormwater runoff are profiled in Table 5-6. Although variable from storm to storm, in general the concentrations of pollutants in urban stormwater can be characterized by an event mean concentration (EMC) on an annual basis. Research indicates the EMC is the same for most pollutants, regardless of storm size, intensity, antecedent conditions, or other factors (EPA, 1983). Consequently, most models of urban stormwater runoff have pollutant loads increase in direct proportion to the amount or percentage of impervious cover in the watershed (Schueler, 1987). It should be noted that this relationship does not hold true for other types of stormwater (e.g., agricultural or forestry runoff).

There are few direct measurements of pollutant concentrations in urban stormwater in the Catskill/Delaware watershed. Efforts have been made to compare the incidence of polluted water samples from "urban" areas to those from "pristine" areas. Monitoring of *Giardia* and *Cryptosporidium* in stream samples across the watershed has revealed that "urban" sources are enriched in both cysts

TABLE 5-6 Average Event Mean Concentrations (EMC) Found in Urban
Stormwater

Pollutants	EMC[a]
Total Suspended Solids	80 mg/L
Total Phosphorus	0.30 mg/L
Total Nitrogen	2.0 mg/L
Total Organic Carbon	12.7 mg/L
Fecal Coliform Bacteria	15,000–20,000 MPN/100 mL
E. coli	1,450 MPN/100 mL
Petroleum Hydrocarbons	3.5 mg/L
Cadmium	2 µg/L
Copper	10 µg/L
Lead	18 µg/L
Zinc	140 µg/L
Chlorides (winter only)	230 µg/L
Insecticides	0.1–2.0 µg/L
Herbicides	1–5.0 µg/L

[a]These concentrations represent *mean* or *median* storm concentrations measured at typical suburban
sites, and may be greater during individual storms. Also note that mean or median runoff concentrations from s*tormwater hotspots* are 2–10 times higher than those shown here.
Source: MDE (1999).

and oocysts as compared to stream samples from undisturbed areas (NYC DEP,
1998a). In fact, *Cryptosporidium* was detected in urban watersheds more frequently than in agricultural watersheds and in effluent from sewage treatment
plants. These studies clearly demonstrate the importance of urban areas and
impervious surfaces in contributing to pollutant loading. Urban sources of pollution are expected to be more significant in the Kensico and West Branch watersheds, which have a much higher percentage of urban land than the West-of-
Hudson watersheds.

Forestry

The Catskill/Delaware watershed includes 667,517 acres of forest land or 68
percent of the total area. NYS DEC lands comprise about 22 percent of the forest
land, mostly in the Catskill Forest Preserve. The largest contiguous block of
forest preserve includes high-elevation and mountainside lands in the Esopus
Creek (Ashokan Reservoir) watershed. NYC DEP owns a small percentage,
largely adjacent to reservoirs, while private landowners hold the remaining acreage, usually lower slope and valley bottomland between state and city holdings.
Ephemeral and intermittent streams link these lands to perennial tributaries and

downstream reservoirs. Hence, inappropriate land use in these areas can directly affect water quality.

It has been known for decades that without the careful application of best management practices (BMPs), the effects of forest harvesting on site stability and aquatic ecosystems can be severe. Numerous studies of traditional logging and farming practices have demonstrated that removal of riparian vegetation, haphazard road and skid trail construction, careless clearcutting and high-grading (cutting only large, valuable trees), and little or no professional supervision can lead to significant degradation of terrestrial and aquatic resources (Anderson et al., 1976; Satterlund and Adams, 1992; Swank and Crossley, 1988; Verry, 1986). In general, the largest relative impact is associated with roads, followed by skidding, site preparation, and then felling operations (Satterlund and Adams, 1992). The felling operation may cause appreciable short-term increases in water yield without corresponding increases in soil erosion and sediment transport (Bosch and Hewlett, 1982).

Reducing the density and biomass of forest cover increases rainfall reaching the soil surface, snowmelt rate in openings, and soil water content in the root zone. All three changes combine to increase water yield. The magnitude of the change is proportional to the area harvested, the biomass removed, and species composition of the stand (whether coniferous, deciduous, or mixed). However, the response rate is not uniform or incremental. Partial cuts produce comparatively smaller water yield increases than do seed-tree or overstory removal cuts. That is, although the same quantity of wood is harvested, a 100-acre clearcut generates more water than do 300 acre thinned by one-third of the original biomass. Because of reduced competition for water, light, and nutrients, partially cut stands tend to recover more quickly, as forest regeneration, herbaceous plants, and changes in microclimate (greater solar radiation and wind velocity at the surface) combine to restore evapotranspiration to pre-harvest levels in as little as three years (Brooks et al., 1997).

It has been more difficult to determine water yield response in relation to the fraction of the watershed subject to harvesting or conversion to another land use. Because of the biophysical and financial limitations associated with classical paired watershed experiments, most of what is known or hypothesized comes from retrospective analyses of streamflow and land use data for very large watersheds. The few convincing studies suggest 30 percent to 40 percent of a forested watershed must be clearcut before a substantial water yield or peak discharge increase is noted. The response rate increases between 30 percent and 60 percent to 70 percent harvested until stabilizing at 70 to 100 percent removal (Verry, 1986).

Due diligence in the application of forestry BMPs is the key to preventing adverse environmental impacts. These BMPs are discussed in more detail in Chapter 9.

On-site Sewage Treatment and Disposal Systems

Individual septic systems (on-site sewage treatment and disposal systems or OSTDS) are frequent alternatives to sewage treatment plants in less densely populated areas where the costs of constructing sewage treatment systems are prohibitive. Properly sited and functional OSTDS receive, treat, and dispose of wastes in a manner that does not contaminate the environment or expose humans to pathogens. OSTDS are often considered nonpoint sources of pollution because they are small-scale and widely dispersed, and they discharge to relatively large subsurface areas. As with other nonpoint sources of pollution, measuring the impact of these systems on nearby water quality can be difficult. Performance monitoring of OSTDS effluent in the Catskill/Delaware watershed has not occurred on a regular basis and has only recently become a goal of NYC DEP's septic siting study (see Chapter 11).

The term OSTDS encompasses several technologies ranging from cesspits to aerobic treatment units. The most common OSTDS is the septic tank and drainfield shown in Figure 5-3. The septic tank is a watertight container with a typical capacity of 1,900–4,550 L, and it is designed to detain raw sewage discharged from a home or building. The septic tank is essentially a settling basin, in which the suspended solids are separated from the liquid fraction (EPA, 1980). The solids settle to the bottom of the tank where they are degraded by anaerobic bacteria. Lighter material, including fats, oils, and grease, accumulates at the

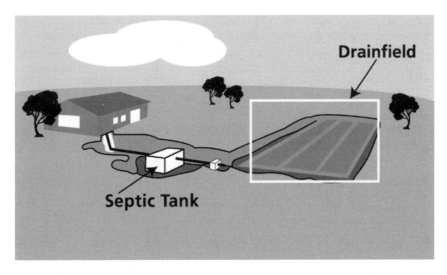

FIGURE 5-3 Septic tank and drainfield. Both parts of the system are generally underground.

liquid surface. The liquid portion of the waste flows from the tank through an outlet near the top. This liquid is then distributed through perforated pipes into a subsurface drainfield or infiltration system. The removal of solids in the tank allows the effluent to be disposed to the subsurface drainfield without clogging the perforated pipes and soil, which would cause failure of the system (Canter and Knox, 1986).

The purpose of the drainfield is to distribute wastewater evenly to the soil. The soil beneath the drainfield filters out pathogenic microorganisms from the OSTDS effluent before it reaches groundwater. Once in the soil, pathogens on soil particles may succumb to unfavorable environmental conditions or be destroyed by aerobic microorganisms. Because aerobic conditions enhance destruction of pathogens, drainfields are usually placed only 61–76 cm beneath the surface. Properly sited and maintained septic tanks and drainfields remove approximately 80 percent of the BOD and virtually all of the total suspended solids. Removing pathogens from OSTDS effluent before it reaches groundwater is important because various studies have shown that bacteria and especially viruses can travel long distances (15–60 m) in groundwater down gradient of a properly sited and functioning OSTDS (Hagedorn et al., 1978; Moe et al., 1984; Parker et al., 1978; Paul et al., 1995; Vaughn et al., 1983; Yates et al., 1986).

The aerobic treatment unit (ATU) is the most effective OSTDS technology that is backed by third-party certification. It uses mechanical devices to mix atmospheric oxygen into the tank, which allows aerobic microorganisms to colonize the tank rather than the anaerobic microorganisms found in the conventional septic system. This results in more effective removal of pathogens and higher treatment efficiencies (Derge, 1983). Other innovative types of OSTDS that make use of aerobic environments include the Ruck system, which uses aquatic vegetation to aid degradation of the effluent, the Peat Moss filter, the recirculating sand filter, and a variety of waterless and composting toilets. However, unlike the ATU, these technologies have received limited testing and are not yet endorsed by most regulatory agencies across the country, including the New York State Department of Health (NYS DOH). Thus, they are not considered further in this report.

Among other types of OSTDS, the least efficient is the cesspit. This is simply an underground tank without a bottom that allows the liquid fraction of the wastewater to percolate into the soil with little or no degradation by soil organisms. Cesspits are not legal in the New York City watershed, although old or illegally installed cesspits may be of concern.

One of the primary limitations of OSTDS is that they provide little removal of nutrients (details provided in Chapter 11). A variety of experimental systems that remove nutrients are being studied, including those that use small rotating biological contactors or recirculating sand filters. The latter system attempts to maintain colonization of the treatment media for vacation homes or other facilities not subject to regular dosing with domestic sewage. As with WWTPs, solid waste that accumulates in OSTDS must be properly disposed of on a regular basis. The six New York City-owned WWTPs in the Catskill/Delaware watershed can accept and treat the sludge generated by OSTDS (known as septage), but the extent to which this is carried out is not known (D. Warne, NYC DEP, personal communication, 1998).

The 31,270 OSTDS in the Catskill/Delaware watershed (NYC DEP, 1993a, Table VIII.F-1) represent a significant potential source of contamination to the reservoirs. Individual systems (22,454) have flows of less than 1,000 gpd and are regulated by NYS DOH and the County Health Departments under 10 NYCRR Appendix 75A. Intermediate systems (59) discharge more than 1,000 gpd and thus operate under SPDES permits, necessitating oversight by either NYS DOH or NYS DEC. Small, nonresidential systems (such as restaurants) with flows of less than 1,000 gpd (7,754) are classified as "other" systems and are regulated by NYS DOH. Regardless of the classification, failing OSTDS must be detected and repaired rapidly through a vigorous enforcement effort if contaminants are to be prevented from degrading surface water quality.

Atmospheric Deposition

Atmospheric deposition is a source of nonpoint pollution that is difficult to quantify and control. Both wet and dry deposition of chemicals, such as acids and nitrogen compounds, can result from industrial activities that produce large amounts of airborne pollutants. Atmospheric deposition onto land areas can be transported to waterbodies via surface runoff. For this reason, some atmospheric deposition may be treated by the BMPs primarily designed to treat land-based sources of nonpoint pollution. However, direct deposition into reservoirs also occurs, and this source of pollution is much more difficult to control.

As shown in Table 5-7, dry and wet deposition are relatively important sources of nitrogen, sulfate, and acids. In particular, atmospheric deposition is the primary mechanism for transporting SO_2 (gas), sulfate particles, and sulfate-containing aerosols to the watershed region. When dissolved in precipitation, sulfate is a major component of acid rain. Sulfate ions do not substantially influence the total alkalinity and stability of the water that New York City receives. However, the associated acid rain may result in changes in species composition within forested areas. Fortunately, atmospheric deposition of sulfate has been decreasing in the watershed region since the late 1980s, probably a result of sulfur emission controls on power plants in the Midwest and East following the 1990 Clean Air Act Amendments (Murdoch et al., 1998).

Nitrogen, and the associated acidity, is the primary pollutant of concern in atmospheric deposition in the Catskill/Delaware watershed, as discussed previously. In contrast to the atmospheric deposition of sulfur, the atmospheric deposition of nitrogen has been increasing in the watershed region because of increasing vehicular emissions (number of cars and number of miles driven), despite the

TABLE 5-7 Speciation of Atmospheric Deposition in the Northeastern United States

Processes	Nitrogen Species	Sulfur Species	Hydrogen Ions	Other Inorganics
Wet Deposition	NO_3^-, NH_4^+	SO_4^{2-}	H^+	Ca^{2+}, Mg^{2+}, K^+, Na^+, Hg^{2+}
Dry Deposition, Particles and Aerosols	NH_4NO_3, $(NH_4)_2SO_4$ NH_4HSO_4	NH_4HSO_4 $(NH_4)_2SO_4$ $CaSO_4$ $MgSO_4$	H_2SO_4	Ca^{2+}, Mg^{2+}, K^+, Na^+, **carbonates / oxides (dust)**
Dry Deposition, Gases	NO_x, NH_3	SO_2	HNO_3	$Hg(0)$

Note: Largest contributions are in bold type.

Clean Air Act Amendments of 1990. Wet deposition of nitrogen falls primarily as nitrate ions, while dry deposition occurs as NO_x gas, HNO_3, and ammonium nitrate and ammonium sulfate aerosols. Deposition of nitrate and ammonia are important contributions to overall watershed nitrogen budgets in forested watersheds, but are relatively insignificant in agricultural settings where nitrogen fertilizers are often used. Total nitrogen modeling and load allocation in the New York City watershed region has not been performed because the reservoirs have been shown to be primarily phosphorus-limited and because nitrogen water quality standards have not been exceeded.

Finally, there is the potential for atmospheric deposition of mercury into the water supply reservoirs and associated detrimental effects on fish health. A few fish caught in the Neversink and Rondout reservoirs have been found to have high mercury levels in their tissues, most likely attributable to mercury deposition into these acidic watersheds (Murdoch, 1999). NYC DEP monitoring of the water supply has detected mercury in concentrations near the detection limit, suggesting that it is not a cause for concern at this time (Warne, 1999b).

WEIGHING POINT AND NONPOINT SOURCES

Assessing the contributions of point and nonpoint sources to overall pollutant loading is a critically important watershed management task. Such an analysis can direct monetary, personnel, and other resources toward the most polluting sources. In the New York City watershed, this type of evaluation is easily conducted using models developed for the Total Maximum Daily Load (TMDL) program. A thorough description of the TMDL program, including data requirements, modeling efforts, and evaluation, is found in Chapter 8.

This discussion is limited to total phosphorus because, as of 1999, TMDLs have been calculated only for that priority pollutant. Table 5-8 and Figure 5-4 show TMDL model predictions of the relative contributions of point and nonpoint sources of phosphorus for each Catskill/Delaware basin. For the purposes of this analysis, point sources include WWTPs and the contributions from upstream reservoirs. Nonpoint sources include agriculture, urban runoff, OSTDS, ambient conditions emanating from pristine areas (forestland), groundwater (which reflects other nonpoint sources), and atmospheric deposition over water. For the six West-of-Hudson reservoirs, the predictions were generated by the Generalized Watershed Loading Function, a terrestrial runoff model that is discussed in detail in Chapter 8. The Reckhow model was used for the Kensico and West Branch watersheds.

Table 5-8 clearly demonstrates that sources of pollution in the six West-of-Hudson basins, West Branch, and Kensico are very different from one another. The more terminal reservoirs (Ashokan, West Branch, Kensico, and Rondout) receive almost their entire phosphorus loading from upstream reservoirs. This fact underscores the importance of watershed management in upstream areas to the maintenance of high water quality in these reservoirs. Schoharie and Pepacton reservoirs receive phosphorus from multiple sources of approximately equal magnitude: groundwater, agriculture, and forest land. Cannonsville, the most developed of the West-of-Hudson watersheds, is affected primarily by agriculture and then by forest land, groundwater, and WWTPs including several dairy-processing plants. Finally, runoff from undeveloped areas (forests) comprises the majority of phosphorus loading to the Neversink Reservoir. Impervious surfaces, septic systems, and atmospheric deposition over water contribute relatively little total phosphorus to the reservoirs. It should be noted that the groundwater category in the Generalized Watershed Loading Function may contain dissolved phosphorus derived from urban, agricultural, and forest sources as well as OSTDS. Because of its relatively significant contribution, further differentiation of the groundwater category should be a goal of future modeling efforts. The implications of the relative contributions of point and nonpoint source pollution for New York City's overall watershed management strategy are discussed in Chapter 12.

CURRENT STATE OF HEALTH OF THE WATERSHED AND WATER SUPPLY

This chapter closes with an overview of current conditions in the New York City watershed and water supply. As introduced in Chapter 3, there are important ecological and human health concerns related to drinking water supply systems. Water quality in the water supply reservoirs has a direct impact on aquatic ecosystems and habitats, particularly during eutrophication events. Although few epidemiological studies have conclusively linked waterborne DBPs to human

TABLE 5-8 Predicted Range of Percent Contributions of Point and Nonpoint Source Total Phosphorus Loading into Individual Reservoirs from 1993 to 1996

Reservoir	WWTP	Upstream Reservoirs	Urban Areas	Septic Systems	Agriculture[a]	Forest Land[b]	Groundwater[c]	Atmospheric Deposition over Water
Ashokan W	<1	5–30	<1	1–3	3–4	38–74	12–25	<1
Ashokan E	<1	84–94	<1	1–4	1–3	2–7	1–3	1–4
Cannonsville	7–20	0	<1	1–2	43–57	13–20	13–21	<1
Neversink	0	0	<1	<1	15–19	48–61	19–33	<1
Pepacton	1–2	0	<1	1–2	29–34	39–51	14–27	<1
Rondout	0–1	26–58	<1	1–2	14–29	16–33	6–10	<1
Schoharie	2–5	0	1	2–4	25–30	31–45	19–32	<1
Kensico[d]	0	95	2	<1	2	<1	ND	<1
West Branch[d]	<1	93	2	1	1	2	ND	1

Note: All values are percentages (%) of basinwide phosphorus loading (mass/time). The West-of-Hudson contributions are derived from the Generalized Watershed Loading Function.

[a]Agriculture is the combination of six subgroups: grass/shrub, grass, bare soil, corn, alfalfa, and barnyard.

[b]Forestland is the combination of three subgroups: deciduous forest, coniferous forest, and mixed forest.

[c]Sources of pollutants in groundwater are not defined in the GWLF. They may derive from multiple other nonpoint sources.

[d]Kensico and West Branch point and nonpoint source contributions were calculated using the nested Reckhow model, which does not include groundwater as a category.

Source: NYC DEP (1999c–j).

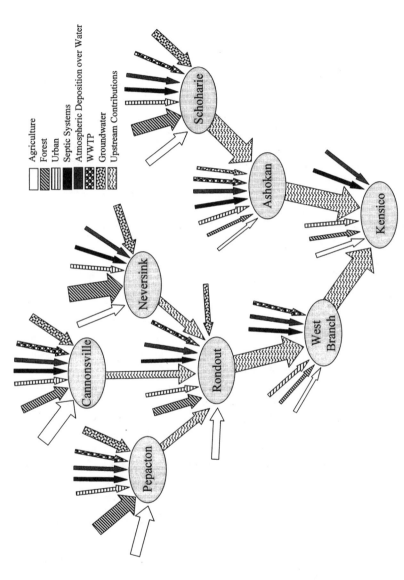

FIGURE 5-4 Relative contributions of point, nonpoint, and upstream sources to phosphorus loading in the Catskill/Delaware reservoirs. The thickness of the arrow represents the relative contribution to total inflow (Table 5-8).

health problems, some of these compounds are known carcinogens and are regulated by the SDWA. Finally, waterborne microbial pathogens, most notably *Cryptosporidium* and *Giardia,* are a constant concern, particularly for systems using disinfection as the sole treatment process.

Reservoir Water Quality

Direct human access to the New York City water supply reservoirs is limited to recreational fishing from row boats, shores, or bridges. However, human activities within the drainage basins have significantly influenced reservoir water quality. The poorest water quality is associated with areas that have significant population growth (see Chapter 2), particularly east of the Hudson River in the Croton system. Water quality in the Catskill and Delaware systems is better, in relative terms, because population densities are lower and a larger proportion of the watersheds is forested. The greatest threats to the reservoirs from watershed uses are microbial contamination and eutrophication caused by nutrient enrichments.

The inflow of nutrients, particularly phosphorus, to the New York City reservoirs is sufficient to promote moderate to abundant phytoplankton growth. Average chlorophyll *a* concentrations are a reasonable estimate of phytoplanktonic biomass, and these data have been coupled to the average total phosphorus (TP) concentrations contained in reservoir water. As shown in Table 5-9, phosphorus concentration data indicate all reservoirs are either moderately productive (mesotrophic) or productive (eutrophic). To date, none of the reservoirs has reached levels of productivity that raise serious health problems related to algal toxicity. However, decomposition of phytoplankton can increase dissolved organic carbon and subsequent DBP formation.

Long-term analyses suggest average total phosphorus concentrations within some of the reservoirs are slowly increasing (NYC DEP, 1993b). In some of the reservoirs, the average annual total phosphorus concentration exceeds 20 µg/L, a concentration known to induce eutrophic development of phytoplankton. The problem is particularly severe in the Cannonsville Reservoir, which is routinely taken offline during the summer and fall because of high algae levels (NYC DEP, 1993b). Current phosphorus levels and the associated formation of DBPs meet regulatory requirements in the other Catskill/Delaware reservoirs. However, reductions of phytoplankton in all reservoirs may be essential to meet regulatory standards for DBPs by 2002.

Another significant water quality impairment in the Catskill/Delaware watersheds is sediment loading and turbidity. The Ashokan and Schoharie reservoirs intermittently evince problems from erosional loading of quartz and clay particles that cause unacceptably high turbidity levels (Effler et al., 1998). Although this problem is largely attributed to geologic characteristics of some tributary streams (e.g., Stony Clove in the Esopus Creek watershed), opportunities to

TABLE 5-9 Estimates of Phosphorus Concentration, Chlorophyll *a* Concentration, and Watershed Phosphorus Loading to New York City Reservoirs

Reservoir	Average[a] P µg/L	Average[a] Chl *a* µg/L	Vollenweider RM[b] Current Loading g/m^2 y	Reckhow RM[b] Current Loading g/m^2 y	Trophic State
Catskill/Delaware System					
Ashokan	13	4.3	0.50	0.52	Mesotrophic
Cannonsville	22	11.6	1.49	1.55	Eutrophic
Neversink	6	3.5	0.39	0.31	Mesotrophic
Pepacton	12	5.3	0.80	0.61	Mesotrophic
Rondout	10	4.5	2.04	1.91	Mesotrophic
Schoharie	14	4.4	2.13	1.78	Mesotrophic
Kensico	12		3.09	3.14	Mesotrophic
Croton System					
Amawalk	19	8.9	0.53	0.58	Eutrophic
Bog Brook	16	4.6	0.19	0.30	Eutrophic
Boyd Corners	18	4.5	0.67	0.75	Eutrophic
Cross River	15	7.4	0.26	0.34	Eutrophic
Croton Falls	22	9.5	0.75	0.72	Eutrophic
Diverting	30	13.7	3.92	4.43	Eutrophic
East Branch	22	12.0	1.27	1.29	Eutrophic
Middle Branch	17	8.1	0.48	0.54	Eutrophic
Muscoot	22	11.5	2.02	2.08	Eutrophic
New Croton	20	8.7	1.34	1.24	Eutrophic
Titicus	14	6.9	0.35	0.38	Eutrophic
West Branch	14	5.0	0.49	0.51	Eutrophic

[a]Average of data measured in the epilimnion of the lacustrine zone from May through October (growing season) from the period 1988–1996.
[b]RM = reservoir model. The Vollenweider and Reckhow reservoir models use data on in-reservoir phosphorus concentration to determine phosphorus loading. Phosphorus concentration data from the period 1984–1991 were used.
Sources: NYC DEP (1993b, 1999b).

control sediment emanating from unpaved roads, stormwater from impervious surfaces, and unstable streambanks should not be neglected. Water diversions from the Schoharie Reservoir to the upper Esopus Creek via the Shandaken Tunnel also may contribute to sediment transport.

Finally, accidental spills of hazardous or other materials in the watershed can have acute, short-term effects on water quality in the water supply reservoirs. During the last five years, there have been no substantial spills of hazardous material that resulted in a discharge to the water supply reservoirs (Warne, 1998b).

Compliance with the Safe Drinking Water Act

The ambient quality of New York City's drinking water is also indicated by its ability to comply with all provisions of the SDWA and the Surface Water Treatment Rule (SWTR). This includes meeting regulatory standards for chemical and microbial parameters, as well as demonstrating an absence of waterborne disease. This section discusses both current and potential future compliance with specific provisions of the SDWA.

Surface Water Treatment Rule

Disinfectant Residual. Because disinfection with chlorine is the only regular treatment given to New York City's drinking water, NYC DEP ensures chlorine residuals are sufficient to comply with federal regulations. The SWTR currently requires water at the entry point to the distribution system to have a disinfectant concentration of at least 0.2 mg/L. As shown in Figure 5-5 (chlorine residual at two entry point locations in the distribution system during 1998), New York City is currently meeting this requirement. These data indicate chlorine residuals so far above the minimum requirement that little difficulty is expected in meeting this requirement in the future as long as the City maintains its current disinfection practice.

CT. The chlorine concentration multiplied by the contact time, or CT, is the metric used to comply with the SWTR requirement for 3-log removal of *Giardia* and 4-log removal of viruses. NYC DEP calculates actual CT values for the City's distribution system and divides them by the CT value dictated by EPA regulations, generating an Inactivation Ratio (IR). New York City is required to keep this ratio above 1.0; however, NYC DEP strives to achieve an IR of 2.0. Vigilance is most necessary during winter, when low temperatures reduce the effectiveness of chlorine. New York City drinking water is currently meeting CT requirements. Typical data, collected during January 1997, are shown in Figure 5-6.

Total Trihalomethanes. DBPs such as trihalomethanes are of considerable concern for water supplies that rely on chlorine as the sole treatment process. The current MCL for total trihalomethane (TTHM) in drinking water is 0.10 mg/L. Since NYC DEP began measuring this parameter in 1993, this standard has been met (Figure 5-7). As discussed in Chapter 3, the Stage 1 Disinfectants/Disinfection By-Products (D/DBP) Rule MCL for TTHM is 0.080 mg/L, although this standard will not become effective until November 2001. Based on the quarterly compliance data from the Catskill/Delaware system (March 1993 through December 1998) New York City drinking water should meet this TTHM requirement.

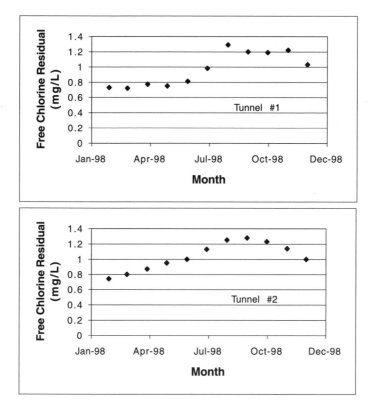

FIGURE 5-5 Monthly average free residual chlorine levels in the entry points to Tunnels 1 and 2 during 1998. Source: NYC DEP (1999k).

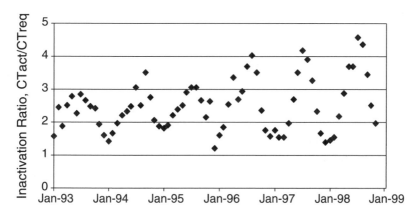

FIGURE 5-6 Inactivation ratios measured in Tunnel No. 2 during January 1997. Source: NYC DEP Compliance Reports.

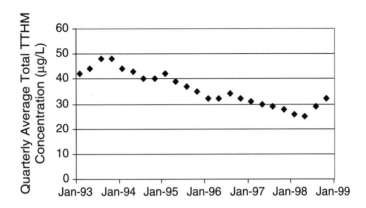

FIGURE 5-7 New York City quarterly running averages for total trihalomethanes within the distribution system of the Catskill/Delaware supply. (Data are averages of multiple locations.) Source: NYC DEP Compliance Reports.

Site-Specific Criteria: Waterborne Disease. The ultimate purpose of watershed management is to protect drinking water quality and public health by reducing the amount of contaminants entering the water source. Thus, the filtration avoidance criteria of the SWTR mandate no waterborne disease outbreaks shall occur among the population served by an unfiltered water supply. A disease outbreak can be defined as an increase in the observed number of cases of disease relative to the expected number of cases (background level) over a specific time period. Waterborne disease transmission refers to human exposure to a pathogenic agent (microbial or chemical) via consumption of or contact with contaminated water.

Investigations of recognized waterborne disease outbreaks by state and local health departments are reported voluntarily to CDC and EPA. This system, based on voluntary reporting by state health departments, clearly represents only a fraction of the true incidence of waterborne disease outbreaks. There have been only three reported waterborne disease outbreaks in New York City since 1941 (Table 5-10). All these outbreaks were due to cross-connections or back-siphonage rather than to contaminated source water. One outbreak of unidentified etiology in 1942 resulted in 225 cases; another outbreak in 1949 resulted in 31 cases of shigellosis in an apartment building; and in 1974 an outbreak of 20 cases of illness related to high levels of chromium occurred in an office building.

These data indicate New York City is in compliance with the SWTR relating to waterborne disease outbreaks. However, it is possible that additional waterborne disease outbreaks may have occurred and were not recognized. There also may be unrecognized endemic disease associated with the New York City water

TABLE 5-10 Waterborne Outbreaks Reported in Community Water Systems in New York State and New York City: 1941–1994

Year	New York State excluding NYC	New York City	New York State Total
1941–1950	27	2	29
1951–1960	2	0	2
1961–1970	3	0	3
1971–1980	3	1	4
1981–1994	6	0	6
Total	41	3	44

Source: Reprinted, with permission, from Craun (1998).

supply. The City's surveillance systems and epidemiologic studies are reviewed in Chapter 6.

Total Coliform Rule

Because Kensico Reservoir is the terminal reservoir for the Catskill/Delaware system prior to disinfection, much of the compliance monitoring conducted by NYC DEP occurs in this basin. Under the SWTR, Kensico is considered the source water of the Catskill/Delaware systems and is subject to strict standards for fecal coliforms. During 1988–1992, Kensico Reservoir experienced elevated fecal coliform concentrations (Figure 5-8). However, although Kensico has shown seasonal increases in fecal coliform bacteria, usually beginning in October or November and continuing through December or January, NYC DEP has managed to not violate the Total Coliform Rule. During the fall and winter of 1991–1992, 1992–1993, and 1993–1994, this was accomplished by bypassing the Kensico Reservoir.

Because bypassing is not an ideal solution given the operational and water quality benefits Kensico provides, in 1991 the City began a study to identify and eliminate the cause of seasonal coliform increases at Kensico Reservoir. Field investigations and limnological sampling showed that an increase in waterfowl populations coincided both temporally and spatially with increases of fecal coliform bacteria in the reservoir. These data also showed that the seasonal increases occurred only in Kensico Reservoir and not in upstream Catskill and Delaware reservoirs. In response to these findings, the city embarked on a waterfowl management program.

Mitigation (e.g., landscaping changes and fence construction to reduce foraging) and the use of noisemakers to frighten the birds were implemented in 1991 and 1992. In the winter of 1993, the City strengthened its efforts with implemen-

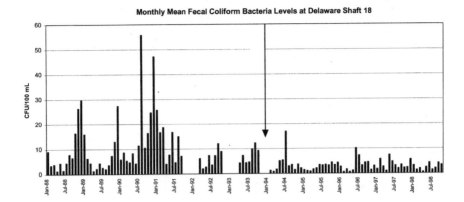

FIGURE 5-8   Total coliform concentrations in Kensico Reservoir from 1991 to 1998. The arrow marks the commencement of the bird harassment program.   Source: Ashendorff et al. (1997).   Adapted from Journal AWWA, Volume 89, No. 3 (March 1997), by permission. © 1997 by American Water Works Association.

tation of a round-the-clock harassment program using boats, hovercraft, and noise-makers.   These efforts simultaneously reduced both the bird populations and coliform densities.   Since waterfowl management began, NYC DEP has not observed seasonal increases in fecal coliform bacteria in Kensico.   Both bird populations and fecal coliform bacteria levels were low in the fall and winter of 1994–1995 and 1995–1996, making the bypass of Kensico Reservoir unnecessary (Ashendorff et al., 1997).

Disinfectants/Disinfection By-Products Rule

Although the D/DBP Rule will not be promulgated until 2002, it is worthwhile to evaluate this regulation to determine whether the New York City water supply system would be in compliance based on current conditions.

Bromate.   In waters containing bromide, ozonation can produce bromate through the oxidation of bromide to hypobromous acid.   Bromate is not expected to be a problem for New York City because the water supply is not currently ozonated. However, trial studies with ozone conducted by NYC DEP indicate an interest in switching to this disinfectant. Unfortunately, bromide concentrations were not measured as part of these studies. If ozone is eventually used and bromide is detected in the water supply, bromate may become a pollutant of concern in the New York City drinking water system.

Chlorite. Chlorite ion in drinking water has two possible sources: (1) from the use of chlorine dioxide and (2) from the breakdown of sodium hypochlorite (NaOCl). NYC DEP currently uses chlorine gas rather than chlorine dioxide at all disinfection locations except on Staten Island. Hypochlorite is used at that location, raising the possibility of chlorite problems in the future.

Haloacetic Acids. The Stage 1 D/DBP MCL for HAA5 is 0.060 mg/L. HAA5 quarterly data for December 1993 through 1998 (Figure 5-9) suggest that New York City may have difficulty in meeting the new HAA5 MCL for the Catskill/Delaware system.

Note that since June 1994, the quarterly HAA5 concentrations in the Catskill/Delaware system were higher than the quarterly compliance TTHM concentrations (compare Figures 5-7 and 5-9). These trends have been observed in North Carolina waters (Singer et al., 1995) and in New England waters (McClellan et al., 1996) and may result from the characteristics of the DOC. Because the Stage 1 HAA5 MCL is lower than the TTHM MCL, and undoubtedly will continue to be so in the Stage 2 D/DBP Rule, this finding has potentially serious consequences for New York City in the future.

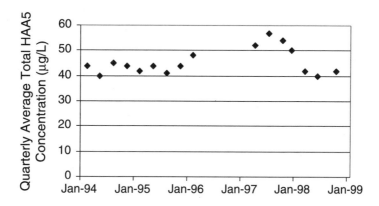

FIGURE 5-9 Quarterly running averages of the sum of five haloacetic acids measured within the distribution system of the Catskill/Delaware supply. The 1994–1997 data were collected from a limited number of sites while the 1998 data were collected from all sites that will be required by the SDWA amendments. Source: NYC DEP Compliance report.

Chlorine. Because disinfectants were not historically thought to pose health risks, upper bounds on chlorine concentration in drinking water have not been included in regulations. Any imposed upper bound on disinfectant residuals was generally determined by the taste and odor of the finished water. However, new evidence indicates some disinfectants can cause harm to humans (EPA, 1994a,b). Thus, the proposed D/DBP Rule includes maximum residual disinfectant levels (MRDLs) for chlorine, chloramine, and chlorine dioxide. The MRDL for chlorine, the primary disinfectant used in New York City, will be 4 mg/L. Based on 1998 data (Figure 5-5), New York City drinking water should not have difficulty meeting this requirement.

Best Available Technology. The proposed D/DBP Rule makes recommendations for improving the performance of best available technologies to achieve new MCLs, MCLGs, MRDLs, and maximum residual disinfectant level goals (MRDLGs). No direct impact on New York City is expected because the treatment technologies recommended for control of TTHMs are enhanced coagulation and granular activated carbon adsorption, which must supplement filtration.

CONCLUSIONS

This chapter has discussed the types and sources of pollution that may have an impact on water quality in the New York City reservoirs. Predictions derived from the Generalized Watershed Loading Function (Table 5-8) indicate that the most important sources of phosphorus differ from basin to basin. The same is likely to be true for other priority pollutants. Some actual measures of reservoir water quality indicate a growing problem with the eutrophic health of the New York City water supply reservoirs (Table 5-9). All are classified as either mesotrophic or eutrophic, and some have variably high concentrations of phosphorus, chlorophyll *a*, and turbidity, particularly after large storm events and during certain seasons. Reservoir conditions are routinely more severe in the Croton reservoirs than in the Catskill/Delaware system reservoirs.

Despite these conditions, source water and drinking water in New York City are in compliance with the SDWA. The Catskill/Delaware water supply currently meets the necessary criteria for disinfectant residual, inactivation ratio, TTHM, and total coliforms. No waterborne disease outbreaks have been recognized and documented. Given current water quality data and upcoming amendments to the SDWA, the Catskill/Delaware system may have difficulty complying with the newly promulgated D/DBP Rule, particularly the MCL for haloacetic acids. It is also likely that EPA will propose new regulations for microbial pathogens, particularly *Cryptosporidium*, in the next few years. Such regulations could cause the City to alter its disinfection process or use additional treatment facilities. Areas where the City may have difficulty maintaining compliance should be given high priority as the watershed management strategy is implemented.

REFERENCES

Anderson, H. W., M. D. Hoover, and K. G. Reinhart. 1976. Forests and water: Effects of forest management on floods, sedimentation, and water supply. USDA Forest Service General Technical Report PSW-18.

Ashendorff, A., M.A. Principe, A. Seeley, J. LaDuca, L. Beckhardt, W. Faber, Jr., and J. Mantus. 1997. Watershed Protection for New York City's Supply. Journal of the American Water Works Association 89(3):75–88.

Auer, M. T., K. A. Tomasoski, M. J. Babiera, M. L. Needham, S. W. Effler, E. M. Owens, and J. M. Hansen. 1998. Phosphorus bioavailability and P-cycling in Cannonsville Reservoir. Journal of Lake and Reservoir Management 14(2-3): 278–289.

Black, R. E., G. F. Craun, and P. A. Blake. 1978. Epidemiology of Common-Source Outbreaks of Shigellosis in The United States, 1961-1975. American Journal of Epidemiology 108(1):47–52.

Bosch, J. M., and J. D. Hewlett. 1982. A review of paired catchment experiments to determine the effect of vegetation changes on water yield and evapotranspiration. Journal of Hydrology 55:3–23.

Briley, K. F., R. F. Williams, K. E. Longley, and C. A. Sorber. 1980. Trihalomethane production from algal precursors. Pp. 117-129 In Jolley, R. L. , Brungs, W. A., and Cumming, R. B. (eds.) Water Chlorination: Environmental Impact and Health Effects, Volume 3. Ann Arbor, MI: Ann Arbor Science Publishers, Inc.

Brooks, K. N., P. F. Ffolliott, H. M. Gregersen, and L. F. DeBano. 1997. Hydrology and the Management of Watersheds. 2nd Edition. Ames, IA: Iowa State University Press.

Brown, M. P., M. R. Rafferty, P. B. Robillard, M. F. Walter, D. A. Haith, and L. R. Shuyler. 1984. Nonpoint Source Control of Phosphorus—A Watershed Evaluation. Albany, NY: New York State Department of Environmental Conservation, Bureau of Water Research.

Butterfield, C. T., E. Wattie, S. Megregian, and C. W. Chambers. 1943. Influence of pH and temperature on the survival of coliforms and enteric pathogens when exposed to free chlorine. U.S. Public Health Reports 58(51):1837–1866.

Canter, L. W., and R. C. Knox. 1986. Septic Tank System Effects on Ground Water Quality. Chelsea, MI: Lewis Publishers, Inc.

Carraway, M., S. Tzipori, and G. Widmer. 1997. A new restriction fragment length polymorphism from *Cryptosporidium parvum* identifies genetically heterogeneous parasite populations and genotypic changes following transmission from bovine to human hosts. Infection and Immunity 65:3958–3960.

Centers for Disease Control and Prevention (CDC). 1998. 1997 Final Foodnet Surveillance Report. Atlanta, GA: Division of Bacterial and Mycotic Diseases.

Chanlett, E. T. 1979. Environmental Protection. New York, NY: McGraw-Hill.

Clancy, J. L., W. Gollnitz, and Z. Tabib. 1994. Commercial labs: How accurate are they? Journal of the American Water Works Association 86(5):89–97.

Clarke, N. A., and S. L. Chang. 1959. Enteric viruses in water. Journal of the American Water Works Association 51:1299–1317.

Cooper, R. C. 1974. Waste water management and infectious disease. I. Disease Agents and Indicator Organisms. Journal of Environmental Health:217–224.

Craun, G. 1998. Presentation at the Microbial Risk Assessment Workshop for the NRC Committee to Review the New York City Watershed Management Strategy. April 14–15, 1998, Atlanta, GA.

Cummins, K. W., M. J. Klug, R. G. Wetzel, K. F. Suberkropp, R. C. Petersen, B. A. Manny, J. C. Wuycheck, F. O. Howard, and R. H. King. 1972. Organic enrichment with leaf leachate in experimental lotic ecosystems. BioScience 22:719–722.

De Graaf, R. M., M. Yamasaki, W. B. Leak, and J. W. Lanier. 1992. New England Wildlife: Management of Forested Habitats. USDA Forest Service GTR NE-144.

Derge, R. E., Jr. 1983. Evaluation of Selected Performance Parameters of an Aerobic Wastewater Treatment Unit for Individual Homes. M.S.E.H. Thesis. East Tennessee State University.

Dodds, W. K., K. R. Johnson, and J. C. Priscu. 1989. Simultaneous nitrogen and phosphorus deficiency in natural phytoplankton assemblages: Theory, empirical evidence, and implications for lake management. Journal of Lake and Reservoir Management 5(1):21–26.

Effler, S. W., M. G. Perkins, N. Ohrazda, C. M. Brooks, B. A. Wagner, D. L. Johnson, F. Peng, and A. Bennett. 1998. Turbidity and Particle Signatures Imparted by Runoff Events in Ashokan Reservoir, NY. Journal of Lake and Reservoir Management 14(2-3):254–265.

Effler, S. W., and A. P. Bader. 1998. A limnological analysis of Cannonsville Reservoir, NY. Journal of Lake and Reservoir Management 14(2-3):125–139.

Environmental Protection Agency (EPA). 1980. Design Manual: Onsite wastewater treatment and disposal systems. Washington, DC: Environmental Protection Agency.

EPA. 1983. Nationwide Urban Runoff Project (NURP) Final Report. Washington, DC: EPA Office of Water.

EPA 1994a. Disinfectants/Disinfection By-Products. Proposed Rule. Federal Register 59:145:38668. July 29.

EPA. 1994b. Enhanced Surface Water Treatment Requirements. Proposed Rule. Federal Register 59:145:38832. July 29.

EPA. 1997. New York City Filtration Avoidance Determination. New York, NY: EPA.

EPA. 1998a. Clean Water Action Plan: Restoring and Protecting America's Waters. Letter to the Vice President from Carol Bownner, EPA Administrator and Dan Glickman, Secretary of Agriculture.

EPA. 1998b. Memorandum from Jeff Gratz to Mark Izeman and Robin Marx. Regarding NRDC's Letter dated April 21, 1998 concerning microfiltration equivalency testing.

EPA. 1999. National Priority Site Fact Sheet: Richardson Hill Road Landfill Site. http://www.epa.gov/region02/superfnd.

Erlandsen, S. L., and W. J. Bemrick. 1988. Waterborne Giardiasis: Sources of *giardia* cysts and evidence pertaining to their implication in human infection. Pp. 227–236 In Wallis, P. M., and B. R. Hammond (eds.), Advances in *Giardia* Research. Calgary, Canada: University of Calgary Press.

Fraser, R. H., P. K. Barten, and D. A. K. Pinney. 1998. Predicting stream pathogen loading from livestock using a geographical information system-based delivery model. Journal of Environmental Quality 27:935–945.

Fraser, R. H. 1999. SEDMOD: A GIS-based delivery model for diffuse source pollutants. Ph.D. Dissertation, Yale University, Graduate School, Department of Forestry and Environmental Studies, New Haven, Conn.

Geldreich, E. E. 1990. Microbiological quality of source waters for water supply. Pp. 3–31 in McFeters, G. A. (ed.) Drinking Water Microbiology. New York, NY: Springer-Verlag.

Gerba, C. P., and J. B. Rose. 1990. Viruses in source and drinking water. In McFeters, G. A. (ed.), Drinking Water Microbiology. New York, NY: Springer-Verlag.

Hagedorn, C., D. T. Hansen, and G. H. Simonson. 1978. Survival and movement of fecal indicator bacteria in soil under conditions of saturated flow. J. Environ. Qual. 7(1):55–59.

Hoehn, R. C., D. B. Barnes, B. C. Thompson, C. W. Randall, T. J. Grizzard, and P. T. B. Shaffer. 1980. Algal as sources of trihalomethane precursors. Journal of the American Water Works Association 72(6):344–350.

Hoehn, R. C., K. L. Dixon, J. K Malone, J. T. Novak, and C. W. Randall. 1984. Biologically induced variations in the nature and removability of THM precursors by alum treatment. Journal of the American Water Works Association 76(4):135–141.

Hoff, J. C., and E. W. Akin. 1986. Microbial resistance to disinfectants: Mechanisms and significance. Environmental Health Significance 69:7–13.

Jordan, M. J., G. E. Likens, and B. J. Petersen. 1985. Organic carbon budget. Pp. 292–301 In Likens, G. E. (ed.) An Ecosystem Approach to Aquatic Ecology: Mirror Lake and Its Environment. New York, NY: Springer-Verlag.

Longabucco, P., and M. Rafferty. 1998. Analysis of material loading to Cannonsville Reservoir: Advantages of event-based sampling. Journal of Lake and Reservoir Management 14(2-3):197–212.

Marx, R., and E. Goldstein. 1993. A Guide to New York City's Reservoirs and Their Watersheds. New York, NY: Natural Resources Defense Council.

Maryland Department of Environment (MDE). 1999. Stormwater Design Manual. Volume I. Baltimore, MD: Maryland Department of Environment.

May, C., R. Horner, J. Karr, B. Mar, and E. Welch. 1997. Effects of urbanization and small streams in the Puget Sound lowland eco-region. Watershed Protection Techniques 2(4):483–494.

Maxted, J., E. Dickey, and G. Mitchell. 1994. Habitat Quality of Delaware Non-tidal Streams. Dover: Delaware Department of Natural Resources, Division of Water Resources.

McClellan, J. N., D. A. Reckhow, J. E. Tobiason, J. K Edzward, and A. F. Hess. 1996. Empirical models for chlorination by-products. Pp. 26–47 in Minear, R. A., and G. L. Amy (eds.) Water Disinfection and Natural Organic Matter. Washington, DC: American Chemical Society.

Melnick, J. L., C. P. Gerba, and C. Wallis. 1978. Viruses in water. Bulletin of the World Health Organization 56(4):499–508.

Mickle, A. M., and R. G. Wetzel. 1978. Effectiveness of submersed angiosperm-epiphyte complexes on exchange of nutrients and organic carbon in littoral systems. II. Dissolved organic carbon. Aquat. Bot. 4:317–329.

Moe, C. L., C. G. Cogger, and M. D. Sobsey. 1984. Viral and Bacterial Contamination of Groundwater by On-Site Wastewater Treatment Systems in Sandy Coastal Soils. In Proceedings of the 2nd International Conference on Groundwater Quality Research. U.S. Environmental Protection Agency, Ada, OK.

Morris, D. P., and W. M. Lewis. 1988. Phytoplankton nutrient limitation in Colorado Mountain lakes. Freshwater Biology 20:315–327.

Murdoch, P. S., D. A. Burns, and G. B. Lawrence. 1998. Factors that Inhibit Recovery of Acid Neutralizing Capacity (ANC) in Catskill Mountain Streams, New York. American Geophysical Union Spring Meeting Abstracts H22G-04. Published as a supplement of EOS, April 28.

Murdoch, P. 1999. U.S. Geologic Survey. E-mail Memorandum to the NRC. April 1999.

New York City Department of Environmental Protection (NYC DEP). 1993a. Final Generic Environmental Impact Statement for the Proposed Watershed Regulations for the Protection from Contamination, Degradation, and Pollution of the New York City Water Supply and its Sources. Corona, NY: NYC DEP.

NYC DEP. 1993b. Implications of Phosphorus Loading for Water Quality in NYC Reservoirs. Corona, NY: NYC DEP.

NYC DEP. 1997a. The Relationship between Phosphorus Loading, THM Precursors, and the Current 20 µg/L TP Guidance Value. Bureau of Water Supply, Quality & Protection. Valhalla, NY: NYC DEP.

NYC DEP. 1997b. Kensico Watershed Study Annual Research Report: April 1997–March 1998. Valhalla, NY: NYC DEP.

NYC DEP. 1997c. Water Quality Surveillance Monitoring. Valhalla, NY: NYC DEP.

NYC DEP. 1997d. Watershed Agricultural Program Preliminary Evaluation.

NYC DEP. 1998a. DEP Pathogen Studies of Giardia spp., Cryptosporidium spp., and Enteric Viruses. January 31. Valhalla, NY: NYC DEP.

NYC DEP. 1998b. Wastewater Treatment Facility Compliance Inspection Report and Year-End Summary. Valhalla, NY: NYC DEP.

NYC DEP. 1998c. Quarterly Report on the Status of Implementing Projects Designed to Reduce Nonpoint Source Pollution. Valhalla, NY: NYC DEP.

NYC DEP. 1999a. Waterborne Disease Risk Assessment Program 1998 Annual Report. Corona, NY: NYC DEP.

NYC DEP. 1999b. Development of a water quality guidance value for Phase II TMDLs in the New York City Reservoirs. Valhalla, NY: NYC DEP.

NYC DEP. 1999c. Proposed Phase II Phosphorus TMDL Calculations for Ashokan Reservoir. Valhalla, NY: NYC DEP.

NYC DEP. 1999d. Proposed Phase II Phosphorus TMDL Calculations for Cannonsville Reservoir. Valhalla, NY: NYC DEP.

NYC DEP. 1999e. Proposed Phase II Phosphorus TMDL Calculations for Neversink Reservoir. Valhalla, NY: NYC DEP.

NYC DEP. 1999f. Proposed Phase II Phosphorus TMDL Calculations for Pepacton Reservoir. Valhalla, NY: NYC DEP.

NYC DEP. 1999g. Proposed Phase II Phosphorus TMDL Calculations for Rondout Reservoir. Valhalla, NY: NYC DEP.

NYC DEP. 1999h. Proposed Phase II Phosphorus TMDL Calculations for Schoharie Reservoir. Valhalla, NY: NYC DEP.

NYC DEP. 1999i. Proposed Phase II Phosphorus TMDL Calculations for Kensico Reservoir. Valhalla, NY: NYC DEP.

NYC DEP. 1999j. Proposed Phase II Phosphorus TMDL Calculations for West Branch Reservoir. Valhalla, NY: NYC DEP.

NYC DEP. 1999k. Filtration Avoidance Report for the Period January 1 to December 31, 1998. Valhalla, NY: NYC DEP.

New York State Department of Environmental Conservation (NYS DEC). 1996. New York State Water Quality 1996. 305(b) report to the EPA. Albany, NY: NYS DEC.

New York State Water Resource Institute (NYS WRI). 1997. Science for Whole Farm Planning. Cornell University Phase II Twelfth Quarter and Completion Report. Ithaca, NY: NYS WRI.

O'Donoghue, P. J. 1995. *Cryptosporidium* and cryptosporidiosis in man and animals. International Journal for Parasitology 25(2):139–195.

Parker, D. E., J. H. Lehr, R. C. Roseler, and R. C. Paeth. 1978. Site evaluation for soil absorption systems. Pp. 3-15 In Proceedings of the second national home sewage treatment symposium. American Society of Agricultural Engineers pub. 5-77. St. Joseph, MI: ASAE.

Patric, J. H. 1976. Soil erosion in the eastern forest. Journal of Forestry 74(10):671–677.

Patric, J. H., J. O. Evans, and J. D. Helvey. 1984. A summary of sediment yield data from forested lands in the United States. Journal of Forestry 82:101–104.

Paul, J. H., J. B. Rose, J. Brown, E. A. Shinn, S. Miller, and S. H. Farrah. 1995. Viral Tracer Studies Indicate Contamination of Marine Waters by Sewage Disposal Practices in Key Largo, Florida. Applied and Environmental Microbiology 61:2230–2234.

Pinney, D. A. K., and P. K. Barten. 1997. Characterization of livestock management practices in the Tivoli Bays watersheds. Pp. VIII–1–26 In Waldman, J. R., W. C. Nieder, and E. A. Blair (eds.) Final Reports of the Tibor T. Polgar Research Fellowship Program. New York, NY: Hudson River Foundation.

Pinney, D. A. K., and P. K. Barten. 1998. Characterization of demographics and attitudes of farmers in Dutchess County, New York. Pp. VI–1–32 In Waldman, J. R., W. C. Nieder, and E. A. Blair (eds.) Final Reports of the Tibor T. Polgar Research Fellowship Program. New York, NY: Hudson River Foundation.

Purdom, W. P. 1971. Environmental Health. New York, NY: Academic Press.

Quentin, D. H. 1996. Pesticide Concentrations Within Streams of Four New York City Reservoir Drainage Basins. Proceedings of the Symposium on Watershed Restoration Management. July 14–17, Syracuse, NY.

Robillard, P. D., and M. F. Walter. 1984. Phosphorus losses from dairy barnyard areas. In Brown, M., et al. (eds.). Nonpoint Source Control of Phosphorus—A Watershed Evaluation. Albany, NY: New York State Department of Environmental Conservation, Bureau of Water Research.

Rose, J. B. 1997. Environmental ecology of *Cryptosporidium* and public health implications. Annual Reviews in Public Health 18:135–161.

Satterlund, D. R., and P. W. Adams. 1992. Wildland Watershed Management. 2nd Ed. New York, NY: John Wiley & Sons.

Schueler, T. 1987. Controlling Urban Runoff—A Manual for Planning and Designing Urban Best Management. Washington, DC: Metropolitan Washington Council of Governments.

Schueler, T. 1995. The importance of imperviousness. Watershed Protection Techniques 1(3):100–112.

Singer, P.C., A. Obolensky, and A. Greiner. 1995. DBPs in chlorinated North Carolina drinking water. Journal of the American Water Works Association 87(10):83–92.

Smith, H. V. 1992. *Cryptosporidium* and water–A review. Journal of the Institution of Water and Environmental Management 6(4):443–451.

Stepczuk, C. L., A. B. Martin, P. Longabucco, J. A. Bloomfield, and S. W. Effler. 1998. Allochthonous contributions of THM precursors in a eutrophic reservoir. Journal of Lake and Reservoir Management 14 (2-3):344–355.

Swank, W. T., and D. A. Crossley, Jr. 1988. Forest Hydrology and Ecology at Coweeta. New York, NY: Springer-Verlag.

Toxics Release Inventory. 1996. http://www.epa.gov/opptintr/tri/index.html.

Tranvik, L., and S. Kokalj. 1998. Decreased biodegradability of algal DOC due to interactive effects of UV radiation and humic matter. Aquat. Microb. Ecol. 14:301–307.

Vaughn, J. M., E. F. Landry, and M. Z. Thomas. 1983. Entrainment of viruses from septic tank leach fields through a shallow, sandy soil aquifer. Applied and Environmental Microbiology 45(5):1474–1480.

Veneman, P. L. M. 1996. Principles of wastewater treatment. Pp. 5–11 In Sturbridge, M. A., and P. L. M. Veneman (ed) in Proceedings on-site sewage systems conference. Society of Soil Scientists of Southern New England.

Verry, E. S. 1986. Forest harvesting and water: The Lake States experience. Water Resources Bulletin 22(6):1039–1047.

Watershed Agricultural Program (WAP). 1997. Pollution Prevention through Effective Management. Walton, NY: Watershed Agricultural Program.

Warne, D. 1998a. NYC DEP. Memorandum to the National Research Council dated August 1998.

Warne, D. 1998b. NYC DEP. Memorandum to the National Research Council dated November 1998.

Warne, D. 1999a. NYC DEP. Memorandum to the National Research Council dated April 1999.

Warne, D. 1999b. NYC DEP. Memorandum to the National Research Council dated July 1999.

Wattie, E., and C. Butterfield. 1944. Relative resistance of *Escherichia coli* and *Eberthella typhosa* to chlorine and chloramines. U.S. Public Health Service Reports 59:1661.

West, P. A. 1989. The human pathogenic vibrios-A public health update with environmental perspectives. Epidemiology and Infection 103:1–34.

Wetzel, R. G., and B. A. Manny. 1972. Decomposition of dissolved organic carbon and nitrogen compounds from leaves in an experimental hardwater stream. Limnol. Oceanogr. 17:927–931.

Wetzel, R. G., and A. Otsuki. 1974. Allochtonous organic carbon of a marl lake. Arch. Hydrobiol. 73:31–56.

Wetzel, R. G., P. G. Hatcher, and T. S. Bianchi. 1995. Natural photolysis by ultraviolet irradiance of recalcitrant dissolved organic matter to simple substrates for rapid bacterial metabolism. Limnol. Oceanogr. 40:1369–1380.

Yates, M. V., S. R. Yates, A. W. Warrick, and C. P. Gerba. 1986. Use of geostatistics to predict virus decay rates for determination of septic tank setback distances. Applied and Environmental Microbiology 52(3):479–483.

6

Tools for Monitoring and Evaluation

The previous chapters have laid the groundwork for analyzing several aspects of New York City's watershed management strategy as embodied in the Memorandum of Agreement (MOA). In conducting its analysis, the committee identified components of New York City's strategy that correspond to the necessary components of a source water protection program described in Chapter 4. In some cases, this was relatively straightforward; in others, corresponding activities were more difficult to identify. The following six chapters consider how effectively New York City is carrying out source water protection and other related activities. Because of the study's broad statement of task, it was necessary to group particular programs and issues for analysis and discussion. Whenever possible, programs were grouped to correspond with specific necessary elements of a source water protection program.

The first step in any watershed management program is to set goals and objectives, both numeric and narrative. The stated goals of the New York City MOA are many. First, as with all other source water protection programs, one goal of the MOA is to comply with local, state, and federal statutes that protect drinking water quality. Thus, the City has developed an extensive water quality monitoring program, active disease surveillance, a Total Maximum Daily Load (TMDL) program, and a variety of programs for controlling and treating pollution. These programs, when carried out successfully, contribute toward the delivery of clean drinking water and the maintenance of healthy water supply reservoirs. However, the City's overall goals clearly go beyond those mandated by environmental laws. Supporting economic development within the watershed region is desired, as evidenced by the Watershed Agricultural Program, Watershed Forestry Program, and Watershed Protection and Partnership Programs.

Finally, a primary motivation for New York City to draft new watershed rules and regulations was to avoid filtration, which is considered to be an expensive water treatment option for this system. Although cost estimates of filtration differ greatly, all exceed the estimated costs of fully implementing the MOA.

This chapter considers two fundamental activities that inform decision-makers about the quality and safety of water from the Catskill/Delaware system. The first activity is water quality monitoring of all sections of the water supply system, including groundwater, streams, reservoirs, and the delivery system. The physical, chemical, and biological parameters being measured, their importance in assessing the condition of the water supply, and their role in water quality modeling are discussed. The Geographic Information System (GIS) developed for watershed inventory and other purposes is critically evaluated. Second, the safety of the drinking water system is considered by reviewing the role of active disease surveillance in New York City and by conducting a microbial risk assessment on the source water. Current activities that are evaluated include the detection of waterborne disease outbreaks and epidemiological studies for determining the proportion of illness attributable to drinking water.

WATER QUALITY MONITORING PROGRAM

The quality of the drinking water in New York City depends highly on the water quality of the 19 reservoirs of the Catskill/Delaware and Croton watersheds. Reservoir water quality is directly coupled to, and dependent upon, the loadings of pollutants from the individual drainage basins and from atmospheric deposition. Each of the drainage basins of the individual reservoirs combines a unique set of physical, chemical, and biological characteristics. These characteristics—including elevation, geomorphology, rock and soil composition and distribution, soil chemistry, rates of runoff and water residence times, types and extent of plant cover, and human modifications by land use, development, and waste releases—can vary markedly from basin to basin.

Because the volume of drinking water required by New York City is so large and entirely dependent upon the aggregate sources of these reservoirs, users rapidly realize changes in reservoir water quality. Therefore, any response by water managers and treatment operators requires rapid acquisition of information concerning changes to the reservoir ecosystems and their drainage basins. The regulatory steps that are taken to control various water quality parameters may be different. In addition, management steps may differ with varying seasonal, meteorological, and human influences. Thus, monitoring frequency must be flexible and respond to the rates of change that are observed for individual water quality parameters.

Four types of monitoring are discussed in Chapter 4: compliance monitoring, operational monitoring, performance monitoring, and monitoring to support modeling efforts. Efforts of the New York City Department of Environmental

Protection (NYC DEP) in most of these areas are analyzed below, and recommendations for improvement, when necessary, are given. A full description of all monitoring efforts conducted by NYC DEP is available in the Water Quality Surveillance Monitoring report (NYC DEP, 1997a).

Compliance Monitoring

In order to comply with the Safe Drinking Water Act (SDWA) and the filtration avoidance determination, a variety of biological, chemical, and physical parameters are measured in the New York City source water reservoirs and the water distribution system (Table 6-1). In almost all cases, the New York City water supply has met the requirements for the physical and chemical parameters up to the present time. Compliance monitoring of fecal and total coliform measurements revealed increases in these parameters during the winters of 1991–1992 and 1992–1993 in the Kensico Reservoir. However, violations of the SDWA were avoided by bypassing the Kensico Reservoir (see Chapter 5 for details).

Operational Monitoring

Operational monitoring refers broadly to activities that are necessary for both short-term and long-term operation of the water supply system. This category, which encompasses much of NYC DEP's efforts, includes both routine activities and special projects (1) to follow changes in water quality and (2) to

TABLE 6-1 Frequency with which Parameters are Measured during SDWA Compliance Monitoring

Parameter	Catskill System		Delaware System	
	CAT(LEFF)[a]	CAT(EV)[b]	DEL18[a]	DEL19[b]
Turbidity	Continuous	Continuous	Continuous	Continuous
pH	Daily	Continuous	Daily	Continuous
Free Chlorine Residual	Not determined	Continuous	Not determined	Continuous
Total Coliform	Daily	Daily	Daily	Daily
Fecal Coliform	Daily	Daily	Daily	Daily
Temperature	Daily	Continuous	Daily	Continuous

[a]CAT(LEFF) and DEL18 are the effluent locations for the Catskill and Delaware systems, respectively, at Kensico.
[b]CAT(EV) and DEL19 are within the Catskill and Delaware aqueducts, respectively, just downstream of Kensico.
Note: SDWA compliance monitoring also includes some pesticides within the distribution system, which are not listed in this table.
Source: NYC DEP (1997a).

identify sources of pollution that affect reservoir water quality. Monitoring of physical and chemical parameters is discussed first (organized by waterbody type) followed by a review of the microbial monitoring efforts that occur throughout the watersheds.

Regional Meteorology

Meteorological data, including air temperature, relative humidity, rainfall, snow depth, solar irradiance, photosynthetically active radiation (selected sites), and wind speed and direction, are measured at 22 stations throughout the Catskill/ Delaware and Croton watersheds at one-minute intervals. This new (1998) network of meteorological stations has been established based on a reasonable set of criteria, including precipitation patterns, elevational gradients, and modeling requirements. Each station contains instrumentation for a large and complete set of meteorological parameters, and the one-minute interval frequency is satisfactory.

One could champion for greater meteorological data collection in a region such as the Catskill/Delaware watershed, where topography is quite heterogeneous. However, the new network is greatly improved over the previous system, which measured precipitation only, and it is adequate for most of the eutrophication and public health questions of concern.

Groundwater and Shallow Subsurface Monitoring

Regular groundwater monitoring has only occurred in the Kensico watershed, as this area has a high potential for contamination related to urbanization (NYC DEP, 1997b). Eighteen monitoring locations exist, consisting of 13 wells, some of which have multiple depths (ranging from 3.5 to 120 ft). These locations, which were selected in relation to geology, land and chemical use, and proximity to sewer lines, are reasonable for the Kensico watershed. All 13 wells were monitored for turbidity, pH, alkalinity, conductance, total and fecal coliforms, and nutrients on a monthly basis between April 1995 and April 1997, and static water levels were measured weekly. Analysis of monitoring data collected during that time period led to biannual sampling of all wells, which has continued to the present time. This frequency of sampling is sufficient for the deep subsurface but not for the shallow subsurface, which is influenced by seasonal variations. Other groundwater monitoring activities associated with special projects are discussed in a later section on performance monitoring.

Stream Monitoring

Monitoring of stream inputs is critical for determining reservoir water quality and managing reservoir operations. One of the first steps in understanding pollutant dynamics within short-detention-time reservoirs is to sample their tributaries frequently and to construct input–output budgets of key physical and chemical parameters. There are two essential monitoring parameters for streams: (1) volume of influent water to compute loadings and for use in nutrient loading and productivity models and (2) concentrations of critical chemical parameters to evaluate the loadings of nutrients, potential toxic heavy metals, and organic compounds.

The locations of the 145 sampling sites throughout the Catskill/Delaware and Croton watersheds include primary river inflows, sites above and below selected wastewater treatment plants (WWTPs) and towns, and outflows from subcatchment basins. Grab samples are generally collected at biweekly intervals and are analyzed for temperature, pH, alkalinity, conductance, dissolved oxygen, major cations (Ca^{2+}, Mg^{2+}, Na^+, K^+), turbidity, color, suspended solids (SS), nutrients, total organic carbon (TOC), silica, chloride (Cl^-), trace elements, and total and fecal coliform, among others. Forty-eight (48) U.S. Geological Survey gauging stations measure water level (stage) continuously. The computerized data acquisition system being developed at present appears to be adequate.

Because samples are collected at fixed time intervals rather than on the basis of discharge, it is certain that major fluctuations in the loading of chemical and biological parameters are not captured. This issue, which pertains to stream sampling, precipitation measurements, and sampling of shallow subsurface flows, is discussed below with regard to certain parameters, and it is generally addressed later under a separate section titled Flow Proportional Monitoring.

Physical Parameters and Cations. Automation of stream monitoring can ease the transition from fixed frequency sampling to event-based sampling. Within the New York City watersheds, the monitoring of some stream parameters, most notably temperature, conductance, pH, and dissolved oxygen, could easily be automated. This is possible even at remote sites via telemetry of data to data acquisition centers. Although the initial expense to install automation would be high, costs could subsequently be reduced by decreasing the required personnel and by eliminating other analyses. For example, rather than being measured directly, the concentrations of Ca^{2+}, Mg^{2+}, Na^+, and K^+ could be obtained from strong correlations with conductance within 5 percent to 10 percent accuracy, which would be quite adequate for the purpose of determining hardness and reactivity (Otsuki and Wetzel, 1974). Alkalinity could also be continuously estimated with reasonable accuracy from the parameters measured automatically.

Turbidity, Color, Suspended Solids, Nutrients, Total Organic Carbon, and Silica. These parameters are currently measured at stream sites on a fixed biweekly or longer interval. The usefulness and validity of these data, particularly for use in nutrient loading models, are unclear. A fixed biweekly or longer sampling interval is marginally satisfactory for the monitoring of large reservoirs with moderately long (> six months) residence times. However, these sampling intervals are not satisfactory for surface water influents and nonpoint shallow subsurface inflows. Fixed interval sampling will miss significant loading events and will substantially underestimate true loadings. It would be better to thoroughly sample stream inflows at fewer stations with automated discharge-mediated samplers, particularly close to the reservoir inlet mouth, than to sample many stations at infrequent, fixed intervals. In addition to the parameters currently measured, dissolved organic carbon (DOC) should also be measured in streams on a flow proportional basis, as suggested by others (ILSI, 1998).

Chloride, Trace Elements, and Toxic Compounds. Chloride and trace element analyses are adequate at present frequencies. Toxic compounds are not measured on a regular basis at stream sites. In 1997, concentrations of several pesticides and polychlorinated biphenyls (PCBs) were monitored at two stream sites in the Kensico watershed; compounds were detected at very low levels (NYC DEP, 1998a). Monthly sampling of streams for pesticides is a planned future activity of NYC DEP that is strongly supported by the committee as a way to determine presence/absence, establish ambient concentrations, and better pinpoint sources. Sampling should be timed to correspond with the application of pesticides and expected pesticide transport by stormwater from rainfall and snowmelt. This activity is needed most in the Cannonsville and Pepacton watersheds because of the density of pesticide application and in the Kensico and West Branch watersheds because of their proximity to the distribution system.

Macroinvertebrates. Monitoring of stream macroinvertebrates is done annually in August and September at 14 regular stream sites throughout the entire watershed region and at additional sites that vary in location. This occasional macroinvertebrate sampling is being used as a biotic index of relative stream "health" based on indicator species. Based on monitoring results, the stream sites are classified as nonimpaired, slightly impaired, or severely impaired. Between 1994 and 1997, 29 sites were nonimpaired, 19 sites were slightly impaired for at least one year, and no sites were severely impaired (NYC DEP, 1999a). Because the present frequency of sampling is too low and the quantitative measures are too marginal to overcome high natural variance, the usefulness of these data in relation to the information accrued and the effort expended is questionable.

Aqueduct Monitoring

The aqueducts (or tunnels) are sampled at ten locations west of the Hudson River and at 11 locations east of the Hudson River, generally where water enters and exits the tunnels. The sampling interval is daily or weekly, depending on the proximity of the tunnel to Kensico Reservoir. Twenty-seven (27) different parameters are measured within the aqueducts for both operational and compliance monitoring purposes and to support the Process Control-Remote Monitoring program. Those measured on a daily basis include odor, color, turbidity, temperature, specific conductivity, pH, free chlorine residual, fluoride, and total and fecal coliforms. In addition to the 27 parameters, some pesticides are being monitored on an annual basis at six aqueduct (key point) locations.

As part of the Process Control-Remote Monitoring Program, sampling of turbidity, pH, conductivity, temperature, free chlorine, and fluoride is automated at 13 aqueduct locations east of the Hudson River. NYC DEP plans to extend automated sampling to aqueduct key points west of the Hudson River when resources are available. The automated sampling is conducted at a significantly higher frequency than grab samples, and the results are continuously downloaded to chart recorders that display the data. The frequency and analytical techniques used for within-aqueduct analyses of water quality are adequate.

Reservoir Monitoring

Reservoirs receive collective loadings both from the atmosphere and from their drainage basins. The effects of these pollutant loadings on reservoir water quality are relatively low because loading volumes are significantly less than total reservoir volume. In addition, the residence time of reservoirs is increased compared to streams. Therefore, assuming that withdrawal volumes are relatively small in proportion to total reservoir volume, the frequencies of monitoring for most water quality parameters can be reduced from that of stream monitoring. This reduction, however, should not exceed the generation times of controlling processes, including biological processes.

The water supply reservoirs are sampled monthly from late March to early December, with biweekly sampling occurring at some reservoirs. Temperature, pH, dissolved oxygen, and specific conductivity are measured *in situ* with automated samplers at 1-m depth intervals. Other measured parameters can be found in NYC DEP (1997a). Present sampling includes a depth-integrated water sample (e.g., 1–3 m) from the euphotic or light-penetrating zone, the depth of which may or may not represent an integrated sample of the epilimnion[1] during periods when the reservoirs are thermally stratified. Metalimnetic and hypolimnetic samples are taken in the deeper portions of the reservoirs. For the general purposes of assessing water quality in these moderately impacted reservoirs, the spatial sampling sites for monitoring are generally adequate.

---

[1]Reservoirs with moderate to long retention times stratify thermally, with less dense, warmer water overlying cooler, more dense water in summer. The water of the upper stratum, the *epilimnion*, is uniformly warm, circulating, and fairly turbulent. The epilimnion essentially floats upon a cold and relatively undisturbed bottom stratum, the *hypolimnion*. The intervening stratum between the epilimnion and the hypolimnion is the *metalimnion*, characterized by a steep thermal gradient from warm to cool water (decreasing at least 1°C per meter of increasing depth).

Growing Season Considerations. The in-reservoir "growing season" (March to December) is assumed to be the period during which the reservoir is most susceptible to water quality changes. Thus, there is little reservoir sampling during the winter, except for sampling of the aggregated outflows within aqueducts. This strategy may be acceptable at the present with low mesotrophic to moderate eutrophic conditions in the reservoirs. However, if the productivity increases further, major chemical and biological changes will occur in winter periods that must be monitored. For example, at this latitude 25 percent of the annual primary productivity can easily occur under ice cover of reservoirs.

Color, Secchi Depth, and Euphotic Depth. The euphotic depth refers to the depth of water that receives sufficient solar radiation to support net photosynthetic growth of phytoplankton. This depth is usually limited to the penetration depth of 0.1 percent to 1 percent of light reaching surface waters. Secchi depth is an estimate of water transparency and is equal to the approximate penetration depth of 15 percent of surface light. These parameters, along with color, give an indication of algal levels and other color-producing compounds. The color and clearness of the New York City reservoirs is often governed by dissolved organic matter (DOM). In the case of the Catskill/Delaware reservoirs, the DOM responsible for color and clearness originates largely from terrestrial and wetland sources of higher plant decomposition (allochthonous sources) rather than from in-reservoir sources (autochthonous sources). This is because only under hyper-eutrophic conditions are light penetration and euphotic depth appreciably influenced by the densities of algae and cyanobacteria, and such conditions are not found in most of the Catskill/Delaware reservoirs.

Much of the loading of allochthonous DOM is directly correlated with precipitation events within the drainage basin. Once in the reservoir, this recalcitrant pool of DOM is biologically degraded at a relatively slow rate. Therefore, biweekly sampling of euphotic depth, Secchi depth, and color should be adequate.

Dissolved Organic Matter. Because DOM can serve as a precursor of trihalomethanes (THMs) and other chlorination byproducts, NYC DEP has spent considerable energy investigating the sources of DOM in the water supply reservoirs. These efforts have consisted of continuous reservoir monitoring as well as special activities to measure a variety of parameters, including DOC and THM formation potential (NYC DEP, 1997c; Stepczuk et al., 1998a–c). Although preliminary evidence suggests that autochthonous sources may predominate during certain times of the year (NYC DEP, 1997c), the slower degradability of allochthonous sources means they will be dominant for significant periods, especially during the winter (December through March). This potential switch in the dominance of different sources of DOM implies that monitoring must take place year round rather than just during the growing season.

DOM is almost always quantitatively expressed as DOC. In order to determine the pool of DOC available for reaction with chlorine, quantitative assays should be conducted using modern DOC analyses (heat oxidation followed by IR or coulometry to measure the CO_2). Measurements of DOC made by the Carlo Erba (EA1108) instrument for CHN analyses (as used by NYC DEP in Stepczuk et al., 1998a) are not satisfactory because the instrument was not designed for such analyses. DOC measurements should occur in depth profiles at regular intervals with a frequency of at least every 2 weeks in the reservoirs. For making such measurements in streams and shallow subsurface sources, the frequencies for such analyses should be based on discharge, not time. It should be kept in mind when conducting such analyses that DOC is, in general, a poor surrogate measure of THM precursor concentration.

To complement DOC measurements, selective ultraviolet absorption spectrophotometry has been employed as a general index of DOM concentration in natural waters (American Public Health Association, 1998; Thurman, 1985; Wetzel and Otsuki, 1974). Ultraviolet absorption (UV254) permits a relative measure of the stable recalcitrant dissolved organic components derived from allochthonous sources. However, it is not particularly useful for detecting variations in autochthonous sources because DOM produced by phytoplankton is more labile by one or two orders of magnitude than allochthonous DOM. UV254 measures tend to exhibit little variation over time because the decomposition rates of allochthonous DOM of about one percent per day approximately balance allochthonous influxes. This trend has been borne out by NYC DEP studies; UV254 data collected in Kensico Reservoir effluents on an irregular basis were found to average around 0.036 OD with little variation (S. Freud, NYC DEP, personal communication, 1998). No correlations were found between UV254 and TOC or trihalomethane formation potential (THMFP) values.

Total DOC and spectral data yield no information about the sources of DOM. Only a detailed structural analysis of the compounds found in water samples can reveal the relative levels of autochthonous and allochthonous DOM. Allochthonous DOM is in large part composed of yellow organic humic acids of plant origin and consists of a mixture of fulvic acids, aromatic polyhydroxy carboxylic acids, and phenolic residues and polymers of these acids. Such compounds do not originate from phytoplankton algae and cyanobacteria.

The more complex organic compounds have been variously categorized on the basis of their structure and their solubility characteristics in acids and bases. Quantitative differentiation of the relative amounts of autochthonous and allochthonous DOM can be estimated by detailed organic chemical analyses using solid-state ^{13}C-nuclear magnetic resonance and gas chromatography–mass spectrometry analyses (e.g., Wetzel et al., 1995; McKnight et al., 1997). Less demanding analyses can be used to estimate the likely proportions of aliphatic, aromatic, and "excess" carbon in a complex mixture of DOM from different sources by evaluating its elemental composition and carboxyl content (Purdue, 1984; Wilson, et al., 1987). At best, these methods yield approximations of ratios between autochthonous and allochthonous sources and should be coupled to more accurate estimates of external DOM loading to reservoirs. If the NYC DEP wants to determine the relative sources of DOM in the water supply reservoirs, these methods must be employed.

In addition to determining sources of DOM, NYC DEP may want to consider regularly measuring disinfection byproduct (DBP) formation potential in the reservoirs, particularly Kensico Reservoir. DBP yields (mg DBP/mg DOC) vary from source to source. Determining the seasonal DBP yields from the outflows of each reservoir would provide information that could help focus control efforts on the most important pollution sources.

Temperature, pH, Specific Conductance, and Dissolved Oxygen. In medium to large natural lakes in the temperate zone with depths greater than 10–15 m, temperature, pH, specific conductance, and dissolved oxygen can be evaluated adequately at biweekly intervals. However, caution should be used when applying these monitoring frequencies to reservoirs because they are subject to more rapid changes than natural lakes. Such rapid changes are most common when the collective reservoir volume is moderate or small in relation to inflows and outflows. In most of the New York City reservoirs, changes in inflow and outflow volume can be large in relation to total reservoir volume. Hence, two-week sampling intervals for these four parameters may not be adequate. In particular, sampling might be increased during the week of autumnal turnover,[2] which is often quite predictable and is a time of major chemical redistribution.

Total Suspended Solids, Volatile Suspended Solids, and Turbidity. Total suspended solids, volatile suspended solids, and turbidity are moderately useful metrics to approximate the loading of organic and inorganic particles. Total suspended solids indicate the presence of inorganic and organic particulate matter, while volatile suspended solids reflect organic particulate matter. Both provide information about particle composition that cannot be derived from turbidity measurements.

In most natural lakes, levels of inorganic particulate matter are low in comparison to organic particulate matter. In reservoirs (including the New York reservoirs), however, inorganic particulate matter such as clays is sometimes found in high concentrations, reversing the ratio toward a dominance of inorganic matter. Whether this is true depends upon geomorphology, precipitation events, and other factors. Because levels of total suspended solids, volatile suspended solids, and turbidity are heavily dependent on precipitation events, the biweekly sampling of these parameters that is currently taking place is likely to be of

---

[2]Autumnal turnover refers to loss of thermal stratification and complete water circulation.

marginal operational value. However, the sampling will indicate which of the reservoirs are routinely problematic. The more rigorous monitoring done at the aqueducts is more responsive to the aggregate loadings from all reservoir sources and therefore should be continued.

Odor. It is assumed that odor measurements are precautionary and that odor is generally not a problem within the reservoirs because of low to moderate production of algae and cyanobacteria. Thus, biweekly sampling is adequate at this time.

Nutrients. Chemical analyses, particularly of nutrients, should be performed at frequencies commensurate with changes induced by loadings, biotic utilization and recycling, and losses. In the case of phosphorus, more than 90 percent to 95 percent is found within living and dead particulate organic matter. Soluble reactive phosphorus (SRP) cycles very rapidly (minutes to a few hours), as does much of the soluble organic phosphorus. Hence, the present biweekly measures of total phosphorus and monthly measures of SRP and total dissolved phosphorus (mostly organic) are likely to be adequate for the predictive modeling purposes.

Total nitrogen (inorganic and organic, particulate and dissolved) tells one relatively little in any functional sense because of the complexities of the oxidative–reductive interactions among the different chemical compounds. It is assumed (and it is likely correct, although algal bioassays have not been conducted) that nitrogen is not limiting phytoplanktonic productivity in these reservoirs except when phosphorus loading is excessive and the availability of phosphorus exceeds demand. Therefore, determinations of combined nitrogen (nitrate/nitrite and ammonium ion concentrations) are useful in relation to vertical intensities of bacterial metabolism, rates of hypolimnetic oxygen reduction, anoxia and related problems (such as sulfide production and iron reduction), and potential nitrogen limitation under eutrophic conditions caused by excessive phosphorus loading. Biweekly sampling frequency is generally adequate to follow seasonal changes in stratified reservoirs.

Chemicals. Chloride is a highly conservative ion of relatively minor limnological interest in freshwater inland lakes. Concentrations nearly always exceed biological requirements, they change little either spatially or temporally, and they are not a problem. Depending on the composition of road salt used in the watershed region, however, the reservoirs may have a chloride gradient. In order to assess this, the present sampling schedules are adequate. Bromide is currently not measured in the reservoirs. However, should New York City decide to install ozonation, bromide should be measured on a regular basis in the Kensico Reservoir using the IC method with a detection limit of 10 μg/L or less.

Sulfate ions can become reduced and depleted in the hypolimnia of productive stratified reservoirs, with the production of hydrogen sulfide and related problems. Monthly sampling is likely to be adequate in the least productive reservoirs but should be increased to biweekly in the more productive reservoirs.

Silica (SiO_2) concentration is an important parameter because of the dependence of desirable diatom and certain crysophyte algae on this substance. With increasing phosphorus loading and eutrophication, silica concentrations can be reduced to concentrations that are limiting to competitive growth of diatoms (<0.5 mg/L). Reduction of silica concentrations in the euphotic zone during productive thermally stratified periods to limiting levels is an important predictive parameter that should be monitored. Monthly sampling is adequate if silica concentrations exceed 5 mg/L, but sampling should be increased to biweekly when concentrations are less than 5 mg/L.

The present frequency of analyzing for major cations and trace metals appears to be adequate. As mentioned above for stream sampling, major cation concentrations can be estimated quite accurately by correlations with specific conductance to help reduce the frequency of direct analyses.

There is no regular monitoring of toxic chemicals in the water supply reservoirs except for pesticide monitoring that occurs annually at six reservoir locations immediately adjacent to aqueducts. Because of the expense associated with monitoring toxic chemicals in the reservoirs on a regular basis, it has been suggested that as a substitute measure, New York City sample fish tissues for bioaccumulating compounds (ILSI, 1998). The committee concurs that this type of activity should be undertaken, as long as the exposure of fish can be well characterized. Large native fish are highly preferable to stocked species. It should be noted that this type of assay is only effective for bioaccumulating organic compounds and mercury. It is not likely to detect pesticides (most of those used in the watershed do not bioaccumulate) and metals other than mercury. For this reason, the committee encourages the planned monthly sampling of pesticides in the reservoirs, particularly at Kensico Reservoir (NYC DEP, 1997a).

Chlorophyll. Chlorophyll concentrations can function effectively as a general indicator of phytoplanktonic biomass, and they can also be used as parameters for predictive models. The average generation times for algae in nature are 1–3 days, depending on the environmental conditions. Hence, it is reasonable to anticipate significant alterations of phytoplankton growth and productivity within a five- to seven-day period under good growing conditions. More frequent sampling (e.g., seven-day intervals at a minimum) would be an improvement, but for general modeling and evaluation purposes, the biweekly sampling is adequate. Monthly sampling is too long an interval and is not recommended for any reasonable evaluation, even in "nonkey" reservoirs.

Algae. The sampling of algae, cyanobacteria, and zooplankton at the generic levels is sufficient for a general overview of microbial developments, and biweekly sampling is adequate for the evaluation processes under way at this time.

Sedimentation. Since 1993, sediment samples have been collected in the Ashokan and Schoharie reservoirs in order to calculate a sediment mass balance in these reservoirs and to determine particle settling rates. Sediment traps, which are outfitted with between one to three sampling depths, are set early in the year and are collected on a weekly or biweekly basis. Samples are analyzed for total and volatile suspended solids. A sediment trap that can measure resuspension is currently being designed for the Ashokan Reservoir and should be operational soon.

The sediment trap data, as presently collected, will not permit an estimation of a sediment mass balance in the reservoirs. Rather, a mass balance calculation requires sediment loading data from the influents and effluents to the reservoirs collected during storm events. Sedimentation data can be of some use in estimating filling rates, but such accretion data can be obtained more effectively by paleolimnological methods. The current sedimentation data permit only minimal interpretations of sediment accretion over the long term. For these reasons, the present use of sediment traps in the monitoring program has no useful application.

Zebra Mussels. Zebra mussel monitoring occurs at 60 sites in the reservoirs by way of settling plates, water quality sampling, and shoreline and structure inspection. Although they have yet to be detected, should zebra mussels appear, they could cause major water quality problems. New York City's primary goal with its monitoring program is detection of zebra mussels, and its efforts are adequate for this purpose.

Distribution System Monitoring

New York City has two types of sampling sites in its distribution system: (1) compliance sites, which are located on distribution mains equal to or smaller than 20 inches with connections directly serving the public, and (2) surveillance sites from reservoirs, shafts, pumping stations, trunk mains, and wells within the distribution system. Table 6-2 presents the distribution of sites in the boroughs of New York City.

Each month, NYC DEP collects on average 946 compliance samples and 473 surveillance samples, resulting in the sampling of individual sites every 9–14

TABLE 6-2 Sampling Sites in New York City Distribution System

Type	Bronx	Brooklyn	Manhattan	Queens	Staten Island	Total
Compliance	46	70	56	85	29	286
Surveillance	31	27	32	113	19	222
Total	77	97	88	198	48	508

Source: NYC DEP (1998b).

days under normal conditions. Because most chemical parameters would not be expected to change in concentration very rapidly in a system as large as New York City's, sampling every 9–14 days is adequate. In addition to those compounds measured for SDWA compliance purposes (Table 6-1), the distribution system is also sampled for a variety of other important parameters, including *E. coli*, orthophosphate, total organic carbon, total organic halides, haloacetic acids, additional pesticides, and PCBs, many of which are not yet regulated.

It should be noted that because of their location, the surveillance sites may not be representative of the water actually consumed (unlike the compliance sites). Furthermore, all sampling devices are subject to inspection, predisinfection flushing, and disinfection prior to sample collection—treatment that consumers do not generally provide to their tap water (NYC DEP, 1998b). Such treatment is performed to remove stagnant water from the sampling device and eliminate the possibility of external contamination.

Wastewater Treatment Plant Monitoring

It is apparent from available data that a major source of pollutant loading to the water supply reservoirs emanates from wastewater treatment plant (WWTP) effluents (NYC DEP, 1998c, 1999a). According to 1997 and 1998 end-of-year compliance reports for the individual plants, WWTP effluent in the Catskill/ Delaware watershed often contains high (noncomplying) levels of phosphorus, coliform bacteria, and other pollutants. Phosphorus is the primary concern at plants utilizing secondary sewage treatment because these processes, when functioning properly, almost always reduce pathogenic microorganisms and particulate and dissolved organic matter to acceptable levels. However, they are not particularly effective in the removal of nutrients.

At the present time, monitoring of WWTP effluent occurs weekly at all six City-owned plants in accordance with State Pollutant Discharge Elimination System (SPDES) permits and twice monthly via grab samples at all 35 non-City-owned plants. Flow rates are generally measured by the plant operator. Measured parameters include the common physical, chemical, and biological parameters, but do not include metals and other toxic compounds (NYC DEP, 1997a). It has been suggested that plant effluents be monitored for toxic organic compounds and metals on an annual basis (ILSI, 1998). In the committee's opinion, such an activity is expensive and of limited usefulness, given that toxic contaminant input is infrequent and usually occurs in slugs (which can be identified if they cause plant upsets). Such compounds often end up in plant residuals, further limiting the information to be gained by effluent monitoring.

As with many of the monitoring activities previously discussed, monitoring of WWTP effluents should be based on volume discharges rather than on a fixed frequency. Although there is some disagreement about the response of WWTP effluent volumes to storm events, the committee is not convinced that precipitation events are unrelated to larger WWTP pollutant loadings because the necessary data to resolve this issue are not available. Depending on the extent of stormwater infiltration into nearby sewer lines, WWTPs could play a significant role in initiating, stimulating, or supporting algal blooms or other polluting events in nearby reservoirs. Only through event-based sampling can the monitoring program accurately assess this possibility.

Microbial Monitoring

The microbial monitoring programs being conducted by NYC DEP (beyond compliance monitoring) are extensive, although they are focused on a few parameters (NYC DEP, 1998b). Total and fecal coliform bacteria, enteric viruses, *Giardia, Vibrio,* and *Cryptosporidium* samples are regularly collected at over 50 locations in the watershed, including tributaries, WWTP effluents, reservoirs, aqueducts, and the compliance points for raw and chlorinated water (which are also used for coliform sample collection). The sampling frequency varies from daily to monthly, and attempts are made to monitor at least one storm event per month. Storm event sampling has been automated at some sites located near the discharge of relatively urban subbasins. In addition to routine pathogen monitoring noted above, New York City has also occasionally measured other pathogens in water samples.

Methods. The procedures being used for sample collection and analysis of virus, bacteria, and protozoan samples are in general accordance with methods employed elsewhere in the United States. However, there are some important limitations to New York City's current methodology, especially for the protozoa. The City's standard method to generate (oo)cyst concentration data, which has not changed significantly since sampling began in 1992, is a slightly modified version of the ASTM P229 method and is similar to the ICR method (NYC DEP, 1997d). The City routinely samples approximately 300 gallons of source water to measure (oo)cysts. Whether the entire pellet generated from this sample is examined depends on water quality; however, enough of the pellet is always examined to result in a detection limit of at least 0.7 or less (oo)cysts/100L (Stern, 1998). Overall recovery efficiency is measured quarterly and varies between 30 percent and 70 percent.

Data from the pathogen studies suggest that improvements to the methods are needed. First, the very high number of nondetects makes it difficult to determine characteristic (oo)cyst concentrations emanating from each catchment in the Catskill/Delaware watershed. Ideally, at least 12 monthly nonzero or 24 twice-monthly nonzero measurements are required in order to establish characteristic concentrations for each source. The volumes required to do this may be large. Second, a determination of recovery efficiency should be a routine part of the weekly pathogen sampling in the watershed. Although the 30 percent to 70 percent recovery efficiencies measured by New York City are possible given the purity of the source water, other studies suggest that recovery efficiencies are substantially lower (perhaps as low as 5 percent) (Clancy et al., 1994; J. Ongerth, University of South Wales, personal communication, 1999; Shepherd and Wyn-Jones, 1995), and this should be investigated further by NYC DEP. Third, New York City compares counts of "confirmed" and "presumed" (oo)cysts, based on the microscopic observation of internal cell structures. This method has not been proven to be a true measure of pathogen viability. Finally, in the documentation examined, there is little evidence of determining comparability between laboratories based on the use of split samples (NYC DEP, 1997d; Stern, 1998).

Fortunately, there is ongoing work in methods development for pathogen detection (Clancy et al., 1997; Falk et al., 1998), and New York City should take a higher profile in testing and using these approaches in order to address the deficiencies described above. Some of the new techniques, such as cell culture (Slifko et al., 1997), offer more promise in terms of increased sensitivity, specificity, and assessment of viability. For example, most current viability work on oocysts has used infectivity as an endpoint, and there is a growing sense that tissue culture methods may represent a more realistic approximation of viability. The use of these alternative methods in parallel with existing methods should allow New York City to test more samples with better precision (i.e., reduced variability among split samples) and accuracy (i.e., improved recovery and specificity of viable human infectious organisms). The City's involvement in the Environmental Protection Agency's (EPA) Supplemental Information Collection Rule Survey and evaluation study of the newly proposed Method 1622 for *Cryptosporidium* should be of considerable benefit in improving recovery efficiencies and increasing the number of positive (nonzero) samples (EPA, 1999).

Source of Pathogens. As discussed in greater detail in Chapter 12, the primary goals of the NYC DEP pathogen monitoring program should be to (1) determine sources of pathogens in the Catskill/Delaware watershed and develop quantitative source terms and compare the relative contributions of different sources or catchments, (2) measure the effect of best management practices on pathogen loadings, and (3) direct resources to the most polluting sources. In support of the first two goals, over 5,000 pathogen samples have been collected to date from watershed sites that are categorized as urban, agricultural, or undisturbed and from the effluent of WWTPs. Viruses have been detected in only five percent of all samples, while *Giardia* and *Cryptosporidium* are more common. *Giardia* is detected most frequently in the effluent of WWTPs (21 percent of all samples), followed by agricultural, urban, and undisturbed areas. *Cryptosporidium* is found most frequently in urban samples (18 percent), followed by agricultural areas, WWTP effluent, and undisturbed areas. Concentrations of these parameters vary widely, but average about 20 *Giardia* cysts/100 L and 1 *Cryptosporidium* oocyst/100 L. Although the data are limited in terms of their usefulness when presented as percent positive, the (oo)cyst concentration data will eventually be needed for evaluation purposes and should continue to be collected.

The pathogen studies have been complemented by attempts to pinpoint actual animal or human sources of pathogens. NYC DEP and Cornell University have investigated both farm animals and wildlife (more than 6,000 specimens) for infection with *Giardia* and *Cryptosporidium*. To date, the prevalence of *Cryptosporidium* in farm animals and wildlife varies between 0.66 percent and 2 percent (NYS WRI, 1997). Research results also indicate that previously infected animals are more likely to become reinfected than previously uninfected animals (NYS WRI, 1997). There are currently no wildlife population data available to assess the relative contributions of different animal source terms to the overall pool of pathogenic protozoa in the Catskill/Delaware watershed. A study to determine the population of mammals with high protozoan infection rates has recently been initiated (NYC DEP, 1999a).

Although progress has been made, current efforts only minimally address the goals described above. In addition to identifying sources, the development of quantitative source terms (e.g., (oo)cysts per animal per time, or (oo)cysts per area per time) is greatly needed, including a determination of the variability of these source terms. Initially such terms could be developed for whole catchments by sampling reservoir effluents, with the eventual goal of developing such terms for urban, agricultural, and wildlife sources. It might be useful for NYC DEP to review and expand upon other attempts to "mass balance" a watershed for protozoa and to quantitatively compare the results in the Catskill/Delaware watershed with other investigations (Hansen and Ongerth, 1991; Ong et al., 1996; Ongerth, 1989).

A first approach to this problem for the Catskill/Delaware watershed has just been published by Walker and Stedinger (1999). The authors utilized local data on WWTP effluent volumes, farming and dairy practices, and dairy populations along with literature values of oocyst levels in sewage effluents and manures and decay rates of organisms under various land management practices. These data were used in watershed runoff models and reservoir water quality models to determine the impact of different oocyst sources on reservoir water quality in the New York City watershed. Although there are a number of data gaps and limitations in their analysis (which could alter the details of their conclusions), the study indicated that the contribution of oocysts in sewage far exceeded the contribution of oocysts in runoff emanating from dairy farms. As additional data are collected, it would be useful for NYC DEP to update and revise this approach and also to use this framework as a method for locating key sampling locations where model validation could occur.

Other Microbial Parameters. It is somewhat surprising that in an ongoing activity of the scope being carried out by New York City, only relatively few microbial parameters are under investigation. There are a number of other pathogens and indicator organisms that should also be considered, as has been suggested previously (ILSI, 1998). *E. coli*, a more fecal-specific indicator than fecal coliform, should be regularly measured at all locations where fecal coliform is sampled. (*E. coli* is regularly measured in distribution system samples that are positive for total coliform.) Use of *Clostridium perfringens* as a water quality indicator, initially developed by Cabelli (1977), may also have merit. The relative insensitivity (Payment and Franco, 1993) of clostridia spores to die-off may make it a most advantageous indicator to track human and animal waste movement in the watershed. It has been used in this regard to track septic discharges in the marine environment (Paul et al., 1995).

The committee anticipates that with the development of techniques founded on modern molecular biology, there will be a number of different and complementary approaches to address source attribution. New York City should keep abreast of these technologies and become early adopters of such methods. In particular, work of Sobsey and colleagues indicates that genotyping F[+] coliphage may provide a useful means of distinguishing between animal and human sources of contamination with specific typing indicators (Hsu et al., 1995). NYC DEP has been measuring for fecal streptococci in the Kensico Reservoir to help determine the relative contributions of human and avian sources of fecal bacteria to this important reservoir. It is not clear from available reports (NYC DEP, 1999b) whether the ratio of fecal coliform to fecal streptococci is being calculated to make determinations about bacterial sources. If so, this practice should be discontinued because this ratio is useless (APHA, 1998, pp. 9-74 to 9-75).

Other organisms that should also be considered for microbial monitoring include cyanobacteria capable of producing toxins. In mesotrophic and particularly eutrophic reservoirs, cyanobacterial blooms are common and during these blooms, population growth of other phytoplankton can be strongly suppressed. For example, the hydroxamate sidechrome compounds released by cyanobacteria to scavenge iron have growth-suppressing effects on algae. Certain cyanobacteria, particularly certain strains of *Microcystis aeruginosa,* produce microcystins and nodularins, a series of some 40 very toxic cyclic heptapeptide compounds. Effects of these compounds in humans are numerous and include gastroenteritis, liver damage, nervous system damage, pneumonia, sore throat, earache, and contact irritation of skin and eyes (Falconer et al., 1983; Carmichael, 1986; Codd et al., 1989; Turner et al., 1990; Harada et al., 1996; Ueno et al., 1996). Cyanobacterial toxins have been implicated in waterborne disease outbreaks in Pennsylvania (Carmichael et al., 1985) and Australia (El Saadi et al., 1995; Hayman, 1992; Soong et al., 1992). Although the potential chronic health effects of long-term exposure to cyanobacterial toxins in drinking water are unknown, the World Health Organization has recently proposed a provisional guidance value of 1 µg/L for the microcystin-LR microtoxin for the protection of drinking water (World Health Organization, 1998).

Flow Proportional Monitoring

This chapter has repeatedly stated that fixed-frequency sampling for both chemical and microbial parameters will not capture representative values for pollutant loading from streams, diffuse subsurface sources, and WWTPs. Support for this statement is abundant among limnological and environmental surveying studies. Box 6-1 discusses the inadequacies of fixed-frequency sampling specifically for pathogenic microorganisms. Numerous examples exist for other types of pollutants, particularly those in the particulate phase. Particulate-phase pollutants tend to increase in concentration during storm events because of the increased soil and sediment erosion associated with high flows.

Data collected from the Cannonsville watershed strongly support the superiority of event-based sampling over fixed-frequency sampling. Intensive sampling of the West Branch of the Delaware River during both base flow and storm events demonstrated that stream flow during storm events contains significantly higher concentrations of several pollutants than base flow, including all measured forms of phosphorus, total suspended solids, and total ammonia nitrogen (Longabucco and Rafferty, 1998). In addition to increased pollutant concentrations, storm events were also found to deliver the bulk (from 55 percent to 95 percent) of the annual load of almost all measured parameters. The study demonstrated that sampling based on a fixed frequency would considerably underestimate pollutant loading to the Cannonsville Reservoir via the West Branch of the Delaware River.

The extent of the discrepancy between fixed-frequency and event-based sampling regarding pollutant loadings depends on the particular pollutant, the length of the storm event, and the interval of fixed-frequency sampling. In general, errors in fixed-frequency sampling seem to increase nearly exponentially as the time interval between sampling events increases. The case study in Box 6-2 demonstrates the effect of increasing the frequency of streamflow sampling in three subwatersheds in Connecticut.

State-of-the-art monitoring that includes event-based sampling can substantially improve the effectiveness of watershed management. Fortunately, event-based sampling can be partially automated and coupled with nearly continuous automated monitoring of certain parameters (dissolved oxygen, pH, temperature, conductance, turbidity), although the initial expenses are high. Data from some automated monitoring can be downloaded by radio transmission to central processing stations. However, in other cases, collection and analysis of samples from proportional collectors requires personnel maintenance based on precipitation and snowmelt periods, which are unpredictable. A commensurate database management system is needed to make full and timely use of new, more detailed information about water flow and quality. Otherwise the volume of data generated by automated equipment can quickly overwhelm watershed managers and simply be archived instead of being actively used to support field operations. In

BOX 6-1
Storm Event Sampling of Pathogenic Protozoans

At the Metropolitan Water District of Southern California, use of innovative grab and "first flush" samplers was found to yield a substantially
greater frequency of positive hits than the current Information Collection
Rule (ICR) methodology (see Table 6-3). Although some of this may be
due to the poor recovery associated with the classical ICR methods
(Clancy et al., 1994), a substantial cause of this difference may be due to
the "spiky" nature of oocyst (and cyst) prevalence in surface waters, particularly during storm events.

This finding is further supported by work of two other research groups.
LeChevallier et al. (1997) conducted an intensive sampling campaign on
a watershed and reported relationships between flow rates, turbidity

TABLE 6-3 Results of Different Types of Sampling Methods on
Isolation Frequency for Protozoans in Water Samples

	Sampling Method					
	Filter (ICR)[a] N=87		First flush[b] N=20		Grab[c] N=21	
Protozoan	% Positive	Range[d]	% Positive	Range[d]	% Positive	Range[d]	Total
Crypto	10	3–415	35	46–41,666	19	3.4–647	16
Giardia	29	2–119	60	25–16,666	19	42–2,428	32
Crypto or *Giardia*	28		45		19		38
Crypto and *Giardia*	6		25		10		9

[a]Sampled using current EPA methods (large-volume samples) during normal meteorological and flow conditions.
[b]Small-volume samples were collected during the periods of rapid initial flow increase following storm events.
[c]Small-volume samples were collected during normal meteorological and flow conditions.
[d]Range measured in 100 (oo)cysts /100L.
Source: Stewart et al. (1997). Reprinted from *Proceedings of the 1997 AWWA Water Quality Technology Conference,* by permission. Copyright © 1998, American Water Works Association.

continued

BOX 6-1 Continued

spikes, and peak levels of oocysts. As shown in Figure 6-1, peak oocyst levels may be found in proximity to peak flow and turbidity levels, in at least certain watersheds. Findings of protozoan peaks during wet weather periods are in accord with well-established findings with respect to indicator organism occurrences in runoff flows (Geldreich, 1978).

The "flashiness" of oocyst levels indicated by these two investigations shows that for the purposes of estimating the total loading of oocysts (number/day) to a waterbody, or estimating the total flux past a location, it is necessary to obtain samples during storm events (event-based sampling). For example, had samples not been collected during days 107–109 in Figure 6-1 (the storm event period), an underestimate of the maximum range of oocyst levels (and thus the flux) would have been made.

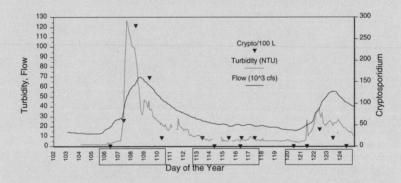

FIGURE 6-1 Variation in turbidity, river flow, and *Cryptosporidium* levels during a spring sampling campaign. Boxes highlight days on which samples were tested for *Cryptosporidium*. Source: LeChevallier et al. (1997). Reprinted from *Proceedings of the 1997 AWWA Water Quality Technolgy Conference,* by permission. Copyright ©1998, American Water Works Association.

BOX 6-2
Integration of Fixed-Interval and
Event-Based Water Quality Sampling:
An Example from Current Watershed Research

This case study provides an example of how rapid advances in field, laboratory, computing, and communications equipment during the last decade have transformed the idea of event-based watershed monitoring from a logistically and financially impossible dream to an operationally feasible, albeit challenging, goal. This example shows part of the water quality sampling system used for the New Haven Watershed Project, which has a hydrogeologic setting and climate and watershed characteristics roughly comparable to the watersheds of the New York City water supply system.

Table 6-4 illustrates the effect of sampling frequency on the information content of water quality measurements. Stream sampling of three subwatersheds (rural, intermediate, and urban) was accomplished using an automated probe connected to a data logger. Four parameters—discharge, turbidity, dissolved oxygen, and nitrate-nitrogen—were compiled at 20 minutes, 24 hours, 7 days, and 14 days. The original 20-minute time interval was resampled to simulate the daily, weekly, and biweekly data collection.

As expected, the mean, standard deviation, maximum, and minimum reflect differences in the three classes of watershed and sampling frequency. Most notable are the maximum values for discharge and turbidity for all three sites. At the 20-minute time interval, discharge from the urban watershed exhibits the flashy response commonly associated with impervious cover. However, the observed response is an order of magnitude lower when data are collected at 7- and 14-day intervals. Similar trends are exhibited by turbidity data for all three subwatersheds.

Also note the difference between decreases in mean and maximum values as the sampling interval becomes larger and sample size is reduced ($N = 1,800, 25, 4$, and 2 for 20-minute, 24-hour, 7-day, and 14-day sampling intervals, respectively). If, by chance, the rising limb of a major storm event was sampled during routine biweekly data collection, it is likely to be eliminated as an outlier or influential case during subsequent statistical analyses. Turbidity is directly correlated with suspended particulate matter including pathogens and nutrients adsorbed to sediment particles. Hence, the ability to detect and respond to water quality problems may be seriously limited by the present monitoring system.

continued

Box 6-2 Continued

TABLE 6-4 A Comparison of Stream Flow and Selected Water Quality Data (June 4–29, 1998) at Four Sampling Frequencies for Three Sub-watersheds of the Quinnipiac and West Rivers near New Haven, Connecticut

Sampling Interval	Stats	Discharge (mm/day)			Turbidity (NTU)		
		Rural	Inter.	Urban	Rural	Inter.	Urban
20 mins.	Mean	1.6	1.8	4.2	2.4	8.2	7.2
	SD	0.6	0.8	2.3	4.4	9.5	3.8
	Max	5.2	6.1	38.1	80.9	93.6	99.3
	Min	0.8	0.9	1.6	0.6	0.6	3.8
24 hours	Mean	1.6	1.9	4.4	1.9	7.7	7.0
	SD	0.6	0.9	2.6	1.2	7.8	1.6
	Max	3.0	4.1	14.3	6.0	37.9	11.1
	Min	0.9	1.1	2.2	0.8	1.0	5.0
7 days	Mean	1.4	1.5	4.3	1.8	4.6	6.8
	SD	0.4	0.2	1.7	0.7	3.4	1.6
	Max	1.7	1.7	6.8	2.6	8.0	8.2
	Min	1.0	1.4	3.0	1.0	1.2	5.1
14 days	Mean	1.68	1.6	5.4	1.4	4.6	6.9
	SD	0.02	0.1	1.9	0.6	4.8	1.8
	Max	1.69	1.7	6.8	1.8	8.0	8.2
	Min	1.67	1.5	4.1	1.0	1.2	5.7

Notes: General land cover statistics: RURAL subwatershed (#1050, 5.00 km^2) is 91 percent forest, 4 percent developed, 3 percent fields, and 2 percent water and wetlands; INTERMEDIATE subwatershed (#2020, 5.21 km^2) is 70 percent forest, 20 percent developed, 7 percent fields, and 3 percent water and wetlands; URBAN subwatershed (#1005, 3.92 km^2) is 1 percent forest, 85 percent developed, 6 percent fields, and 8 percent water and wetlands. The 20-minute interval data were generated with automated probes. The 24-hour, 7-day, and 14-day data were resampled from the 20-minute interval set.

addition, a backup methodology is needed because failure of automated systems can cause gaps in data needed for predictive modeling. For these reasons, flow proportional monitoring can require a substantial investment of personnel and monetary resources.

Samples collected by flow proportional monitoring must be properly preserved and retrieved in accordance with the vulnerability of the assays being

D.O. (mg/L)			NO_3–N (mg/L)		
Rural	Inter.	Urban	Rural	Inter.	Urban
7.3	9.0	5.7	0.19	1.15	1.24
1.3	1.4	1.8	0.05	0.29	0.23
11.3	11.4	11.0	0.27	1.53	1.70
5.1	6.4	2.9	0.11	0.47	0.74
7.2	9.1	5.3	0.11	1.16	1.25
1.3	1.4	1.6	0.05	0.32	0.24
8.9	10.9	8.7	0.27	1.53	1.70
5.2	6.6	3.2	0.11	0.47	0.74
7.3	9.2	5.6	0.19	1.28	1.27
1.4	1.7	2.2	0.06	0.17	0.19
8.4	10.9	8.8	0.27	1.51	1.42
5.4	7.0	3.5	0.12	1.13	1.00
7.6	9.8	4.4	0.14	1.15	1.13
1.1	1.5	1.2	0.03	0.02	0.18
8.4	10.9	5.3	0.16	1.16	1.25
6.9	8.8	3.5	0.12	1.13	1.00

performed. For example, samples collected for phosphorus analysis should be stored on ice and processed for analyses within 8–12 hours of collection. The use of flow proportional sampling for microorganisms (indicators such as coliforms and pathogens such as *Cryptosporidium*) is technically feasible with the use of refrigerated field samplers. However, there are a number of logistical and economic issues that may limit the applicability of this approach. For example, the storage of samples for more than 24 hours is not generally acceptable, and so field samplers would need to be serviced on a daily basis. Sources of electric power (for refrigeration) for the sampler would be required. There are approaches that have been used for "first-flush" collection devices that may be attractive for use at a limited number of sites (Stewart et al., 1997).

It is suggested that fixed-frequency sampling be continued for a reasonable length of time (minimally one year) after the proportional sampling system is operational. By this means, interpretation of old fixed-frequency sampling data can be analyzed for evaluations of comparative reliability. In some cases, where the relation between discharge and the water quality constituent is well established, it may be reasonable to measure some parameters at a longer time interval. Once a sufficient concurrent record of fixed-frequency and event-based monitoring exists, a wide range of advanced statistical techniques can be used for retrospective modeling and analyses of historical data (Clarke, 1994; Haan, 1977; Hosking and Wallis, 1997).

For some time, NYC DEP has recognized the need to include event-based sampling in their water quality monitoring program (NYC DEP, 1997a), and its importance has also been repeatedly stressed by other sources (ILSI, 1998; Longabucco and Rafferty, 1998). Event-based sampling is on the rise in the streams of the New York City watersheds. Between April 1995 and April 1996, storm event sampling was conducted in a tributary of the Cannonsville Reservoir to determine nutrient loading for calibration and validation of water quality models. Samples from 18 storm events were collected and analyzed. In 1996, turbidity and suspended solids were measured during four storm events in the Ashokan basin, including the Esopus Creek, which is known to contribute to turbidity in the Catskill System. This increased to sampling eight events at 25 sites in 1998. To support further development of reservoir models, storm event sampling for nutrients is being conducted in all of the major tributaries west of the Hudson River. Finally, streams in the Kensico Reservoir basin are being sampled intensively for coliforms and turbidity, as mentioned below under the special section on the Kensico Reservoir. It is hoped that the efforts described above will lead to the rapid establishment of automated (when possible) event-based sampling, especially for the Kensico watershed and for those reservoir basins with impaired water quality (such as Cannonsville).

Performance Monitoring

In many water supply systems, compliance and operational monitoring make up the vast majority of all monitoring activities. However, in large, complex systems in which watershed land use varies, additional monitoring is needed to determine the effectiveness of management practices and policies for reducing nonpoint source pollution. For the purposes of this report, this additional monitoring is roughly classified as performance monitoring.

Performance monitoring is usually conducted on a limited basis to determine the effectiveness of a practice or a technology. It often consists of multiple data sets that are analyzed by conducting a mass balance around the particular practice or area of interest. Performance monitoring can occur over a wide range of scales, from a single farm best management practice (BMP) to an entire reservoir or watershed. Because of natural variability and the time lag between BMP implementation and system response, substantial time may pass before performance monitoring can detect changes in water quality around large (watershed-scale) areas. Performance monitoring results accrue more quickly as the monitored area diminishes in size. The types of performance monitoring that are most important for the New York City watersheds are discussed below, with indications as to whether this monitoring is taking place. Suggestions for improved performance monitoring are found in this chapter and in the discussion of individual programs in Chapters 8–10.

Shallow Subsurface Monitoring

Perhaps the most important type of performance monitoring for a source water protection program is monitoring of shallow subsurface flow. This monitoring should occur in multiple locations throughout the watershed, but it is particularly important in the riparian zones surrounding the reservoirs and major tributaries. These areas are critical for sequestering nonpoint source pollution from stormwater runoff and are often the sole barrier of protection to nearby waterbodies. The goal of this performance monitoring is to determine the retention efficiencies of the shallow subsurface for a variety of pollutants. Such analyses should be performed under a spectrum of differences in soil types, vegetation cover, precipitation, elevational gradients, and topography. In addition to riparian zones, performance monitoring of the shallow subsurface should be conducted around on-site sewage treatment and disposal systems (OSTDS) and their drainfields, above and below certain agricultural BMPs, and above and below urban stormwater BMPs.

Although some performance monitoring of the shallow subsurface has occurred in the New York City watersheds, it has been relatively infrequent. Given data on groundwater flow rates in the Kensico watershed, there is a general assumption that subsurface flows and their attendant pollutant loadings to the reservoirs are minor in comparison to surface (river) runoff. Such assumptions are likely invalid given quantitative studies in regions of similar geology and topography [e.g., Mirror Lake ecosystem of Hubbard Brook, NH, studies (Likens, 1985; Likens and Bormann, 1995)]. These studies, which are based on real data, not on models, have shown that composite pollutant loadings cannot be based on river discharge volumes alone.

To our knowledge, ongoing performance monitoring of the shallow subsurface is associated with five main activities in the New York City watersheds.

Eight of the groundwater monitoring wells in the Kensico watershed are being used to determine whether exfiltration from sewer lines is affecting water quality in the shallow subsurface (NYC DEP, 1999a). Monitoring of total and fecal coliforms from these wells has not detected significant increases in these parameters over other groundwater wells in the watersheds, suggesting that exfiltration of sewage is not important. The Septic Siting Study, which is designed to determine the fate and transport of pathogens introduced into septic systems, is measuring pollutant concentrations in septic system drainfields and control plots at six locations (two in each watershed). A more detailed description of this MOA-mandated study is found in Chapter 11. Researchers at the U.S. Geological Survey are determining how pollutant loadings in shallow subsurface flow respond to a complete removal of forest vegetation in a small tributary of the Neversink. Cornell University research in support of the Watershed Agricultural Program has measured phosphorus concentrations in shallow subsurface affected by manure application. And finally, performance monitoring of wetlands east of the Hudson River (in the West Branch and Boyd's Corner watersheds) is planned for the near future (NYC DEP, 1999a). Phosphorus, suspended solids, dissolved organic carbon, and other parameters will be measured in the inflow and outflow of 9–15 reference wetlands and in the shallow subsurface. These activities should continue and be supplemented by more wide-scale performance monitoring of riparian buffer zones, septic system drainfields (particularly in the Kensico watershed), and the shallow subsurface above and below agricultural and urban stormwater BMPs.

Other Performance Monitoring

The shallow subsurface is not the only important target of performance monitoring. Any individual BMP is a logical candidate for performance monitoring, particularly BMPs that have not been used extensively or for which there are no published pollutant removal rates. The 44 urban stormwater BMPs that are being installed in the Kensico watershed (described in Box 6-3) should undergo such performance monitoring subsequent to their installation. Performance monitoring can also be conducted on a larger-scale, in order to assess the "performance" of a region or subwatershed. This larger-scale strategy is being used by NYC DEP in monitoring agricultural areas. Two locations, the Robertson farm and Shaw Road (a control site), are undergoing stream monitoring to compute the difference in pollutant loading emanating from these sites. Similarly, a comparison of conditions at six pristine or disturbed wetland sites located partially within the Croton watershed is being conducted to determine the relationship between macroinvertebrate communities and wetland characteristics (NYC DEP, 1999a).

Performance monitoring does not necessarily require the acquisition of new data. In some cases, data collected during compliance and operational monitoring can be used to determine the effectiveness of a pollution reduction scheme.

BOX 6-3
The Kensico Reservoir and Watershed

About 90 percent of the total New York City water supply is derived from the City's six West-of-Hudson reservoirs in the Catskill Mountains and the Upper Delaware River Basin (the Catskill/Delaware system). Water from those reservoirs is delivered via the Catskill and Delaware aqueducts, which cross beneath the Hudson, to a common destination at Kensico Reservoir, 15 miles north of New York City. There, water from both aqueducts mixes and, after 15–25 days of storage time, it is chlorinated and delivered to the City and other user communities. The state of water quality in Kensico Reservoir is thus critical: if it becomes degraded, it would contaminate the high-quality water derived from the other Catskill/ Delaware reservoirs. Although both aqueducts are designed to permit bypassing of Kensico, bypassing is undesirable because the settling time, mixing, and flow control that Kensico provides would be lost.

Kensico Reservoir, with a capacity of 30 billion gallons, was placed in service in 1915 with the damming of the Bronx River. Although most water in Kensico Reservoir is delivered by the two trans-Hudson aqueducts, it also receives runoff from its immediate drainage area of 13.1 square miles. Largely forested, this area contains about 1,500 dwelling units, a number of office parks and other commercial facilities, as well as transportation facilities. The latter include Westchester County Airport, Interstate 684, and state highway 22, which borders the reservoir and crosses it at one point. Stormwater from each of these facilities and structures drains into Kensico.

Recent violations in stormwater runoff quality from the airport, the proposed widening of Routes 22 and 120, and new office construction have made the Kensico watershed the subject of much discussion (Marx and Goldstein, 1999). These projects pose serious threats to water quality in Kensico Reservoir if stormwater management is not undertaken to counteract associated increases in pollutant loading. In particular, wetlands bordering the reservoir will likely be removed to accommodate road-widening, thereby reducing the ability of the surrounding buffer land to diminish pollutant loading from stormwater. If such projects go forth, stormwater management must (1) control all new pollutant loadings, (2) compensate for losses in nonpoint source pollutant control via wetland functioning, and (3) mitigate current problems with runoff from the Westchester County Airport.

To confront these challenges, NYC DEP has developed the Kensico Stormwater Management Plan, which currently includes 44 stormwater facilities in 16 subbasins to reduce loadings of turbidity and fecal coliform bacteria (and associated pollutants) to the reservoir. The stormwater

continued

BOX 6-3 Continued

BMPs will be constructed first in the vicinity of Malcolm Brook and Young Brook, which are major tributaries to the reservoir near the Catskill aqueduct effluent chamber. All facilities should be operational by the year 2000. Performance monitoring of these BMPs and the shallow subsurface around the airport and the proposed road widenings should be one of NYC DEP's highest priorities.

In addition to stormwater runoff control, several other preventive measures are being taken to reduce the impact of watershed activities on Kensico reservoir water quality. An 850-ft turbidity curtain was installed in 1995 at the mouths of Malcolm and Young brooks to prevent direct transit of sediment to the Catskill aqueduct effluent chamber. So far it has been effective in preventing water at that effluent chamber from exceeding 5 NTU. As mandated by the MOA, maintenance dredging is occurring in the Kensico Reservoir to remove accumulated sediments near both aqueduct effluent chambers. Approximately 980 cubic yards of sediment will eventually be removed, dewatered, and shipped to a facility in Ohio.

Although there are no wastewater treatment plants in the Kensico watershed, the area is crossed by various sewer lines leading to treatment plants elsewhere. NYC DEP is conducting video inspection of sewers to find and repair leaks. As described elsewhere in this report, groundwater monitoring has not detected any exfiltration of sewage from these pipes to the surrounding subsurface. There are approximately 580 septic systems in the watershed, of which 210 lie within 300 ft of the reservoir or 100 ft from tributary streams. NYC DEP, as mandated by the MOA, is responsible for detecting failing septic systems within the watershed and in assisting those using septic systems to connect to sewers where feasible.

For example, operational monitoring data could be used to determine the reduction in pollutant concentration between the influent and effluent of an entire reservoir. Although there appears to be a large database taking shape that would permit this type of analysis, the only known attempts have been conducted for the West-of-Hudson reservoirs using percent detection data instead of concentration and flow data (Bagley et al., 1998; Stern, 1996). If NYC DEP wants to conduct this type of large-scale mass balance with accuracy, data need to be collected over long time periods and during storm events to reduce both variability and the risk of systematic sampling bias.

Because of the serious consequence of hazardous waste spills within the watershed, NYC DEP has developed a Kensico Spill Response Program. Four incidents in 1998 prompted action under the program; none were found to negatively affect water quality (NYC DEP, 1999a). Part of the success of this program is dependent on alert residents who notified NYC DEP of potential problem activities. Citizen involvement in monitoring and preventing pollution in the watershed is the backbone of the Kensico Environmental Enhancement Program (KEEP), a joint effort of NYC DEP and community groups. In 1998, KEEP held workshops on environmental protection, it patrolled the reservoir and its perimeter, and it distributed educational information to local residents suggesting how they can help protect water quality in the Kensico Reservoir.

These activities and others (such as the gull mitigation efforts discussed in Chapter 5) contribute to high water quality in the Kensico Reservoir and subsequently in the finished drinking water. This is demonstrated by the fact that the Kensico Reservoir has not been bypassed since 1994 (Delaware aqueduct) and 1995 (Catskill aqueduct). Because West Branch Reservoir is similarly positioned within the Catskill/Delaware system, these special types of activities should be considered for possible implementation in the West Branch watershed as well. Under normal conditions, water from the Delaware aqueduct (55 percent of the total City supply) passes through the West Branch Reservoir before reaching the Kensico Reservoir. As discussed in Chapter 2, population (and, by association, development) is increasing more rapidly in the West Branch watershed than in the Kensico watershed, highlighting the importance of extra protection efforts for West Branch.

Monitoring in Support of Modeling Efforts

Water quality parameters vary significantly among reservoirs and their tributaries. Spatial variations arise from differences in the drainage basin characteristics of physiography, land use practices, and meteorology. In addition, water quality within individual reservoirs is uncertain because of variable basin morphometry, natural and managed hydrology, climatic conditions, and fluctuations in pollutant loadings. As noted earlier, temporal variations in water quality typically occur over small time scales. One of the purposes of water quality modeling is to follow and predict these patterns of spatial and temporal variability in water quality parameters. If successfully designed, calibrated, and verified, water quality models can lead to reduced monitoring requirements over the long term.

New York City has developed and is using a large number of terrestrial runoff and water quality models for both the Catskill/Delaware and Croton reservoirs. Table 6-5 lists some of the models being used by New York City and the data requirements for each. The table includes models of reservoir water hydrodynamics and quality, terrestrial models of surface water runoff and nonpoint source pollution, and precipitation and atmospheric deposition models. A long-term goal of these NYC DEP modeling efforts is to link models representing different processes into one integrated model of watershed functioning. It should be noted that pathogen models and agricultural models (as suggested earlier in this chapter) are not yet being developed to interface with the current runoff and water quality models. Development of pathogen fate and transport models will be critical to moving beyond the simplified presence/absence evaluation that characterizes the NYC DEP pathogen studies.

This report does not review and critique each water quality model being used by New York City. However, when a model is being used to comply with a regulatory program, the quality of the data used as input to the model must be taken into consideration. Thus, the report addresses the quality of the data required for calibration and verification of the Generalized Watershed Loading Function (GWLF), the Vollenweider model, and the Reckhow model, all of which are used for the Total Maximum Daily Load (TMDL) Program. Because a significant modeling effort is necessary to calculate TMDLs, the data acquired to support these models must be of sufficient quantity and high quality. In addition, the report reviews the hydrothermal and nutrient/phytoplankton models being considered for possible use during the Phase III TMDL process. These discussions are found in Chapter 8.

Monitoring of the Kensico Watershed

As the collection point for Catskill/Delaware system water prior to chlorination and distribution, the Kensico Reservoir is critical in controlling the quality of the New York City drinking water supply (Figure 6-2). The residence time afforded by travel through the Kensico Reservoir allows particulate phase pollutants to settle, and it supports greater degradation or inactivation of microbial pathogens. Because polluted runoff and other contamination from the Kensico watershed (and to a lesser extent the West Branch watershed) has the potential to adversely affect the quality of water emanating from the six West-of-Hudson reservoirs, a significant fraction of all monitoring efforts and pollution prevention activities are conducted in Kensico. Box 6-3 describes this unique reservoir and watershed in more detail, discussing the special activities being undertaken to

TABLE 6-5 Selected Water Quality Models Used in New York City

Model Name	Model Type	Data Requirements
Reckhow Model	Surface Runoff Water Quality	• Stream and WWTP flows • Average pollutant concentrations in stream flows (for validation), WWTP effluents • Land cover/land use • Number of septic systems • Export coefficients for various nonpoint sources • Soil retention coefficient
Generalized Watershed Loading Function (GWLF)	Surface and Subsurface Runoff Water Quality	• Precipitation • Air temperature • Stream and WWTP flows • Average pollutant concentrations in stream flows, WWTP effluents • Septic system pollutant loads • Land cover/land use • Mean elevation and slope • Soil chemistry data
Vollenweider Model	Reservoir Water Quality	• Pollutant concentration in reservoir • Retention rates for pollutants in reservoir • Reservoir residence time • Mean reservoir depth
Hydrothermal Model	Reservoir Water Quality	• Meteorological data[a] • Geometric data (reservoir topography) • Stream flows, all outflows, and water surface elevation • Reservoir temperature profiles with depth, stream temperatures, aqueduct temperature • Hydrodynamic and kinetic coefficients • Structural details
Eutrophication Model (also known as the nutrient/ phytoplankton model)	Reservoir Water Quality	• Meteorological data[a] • Stream flows, all outflows, and water surface elevation • Reservoir temperature profiles with depth, stream temperatures, aqueduct temperature • Light extinction • Average pollutant concentrations in the epilimnion and the hypolimnion, in stream flows, and in outflows

[a]Meteorological data include air temperature, dew point temperature, wind velocity, and cloud cover (or solar radiation).
Sources: Doerr et al. (1998), NYC DEP (1998d), and Owens (1998).

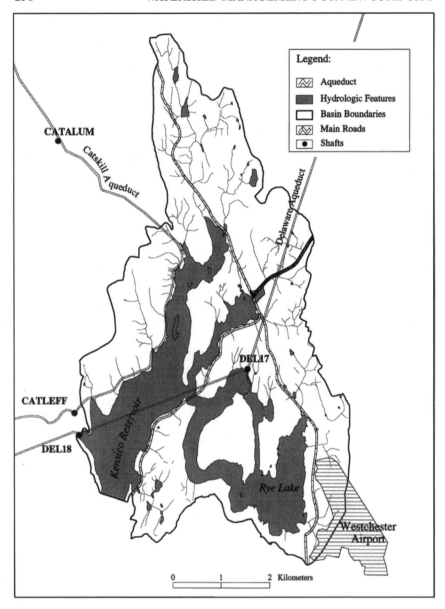

FIGURE 6-2 Kensico Reservoir and watershed. Courtesy of the NYC DEP.

identify, quantify, evaluate, and mitigate sources of chemical and biological pollution within the Kensico watershed that affect its water quality.

Monitoring in the Kensico basin is intensive and frequent and encompasses some 35 physical, chemical, and biological water quality parameters. A significant quantity of all compliance and operational monitoring for the Catskill/Delaware system occurs in the Kensico watershed because of its proximity to the distribution system. Performance monitoring also takes place at a high frequency in the Kensico watershed in comparison to the other New York City watersheds. For example, Kensico is the only watershed in which groundwater monitoring is conducted on a regular basis. Table 6-6 lists all of the special monitoring activities occurring in the Kensico watershed beyond the compliance monitoring required by the Surface Water Treatment Rule (SWTR) and the standard operational monitoring that takes place in all Catskill/Delaware reservoir basins.

Given its influence on the quality of drinking water in New York City, Kensico Reservoir and its watershed should continue to be the focus of intensive monitoring by NYC DEP. As discussed in Box 6-3, the most serious local water quality threat to the reservoir appears to be pollutant-laden stormwater from roads, bridges, and the Westchester County Airport. Monitoring the performance of urban stormwater BMPs in the vicinity of these pollution sources should be one of NYC DEP's highest watershed management priorities.

Conclusions and Recommendations for the Monitoring Program

The complexity of the multiple interacting reservoir ecosystems of the New York City water supply imposes major monitoring demands to allow for effective management and responses to problems. In general, NYC DEP has been performing these formidable tasks excellently. Several recommendations for improvement are made below.

1. Monitoring should be conducted on the basis of discharge-mediated volume rather than on fixed-frequency intervals for stream, shallow subsurface groundwater, WWTP effluent, and precipitation analyses. Event-based or flow proportional sampling is needed to capture rapid variation in water flow and quality. Some parameters of sampling can be automated, but care is needed to avoid problems associated with storage of samples collected by flow proportional sampling.

2. Shallow subsurface and groundwater parameters should be monitored regularly throughout the reservoir watersheds. This is particularly important given the prominent role of agriculture in some areas, the high density of OSTDS, and the potential for leaking sewer lines through the watersheds. Biannual sampling is sufficient for deep groundwater but not for shallow subsurface monitoring. Such routine groundwater sampling should be integrated with

TABLE 6-6 Special Monitoring Efforts in the Kensico Watershed Beyond Regular Compliance and Operational Monitoring[a]

Type of Monitoring	Parameters Measured	Frequency of Sampling
Reservoir	• Sediment cores and sediment samples: metals, pesticides, polyaromatic hydrocarbons, radioisotope dating	• One-time events in several locations
	• Fecal streptococci	• NA
Stream	• Flow • Fecal coliforms • Protozoan pathogens • Turbidity • Suspended solids • Total phosphorus	• Event-based
	• Sediment and river water sampled for pesticides and hydrophobic organic compounds	• June 1997, June 1998
Groundwater/ Shallow Subsurface	• Total and fecal coliforms • Physical parameters • Nutrients	• Biannual
	• Synthetic organic compounds	• One time event with followup
Wetlands in the Malcolm Brook Subwatershed	• Dye test study to determine travel time through wetlands	• Biannual
Bird Populations and Bird Fecal Matter	• Fecal coliforms • Fecal streptococci	• Weekly • NA
Forest Regeneration Study	• Seedling survival	• Biannual between 1995 and 1997

Notes: Entries indicate both aquatic and terrestrial monitoring efforts beyond those made for the other basins in the Catskill/Delaware system.

NA = not available

[a] Compliance monitoring is indicated in Table 6-1 and operational monitoring is discussed previously in the chapter.

Source: NYC DEP (1997a).

direct experiments (performance monitoring) on the efficacy of OSTDS and riparian buffer zone development and management.

3. With the exception of dissolved organic carbon, the analytical methods employed for physical and chemical parameter monitoring in both reservoirs and streams are generally adequate. Some additions to the evaluations of dissolved organic matter are suggested. The quality assurance/quality control methods are generally adequate for chemical and biological limnological parameters. They should be rigorously maintained, since they are critical for long-term analyses and continuity. If methods are changed or improved, both the old and new techniques should be performed simultaneously on the same samples for a period (e.g., two years) in order to evaluate the comparability of old and new data sets.

4. Although methods currently used for pathogen monitoring are acceptable, New York City should take a more active role in the development and use of new and improved methods for pathogen detection. The goal of these efforts should be to achieve significantly higher numbers of nonzero measurements for (oo)cysts, to assess overall recovery efficiency on a weekly basis, to determine viability, and to build into the routine sampling plan split samples for these developing protocols. Methods for pathogen monitoring will change in the future, and the use of split samples will enable future analysts and managers in accurately interpreting historical data.

5. The Pathogen Studies should have as a goal the estimation of source terms [e.g., (oocysts/acre)/day or (oocysts/calf)/day] for various catchments, animals, agricultural and urban activities, and farm waste management. The Pathogen Studies as currently conducted (and expressed as percent detection from different land uses) do not obviously lead to a greater understanding of pathogen dynamics at the watershed level. In this regard, NYC DEP could benefit by assimilating work done on pathogen inactivation under variable environmental conditions and by making use of approaches to watershed modeling developed for nonpathogens (e.g., nutrients).

6. Additional microbial parameters should be considered for inclusion in routine water quality monitoring programs, in particular *E. coli*, coliphage, *Clostridium perfringens*, and cyanobacteria that produce toxins. Coliphage may be particularly useful for source typing in reservoirs and streams and for assessing potential groundwater contamination by human fecal pollution emanating from leaking sewerage. *Clostridium perfringens* can be used to assess the potential viability of protozoa in surface waters distant from the point of contamination. Fecal streptococci or enterococci may be useful as supplements to coliforms as indicators of pollution. However, the outmoded and ambiguous ratio of fecal coliform to fecal streptococci should not be used.

7. If NYC DEP wants to determine the effectiveness of managed riparian buffer zones, septic system drainfields, and agricultural and urban stormwater BMPs, performance monitoring using paired measurements is needed. The Septic Siting Study is an example of this type of monitoring currently under way in the watershed. Performance monitoring of septic systems and new urban stormwater BMPs is strongly recommended in the Kensico watershed. As suggested in Chapter 9, performance monitoring of agricultural BMPs for pathogen and nutrient removal would greatly enhance the Watershed Agricultural Program.

8. Monitoring and other special pollution prevention activities in the Kensico Reservoir and watershed should continue at their present high intensity given the importance of this basin in controlling Catskill/Delaware drinking water quality. The goal of these activities should be to demonstrate no net increase in pollutant loading to the reservoir in the face of future development, road expansion, and airport use. Similar measures should be considered for the West Branch Reservoir and watershed, which is experiencing population growth at a rate greater than that in the Kensico watershed.

GEOGRAPHIC INFORMATION SYSTEMS

A Geographic Information System (GIS) is a relational database that supports mapping and spatially distributed modeling. Such systems can store data on a wide variety of watershed characteristics, including topography, waterbodies, roads, vegetation and soil type, land use, and such discrete points as buildings, wells, and septic systems. GIS also allows new data sets (e.g., flow path) to be derived from available data sets (e.g., slope), which is a valuable function for mapping of watershed areas and for a variety of NYC DEP watershed programs.

Over the past several years, NYC DEP has developed a GIS program at its Valhalla, NY, office with network links to field offices. Primary data layers have been obtained from the U.S. Geological Survey National Mapping Center or have been developed under contract. NYC DEP has the required hardware, software, and, most importantly, the staff expertise to make full use of state-of-the-art GIS capabilities to support the complex management actions included in the MOA. In addition, water quality monitoring data can be analyzed using the GIS to determine if management strategies are meeting their objectives.

Notwithstanding its considerable accomplishments to date, there are opportunities within NYC DEP to improve the application of GIS to watershed monitoring and management. As currently applied, land cover (e.g., forest, grassland, and wetland) and land use (e.g., residential, pasture, or row-crops) data, compiled from Landsat imagery, are combined in one layer. This limits the ability of analysts and managers to differentiate between areas with similar appearances (spectral reflectance in a satellite image) but markedly dissimilar uses and pollution potential (e.g., alpine meadow, hayfield, pasture, or lawn). Additional field and integrated image analyses are needed to refine the characterization of human use and potential impacts. Such analyses could employ more Global Positioning System (GPS) data as well as conventional aerial photographs, color infrared aerial photographs, digital orthophotography, and multiseason/multispectral satellite imagery. Land cover/land use data are currently used to support several watershed management programs, including the Land Acquisition Program and the TMDL Program.

GIS is a particularly cost-effective means of evaluating wastewater treatment options. Soil, terrain, and flow path data can be used to assess the suitability of areas for residential development with OSTDS. Similarly, the location and spatial pattern of existing or potential residential development, again combined with other watershed features, can be used to assess the feasibility and costs of centralized wastewater treatment, package plants, or advanced OSTDS. A cooperative study by NYC DEP and the Catskill Watershed Corporation (CWC), fully supported by the GIS group, would greatly advance this key task.

As part of watershed planning and management efforts, citizens need to be more engaged. Because the debate about how to manage land is more substantive and productive when all stakeholders have the same information, the GIS database should be accessible to the public. A web site with "read-only" layers, a database directory, and basic documentation would enable interested residents to study watershed characteristics and formulate ideas and alternatives.

Conclusions and Recommendations for the GIS

1. In light of its central importance to watershed modeling and management, land cover and land use data should be separated, refined, and regularly updated. As currently operated, NYC DEP's GIS does not differentiate between these two types of data sets.

2. Management of domestic wastewater should be fully supported by spatially distributed modeling and GIS. A cooperative effort involving NYC DEP, CWC, and state and county health departments would advance this goal.

3. The GIS database should be made available, perhaps via a web site, to interested citizens, communities, nonprofit organizations (e.g., CWC, Catskill Center for Conservation and Development, and many others), and scientists (e.g., the Catskill Institute for the Environment). GIS maps, which can display varying degrees of complexity, can illustrate how individual actions might affect downstream water quality.

DISEASE SURVEILLANCE AND PUBLIC HEALTH PROTECTION

Water quality monitoring is a powerful and direct way of revealing the presence of contaminants in drinking water. However, it is not a particularly good predictor of the likelihood of human illness as a result of drinking water consumption because (1) pollutants in drinking water often exist at extremely low concentrations that are below detection limits, (2) it is difficult to know the quality and quantity of water actually ingested, (3) individual responses to contaminated water vary considerably, and (4) it is hard to distinguish the harmful effects of contaminated water from those of other vehicles such as food. For these reasons, water quality monitoring programs should be complemented by active disease surveillance and monitoring for disease outbreaks to determine whether water consumption presents a disease risk for the public.

The New York City Department of Health (NYC DOH) and NYC DEP maintain a Waterborne Disease Risk Assessment Program that encompasses both active disease surveillance and outbreak detection. Acute diseases of microbial origin are the focus of this program, rather than chronic conditions that have been linked to long-term exposure to chemical contaminants (such as trihalomethanes). The objectives of New York City's program are to (1) determine rates of giardiasis and cryptosporidiosis, along with demographic and risk factor information on case patients, (2) track diarrheal illness to assure rapid detection of outbreaks, and (3) determine the contribution (if any) of tap water consumption to gastrointestinal disease. In addition to active disease surveillance and outbreak detection, the program includes special studies, information sharing, and public education.

The Waterborne Disease Risk Assessment Program is meant to provide evidence of the lack of waterborne disease outbreaks in New York City, as required under the SWTR. The program also compares reported cases of giardiasis and cryptosporidiosis or surrogate measures of diarrheal disease incidence in the City's population with measures of water quality. However, although temporal trends in diarrheal disease may coincide with trends in water quality parameters, determining whether an observed pattern of disease is associated with exposures to drinking water contaminants requires a specifically designed epidemiological study.

Active Disease Surveillance

NYC DOH staff regularly contact 84 clinical laboratories to collect case reports on stool specimens positive for *Giardia* and *Cryptosporidium*. For each case, an interview is conducted and/or medical charts are reviewed for information on demographic features and potential risk factors. Active surveillance for giardiasis in New York City began in July 1993, and case interviews were discontinued in August 1995. Prior to 1993, there was a passive surveillance program for giardiasis, and responsibility for reporting recognized cases rested with the health care provider or laboratory. Active disease surveillance resulted in a marked increase in the reported number of giardiasis cases—from less than 300 cases per year during 1982–1988 to 2,456 cases in 1994 (the first year of active surveillance). Active surveillance of cryptosporidiosis in New York City began in November 1994, and case interviews and chart reviews have been conducted since January 1995. Active surveillance for cyclosporiasis began in June 1996 after a series of foodborne outbreaks were associated with this agent.

Active disease surveillance provides information on endemic rates of infection and demographic patterns in infection rates (such as residence, age, gender, and ethnicity). For cryptosporidiosis cases, additional information is collected on the immune status of the case and on known or suspected risk factors for infection, such as animal contact, high-risk sexual behavior, contact with a child enrolled in daycare, or international travel. Cases are also queried about their source of drinking water (e.g., plain tap water, filtered tap water, boiled tap water, or bottled water).

Endemic Rates of Giardiasis and Cryptosporidiosis

Active surveillance data for both giardiasis and cryptosporidiosis indicate that case rates per 100,000 have been decreasing since these data were first collected in 1994–1995 (see Figure 6-3 for cryptosporidiosis rates). Giardiasis

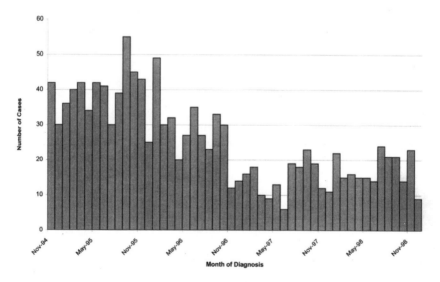

FIGURE 6-3 Reported cases of cryptosporidiosis in New York City, 1994–1998. Source: NYC DEP (1999c).

has declined from 33.5 reported cases per 100,000 in 1994 to 25.7 cases per 100,000 in 1998. A total of 1,881 cases of giardiasis were reported in 1998, with children less than 10 years of age experiencing the highest rates of infection. Geographically, the highest case rate was in Manhattan. These age and geographic patterns have been observed for several years (NYC DEP, 1999c). Interview data from giardiasis cases during July 1994 to May 1995 indicated that the most commonly reported exposure risk was travel outside the United States, with other prevalent risk factors being contact with a household member having diarrhea, recent immigration, and adult male homosexual activities (NYC DEP, 1997e). Because most of the cases were in children, it is surprising that contact with childcare centers was not reported as an exposure risk.

Between 1995 and 1998, cryptosporidiosis rates dropped from 6.5 reported cases per 100,000 to 2.8 cases per 100,000 (NYC DEP, 1999c). In 1995 and 1996, the majority of reported cryptosporidiosis cases were among persons with AIDS, and the overall decline in cryptosporidiosis between 1995 and 1998 is due to a decline in cases in this subpopulation. This trend coincides with the increased use nationwide of newer antiretroviral therapies (Miller, 1998). Reported cryptosporidiosis among immunocompetent persons in New York City has increased from 71 cases in 1995 to 116 cases in 1998; the reasons for this rise are unclear. In addition, case rates for immunocompetent persons peak each year in late summer, a seasonal pattern observed elsewhere (MacKenzie, 1998). From 1995 to 1998, the most commonly reported exposure for both immuno-compromised and immunocompetent persons was contact with an animal. For cryptosporidiosis cases among immunocompetent persons, foreign travel and recreational water contact were the second most commonly reported exposures.

For both giardiasis and cryptosporidiosis cases, the vast majority of cases reported drinking unboiled tap water as their primary source of drinking water. However, the significance of this exposure cannot be determined because there is no information on exposure histories in a suitable control population, such as persons who consume only bottled water. In addition, it is difficult to compare the New York City case rates with rates of giardiasis and cryptosporidiosis in other parts of the United States because there are few communities that have similar active surveillance for these diseases. Preliminary laboratory-based active surveillance data from six states in 1998 showed that cryptosporidiosis rates ranged from 1.3 to 3.5 per 100,000 (V. Dietz, CDC, personal communication, 1998), suggesting that the endemic rates of cryptosporidiosis in New York City are similar to those observed elsewhere.[3]

---

[3]It should be noted that these 1998 case rates differ from those found in Chapter 5, Table 5-2, which relied on 1997 data.

Limitations of Active Disease Surveillance

Although much more sensitive than passive surveillance, active surveillance suffers from significant underdiagnosis, making comparisons among different localities problematic. Underdiagnosis is caused by the combined effects of several factors, illustrated in Figure 6-4. The net result of these factors is an underestimation of the true disease rate by several orders of magnitude. The greatest loss of information is due to the fact that most cases (94 percent in the Milwaukee *Cryptosporidium* outbreak) do not seek medical attention, and most health care providers do not routinely collect and test stool specimens from patients with gastroenteritis. Because most cases of gastroenteritis are only given supportive care, knowledge of the etiology of the disease does not affect treatment and is therefore rarely sought. These limitations apply to active disease surveillance programs nationwide.

To illustrate these effects, Table 6-7 compares the expected number of endemic *Cryptosporidium* cases reported per month among immunocompetent persons in New York City to the expected number of reported *Cryptosporidium* cases in a situation where there is an outbreak affecting one percent of the population. Reporting-loss estimates used in this table are based on the 1993 Milwaukee cryptosporidiosis outbreak, during which only 1 in every 22,000 cases of cryptosporidiosis were reported to the Health Department prior to recognition that the outbreak was waterborne (Juranek, 1999). (The Foodnet surveillance system of the Centers for Disease Control and Prevention (CDC) has made similar observations of the percent loss in this series of steps.) The estimate of six reported cases of endemic cryptosporidiosis per month is similar to that observed by the NYC DOH active surveillance system. That is, for 37 months of surveillance data, the number of reported cryptosporidiosis cases per month ranged from 1 to 17 with an average of 6.4. Table 6-7 shows that an outbreak affecting one percent of the population would result in the recognition of an additional nine cryptosporidiosis cases. It is clear that a large portion of the population (perhaps as much as ten percent) would have to become infected during an outbreak in New York City before a significant rise in the number of reported cryptosporidiosis cases would be detected by active surveillance (MacKenzie, 1998).

In addition to underestimating cases, active disease surveillance suffers from problems of timing. In the case of cryptosporidiosis, there are approximately 7–14 days between time of exposure and occurrence of symptoms. Delays in seeking medical care, and the time needed for laboratory diagnosis and reporting, would cause any observed peak in cryptosporidiosis rates detected by active surveillance to occur at least two weeks after a problem in water quality (assuming water is the causative agent). For example, surveillance data collected from the Milwaukee cryptosporidiosis outbreak showed that the peak in laboratory-reported cases occurred 15 days after the peak in turbidity reported at the faulty

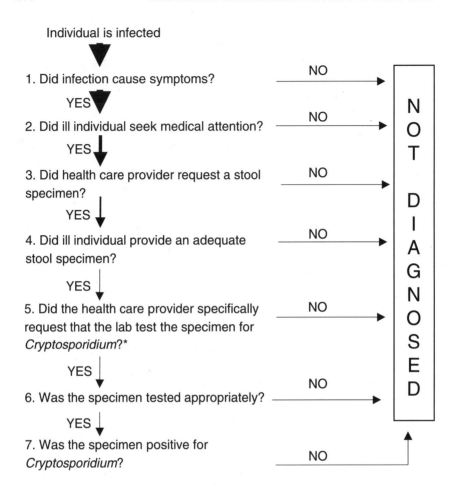

FIGURE 6-4 Sources of loss of information during active disease surveillance.
Note: In the case of cryptosporidiosis, the health care provider must specifically request that a stool specimen be tested for *Cryptosporidium* because this is not included in routine parasitic examinations of stool specimens. *A survey of U.S. laboratories in 1994–1995 indicated that only five percent of stool specimens submitted for routine "Ova and Parasite" exam are tested for *Cryptosporidium* (Boyce et al., 1996). Recent data from a survey of clinical labs in New York State indicate that about 18 percent of laboratories now include *Cryptosporidium* testing as part of the routine "Ova and Parasite" exam (J. Miller, NYC DOH, personal communication, 1998). Adapted from Frost et al. (1996) and Juranek (1997).

TABLE 6-7 Comparison of Expected Number of Reported Endemic Cryptosporidiosis Cases vs. Expected Number of Reported Cryptosporidiosis Cases in a Simulated Outbreak

Assumptions[a]	Endemic Situation	Simulated 1% Outbreak
Total population = 7,000,000		
Baseline diarrhea rate in immunocompetent persons = 11% prevalence monthly	770,000 cases of diarrhea	
1% of population infected with *Cryptosporidium* in simulated outbreak		70,000 cases of cryptosporidiosis
2% of cases tested for any enteric pathogen	15,400 diarrhea cases	1,400 cryptosporidiosis cases
25% of these tested for "O&P" (ova and parasites)	3,850 diarrhea cases	350 cryptosporidiosis cases
5% of these tested for *Cryptosporidium*	193 diarrhea cases	18 cryptosporidiosis cases
3% of gastroenteritis cases test positive for *Cryptosporidium*	6 reported cryptosporidiosis cases per month in immunocompetent persons	
50% of outbreak-associated cryptosporidiosis cases test positive		9 recognized cryptosporidiosis cases

[a] Estimates by MacKenzie (1998).

water treatment plant (Proctor et al., 1998). This time lag reinforces the argument that this type of active surveillance system is limited in recognizing and managing an outbreak in real time and can only provide retrospective information.

Outbreak Detection

Aware of the fact that active, laboratory-based surveillance is neither sensitive enough nor rapid enough to detect waterborne disease outbreaks, New York City has developed three independent and complementary diarrheal disease-monitoring systems as part of an Outbreak Detection Program. These systems collect information during the earliest steps in Figure 6-4 and thus detect more potential cases of gastrointestinal disease than does the active surveillance program. Because the three systems are independent, their results can be compared to determine if an observed trend in a single system is confirmed by the other systems or is a reporting artifact. Although the sensitivity of these systems is potentially far greater than that of active surveillance, in some cases (monitoring of medication sales and number of stool specimens examined), the data collected are surrogate measures for cases of gastroenteritis and do not have the specificity of laboratory-confirmed infections.

Monitoring of Antidiarrheal Medications

Several reports in the literature indicate that outbreaks of enteric disease have been accompanied by increased sales of antidiarrheal medications (Rodman et al., 1998; Sacks et al., 1986). For example, Angulo et al. (1997) found that sales increased by 600 percent during an outbreak of salmonellosis. Monitoring antidiarrheal medication sales has been embraced by New York City as a sensitive, real-time indication of increased gastrointestinal disease and as one approach for detecting disease outbreaks. One of its advantages is that it provides some information on the total population experiencing diarrheal symptoms (step 1 in Figure 6-4), most of whom will not seek medical attention (step 2 in Figure 6-4).

In May 1995, New York City started collecting information on weekly shipments of Imodium from a major distributor to 1,265 pharmacies including 564 (about one-third of all pharmacies) in New York City. In addition, the weekly sales of 22 antidiarrheal medicines at a chain of 38 pharmacies located in the City have been monitored since February 1996. Both these monitoring activities provide information on the annual patterns of antidiarrheal medication sales at the distribution and retail level.

According to information provided by NYC DOH, the weekly sales of antidiarrheal medication have been relatively stable since 1996 (Miller, 1998). Peaks have been noted and attributed to promotional sales, including a large sales peak in November 1997.

Monitoring of Clinical Laboratories

To complement monitoring of antidiarrheal medicine sales, since November 1995 NYC DOH has received daily information by fax from three clinical laboratories (including the largest laboratory in the metropolitan area) on the number of stool specimens submitted for bacterial and parasitic testing. One of the three labs notes whether the stool specimens were submitted specifically for *Cryptosporidium parvum*. The reported number of routine stool specimens examined is relatively constant with modest seasonal variation, as shown in Figure 6-5. Because this outbreak detection system addresses steps 5 and 6 illustrated in Figure 6-4, it is less likely to detect increased disease rates com-

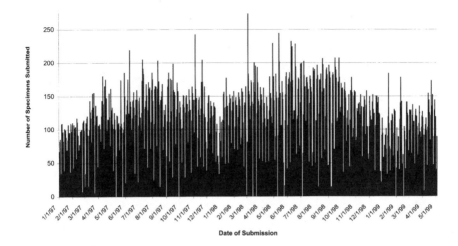

FIGURE 6-5 Number of stool specimens submitted for ova and parasite testing at one New York City laboratory. Courtesy of the NYC DEP.

pared to monitoring of antidiarrheal medication sales, which addresses steps 1 and 2 in Figure 6-5.

Nursing Home Monitoring

The third outbreak detection system is a nursing home surveillance system that involves 11 nursing homes in all five boroughs. Like monitoring of antidiarrheal medicine sales, this system provides information on step 1 in Figure 6-4. A pilot program, begun in March 1997, has continued to expand to include approximately 1,800 nursing home residents. Each nursing home provides daily information by fax on the number of new gastrointestinal illness cases among the residents. The population served by these homes includes elderly and AIDS patients who drink tap, filtered, or bottled water. Tap water drinkers from both the Croton and Catskill/Delaware systems are represented. Although surveillance data are limited because of the newness of the program, results so far indicate very low numbers of new gastrointestinal illness cases per day (NYC DEP, 1999c).

Validating the Outbreak Detection Programs

When a peak is observed in one of these three systems, the reporting source is contacted about the possibility of a reporting artifact. If a reporting artifact is ruled out, then the results of the different outbreak detection systems are compared. If similar increases are observed, then NYC DOH contacts childcare centers, school nurses, and hospital emergency departments to determine if a corresponding rise in the number of clinical cases of gastroenteritis at these institutions has been noticed. Data from the outbreak detection program are also compared to water quality data to observe any simultaneous trends.

Two such follow-up investigations occurred in 1998. First, in March 1998, there was an increase in the sales of antidiarrheal medication and in the number of stool specimens submitted to clinical laboratories for examination. The timing of this rise coincided with the annual spring rotavirus season, although the etiology of this illness could not be confirmed because stool examination for rotavirus is rarely performed in clinical settings. During the same time period, three outbreaks of gastrointestinal illness were reported at New York City nursing homes, two of which were confirmed to be caused by rotavirus. Second, in October 1998, low levels of *Cryptosporidium* were detected in the Catskill effluent chamber of the Kensico Reservoir. However, the outbreak detection systems did not document any increased diarrheal disease in the population.

Because trends in all three outbreak detection datasets have been relatively stable, there have been few opportunities to determine their sensitivity to increases in gastrointestinal disease and compare the three systems to one another. More time must pass before the adequacy of these systems can be fully evaluated. In addition, because the actual sensitivity of these outbreak detection programs is currently unknown, it is premature to evaluate the adequacy of the response of the NYC DOH. Criteria for such evaluations should include (1) how peaks in these reporting systems are operationally defined (e.g., a certain percent rise in incidence over a specified time period); (2) how quickly, aggressively, and consistently peaks are investigated; and (3) under what circumstances these investigations include checking source water quality and informing the public.

Response to Consumer Complaints

As in many other water systems, NYC DEP and NYC DOH respond to consumer complaints about water quality by collecting and testing a water sample from the consumer's home. Although New York City performs a number of microbial and chemical tests, it does not analyze for the presence of cysts or oocysts, which would be infeasible given the needed volumes of tap water. If a consumer reports having diarrhea and thinks water may be the cause, arrangements can be made to obtain a stool specimen and test it at the NYC DOH for ova and parasites and bacterial enteric pathogens (A. Ashendorff, NYC DEP, personal communication, 1998).

There does not appear to be any systematic tracking of consumer complaints of water-related illness or water quality. This is valuable information that should be recorded and reviewed, preferably using the GIS, to determine if any spatial or temporal patterns are evident. Consumer complaints in Milwaukee were one of the first indications that a serious problem with water quality and waterborne disease was occurring.

Boil-Water Advisories

A boil-water advisory is a public health measure that, if implemented promptly, can successfully reduce the risk for potentially serious diarrheal and other waterborne diseases among persons whose water supply has been contaminated by microbial pathogens (CDC, 1995). In New York City, it is not clear what concentration of contaminants triggers a boil-water advisory or other public health response. According to NYC DEP, should any elevated level of *Cryptosporidium* be detected in the water supply, NYC DEP, working with the City and State Departments of Health and EPA, will evaluate factors included in the 1995 CDC guidance to determine what action, if any, to take in response (Ashendorff, 1999). Developing a plan of action for issuing boil-water advisories and responding to changes in pathogen concentrations in source waters are concerns for water supplies across the country. New York City should take the lead in creating such plans by working with state and federal partners to formalize a decision-making process.

Information Sharing and Public Education

A fundamental element of any public health surveillance system is to share its findings with those who provide the data. Active disease surveillance data are regularly compiled into quarterly and annual reports. In addition to these reports, presentations on the surveillance system findings are made to health care professionals in the New York City area. NYC DOH staff also participate in national and international conferences that focus on drinking water and waterborne disease issues.

Two public outreach efforts by NYC DOH deserve mention. Special reports of the *City Health Information* publication that focus on *Cryptosporidium* have been mailed to health care professionals to draw attention to the need to request specific laboratory testing for *Cryptosporidium* when this infection is suspected. In addition, announcements have been faxed to area hospitals, health care providers of HIV/AIDS patients, and organizations serving persons with HIV/ AIDS informing them of the findings of low levels of *Cryptosporidium* in the source water and the significance of these findings. The CDC has encouraged local public health authorities and water utility officials to develop coalitions with health care providers and advocacy groups for immunosuppressed persons in order to communicate important public health information and decide what specific actions should be taken when oocysts are detected in municipal water (CDC, 1995).

Epidemiologic Studies

Active disease surveillance measures rates of specific infections, both endemic and epidemic. Outbreak detection systems monitor numbers of cases of gastrointestinal illness. In both instances, it can be difficult, if not impossible, to demonstrate that drinking water is the vehicle of disease transmission. In order to link contaminated source water with disease, epidemiologic studies must be conducted. As described below, New York City has supported such studies twice during the 1990s. Both of these studies utilized cases identified from the active disease surveillance program.

Giardiasis Case-Control Study

In June and July 1995, NYC DOH compared 120 patients suffering from giardiasis with a control group of 120 persons (NYC DEP, 1997e). The controls and patients were matched by age, gender, neighborhood of residence, and primary language. Interviews with all participants included questions about high-risk activities, food intake, water intake, and immune status. The study could detect no difference in the proportion of tap water drinkers among giardiasis cases versus controls. Thus, drinking water was not implicated as a source of *Giardia.*

NYC DOH study results contrast with other *Giardia* case-control studies. First, the NYC DOH study did not detect any association between disease and commonly recognized risk factors for giardiasis such as travel outside the United States and swimming in freshwater. Contact with childcare centers, another common risk factor, is not mentioned. Furthermore, the study reported that giardiasis cases were less likely than controls to have household members suffering from diarrhea, a surprising finding. These results directly contrast with those of other researchers. A case-control study of residents in New Hampshire that involved 273 cases and 375 controls found that giardiasis was significantly associated with a recent history of drinking untreated surface water (e.g., during camping or hiking), a history of swimming in a lake, pond, or any natural body of freshwater, and contact with a person thought to have giardiasis (Dennis et al., 1993). An earlier study using 171 giardiasis cases and 684 controls found that having a family member in day care, having a family member with diagnosed giardiasis, travel outside the United States, and camping were significant risk factors (Chute et al., 1987). It is possible that the smaller size of the New York City study did not permit a thorough examination of all risk factors for giardiasis.

Cryptosporidiosis Cross-Sectional Study

A cross-sectional study was conducted to examine the prevalence of cryptosporidiosis among HIV-infected persons in New York City and identify risk factors for cryptosporidiosis in this population. Between October 1995 and July 1997, 405 HIV-positive patients at the New York Hospital–Cornell Medical Center were interviewed, 379 blood samples (sera) were collected, and 331 stool specimens were collected. Only 1.2 percent of the stools tested positive for *Cryptosporidium*, while 28 percent of the blood samples were seropositive for a *Cryptosporidium* antibody (NYC DEP, 1998e).

Study subjects were queried about a variety of potential risk factors. Interviews revealed that the drinking water habits of the HIV patients were variable: 45 percent did not drink tap water at home but few completely avoided tap water, and 21 percent had recently changed their water consumption habits (Soave et al., 1998). Because of the low percentage of stools that tested positive for *Cryptosporidium*, the study was not able to examine the association between drinking water habits and enteric infection. In addition, although a sizable percentage of the study participants had antibodies against *Cryptosporidium,* seropositivity was not correlated with any of the potential risk factors for infection, including drinking water habits. Like the giardiasis case-control study, this study was limited by the small size of the study population.

Additional Epidemiologic Studies

Stool Survey for *Cryptosporidium*. Since September 1995, the NYC DOH has tested all stool specimens collected by Child Health Clinics (which serve approximately 80,000 children) for *Cryptosporidium*. The goal of this study is to collect information on the prevalence of *Cryptosporidium* in children. The number of samples tested per year has ranged from 3,400 to 5,400, and the detection rate has ranged from zero to 0.09 percent. Although it is reassuring to see such a low prevalence, there are some important unanswered questions. The study population should be defined in terms of age distribution, geographic distribution, and current gastrointestinal symptoms. Whether the specimens are fresh or preserved will also affect study results. Finally, these data should be compared to rates of *Cryptosporidium* detection in stool specimens submitted for ova and parasite exams as part of active surveillance. If stool surveys for *Cryptosporidium* are continued, further information about the source and quality of the specimens and how these results compare to other measures of *Cryptosporidium* prevalence in New York City and elsewhere, is needed to interpret the significance of the results from this activity.

Serologic Studies. Additional epidemiologic studies would be useful to help determine whether there is a relationship between consumption of New

York City tap water and endemic enteric illness or infection (as recommended by Craun et al., 1994). A cross-sectional or longitudinal serologic study could be conducted to examine seroprevalence or seroincidence of *Cryptosporidium* in New York City tap water drinkers compared either to residents who drink bottled water or to residents of other communities with filtered surface water or groundwater (Griffiths, 1999, Box 4-2). Although there is still some uncertainty about the best methods for determining seropositivity, recent advances have been made in this area, and the results of other *Cryptosporidium* serologic studies are encouraging (Frost et al., 1998a,b; Griffiths, 1999). Such studies could help explain the difference between the low rates of *Cryptosporidium* detected by the active surveillance program (0.2 per 100 HIV-infected persons in 1998) and the higher rate of seropositivity observed in the study of HIV-infected patients (28 percent).

Time Series Studies. Because of the wealth of water quality monitoring data, NYC DOH and NYC DEP should consider conducting time series studies that compare patterns of illness with various water quality parameters over time (as in Schwartz et al., 1997). This could be easily accomplished with the data that are already collected in the active disease surveillance, outbreak detection, and water quality surveillance monitoring programs.

Cohort Studies. Because it has large amounts of accumulated data and an active professional staff, NYC DOH is in an excellent position to conduct a randomized household intervention trial using some type of in-home water treatment device (as in Payment et al., 1991, 1997). This type of study compares the rate of enteric symptoms among households consuming tap water with those among households consuming tap water that receives additional home treatment (e.g., reverse-osmosis filtration or UV disinfection). Ideally, the study design should allow consideration of the separate contributions of source water quality and distribution system water quality to enteric illness or infection.

Although such studies are complex and expensive, they are the cleanest and most definitive way to examine the role of waterborne disease. Cohort studies compare two distinct populations, one using the water source to be evaluated and the other using some other very high-quality water. The longitudinal nature of such studies allows one to look at health effects over time and examine the temporal relationship between water exposure and disease. Because household intervention studies are conducted in single communities, community-to-community differences in other risk factors are not an issue. Finally, household-to-household differences in risk factors for gastrointestinal illness (such as children in childcare centers and travel) should be randomly distributed between the two study groups. Therefore, any difference in gastrointestinal illness rates can be attributed to differences in water quality.

Conclusions and Recommendations

A panel of experts reviewed New York City's Waterborne Disease Risk Assessment Program in April 1998 and individual panel members suggested several ways to enhance or supplement the current surveillance and outbreak detection programs and further examine the safety of New York City tap water. The following recommendations are targeted at the three primary goals of the program: (1) to determine rates of giardiasis and cryptosporidiosis, (2) to track diarrheal illness to detect outbreaks, and (3) to determine the contribution of tap water to gastrointestinal illness. The results are summarized briefly in Table 6-8 and are explained in greater detail below.

TABLE 6-8 Recommendations for Improving the New York City Waterborne Disease Risk Assessment Program

Program	General Limitations	Recommendations	Goal 1, 2, or 3?
Active Disease Surveillance	1. Low sensitivity because only a small percentage of stool specimens are tested. 2. Slow	All stool specimens should be tested. Peaks should continue to be aggressively analyzed.	1 and 2
Antidiarrheal Medicine Monitoring	Some peaks not associated with illness.	Continue aggressive and timely investigation of peaks.	2
Clinical Laboratories Monitoring	1. Some stools are from patients without infectious disease symptoms. 2. Addresses later steps in Figure 6-4.	Of marginal use for detecting outbreaks, but useful for comparison with other detection systems. Could be useful in determining the percentage of stools positive for *Crypto*.	2
Nursing Home Monitoring	May have a high background of gastrointestinal upsets because of multiple transmission routes in institutions.	Continue aggressive investigation of peaks and comparison of data from one home to data from other homes.	2
Epidemiological Studies	Previous studies limited in sensitivity and design.	Intervention (cohort) study	3
Response to Consumer Complaints	Sampling of water or individual likely to occur long after exposure event.	Do not sample water for protozoa, but do flag health complaints and set up GIS database for tracking complaints.	None

1. NYC DOH and NYC DEP are making strong efforts to conduct active surveillance for giardiasis and cryptosporidiosis and to monitor levels of gastrointestinal illness in New York City. Such efforts are expected given the size of the service population and the fact that this is an unfiltered surface water supply. Since 1994, there has been substantial progress in the development of several complementary outbreak detection systems.

2. Health care providers and laboratories should make *Cryptosporidium* testing part of all routine stool examinations. Active disease surveillance is too insensitive and slow to detect waterborne disease outbreaks on a real-time basis. Data from other communities suggest that only 1 in 22,000 cases are diagnosed and reported via disease surveillance. Sensitivity can best be improved by promoting enhanced collection and testing of stool specimens for *Cryptosporidium*. The median cost of such tests is $52 for those New York City laboratories that do not routinely test for *Cryptosporidium* (D. Warne, NYC DEP, personal communication, 1999). The 140 percent increase in testing stools for *Cryptosporidium*, observed between 1995 and 1997, is encouraging.

3. NYC DOH should determine the lowest incidence of disease that can be detected by the current outbreak detection program and increase the sensitivity by studying specific populations. It is not clear what size of outbreak can be detected using the combined system of monitoring sentinel nursing home populations, sales of antidiarrheal medications, and submission of stool specimens. To increase the sensitivity and population base covered by these systems, the following additional approaches could be used: (1) monitoring school absenteeism related to gastroenteritis in sentinel schools (see Rodman et al., 1998), (2) monitoring Health Maintenance Organization nurse hotline calls to track increased incidence of gastroenteritis, (3) monitoring hospital emergency room visits for gastroenteritis, and (4) monitoring gastrointestinal symptoms in a network of sentinel families.

4. NYC DOH and NYC DEP should develop a plan of action to define spikes in surveillance and outbreak detection data and trigger investigation of these peaks after eliminating the possibility of reporting error. There appears to be no systematic approach for defining peaks in active surveillance or outbreak detection data. Investigation of peaks should continue to include comparisons with water quality data. Substantial resources are currently being directed toward collecting data on water quality and health measures (both disease surveillance and outbreak detection). Linking and critically reviewing these data sets may lead to recognition of relationships between water quality and health. Analysis of the Milwaukee cryptosporidiosis outbreak indicated a strong temporal relationship between trends in finished water turbidity and trends in emergency room visits for gastrointestinal illness, customer complaints, laboratory diagnoses of *Cryptosporidium* infections, prevalence of diarrhea in nursing homes, and absentee rate in schools (Proctor et al., 1998).

5. NYC DOH and NYC DEP should develop a plan of action for tracking and investigating consumer complaints about water quality and water-related illness that goes beyond collecting and testing a household water sample for routine parameters. Such a system, which would optimally be linked to the GIS, may lead to earlier recognition of geographic and temporal problems with water quality both at the source and in the distribution system.

6. To determine the role of tap water as a vehicle of infection, NYC DOH should conduct additional epidemiological studies. Two relatively small epidemiological studies of waterborne disease have been conducted in New York City, neither of which implicated drinking water as a risk factor for disease. It is questionable whether the *Giardia* case-control study had sufficient power to detect an association between exposure and disease for any risk factor of giardiasis. In the cryptosporidiosis cross-sectional study, only four laboratory-confirmed cases of *Cryptosporidium* infection were detected, making it impossible to evaluate the relationship between drinking water and risk of cryptosporidiosis. Several possible study designs, including serologic studies and intervention studies, are described in this chapter.

Surveillance systems alone, no matter how sensitive, cannot provide evidence of the risk of endemic waterborne disease. Well-designed and well-conducted epidemiologic studies can examine disease patterns revealed by the surveillance system (such as the rise of cryptosporidiosis among immuno-competent individuals) and provide valuable documentation of the safety of the water supply.

MICROBIAL RISK ASSESSMENT

Epidemiological methods provide one tool for estimating potential disease impacts from pathogens in water. A complementary tool is the use of quantitative microbial risk assessment. Although risk assessment is not currently being practiced by New York City on a regular basis, there are sufficient data being collected to use this technique to estimate potential disease impacts. The objective of this section is to outline the process and to perform a quantitative risk assessment focusing on *Cryptosporidium*. *Cryptosporidium* is selected as the target organism of interest because it is presently the most resistant pathogen to disinfection, with minimal inactivation by free chlorine alone. In contrast, other pathogens (*Giardia* and enteric bacteria and viruses, for example) would be expected to be substantially reduced by current treatment.

This risk assessment follows the general methods developed by the National

Research Council (NRC, 1983) and provided in Appendix C. The following inputs are needed to conduct this assessment:

- Water ingestion per day (V)
- Oocyst concentration at point of ingestion (C)
- Dose-response relationship for *Cryptosporidium* f(V•C)

In accordance with prior risk assessments, each day of exposure (consumption of water) is considered to result in a statistically independent risk of infection (Regli et al., 1991; Haas et al., 1993).

In performing this risk assessment, it is assumed that the oocyst levels in the Kensico Reservoir raw water (at the CATLEFF and DEL 18 sampling locations) reflect concentrations generally experienced at the point of consumption. This assumption is reasonable because free chlorine has been generally regarded as being of minimal efficacy in inactivating oocysts (Korich et al., 1990; Ransome et al., 1993; Finch et al., 1998). It is possible that at the periphery of the distribution network in New York City, sufficient time of contact with chlorine may occur to provide some inactivation of oocysts.

Determination of Input Variables

Water Ingestion

Tap water ingestion was modeled using the lognormal distribution for total tap water consumption developed by Roseberry and Burmaster (1992). In this study, the natural logarithm of total daily tap water consumption was found to be normally distributed with a mean of 7.492 (corresponding to an arithmetic mean of 1.95 L/day) and a standard deviation of 0.407. The study was based on data collected from 5,605 persons over a three-day period from 1977 to 1978 as part of the Nationwide Food Consumption Survey. Water consumption rates for New York City have not been measured,,and although it does not explicitly reflect the New York City population, the Roseberry and Burmaster analysis is widely used (see the Exposure Factors Sourcebook published by the American Industrial Health Council, Washington DC). It should be noted that the default water consumption generally used by EPA in microbial risk assessment is 2 L/day, which is slightly greater than the mean developed in Roseberry and Burmaster.

Oocyst Concentration

Initial examination of oocyst levels measured at the two source water points (DEL 18 and CATLEFF) indicates a number of interesting features (Figure 6-6). First, the levels of oocysts are quite variable, as is common for many microbial data sets. Second, the densities appear to be higher during the earlier portion of

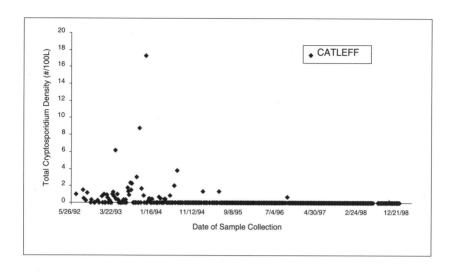

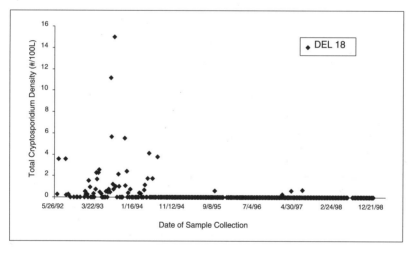

FIGURE 6-6 Total oocyst concentrations in Kensico raw water samples. Data courtesy of the NYC DEP.

the data record than in the more recent part of the data record. The reasons for this difference are not clear. Third, there is a substantial amount of data where no oocysts were detected. The mean detection limit for these nondetects was 0.721 oocysts/100 L.

The overall mean oocyst concentration (treating the "nondetects" as zeros) was 0.26 and 0.31 oocysts/100 L for the CATLEFF and DEL 18 locations, respectively. Of the 292 samples taken at each location, only 45 samples at CATLEFF and 48 samples at DEL 18, respectively, were above individual daily detection limits. Of these samples, only 18 and 21, respectively, were above 0.721 oocysts/100 L (the average detection level for the nondetects). This pattern is not unusual in protozoan monitoring data, and it presents a level of complexity in assessing the risk posed by exposure to these organisms.

Detection Limit Considerations. The significant number of samples with concentrations close to or below the average detection limit must be taken into account when estimating mean oocyst densities and distribution. There are several methods that may be used when dealing with below-detection-limit data (Haas and Scheff, 1990). Two basic approaches are employed here:

1. Observations that are below-detection-limit are treated as if they had values equal to the detection limit, half the detection limit, or zero. The arithmetic mean of the revised data is then computed by simple averaging. These alternatives are called "fill in" alternatives.

2. The method of maximum likelihood is used. In this approach, the data are presumed to come from a particular distribution (e.g., lognormal), and standard methods for analyzing data with a single censoring point are used. A likelihood function is formulated with a contribution equal to the probability density function for all quantified values, and equal to the cumulative distribution function (up to the detection limit) for all below-detection-limit values. The values of the distribution parameters that maximize the resulting likelihood are accepted as the best estimators. If the lognormal distribution is used, and the parameters $m_{\ln(x)}$ and $s_{\ln(x)}$ represent the mean and standard deviation of the log-transformed densities, then the arithmetic mean (m_x) determined by the property of the lognormal distribution is:

$$m_x = \exp\left[m_{\ln(x)} + \frac{\left(s_{\ln(x)}\right)^2}{2} \right] \qquad (6\text{-}1)$$

To develop the distribution for oocyst concentrations at the point of ingestion, all data present in Figure 6-6 for CATLEFF and DEL 18 were examined. Using maximum likelihood, and treating all observations less than or equal to 0.721 oocysts/100 L as being "censored" (for all censored observations, 0.721 oocysts/100 L was regarded as being the detection limit), the parameters of lognormal distributions were determined.

The maximum likelihood method is preferred over the fill-in method because

TABLE 6-9 Mean and Standard Deviation of Best-Fitting Normal Distribution for Natural Logarithm of Oocyst Levels (#/100 L) in Kensico Samples (1992 to July 1998)

Location	Oocyst Levels	
	Mean of Natural Logarithm	Standard Deviation of Natural Logarithm
CATLEFF	−2.752	1.828
DEL 18	−3.210	2.177

it is less subject to bias, especially when a large number of censored observations exist. Numerical studies have shown that this method, and closely related probability regression methods, are reasonably robust to deviations from an assumption of exact log-normality (Helsel and Cohn, 1988; Haas and Scheff, 1990).

Lognormal distributions of oocyst data. Table 6-9 gives the parameters of the best-fitting lognormal distributions to the entire data record at each station. There is some underprediction at the extreme tails of the distribution; however, in general the fit is adequate. Investigation of alternative distributions (gamma, Weibull, and inverse Gaussian) did not yield fits superior to the lognormal distribution. The goodness of fit to the lognormal was acceptable as judged by a chi-squared test. Figure 6-7 provides a plot of the exceedance probabilities for the fitted and observed distributions.

Although the time series plot (Figure 6-6) suggests a potential correlation between the oocyst levels at CATLEFF and DEL 18, the rank correlation coefficient (for observations in excess of the detection limit only) is −0.18. This is lower in absolute value than a correlation that would have substantial influence on the results of a Monte Carlo risk assessment (Smith et al., 1992; Bukowski et al., 1995), and so it has been ignored in this computation.[4]

Comparing Fill-in Methods to the Maximum Likelihood Method. Figures 6-8 and 6-9 summarize the annual arithmetic averages at the CATLEFF and DEL 18 stations, respectively. The first column shows the mean oocyst concentration for the entire 1992–1998 data set, which was computed using the "fill-in" methods. Each subsequent column shows the mean oocyst concentration for an individual year of data. (For 1992 and 1998, these averages are for portions of the year.) The four bars correspond to the method of maximum likelihood and the three different fill-in methods. The "imputed arithmetic mean" (i.e., m_x) is computed from the maximum likelihood estimates of $m_{\ln(x)}$ and $s_{\ln(x)}$; in more

---

[4] The effect of including this slight negative correlation would have been to slightly reduce the estimated risk of infection.

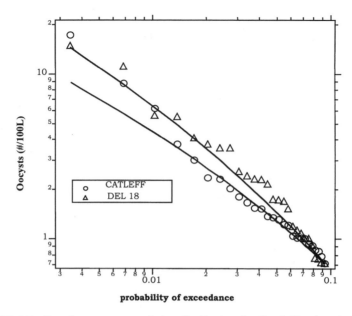

FIGURE 6-7   Complementary cumulative distribution for fitted (lines) and observed (individual points) Kensico oocyst densities.

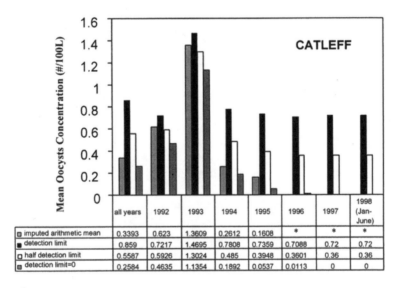

	all years	1992	1993	1994	1995	1996	1997	1998 (Jan-June)
imputed arithmetic mean	0.3393	0.623	1.3609	0.2612	0.1608	*	*	*
detection limit	0.859	0.7217	1.4695	0.7808	0.7359	0.7088	0.72	0.72
half detection limit	0.5587	0.5926	1.3024	0.485	0.3948	0.3601	0.36	0.36
detection limit=0	0.2584	0.4635	1.1354	0.1892	0.0537	0.0113	0	0

FIGURE 6-8   Summary of mean oocyst levels (#/100 L) at CATLEFF estimated by different methods. *The high proportion of nondetects in these years makes it impossible to use the maximum likelihood method.

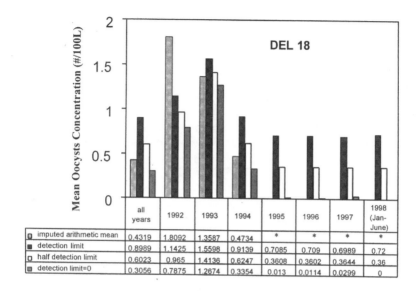

	all years	1992	1993	1994	1995	1996	1997	1998 (Jan-June)
imputed arithmetic mean	0.4319	1.8092	1.3587	0.4734	*	*	*	*
detection limit	0.8989	1.1425	1.5598	0.9139	0.7085	0.709	0.6989	0.72
half detection limit	0.6023	0.965	1.4136	0.6247	0.3608	0.3602	0.3644	0.36
detection limit=0	0.3056	0.7875	1.2674	0.3354	0.013	0.0114	0.0299	0

FIGURE 6-9 Summary of mean oocyst levels (#/100 L) at DEL18 estimated by different methods. *The high proportion of nondetects in these years makes it impossible to use the maximum likelihood method.

recent years, it was not possible to estimate the maximum likelihood mean densities at both locations and all times, since too few (less than two) observations above a detection limit were available.

The bias related to the "fill-in" methods using the detection limit and half the detection limit is quite evident in more recent years, where the oocyst levels were generally below detection. Both these fill-in methods may overestimate total oocyst concentration in the source water. Regardless of the methods used, in 1992 and 1993 average oocyst levels were higher than they were in more recent years.

Assessing Relative Contributions from CATLEFF and DEL 18. The two oocyst concentrations (CATLEFF and DEL 18) represent contributions dominated by Catskill and Delaware flows at Kensico. In combining these contributions to assess exposure, they were flow weighted by the average daily flows to the Hillview reservoir—521 and 800 mgd, for CATLEFF and DEL 18, respectively (NYC DEP, 1997a). In other words, the combined concentration entering the New York City distribution system can be estimated as the sum of 0.394

[which is equal to 521/(521+800)] times the CATLEFF concentration and 0.606 times the DEL 18 concentration.

Dose-Response Relationship

The dose-response relationship for infection of human volunteers with *C. parvum* oocysts has been found to be exponential, as given in Equation 6-2. *V* is the volume of water consumed, *C* is the oocyst concentration, *p* is the risk of infection from a single exposure, and *k* is the best-fit dose response parameter (or median infectious dose—238 oocysts) (Haas et al., 1996).

$$p = 1 - \exp\left(-\frac{VC}{k}\right) \qquad\qquad (6\text{-}2)$$

The confidence distribution to the natural logarithm of *k*, as determined by likelihood theory (Morgan, 1992), was found to be closely approximated by a normal distribution with a mean of 5.48 and a standard deviation of 0.32.

This dose-response relationship, and the variability of *k*, is only for the particular strain used in the underlying human volunteer studies. There is known to be variability in the infectivity of differing strains of *Cryptosporidium parvum* (Chappell, 1999) of perhaps an order of magnitude. However quantitative dose-response relationships for these other strains are not yet available. Hence strain-to-strain variability cannot be accounted for in a quantitative sense at this time.

Risk Assessment Calculation

Given a single value of water consumption (V), oocyst concentration (C), and the dose-response parameter (*k*), the risk of infection to an individual can be computed by application of Equation 6-2. To consider the distribution of risk that incorporates uncertainty and variability in each of the input parameters, this computation needs to be performed a large number of times. For each repetition, a new set of random samples (for water consumption, oocyst concentration at each location, and the dose-response parameter) must be obtained. Individual calculations using these sets of random samples are combined to reveal an esti-mated distribution of risk. This technique is termed the application of Monte Carlo methods to risk assessment, and it has received wide use (Burmaster and Anderson, 1994; Vose, 1996).

Two types of results are presented below. First, the daily risk estimate is calculated for each individual year, given a single water dose, dose-response parameter, and average oocyst concentration. Four oocyst concentrations are used, representing the different methods for considering data points below the detection limit. The purpose of this exercise is to observe trends in the risk estimate over time. The second set of results shows the range of estimated risk, taking into account uncertainty in all of the input parameters. This range is generated using the combined data from 1992 to 1998.

Results: Point Estimates

Point estimates for the daily risk infection from *Cryptosporidium* are presented in Table 6-10. The volume of water consumed was 1.95 L/day and k was 238. Calculations were done using both the total (1992–1998) data set and for each year individually, and they do not take into account variability in any of the input parameters. Table 6-10 presents results of four different methods used to determine the average oocyst concentration: the maximum likelihood method and the three "fill in" methods. Depending on which of the methods was used, the daily risk ranges from zero to 12.2×10^{-5}. The table demonstrates that the risk estimate has dropped between 1993 and 1996.

When fewer than two quantified (above-detection-limit) observations existed, the maximum likelihood method could not be used (as in the cases of data for 1995 and later). The bias inherent in the fill-in methods is particularly evident for 1995 and subsequent data. That is, risk values derived from using the full detection limit and half detection limit methods are likely to be overestimates of the true risk, while those derived from using a zero detection limit method are likely to be underestimates.

TABLE 6-10 Computed Point Estimates for the Daily Risk of Infection from *Cryptosporidium* (all numbers $\times 10^{-5}$)

| Year | Maximum Likelihood Method | Fill-in Methods | | |
		Detection Limit	Half Detection Limit	Zero Detection Limit
all years	3.2	7.1	4.7	2.3
1992	10.7	7.8	6.5	5.3
1993	10.8	12.2	10.9	9.7
1994	3.1	6.9	4.6	2.2
1995	*	5.7	3.0	0.2
1996	*	5.7	2.9	0.09
1997	*	5.6	2.9	0.1
1998 (Jan–June)	*	5.6	2.9	0

*Could not be estimated because fewer than two quantified observations are available.

Results: Monte Carlo Simulation

Although useful, point estimates of risk do not reveal the degree of uncertainty in the risk estimate. Monte Carlo simulations are currently the most rigorous way to take uncertainty into account during risk assessment, assuming high quality data are available. Summary statistics on 10,000 iterations of the Monte Carlo model are shown in Table 6-11. For this computation, the entire oocyst concentration database (1992–1998) was used.

The main result of the analysis is that the mean individual daily risk is estimated as 3.4×10^{-5}. For an exposed population of 7.5 million, this would translate into an estimated 255 infections per day. The range of the 95 percent confidence limits would translate into an estimated range of infections per day from 2.6–1,643. It should be noted that the results of the Monte Carlo analysis bracket the range of point estimates observed by considering each year's data set separately, whether maximum likelihood or "fill-in" methods are used.

As part of this computation, a sensitivity analysis was conducted. The rank correlation of the individual daily risk with the various input parameters was computed (Table 6-12). The densities of pathogens in the two effluent flows

TABLE 6-11 Summary of 10,000 Monte Carlo Trials on Kensico Risk Assessment: Daily Risk of *Cryptosporidium* Infection ($\times 10^{-5}$)

Statistic	Individual Daily Risk	Daily # of Infections[a]
Mean	3.4	255
Median	0.7	53
Standard Deviation	19.8	1,485
Lower 95% confidence limit	0.034	2.6
Upper 95% confidence limit	21.9	1,643

[a]Based on an exposed population of 7.5 million persons.

TABLE 6-12 Rank Correlation of Input Parameters with Daily Risk of Infection[a]

Input Parameter	Rank Correlation with Daily Risk
DEL 18 oocyst density	0.61
CATLEFF oocyst density	0.56
Water consumption	0.24
Dose-response "k" value	−0.20

[a] Rank correlation is the correlation between two sets of data when the individual observations in each set are replaced by the rank in that set.

from Kensico have the greatest correlation with the estimated daily risk. The other inputs (water consumption, dose-response parameter) contribute only a minor amount to the uncertainty and variability of the estimated risk. This suggests that to reduce the degree of uncertainty and variability in the estimated risk, attention should be paid primarily to obtaining better (more precise) estimates of the effluent oocyst concentrations.

Caveats

Dose-Response Relationship

The above risk assessment has a number of caveats that should be taken into account in making a decision based on these results. First, the dose-response relationship was obtained from a study on healthy volunteers who were believed to have no prior history of cryptosporidiosis (Dupont et al., 1995). Prior exposure to *Cryptosporidium* may result in reduced susceptibility (Okhuysen et al., 1998). On the other hand, the elderly, children, and persons with lowered immunity (e.g., those on antirejection drugs after organ transplantation, recipients of cancer chemotherapy, and persons with HIV infection) are in general more susceptible to infectious diseases such as cryptosporidiosis (Gerba et al., 1996). The same populations may also suffer more severe symptoms than the general population, as was demonstrated during the cryptosporidiosis outbreaks in Milwaukee (Hoxie et al., 1997) and Las Vegas (Goldstein et al., 1996).

Secondary Infection

The above risk analysis does not incorporate consideration of community-level impacts, such as the formation of secondary cases. Such cases refer to individuals not directly infected by water exposure but who are exposed to other infected individuals. A consideration of secondary infections can raise the risk estimate. For example, in the 1993 Milwaukee outbreak, 4.2 percent of households with one or more ill persons also contained one or more secondary cases (MacKenzie et al., 1995). In a foodborne (apple cider) outbreak of cryptosporidiosis, for each primary case, there were one-third as many secondary cases (Millard et al., 1994).

Oocyst Viability and Recovery

The above analysis used total oocysts found in the Kensico Reservoir raw water sampling locations to assess exposure. Not all of the total oocysts represent viable, human, infectious forms. Although several methods are available, there is no rapid, inexpensive, and reliable method for determining oocyst viability on a routine basis. One frequently used method is to calculate viability based on the ratio of "confirmed" to "presumed" oocysts, terms that are applied to oocysts during microscopic observation. However, this viability analysis is not scientifically defensible because it has never been conclusively established that oocyst collection and processing steps do not themselves result in loss of microscopic features required for confirmation. Simple shaking of oocysts with sand has been found to cause loss of internal structure of oocysts (Parker and Smith, 1993).

The efficiency with which the sampling methodology can recover oocysts can be quite low (Clancy et al., 1994). Oocyst recovery in New York City currently ranges from 30 percent to 70 percent (Stern, 1998). In prior risk assessments for protozoa, it has been assumed that the two errors (errors related to viability and to recovery) roughly cancel (Regli et al., 1991; Rose et al., 1991).

Endpoints

The focus of this risk assessment is on infections. It should be recognized that the end-point of clinically confirmed human illness may be substantially less than this. Based on human feeding studies, only about half of all infections progress to frank illness (Dupont et al., 1995). Even if illness occurs, in normal healthy individuals symptoms may be mild and not cause medical attention to be sought. If the endpoint of illness is used, the risk assessment would predict a lower number of symptomatic cases.

The endpoint for this risk assessment—infection—is different than the endpoint measured by active disease surveillance—frank illness that is diagnosed and reported. This is one of the reasons that the risk estimate predicts a higher rate of infection than is observed in the active disease surveillance program. Other factors, such as the limited sensitivity of active disease surveillance and the contribution of other vectors such as food, are also responsible for discrepancies between the risk assessment and measured surveillance rates.

Time Period

The exposure estimate in the risk analysis used the entire data record at the two locations. If the lower oocyst levels, which have been seen in more recent years (Figure 6-6), are assumed to be more typical of future oocyst levels, then a lower exposure and consequently a lower risk would be estimated.

Strains of Cryptosporidium

It is noted that the present dose-response relationship derives from a single set of studies on a single oocyst strain (the "Iowa" or "Harley Moon" strain). Information currently being analyzed suggests that other strains of oocysts may have higher and lower infectivities and different dose-response curves than the

"Iowa" strain (Chappell, 1999). This represents an additional source of potential variability in the risk assessment that should be amenable to quantitation.

Other Sources of Water Supply

The above analyses assumed that the sole oocyst loading reaching consumers was from the Catskill/Delaware supply. Although this represents the dominant source of drinking water for New York City, some residents are primarily served by water from the Croton system. To the degree that the (current) quality of raw water from the Croton system with respect to oocysts is different than the Catskill/Delaware, the numbers in Table 6-11 may under- or overestimate the total risk to some consumers of New York City water.

Impact of Treatment and Watershed Management

If the watershed management programs described elsewhere in this report are successfully implemented, the level of oocysts in the Catskill/Delaware supply may decrease. Although it is not yet possible to quantitatively forecast the magnitude of this decrease, a reduction in risk to consumers is expected. A primary motivation for conducting microbial risk assessment in water supply systems that pursue watershed management should be to determine the contribution of watershed management to overall risk reduction.

Quantifying the impacts of other treatment processes on the risk estimate is a more straightforward task. Ozonation and particle removal will decrease the oocyst levels in drinking water, and it is possible to determine the magnitude of this reduction based on pilot-scale testing. Given standard treatment efficiencies, a properly functioning water filtration plant is expected to achieve at least a 2-log removal of *Cryptosporidium* oocysts (Nieminski and Ongerth, 1995).

For both treatment processes and watershed management activities, a 1-log reduction in oocyst concentration translates directly into a 1-log reduction in the risk estimate because the dose-response relationship is linear at low dose. In other words, any process that reduces oocyst levels in the Kensico Reservoir by a factor of ten will reduce the risk estimate to 0.34×10^{-5} per person per day, or 25.5 infections per day, in a population of 7.5 million persons.

Conclusions and Recommendations on Risk Assessment

1. Based on the committee's risk assessment using data from 1992 to 1998, the current daily risk of *Cryptosporidium* infection in New York City is 3.4×10^{-5}, with a 95 percent confidence interval ranging from 3.4×10^{-7} to 21.9×10^{-5}. EPA has stated that less than one microbially-caused illness per year per 10,000 people is a reasonable policy (EPA, 1989).[5] This risk level, which corresponds to 10^{-4} per year or 2.7×10^{-7} per day, is smaller than the lower 95 percent confidence interval of the estimated daily risk for New York City based on the Catskill and Delaware supplies (Table 6-11). It is also below the point estimates for risk during individual years. Hence, based on the assumptions used, the calculated risk of cryptosporidiosis would appear to be in excess of the frequently propounded acceptable risk level.

The calculated risk estimate must be considered in conjunction with the caveats listed above. The risk estimate does not take into account measurements of oocyst viability and recovery, secondary infection, multiple strains of *Cryptosporidium*, or multiple dose-response relationships. The endpoint of the risk assessment was assumed to be infection, several years of data were used, and only the Catskill/Delaware system was considered. Finally, the impacts of watershed management on the risk estimate are not quantified.

2. It is recommended that a *Cryptosporidium* risk assessment be performed on a periodic basis for New York City. The goal of these efforts should be to help determine the contribution of watershed management (vs. other treatment options and management strategies) to overall risk reduction. Data that are sufficient for these purposes are currently collected as part of the NYC DEP Pathogen Studies. As new methods for oocyst recovery, detection, speciation (bird vs. human vs. animal), and viability become available, the risk assessment methods used in this report should be improved upon. Depending on the frequency of monitoring, risk assessment can be calculated for varying time periods to assess potential high-risk exposure times, such as during certain seasons and during storm events.

Prior to commencing this regular effort, a decision must be made as to what level of risk is deemed to be acceptable to the regulatory agencies, the City, and the affected parties. This level should be arrived at after full and open discussion with the various stakeholders. Should an annual risk level of greater than 10^{-4} be regarded as acceptable by NYC DEP or other relevant risk managers, then the risk estimates computed in this report can be compared to such alternate yardsticks.

3. An ongoing program of risk assessment should be used as a complement to active disease surveillance. Risk assessment allows one to ascertain the level of infection implied by a very low level of exposure that would go undetected by active surveillance, thus acting as a complementary source of information

---

[5] It should be noted that although the 10^{-4} risk level was developed for giardiasis, it is the only EPA-endorsed value available with which to compare current risks of cryptosporidiosis. As suggested in the second recommendation, New York City should determine an acceptable risk level before undertaking regular risk assessments.

about public health. In combination, risk assessment and active disease surveillance data could be used to estimate the proportion of gastrointestinal disease cases attributable to drinking water (as in Perz et al., 1998). These estimates could then be validated by comparison with epidemiological data from a cohort study. In general, periodic risk estimates should be examined for concordance with prior computed risks and observed illness rates when formulating subsequent water treatment and watershed management decisions.

REFERENCES

American Public Health Association (APHA). 1998. Standard Methods for the Examination of Water and Wastewater. 20th Edition. Clesceri, L. S., A. E. Greenberg, and A. D. Eaton (eds.). Washington, DC: American Public Health Association.

Angulo, F. J., S. Tippen, D. J. Sharp, B. J. Payne, C. Collier, J. E. Hill, T. J. Barrett, R. M. Clark, E. E. Geldreich, H. D. Donnell, and D. L. Swerdlow. 1997. A community waterborne outbreak of salmonellosis and the effectiveness of a boil water order. Am. J. Pub. Health 87(4):580–584.

Ashendorff, A. 1999. NYC DEP. E-mail memorandum to the National Research Council dated January 6, 1999.

Bagley, S. T., M. T. Auer, D. A. Stern, and M. J. Babiera. 1998. Sources and fate of *Giardia* cysts and *Cryptosporidium* oocysts in surface water. Journal of Lake and Reservoir Management 14(2-3): 379–392.

Boyce, T. G., A. G. Pemberton, and D. G. Adiss. 1996. *Cryptosporidium* testing practices among clinical laboratories in the US. Ped. Infect. Dis. J. 15:87–88.

Bukowski, J., L. Korn, and D. Wartenberg. 1995. Correlated inputs in quantitative risk assessment: The effects of distributional shape. Risk Analysis 15(2):215–219.

Burmaster, D. E., and P. D. Anderson. 1994. Principles of good practice for the use of Monte Carlo techniques in human health and ecological risk assessment. Risk Analysis 14(4):477–481.

Cabelli, V. J. 1977. *Clostridium perfringens* as a Water Quality Indicator. Bacterial Indicators/ Health Hazards Associated with Water. A. Hoadley and B. Dutka. Philadelphia, PA: ASTM.

Carmichael , W. W., C. L. A. Jones, N. A. Mahmood, and W. C. Theiss. 1985. Algal toxins and water-based diseases. Crit. Rev. Environ. Control 15:275–313.

Carmichael, W. W. 1986. Algal toxins. Adv. Bot. Res. 12: 47–101.

Centers for Disease Control and Prevention (CDC). 1995. Assessing the public health threat associated with waterborne cryptosporidiosis: report of a workshop. MMWR 1995; 44(RR-6): 1–19.

Chappell, C. 1999. Presentation at Heath Effects Stakeholder Meeting for the Stage 2 DBPR and LT2ESWTR. USEPA. February 1999, Washington, DC.

Chute, C. G., R. P. Smith and J. A. Baron. 1987. Risk factors for endemic giardiasis. Am. J. Public Health 77(5):585–587.

Clancy, J. L., W. Gollnitz, and Z. Tabib. 1994. Commercial labs: How accurate are they? Journal of the American Water Works Association 86(5):89–97.

Clancy, J. L., C. R. Fricker, C. R. Fricker, and W. Telliard. 1997. New US EPA Standard Method for *Cryptosporidium* Analysis in Water. AWWA Water Quality Technology Conference. Denver, CO: American Water Works Association.

Clarke, R. T. 1994. Statistical Modeling in Hydrology. New York, NY: John Wiley & Sons.

Codd, G. A., S. G. Bell and W. P. Brooks. 1989. Cyanobacterial toxins in water. Water Science and Technology 21:1–13.

Craun, G., G. Birkhead, S. Erlandsen, F. Frost, W. Jakubowski, D. Juranek, R. Soave, C. Sterling, and B. Ungar. 1994. Report of New York City's Advisory Panel on Waterborne Disease Assessment. NYC DEP.

Davis, L. J., H. L. Roberts, D. D. Juranek, S. R. Framm and R. Soave. 1998. A survey of risk factors for cryptosporidiosis in New York City: drinking water and other exposures. Epidemiol. Infect. 121:357–367.

Dennis, D. T., R. P. Smith, J. J. Welch, C. G. Chute, B. Anderson, J. L. Herndon and C. F. vonReyn. 1993. Endemic giardiasis in New Hampshire: a case-control study of environmental risks. J. Infect. Dis. 167(6):1391–1395.

Doerr, S. M., E. M Owens, R. K. Gelda, M. T. Aurer, and S. W. Effler. 1998. Development and testing of a nutrient-phytoplankton model for Cannonsville Reservoir. Journal of Lake and Reservoir Management 14(2-3):301–321.

Dupont, H. L., C. C. Chappell, C. R. Sterling, P. C. Okhuysen, J. B. Rose, and W. Jakubowski. 1995. Infectivity of *Cryptosporidium parvum* in healthy volunteers. New England Journal of Medicine 332(13):855.

El Saadi, O., A. J. Esterman, S. Cameron, and D. M. Roder. 1995. Murray River water, raised cyanobacterial cell counts and gastrointestinal and dermatological symptoms. Med. J. Aust. 162:122–125.

Environmental Protection Agency (EPA). 1989. Federal Register 54:27486 (June 29, 1989)

EPA. 1999. Method 1622: *Cryptosporidium* in Water by Filtration/IMS/FA. EPA-821-R-99-001. Washington, DC: EPA Office of Water.

Falconer, I. R., A. M. Beresford, and M. T. Runnegar. 1983. Evidence of liver damage by toxin from a bloom of the blue-green alga *Microcystis aeruginosa*. Med. J. Aust. 1(11):511-514.

Falk, C. C., P. Karanis, D. Schoenen, and H. M. Seitz. 1998. Bench Scale Experiments for the Evaluation of a Membrane Filtration Method for the Recovery Efficiency of *Giardia* and *Cryptosporidium* from Water. Water Research 32(3):565-568.

Finch, G. R., L. L. Gyurek, L. R. K. Liyanage, and M. Belosevic. 1998. Effects of Various Disinfection Methods on the Inactivation of *Cryptosporidium*. AWWA Research Foundation and American Water Works Association. Denver, C.O.

Frost, F. J., G. F. Craun, and R. L. Calderon. 1996. Waterborne disease surveillance. Journal of the American Waterworks Association 88(9):66–75.

Frost, F. J., A. A. de la Cruz, D. M. Moss, M. Curry, and R. L. Calderon. 1998a. Comparisons of ELISA and Western blot assays for detection of *Cryptosporidium* antibody. Epidemiology and Infection 121(1)205-211.

Frost, F. J., R. L. Calderon, T. B. Muller, M. Curry, J. S. Rodman, D. M. Moss, and A. A. de la Cruz. 1998b. A two-year follow-up survey of antibody to *Cryptosporidium* in Jackson County, Oregon following an outbreak of waterborne disease. Epidemiology and Infection 121(1):213–217.

Frost, F. J., G. F. Craun, and R. L. Calderon. 1996. Waterborne disease surveillance. Journal of the American Waterworks Association 88(9):66–75.

Geldreich, E. E. 1978. Bacterial populations and indicator concepts in feces, sewage, stormwater and solid wastes. In Berg, G. (ed.) Indicators of Viruses in Water and Food. Ann Arbor, MI: Ann Arbor Science.

Gerba, C. P., J. B. Rose, and C. N. Haas. 1996. Sensitive populations: Who is at the greatest risk? International Journal of Food Microbiology 30(1-2):113–123.

Goldstein, S. T., D. D. Juranek, O. Ravenholt, A. W. Hightower, D. G. Martin, J. L. Mesnik, S. D. Griffiths, A. J. Bryant, R. R. Reich, and B. L. Herwaldt. 1996. Cryptosporidiosis: An outbreak associated with drinking water despite state of the art water treatment. Annals of Internal Medicine 124(5):459–468.

Griffiths, J. K. 1999. Serological testing for Cryptosporidiosis. Presentation at U.S. EPA Office of Ground Water and Drinking Water Stage 2 Microbial/Disinfection Byproducts Health Effects Workshop. February 10-12. Washington, DC.

Haan, C. T. 1977. Statistical Methods in Hydrology. Ames, IA: Iowa State University Press.

Haas, C. N., and P. A. Scheff. 1990. Estimation of averages in truncated samples. Environmental Science and Technology 24:912–919.

Haas, C. N., J. B. Rose, C. Gerba, and S. Regli. 1993. Risk assessment of virus in drinking water. Risk Analysis 13(5):545–552.

Haas, C. N., C. Crockett, J. B. Rose, C. Herba, and A. Fazil. 1996. Infectivity of *Cryptosporidium parvum* oocysts. Journal of the American Water Works Association 88(9):131–136.

Hansen, J. S., and J. E. Ongerth. 1991. Effects of time and watershed characteristics on the concentration of *Cryptosporidium* oocysts in river water. Applied and Environmental Microbiology 57(10):2790–2795.

Harada, K., M. Oshikata, H. Uchida, M. Suzuki, F. Kondo, K. Sato, Y. Ueno, S.Z. Yu, G. Chen and G.C. Chen. 1996. Detection and identification of microcystins in the drinking water of Haimen City, China. Nat. Toxins 4(6):277–283.

Hayman, J. 1992. Beyond the Barcoo—probably human tropical cyanobacterial poisoning in outback Australia. Med. J. Aust. 157:794–796.

Helsel, D. R., and T. A. Cohn. 1988. Estimation of descriptive statistics for multiply censored water quality data. Water Resources Research 24(12): 1997–2004.

Hosking, J. R. M., and J. R. Wallis. 1997. Regional Frequency Analysis: An Approach Based on L-Moments. London: Cambridge University Press.

Hoxie, N. J., J. P. Davis, J. M. Vergeront, R. D. Nashold, and K. A. Blair. 1997. Cryptosporidiosis-associated mortality following a massive waterborne outbreak in Milwaukee, Wisconsin. American Journal of Public Health 87(12):2032–2035.

Hsu, F. C., Y. S. Shieh, J. van Duin, M. J. Beekwilder, and M. D. Sobsey. 1995. Genotyping male-specific RNA coliphages by hybridization with oligonucleotide probes. Applied and Environmental Microbiology 61(11):3960–3966.

International Life Sciences Institute (ILSI). 1998. Comprehensive Watershed Monitoring: A Framework for the New York City Reservoirs. Washington, DC: ILSI Risk Science Institute.

Juranek, D. 1997. CDC. E-mail from Dennis Juranek to the National Research Council Committee.

Juranek, D. 1999. CDC. E-mail from Dennis Juranek to the National Research Council Committee, May 1999.

Korich, D. G., J. R. Mead, M. S. Madore, N. A. Sinclair, and C. R. Sterling. 1990. Effects of ozone, chlorine dioxide, chlorine, and monochloramine on *Cryptosporidium parvum* oocyst viability. Applied and Environmental Microbiology 56(5):1423–1428.

LeChevallier, M. W., W. Norton, M. Abbaszadegan, T. Atherholt, and J. Rosen. 1997. Variations in *Giardia* and *Cryptosporidium* in Source Water: Statistical Approaches to Analyzing ICR Data. In AWWA Water Quality Technology Conference. Denver, CO: American Water Works Association.

Likens, G. E., (ed). 1985. An Ecosystem Approach to Aquatic Ecology: Mirror Lake and Its Environment. New York, NY: Springer-Verlag.

Likens, G. E., and F. H. Bormann. 1995. Biogeochemistry of a Forested Ecosystem. 2nd Edition. New York, NY: Springer-Verlag.

Longabucco, P., and M. Rafferty. 1998. Analysis of material loading to Cannonsville Reservoir: Advantages of event-based sampling. Journal of Lake and Reservoir Management 14(2-3):197–212.

MacKenzie, W. R., W. L. Schell, B. A. Blair, D. G. Addiss, D. E. Peterson, N. J. Hozie, J. J. Kazmierczak, and J. P. Davis. 1995. Massive outbreak of waterborne *Cryptosporidium* infection in Milwaukee, Wisconsin: Recurrence of illness and risk of secondary transmission. Clinical Infectious Diseases 21:57–62.

MacKenzie, W. R. 1998. Presentation given at the Risk Assessment Workshop for the NRC Committee to Review the New York City Watershed Management Strategy. Atlanta, GA. April 14–15, 1998.

Marx, R., and E. A. Goldstein. 1999. Under Attack: New York's Kensico and West Branch Reservoirs Confront Intensified Development. New York, NY: Natural Resources Defense Council.

McKnight, D. M., R. Harnish, R. L. Wershaw, J. S. Baron, and S. Schiff. 1997. Chemical characteristics of particulate, colloidal, and dissolved organic material in Loch Vale Watershed, Rocky Mountain National Park. Biogeochemistry 36:99–124.

Millard, P., K. Gensheimer, D. G. Addiss, D. M. Sosin, G. A. Beckett, A. Houck-Jankoski, and A. Hudson. 1994. An outbreak of cryptosporidiosis from fresh-pressed apple cider. Journal of the American Medical Association 272(20):1592–1596.

Miller, J. 1998. Presentation given at the Risk Assessment Workshop for the NRC Committee to Review the New York City Watershed Management Strategy. Atlanta, GA. April 14–15, 1998.

Morgan, B. J. T. 1992. Analysis of Quantal Response Data. London: Chapman and Hall.

National Research Council (NRC). 1983. Risk Assessment in the Federal Government: Managing the Process. Washington, DC: National Academy Press.

New York City Department of Environmental Protection (NYC DEP). 1997a. Water Quality Surveillance Monitoring. Valhalla, NY: NYC DEP.

NYC DEP. 1997b. Results of Groundwater Monitoring in the Kensico Reservoir Watershed. July 1997. Valhalla, NY: NYC DEP.

NYC DEP. 1997c. The Relationship Between Phosphorus Loading, THM Precursors, and the Current 20 µg/L TP Guidance Value. December 1997. Valhalla, NY: NYC DEP.

NYC DEP. 1997d. Pathogen Laboratory: Standard Operating Procedure. January 1997. Valhalla, NY: NYC DEP.

NYC DEP. 1997e. Waterborne Disease Risk Assessment Program 1996 Annual Report. Corona, NY: NYC DEP.

NYC DEP. 1998a. Kensico Watershed Study Annual Research Report: April 1997–March 1998. Valhalla, NY: NYC DEP.

NYC DEP. 1998b. Distribution Sampling Plan. March 26, 1997. Valhalla, NY: NYC DEP.

NYC DEP. 1998c. DEP Pathogen Studies of Giardia spp., Cryptosporidium spp., and Enteric Viruses. January 31, 1998. Valhalla, NY: NYC DEP.

NYC DEP. 1998d. Calibrate and verify GWLF models for remaining Catskill/Delaware Reservoirs.

NYC DEP. 1998e. Waterborne Disease Risk Assessment Program 1997 Annual Report. Corona, NY: NYC DEP.

NYC DEP. 1999a. Filtration Avoidance Report for the Period January 1 through December 31, 1998. Corona, NY: NYC DEP.

NYC DEP. 1999b. Kensico Watershed Study Semi-annual Progress Report. Valhalla, NY: NYC DEP.

NYC DEP. 1999c. Waterborne Disease Risk Assessment Program 1998 Annual Report. Corona, NY: NYC DEP.

New York State Water Resources Institute (NYS WRI). 1997. Science for Whole Farm Planning: Cornell University Phase II Twelfth Quarter and Completion Report.

Nieminski, E. C., and J. E. Ongerth. 1995. Removing *Giardia* and *Cryptosporidium* by conventional treatment and direct filtration. J. Amer. Waterworks Assoc. 9:96–106.

Okhuysen, P. C., C. L. Chappell, C. R. Sterling, W. Jakubowski, and H. L. DuPont. 1998. Susceptibility and serologic response of health adults to reinfection with *Cryptosporidium parvum*. Infection and Immunity 66(2):441–443.

Ong, C., W. Moorehead, A. Ross, and J. Isaac-Renton. 1996. Studies of *Giardia* spp. and *Cryptosporidium* spp. in two adjacent watersheds. Applied and Environmental Microbiology 62(8):2798–2805.

Ongerth, J. E. 1989. *Giardia* cyst concentrations in river water. Journal of the American Water Works Association 81(9): 81–86.

Otsuki, A., and R. G. Wetzel. 1974. Calcium and total alkalinity budgets and calcium carbonate precipitation of a small hard-water lake. Arch. Hydrobiol. 73:14–30.

Owens, E. M. 1998. Development and testing of one-dimensional hydrothermal models of Cannonsville Reservoir. Journal of Lake and Reservoir Management 14(2-3):172–185.

Parker, J. F. W., and H. V. Smith. 1993. Destruction of Oocysts of *Cryptosporidium parvum* by Sand and Chlorine. Water Research 27(4):729–731.

Paul, J. H., J. B. Rose, J. Brown, E. A. Shinn, S. Miller, and S. H. Farrah. 1995. Viral tracer studies indicate contamination of marine waters by sewage disposal practices in Key Largo, Florida. Applied and Environmental Microbiology 61:2230–2234.

Payment, P., and E. Franco. 1993. *Clostridium perfringens* and somatic coliphages as indicators of the efficiency of drinking water treatment for viruses and protozoan cysts. Applied and Environmental Microbiology 59(8):2418–2424.

Payment, P., L. Richardson, J. Siemiatycki, R. Dewar, M. Edwardes and E. Franco. 1991. A randomized trial to evaluate the risk of gastrointestinal disease due to consumption of drinking water meeting current microbiological standards. Am. J. Public Health 81(6):703–708.

Payment, P., J. Siemiatycki, L. Richardson, G. Renaud, E. Franco and M. Prevost. 1997. A prospective epidemiological study of gastrointestinal health effects due to the consumption of drinking water. Int. J. Env. Health Res. 7:5–31.

Perz, J. F., F. K. Ennever, and S. M. Le Blancq. 1998. *Cryptosporidium* in tap water: Comparison of predicted risks with observed levels of disease. American Journal of Epidemiology 147(3):289–301.

Proctor, M. E., K. A. Blair, and J. P. Davis. 1998. Surveillance data for waterborne illness detection: an assessment following a massive waterborne outbreak of *Cryptosporidium* infection. Epidemiol. Infect. 120:43–54.

Purdue, E. M. 1984. Analytical constraints on the structural features of humic substances. Geochim. Cosmochim. Acta 48:1435–1442.

Ransome, M. E., T. N. Whitmore, and E. G. Carrington. 1993. Effect of disinfectants on the viability of *Cryptosporidium parvum* oocysts. Water Supply 11:75–89.

Regli, S., J. B. Rose, C. N. Haas, and C. P. Gerba. 1991. Modeling the risk from *Giardia* and viruses in drinking water. Journal of the American Water Works Association 83(11):76–84.

Rodman, J. S., F. Frost, and W. Jakubowski. 1998. Using nurse hotline calls for disease surveillance. Emerging Infectious Diseases 4(2):329–32.

Rose, J. B., C. N. Haas, and S. Regli. 1991. Risk assessment and the control of waterborne giardiasis. American Journal of Public Health 81:709–713.

Roseberry, A. M., and D. E. Burmaster. 1992. Log-normal distributions for water intake by children and adults. Risk Analysis 12(1):99–104.

Sacks, J. J., S. Lieb, L. M. Baldy, S. Berta, C. M. Paton, M. C. White, W. J. Bigler, and J. J. Witte. 1986. Epidemic campylobacteriosis associated with a community water supply. Am. J. Pub. Health 76(4):424-8.

Schwartz, J., R. Levin and K. Hodge. 1997. Drinking water turbidity and pediatric hospital use for gastrointestinal illness in Philadelphia. Epidemiology 8(6):607–609.

Shepherd, K. M., and A. P. Wyn-Jones. 1995. Evaluation of different filtration techniques for the concentration of Cryptosporidium oocysts from water. Water Science and Technology 31(5-6):425–429.

Slifko, T. R., D. Friedman, J. B. Rose, and W. R. Jakubowski. 1997. An *In vitro* Method for Detecting Infectious *Cryptosporidium* Oocysts with Cell Culture. Applied and Environmental Microbiology 63(9):3669–3675.

Smith, A. E., P. B. Ryan, and J. S. Evans. 1992. The effect of neglecting correlations when propagating uncertainty and estimating the population distribution of risk. Risk Analysis 12(4):467–474.

Soave, R., L. J. Davis, and D. Juranek. 1998. Meeting of the Infectious Disease Society of America.

Soong, F. S., E. Maynard, K. Kirke and C. Luke. 1992. Illness associated with blue-green algae. Med. J. Aust. 156:67.

Stern, D. 1996. Initial Investigation of the Sources and Sinks of *Cryptosporidium* spp. and *Giardia* spp. within the watersheds of the New York City Water Supply System. In New York City Water Supply Studies. Proceedings of a Symposium on Watershed Restoration Management. American Water Resources Association. Herndon, VA.

Stern, D. 1998. NYC DEP. E-mail memorandum to the National Research Council dated November 16, 1998.

Stepczuk, C. L., A. B. Martin, P. Longabucco, J. A. Bloomfield, and S. W. Effler. 1998a. Allochthonous contributions of THM precursors in a eutrophic reservoir. Journal of Lake and Reservoir Management 14(2–3):344–355.

Stepczuk, C. L., A. B. Martin, S. W. Effler, J. A. Bloomfield, and M. T. Auer. 1998b. Spatial and temporal patterns of THM precursors in a eutrophic reservoir. Journal of Lake and Reservoir Management 14(2–3): 356–366.

Stepczuk, C. L., E. M. Owens, S. W. Effler, J. A. Bloomfield, and M. T. Auer. 1998c. A modeling analysis of THM precursors for a eutrophic reservoir. Journal of Lake and Reservoir Management 14(2–3):367–378.

Stewart, M. H., D. M. Ferguson, R. De Leon, and W. D. Taylor. 1997. Monitoring Program To Determine Pathogen Occurrence In Relationship To Storm Events And Watershed Conditions. In AWWA Water Quality Technology Conference. Denver, CO: American Water Works Association.

Thurman, E. M. 1985. Organic Geochemistry of Natural Waters. Martinus Nijhoff/Dr. W. Junk Publishers, Dordrecht.

Turner, P. C., A. J. Gammie, K. Hollinrake and G. A. Codd. 1990. Pneumonia associated with contact with cyanobacteria. Br. Med. J. 300:1440–1441.

Ueno, Y., S. Nagata, T. Tsutsumi, A. Hasegawa, M. F. Watanabe, H. D. Park, G. C. Chen, G. Chen, and S. Z. Yu. 1996. Detection of microcystins, a blue-green algal hepatotoxin, in drinking water sampled in Haimen and Fusui, endemic areas of primary liver cancer in China, by highly sensitive immunoassay. Carcinogenesis 17(6):1317–1321.

Vose, D. 1996. Quantitative Risk Analysis: A Guide to Monte Carlo Simulation Modeling. New York, N.Y.: John Wiley.

Walker, F. R., Jr., and J. R. Stedinger. 1999. Fate and Transport Model of *Cryptosporidium*. Journal of Environmental Engineering 125(4):325–333.

Warne, D. 1998. NYC DEP. Memorandum to the National Research Council dated November 1998.

Wetzel, R. G., and A. Otsuki. 1974. Allochthonous organic carbon of a marl lake. Arch. Hydrobiol. 73:31–56.

Wetzel, R. G., P. G. Hatcher, and T. S. Bianchi. 1995. Natural photolysis by ultraviolet irradiance of recalcitrant dissolved organic matter to simple substrates for rapid bacterial metabolism. Limnol. Oceanogr. 40:1369–1380.

Wilson, M. A., A. M. Vassallo, E. M. Purdue, and J. H. Reuter. 1987. Compositional and solid-state nuclear magnetic resonance study of humic and fulvic acid fractions of soil organic matter. Anal. Chem. 59:551–558.

World Health Organization. 1998. Guidelines for Drinking Water Quality. Second Edition. Addendum to Volume 2: Health Criteria and Other Supporting Information. Geneva, Switzerland: World Health Organization.

7

Land Acquisition and
Land Use Planning

This chapter focuses on the use of nonstructural protection strategies for watershed management. As mentioned in Chapter 4, nonstructural protection strategies specify how watershed land can or should be used to prevent pollution of source waters. In publicly owned watersheds, nonstructural protection strategies are unnecessary for the most part, because the drinking water utility can limit access to critical areas and maintain pristine conditions. However, in watersheds such as New York City's that have a substantial population and where much of the land is in private hands, nonstructural protection strategies play a major role in watershed management.

Nonstructural protection strategies in the Memorandum of Agreement (MOA) consist of two main programs: (1) the Land Acquisition Program and (2) the Watershed Protection and Partnership Programs. These programs were designed to accommodate the desire to maintain high water quality in the water supply reservoirs while allowing economic growth in the watershed area. This chapter reviews and critiques several plans concerning land acquisition and management of the acquired lands, land use, and economic development in the watershed communities. Although these issues were not originally part of the study mandate, land use and economic development planning are critical to the MOA's goals of source water protection, economic development in the watershed region, and filtration avoidance and hence were considered by the committee.

LAND ACQUISITION

Purchasing private land is one of the most important nonstructural tools used to protect a watershed. New York City's Land Acquisition and Stewardship program is a key component of the watershed management strategy (MOA, Article II). Unlike the Watershed Rules and Regulations, which are meant to curb current pollutant loading, this program is designed to prevent future contamination of the water supply. The Land Acquisition Program allows New York City to acquire land through selective outright purchase or through the use of conservation easements. A conservation easement is a covenant that limits or restricts development, management, or use of property, protects important natural features of property, and provides the landowner with certain retained rights (NYC DEP, 1997). As part of its filtration avoidance determination, the City must solicit interest by owners of 355,050 acres of land in the Catskill/Delaware watershed over the next ten years and commit $250 million to this effort. (The City is also contacting owners of land in the Croton watershed, but there are no acreage goals.)

Lands are being purchased only from willing sellers and for full market price, as appraised by independent, certified appraisal companies under contract with the City. This is unusual because the City has a legal right (through the State health codes) to take lands in the watershed region by eminent domain. However, it has agreed not to do so as an explicit concession in the MOA. In order to protect the budgetary base of watershed towns and villages, New York City is paying property taxes on the land it now owns and on all new land and conservation easements it acquires. (For conservation easements, the amount of property tax the City will pay is proportional to the value of the easement relative to the overall property value.) The New York State Department of Environmental Conservation (NYS DEC) will be granted a conservation easement on all lands acquired by the City to ensure lands are held in perpetuity in an undeveloped state. There can be no residential structures on the land purchased, though there may be accessory structures and uninhabitable dwellings.

Prioritization of Watershed Areas

New York City has developed a priority list of lands for acquisition based on their proximity to reservoirs, reservoir intakes, and the drinking water distribution system (Figure 7-1). The five levels of priority for the Catskill/Delaware watershed, and the amount land from each that will be solicited, are given in Table 7-1. West-of-Hudson parcels in Priority Area 1A must be at least one acre, while Area 1B parcels must be at least five acres. The minimum acreage requirement for all other priority areas is ten acres. Both the Kensico and West Branch Reservoir basins, which have no minimum acreage requirements, are included in this prioritization scheme as key elements of the Catskill/Delaware system, although they are physically located in the Croton watershed. Yearly solicitation goals for the program are given in Table 7-2.

Like the Catskill/Delaware watershed, the Croton watershed is divided into priority areas based on reservoir basin and travel time. Priority Area A includes

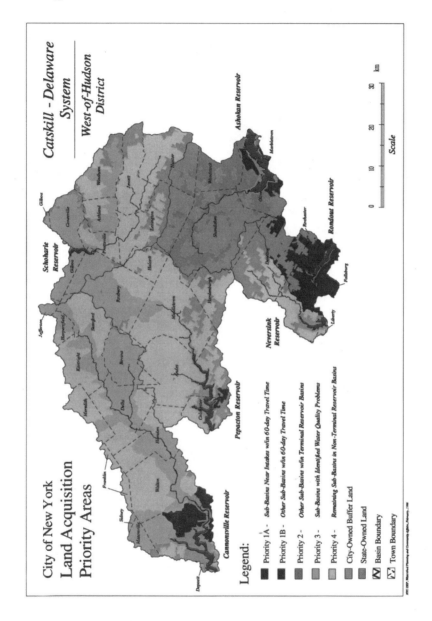

FIGURE 7-1 (A) Land acquisition areas in the Catskill/Delaware watershed. Source: NYC DEP (1997).

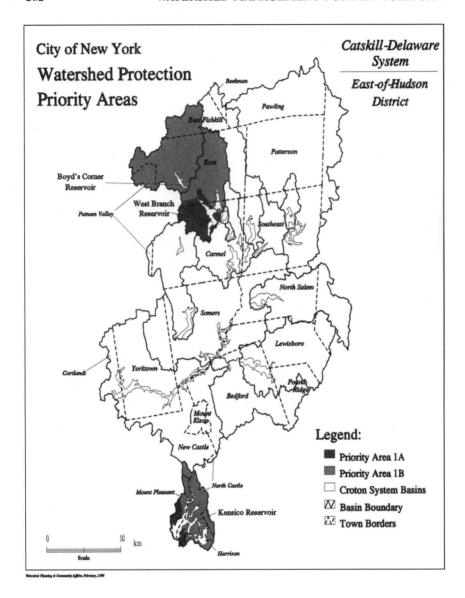

FIGURE 7-1 (B) Land acquisition areas in the West Branch and Kensico watersheds. Courtesy of the NYC DEP.

TABLE 7-1 Priority Categories for Land Acquisition

Priority Areas	Definition of Area	Minimum Acreage	Solicitation Acreage Goals
1A	Subbasins within 60-day travel time[a] to the distribution system that are near intakes	1 (West-of-Hudson)	61,750 (1A and 1B combined)
1B	Subbasins within 60-day travel time to the distribution system that are not near intakes	5	61,750 (1A and 1B combined)
2	Subbasins within terminal reservoir basins that are not within priority areas 1A and 1B	10	42,300
3	Subbasins with identified water quality problems that are not in priority areas 1A, 1B, and 2	10	96,000
4	All remaining subbasins in nonterminal reservoir basins	10	155,000

[a]See Chapter 11 for a thorough discussion of the 60-day travel-time delineation.

Note: Parcels in priority areas 2, 3, and 4 must:
- Be at least partially located within 1,000 feet of a reservoir, or
- Be at least partially located within the 100-year floodplain, or
- Be at least partially located within 300 feet of a watercourse, or
- Contain in whole or in part a federal jurisdictional wetland greater than five acres in size or a NYS DEC mapped wetland, or
- Contain ground slopes greater than 15 percent.

Source: Budrock (1997).

TABLE 7-2 Catskill and Delaware Watershed Land Solicitation Goals Under the MOA

Year	Acres Solicited to Acquire Fee Title or Conservation Easement	
	Annual Goal	Cumulative Total
1	56,609	56,609
2	51,266	107,875
3	42,733	150,608
4	52,846	203,454
5	55,265	258,719
6	48,531	307,250
7	0	307,250
8	47,800	355,050
9	0	355,050
10	0	355,050

the New Croton, Croton Falls, and Cross River Reservoir basins. Area B includes Muscoot Reservoir basin and portions of Amawalk and Titicus Reservoir basins within a 60-day travel time to the distribution system. Area C includes all remaining reservoir basins and subbasins beyond the 60-day travel-time boundary. In the Croton watershed, property that is zoned commercial or industrial as of the date of the City's solicitation is generally (but not always) excluded from acquisition, and all acquired property must be vacant. New York State has also committed $7.5 million to purchase lands or watershed conservation easements in the Croton watershed. After purchase, land ownership is transferred to New York City, which is then responsible for paying property taxes.

Local communities in the Catskill/Delaware watershed were given the option of designating lands around existing population centers to be excluded from acquisition by fee, but not from acquisition of watershed conservation easements. Towns were allowed to delineate the boundaries of existing hamlets and exclude these areas if desired, while villages were allowed to exclude all village land. These exclusions, made available to local communities to allow for reasonable growth and to preserve community character, are shown in Figure 7-2.

Meeting Filtration Avoidance Determination Milestones

The New York City Department of Environmental Protection (NYC DEP) has achieved success in the Land Acquisition Program to date. All the solicitation goals in the filtration avoidance determination have been met. As of June 30, 1999, NYC DEP had 15,380 acres under contract for a purchase price of $40.4 million (NYC DEP, 1999a). This represents about four percent of the total land area that must be solicited under the program. To date, no land has been purchased in the Kensico watershed, the most important target in terms of water quality protection. Reasons for the lack of success in Kensico include the cost of land, the limited amount of land eligible under the program (1,000 acres), and lack of seller interest (Marx and Goldstein, 1999; Miele, 1999). New York City should consider spending more than fair market value to acquire lands in this critical watershed.

Related Programs

Flood Buyout Program

NYC DEP and Delaware County are cooperating in a flood buyout program in which the City can purchase land underlying flood-damaged residences in the watershed as part of the general Land Acquisition Program. Interested landowners are given the opportunity to receive the pre-flood value of their homes and land, while the City pays only for the land. The balance of the purchase price is funded by Hazard Mitigation Grants administered by the Federal Emergency

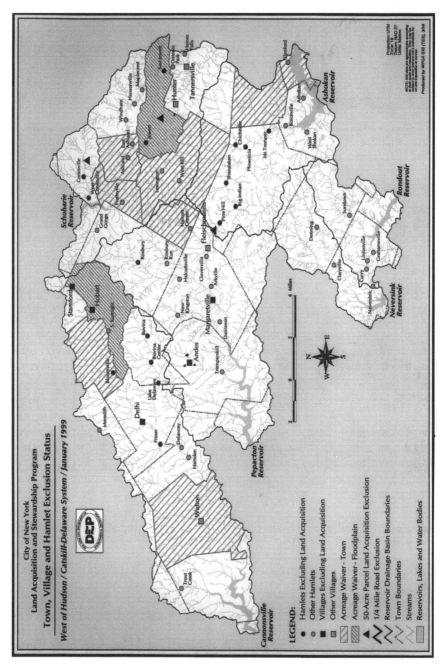

FIGURE 7-2 Towns and villages exempt from the Land Acquisition Program. Courtesy of the NYC DEP.

Management Agency. As of the end of June 1998, 73 interested homeowners had applied to the County to participate in this program. After being prioritized by the County and NYC DEP, the first purchase was completed in July and ten additional purchase contracts have been completed (NYC DEP, 1999a).

Agricultural Easements

A program to acquire watershed agricultural easements has very recently been established by the MOA to protect sensitive agricultural lands and to promote pollution prevention among farmers. It is envisioned that this program would assist in the intergenerational transfer of farmlands and operations. The Watershed Agricultural Council is responsible for landowner outreach and contact, working closely with NYC DEP to survey, appraise, and close agreements associated with agricultural easements. The City has committed $10 million out of the $250 million Catskill/Delaware Land Acquisition Program for watershed agricultural easements and $10 million for forested lands on farms. The easements are to be held in perpetuity by the Watershed Agricultural Council.

Stewardship

Stewardship under the Land Acquisition Program concerns the management of previously acquired lands. The City has agreed to consult with NYS DEC, the Environmental Protection Agency (EPA), local governments, and the Sporting Advisory Subcommittee (established in the MOA for both East- and West-of-Hudson watersheds) concerning appropriate recreational uses of newly acquired and already owned lands. Guidance for recreational uses of New York City-owned land has recently been proposed (NYC DEP, 1999b). Traditional recreational uses such as fishing, hiking, and hunting (currently allowed on City-owned land in the Cannonsville watershed on a limited basis) are likely to be allowed to continue, subject to permits, rules, and regulations. Other activities such as snowshoeing, cross-country skiing, bird watching, and nature study are likely to be allowed in certain areas and under certain conditions. It is highly unlikely that boating (other than rowboats permitted for fishing), snowmobiling, camping, motorcycling, horseback riding, and mountain bike riding will be permitted. Policies are also being developed by NYC DEP related to (1) public access to newly acquired lands and (2) forest and resources management on watershed lands.

Land Trusts

An alternative to outright land purchase is the land trust. Across the United States, tens of thousands of acres of open space are owned by land trusts. A land trust is a nonprofit legal entity established under state law that buys, manages, and occasionally sells or leases interests in undeveloped real estate. Land trusts typically focus on particular communities or regions from which they draw their boards of directors and officers. Some are large enough to maintain a full-time office and staff, while others depend on volunteers to administer their program and manage their properties. Properties may be held in fee simple or some form of less than fee interest (e.g., a conservation easement). Land trusts frequently cooperate with governmental land agencies such as state departments of natural resources or local conservation commissions.

A land trust may be able to act quickly to acquire a key parcel of land and hold it until a governmental agency has the funds on hand to buy it from the land trust, or the land trust may hold the deed or conservation easement in perpetuity. However, nonprofit land trusts would not pay taxes on the land unless special agreements were made with local government.

The Trust for the Public Land (TPL), a signatory to the MOA and a national nonprofit land conservation organization with offices in New York City, is actively acquiring land in the Croton and Catskill/Delaware watersheds. Recently, the TPL purchased 610 acres of undeveloped land in the town of Hunter (west of the Hudson River) and 100 acres for the Catskill Center for Conservation and Development (CCCD). In Putnam County, TPL protected 794 acres of steeply sloping undeveloped land surrounding 30-acre spring-fed Wonder Lake in collaboration with the Open Space Institute. The outfall from the lake is a tributary to the Croton reservoir system. In this case, TPL and the Open Space Institute purchased the land on behalf of the New York State Office of Parks, Recreation, and Historic Preservation. The land will become Wonder Lake State Park.

Land Acquisition and Stewardship Program Analysis

Prioritization Scheme

Because New York City cannot purchase all the private land in the watersheds, a prioritization scheme has been developed. Ideally, prioritization of lands for purchase should depend on two criteria: (1) location of land areas with respect to reservoirs and waterways and (2) land use, with an emphasis of those activities that produce pollutants. NYC DEP has established a prioritization scheme based primarily on the first criterion. (This means first priority has been given to lands in the Kensico and West Branch watersheds, which are important given their locations within the system.) Under the current scheme, land use is not a factor in prioritizing lands for purchase. This scheme may be appropriate for relatively nonmobile pollutants such as particulate phosphorus. However, many other pollutants, such as soluble phosphorus, nitrogen, and solvents, are considerably more mobile and can travel long distances in streams.

The importance of land use in prioritizing lands for acquisition was recently demonstrated in areas of Virginia draining into the Chesapeake Bay. The conventional wisdom in this area was that lands closest to the bay were contributing the most pollution. On that basis, the Virginia Chesapeake Bay Preservation Act was passed, requiring counties east of Interstate-95 to incorporate into their land use plans buffers around agricultural land, wetlands, and waterways. Computer modeling of the Virginia portion of the watershed later showed that lands higher up in the watershed were contributing a much larger proportion of the nutrient loading to the watershed than had been expected (Virginia DEQ, 1998). Unfortunately, the Chesapeake Bay Preservation Act did not regulate activities in these upland and headwater areas. Without computer modeling of nonpoint source pollution, it can be difficult to know which lands are contributing or could potentially contribute the most pollution.

Published information about the Land Acquisition Program does not adequately explain the analytical methods used to set priorities and make decisions at the scale of individual tax parcels. As mentioned in Chapter 6, an advanced Geographic Information System (GIS) has many options for watershed characterization that could be used to better prioritize lands for acquisition. In addition, NYC DEP has recently developed and tested the Generalized Watershed Loading Function (GWLF) model for predicting nutrient loads to the reservoirs based on different land uses (NYC DEP, 1998). This model, or the Reckhow model, could be used to identify and delineate significant polluting areas in the Catskill/Delaware watershed. Such analyses could be used to refine the current prioritization scheme. This would be particularly useful for Priority Area 3 subbasins, which must have "identified water quality problems." This general description, which allows NYC DEP flexibility in deciding which lands to purchase, could be refined by acquiring site-specific information on land use using the computer modeling described above.

Size Requirements

From an administrative point of view, it is probably sensible to pursue only the larger pieces of land. However, it is possible that there are areas critical to water quality that are smaller than the one-acre, five-acre, and ten-acre sizes set as lower limits for participation in the program. It is also questionable to rule out lands for acquisition when the overall goal of the program is to acquire as much land as possible.

Tax Considerations

Because real property taxes are the principal source of revenue for towns, villages, and school districts in New York State, local governments are very concerned about the amount of property removed from the tax rolls when it is purchased by a nonprofit organization or municipality. Real property tax receipts account for about half the revenues of these governmental units and about a fourth of revenues for county government. Real property taxes have the advantage of being a broadly based tax on a resource that is relatively immobile, although it should be noted that property taxes are typically based on the market value of the property, which may have little or no relationship to current uses. To help retain the tax base of the watershed, NYC DEP will be paying taxes on all acquired land. However, the same is not true for land sold to a land trust, unless it transfers the land to New York City.

Removing land from possible future development can also potentially decrease local tax revenues, capital improvements, and economic activity and jobs associated with businesses that might have located on acquired properties. This is a potential loss of funds that is not balanced by fees paid by the City on all acquired lands. It is possible to offset some of these losses in the long run by using tax incentives to help commercial development locate in a particular county, town, or village. Many economic development programs attract new business by waiving or decreasing local property taxes for a set length of time (e.g., 10–20 years) because of the promise of enhanced future revenue from other taxes.

There are other ways that local tax policy and, to some degree, state tax policy can be used to shape the character of future development. In the New York City watersheds, tax policy might be an effective tool for diffusing tensions between economic growth and maintenance of high water quality. This might be accomplished by shifting the share of new development costs from the general taxpayer to the developer, which would be an important political challenge. It should be kept in mind that while such policies can be used to protect important environmental resources, they can also be viewed as antigrowth and thus as inconsistent with some public values within the community.

Conclusions and Recommendations for Land Acquisition

1. NYC DEP should make greater use of the GIS and available land use/ nonpoint source computer models (like the GWLF) to determine more precisely which areas within Priority Area 3 in the Catskill/Delaware watershed are contributing to water quality problems. Although land use modeling in the Catskill/Delaware watershed has been conducted as part of the Total Maximum Daily Load Program, this information does not appear to have influenced the prioritization of lands for the Land Acquisition Program. Thus, there is no way of knowing whether the current prioritization scheme will capture the areas contributing the most pollution. The application of GIS in support of the Land Acquisition Program should include relevant spatial and temporal data and process descriptions. For example, the GIS could be used to integrate data on slope, soil suitability for on-site sewage treatment and disposal systems (OSTDS), erosion potential, and ownership to determine which lands should be targeted for acquisition.

2. Lower limits on the size of parcels that can be solicited may exclude environmentally sensitive lands from acquisition and should eventually be removed. As the number of larger parcels diminishes and as small adjacent parcels become available, NYC DEP should relax and eventually drop the acreage limits.

3. A Catskill/Delaware Land Trust should be created. Currently the methods for managing the lands acquired by the City, both through outright purchase and conservation easement, are complex and require oversight from the City and State. A land trust can move more quickly and quietly than the Land Acquisition Program to acquire or accept gifts of land, and it may be attractive to those landowners who do not wish to sell to New York City.

Although the national organization Trust for Public Land can acquire land within the watershed, the formation of local land trusts or adjuncts to the Catskill Watershed Corporation (CWC) could enhance the ability to purchase and protect critical watershed lands. Ideally, such an organization could promote environmental education and awareness of New York City and local watershed management issues. Property owners in the watershed region should be involved in managing the trusts.

4. In evaluating the tax revenue implications of New York City watershed land acquisitions, the committee recommends that the CWC assist local governments in the review of tax policies and practices to determine their influence on the quality of future development. Opportunities to complement growth-directing regulation with financial incentives should be considered and developed in conjunction with State programs to promote land uses consistent with watershed functions and values.

PLANNING IN THE WATERSHEDS

Nonstructural protection strategies are the core of the Watershed Protection and Partnership Programs in the MOA. The goal of these programs is to maintain and improve water quality while preserving the character and economic vitality of watershed communities. Participation in some of the programs requires that watershed communities adopt land use plans. In providing a comprehensive assessment of the MOA, the committee chose to evaluate whether the adopted plans, and the tools used to implement them, are appropriate and sufficient to provide the level of watershed protection required by the City's filtration avoidance determination.

Table 7-3 lists each program, its intended purpose, and its funding allocation. Those programs that require comprehensive planning by participating local communities and counties are described in detail below.

TABLE 7-3 Watershed Protection and Partnership Programs

Program	Funding Level	Purpose
New Sewage Treatment Infrastructure[a]	$75 million	Construct and install new WWTPs or community septic systems or create septic maintenance districts in up to 22 selected villages and hamlets.
Croton Plan[a, b]	$2 million-plan development $68 million-implementation	Identify sources of pollution to the Croton watershed system, recommend measures to improve water quality, and recommend measures to protect the character and needs of Croton watershed communities.
Catskill Fund for the Future	$59.7 million	Establish a program supporting environmentally sensitive economic development projects in the watershed.
Stormwater Fund	$31.7 million	Design, construct, implement, and maintain new stormwater measures identified in stormwater pollution prevention plans (SPPPs).
Septic System Rehabilitation and Replacement	$13.6 million	Pump out and inspect individual residential septic systems and repair, replace, or upgrade failing or likely-to-fail systems.
Sand and Salt Storage Facilities	$10.25 million	Improve storage of salt, sand, or other deicing materials to better protect water quality.
Sewer Extensions[a]	$10 million	Construct sewer extensions from City-owned WWTPs to nearby communities with water quality problems.
Stormwater Retrofits	$7.625 million	Design, permit, administer, construct, implement, and maintain stormwater best management practices (BMPs) to address existing stormwater problems.
SPDES Upgrades	$5 million	Assist existing WWTPs in replacing and upgrading equipment to help meet requirements of SPDES permits.
Stream Corridor Protection	$3 million	Design, construct, and implement stream corridor protection projects such as stream bank stabilization and fish habitat improvement.
Alternative Design Septic Program	$3 million	Design, construct, and install "alternative design septic systems" that because of site conditions require importation of fill material or the use of pumping equipment.

continued

TABLE 7-3 Continued

Program	Funding Level	Purpose
Public Education	$2 million	Implement multiple projects to educate public on the importance of New York City's water supply system and the role of watershed residents as stewards of water quality. Up to $1 million can be used to establish a regional museum.
Watershed Forestry Program	$500,000	Establish a program to promote forest stewardship and conservation.
Economic Development Study	$500,000	Perform a comprehensive study of economic and community development goals that will be used to guide decision-making for the Catskill Fund for the Future.

Note: Only Catskill/Delaware watershed communities are eligible for participation unless otherwise indicated.

a Denotes those programs that require comprehensive land use planning.

b The Croton Plan is not technically part of the Watershed Protection and Partnership Programs. However, because it contains a requirement for comprehensive planning, it is included here for completeness.

Source: Reprinted, with permission, from Budrock (1997). © 1997 by The Catskill Center for Conservation and Development.

Plans in the MOA

Croton Plans

Although the bulk of the Watershed Protection and Partnership Programs in the MOA focus on the West-of-Hudson watershed, the Croton Plan (MOA, Article V, Paragraph 138) focuses on Westchester, Putnam, and Duchess counties. The purpose of the Croton Plan is to identify significant sources of pollution in the Croton watershed and recommend measures to be taken by watershed municipalities, the counties, and NYC DEP to improve water quality. Recommended measures should protect the character and special needs of the communities within the Croton watershed. The goal of the Croton Plans is to provide a vision of watershed communities that will help guide future residential and commercial development in an environmentally sound manner. Development of the Croton Plan is required for communities to participate in the Phosphorus Offset Pilot Program (described in Chapter 8).

Each county is expected to draft its own version of the Croton Plan, efforts to which New York City will contribute up to $1 million per county. Putnam and Westchester counties have taken quite different approaches to this effort. Putnam

County has employed a town-controlled effort while Westchester County has designed a process that will use both municipal boundaries and subwatershed boundaries for multiple citizen and staff committees (see Box 7-1). A total of $68 million from New York City is available to support projects outlined under the Croton Plans, which must be developed by May 2002. This date reflects both the urgency and complexity of the task.

The Croton Plan is an important effort because water quality has declined substantially in the Croton system as a result of past development. Although there are differences between planning for the Croton and West-of-Hudson watersheds, partly because of a higher population in the Croton watershed, it is possible that planning efforts in the Croton may be relevant to the smaller and more varied planning efforts required for the Catskill/Delaware watershed. The goal of both efforts is the same: to preserve and improve water quality while providing for environmentally sound development.

West-of-Hudson Planning Efforts

The MOA requires several different land use planning activities in order for West-of-Hudson communities to participate in the Watershed Protection and Partnership Programs. These requirements may present a significant challenge because many of these communities are inexperienced in land use planning. Table 7-4 shows how few of the counties, towns, and villages in this region have comprehensive plans, planning commissions, subdivision regulations, and site plan review. Without these tools, it can be difficult for communities to control development. To better understand the relationships between counties, towns, and villages in the West-of-Hudson region, Box 7-2 describes the structure of local government in New York State.

Planning for the New Sewage Treatment Infrastructure Program. The MOA allocates $75 million for new sewage treatment infrastructure. These funds are available under certain conditions to specific towns, villages, or hamlets in the Catskill/Delaware watershed "experiencing water quality problems due to failing septic systems in close proximity to streams and other water courses or where failure is likely to occur" (MOA, Article IV, Section 122). Localities identified as having such problems that participate in the program must "adopt and maintain a comprehensive plan, subdivision regulations and appropriate land use ordinances assuring that future growth within such areas can be adequately serviced by, and will not exceed the capacity of, the sewerage systems and the watershed treatment plant (WWTP) to which it is connected." The MOA identifies 22 communities with water quality problems, all of which are eligible to apply for new infrastructure funds for the construction of community septics or sewerage collection systems or for the creation and funding of septic districts. Seven of the highest-priority communities are also eligible to use the funds for

---

BOX 7-1
Croton Planning

Putnam County. Putnam County places the lead responsibility for plan formulation on watershed towns (Putnam County, 1997). The towns are responsible for identifying their community character and needs, significant sources of contamination, and the measures to be implemented within their own jurisdictions. Putnam County is providing overall direction and information resources and preparing any necessary environmental impact statements (EIS).

Putnam County's Croton Plan is being developed in three phases. During Phase I, the county is compiling resources on manmade, natural, and regulatory features of the watershed to help prioritize areas for protecting community character and meeting community needs in a manner consistent with water quality goals. During Phase II, the towns will identify and choose from a wide variety of water quality control and community development measures. These may include new or expanded WWTPs, improved stormwater management, changes in master plans and zoning laws, and changes in local environmental laws. Final agreement on the plan will be obtained in Phase III during deliberations between the county, towns, and the public.

Westchester County. Westchester County, recognizing that watershed management is both technical and political, has developed an innovative planning process that incorporates both these aspects (Westchester County, 1998a, b). The county is basing its Croton Plan both on

---

the construction of new WWTPs. Only those communities that construct WWTPs or community septics are required to develop a comprehensive plan.

Table 7-5 shows the land use planning mechanisms in effect in these 22 localities. Of particular interest are the mechanisms specifically mentioned in the MOA. A little more than a third of the towns have comprehensive plans in place, while most have subdivision regulations, about the same proportion as for West-of-Hudson towns. Subdivision regulations are important in determining the placement of residences and for providing adequate roadways. However, unlike comprehensive plans, they do not involve local citizens in decisions about the desired quality of life in the region, nor do they guide the extent and location of new growth. In the absence of a comprehensive plan, growth may be determined by a planning commission or by town supervisors. This may not reflect the consensus within the community.

the boundaries of its municipalities and on its seven basin and 27 sub-basin boundaries.

The Croton Watershed Steering Committee, which will direct the planning process, is composed of municipal supervisors from each watershed municipality, the Westchester County Department of Planning, and the Westchester County Department of Health. A second level of committees will be established at the municipal level—one in each of the ten participating towns. Each municipal committee has the same membership, which consists of one member each from the Town Board, the Planning Board, the Conservation Advisory Council, and the Wetlands Board, two town residents, and municipal staff representatives.

Recognizing the importance of public participation, three watershed-wide advisory committees will be formed to assist in developing the plan, one for each of three topics: land management, water management, and information and education. Each advisory committee will include, as appropriate to the topic, professional practitioners, state and New York City representatives, citizens from planning boards, and representatives from chambers of commerce, university faculty, public interest groups, and industry. The work groups addressing the three main topics will be structured around the appropriate geographical boundary: political or watershed. However, regardless of the initial approach used to study a topic, the final analysis will relate to both municipal boundaries and sub-watershed areas.

Planning for the Sewer Extension Program. The MOA also provides for the expenditure of $10 million for the construction of extensions to sewerage collection systems servicing City-owned WWTPs west of the Hudson River. Sewer extension projects are being selected by NYC DEP in consultation with the CWC and the five affected localities: Grahamsville, Margaretville, Grand Gorge, Pine Hill, and Tannersville. As with the new sewage treatment infrastructure program, one of the conditions of the sewer extension program is that localities adopt and maintain a comprehensive plan, subdivision regulations, and appropriate land use laws and ordinances to ensure that future growth will not exceed the capacity of sewerage systems to which they are connected.

The request for proposals (RFP) published for the sewer extension program is primarily for engineering studies, which do not require the level of public involvement expected in a typical comprehensive plan. Thus, it seems unlikely

TABLE 7-4 Planning and Zoning Tools in West-of-Hudson Communities

County	Population (1990)	Comprehensive Plan	Zoning
Delaware County			
Town of Andes	1,291	yes	no
Town of Bovina	550	no	no
Town of Colchester	1,928	yes	no
Town of Delhi	5,015	yes	yes
Town of Deposit	1,824	no	no
Town of Franklin	2,471	no	no
Town of Hamden	1,144	no	no
Town of Harpersfield	1,450	no	no
Town of Kortright	1,410	yes	no
Town of Masonville	1,352	no	no
Town of Meredith	1,513	yes	no
Town of Roxbury	2,388	no	no
Town of Sidney	6,667	yes	yes
Town of Stamford	2,047	no	yes
Town of Tompkins	994	no	no
Town of Walton	5,953	no	yes
Total	37,997	38% yes	25% yes
Schoharie County			
Town of Broome	926	no	no
Town of Conesville	684	no	no
Town of Gilboa	1,207	no	no
Town of Jefferson	1,190	no	no
Total	4,007	0% yes	0% yes
Greene County			
Town of Ashland	803	no	no
Town of Halcott	189	yes	no
Town of Hunter	2,116	yes	no
Town of Jewett	933	yes	yes
Town of Lexington	835	no	yes
Town of Prattsville	774	yes	no
Town of Windham	1,682	yes	no
Total	7,332	71% yes	29% yes
Sullivan County			
Town of Fallsburgh	11,445	no	yes
Town of Liberty	9,825	yes	yes
Town of Neversink	2,951	yes	yes
Total	24,221	66% yes	100% yes
Ulster County			
Town of Hardenburgh	204	no	no
Town of Hurley	6,741	no	yes
Town of Kingston	864	no	no
Town of Olive	4,086	no	yes
Town of Rochester	5,679	yes	yes
Town of Shandaken	3,013	no	yes
Town of Wawarsing	12,348	yes	yes
Town of Woodstock	6,290	yes	yes
Total	39,747	44% yes	77% yes

Source: Updated from the New York State Commission on Rural Resources (1994).

Subdivision Regulations	Site Plan Review	Planning Board
yes	yes	yes
yes	no	yes
yes	yes	yes
yes	no	yes
no	yes	no
no	no	no
yes	yes	yes
yes	no	yes
yes	no	yes
no	no	no
yes	yes	yes
yes	no	yes
yes	no	yes
yes	yes	yes
yes	no	yes
yes	yes	yes
81% yes	44% yes	81 % yes
yes	yes	yes
yes	yes	yes
yes	no	yes
yes	no	no
100% yes	50% no	75% yes
yes	no	yes
yes	yes	yes
yes	yes	yes
yes	yes	yes
yes	yes	yes
yes	no	yes
yes	yes	yes
100% yes	71% yes	100% yes
yes	yes	yes
yes	yes	yes
yes	yes	yes
100% yes	100% yes	100% yes
yes	no	yes
yes	no	yes
no	yes	yes
yes	yes	yes
yes	yes	yes
yes	yes	yes
yes	yes	yes
yes	yes	yes
89% yes	78% yes	100% yes

BOX 7-2
Local Government in the New York City
Watershed Region

New York State is divided into counties that can be run either by an elected county legislature (whose members come from districts that have boundaries different than town boundaries) or by a Board of Supervisors comprised of a town supervisor from each town. For example, Greene, Ulster, and Sullivan counties are run by an elected legislature, while Delaware and Schoharie counties have a board of supervisors. Each county has a planning office and a county planning board that may designate areas outside the limits of any incorporated village as a county subdivision control area if there is no town planning board.

Each county is divided into towns, which are considered the basic unit of local government. Towns are run by an elected town board headed by a town supervisor. Towns are empowered, but not required, to develop a comprehensive plan, zoning, and site review ordinances. All town plans and land use ordinances and actions are sent to the county planning agency for review.

Within a town may be found incorporated villages. An elected board of trustees is responsible for decisions within incorporated village limits, but the villages are still subject to town laws. Incorporated villages may have a comprehensive plan and adopt zoning and site review ordinances. A comprehensive plan, zoning laws, site plans, and special-use permits must be sent to the county planning agency for review.

Hamlets, areas of population concentration within a town that are smaller than villages, do not have their own government and are dependent on town administrators for their services.

TABLE 7-5 Land Use Planning Mechanisms by Priority Ranking in the Towns Considered for New Infrastructure Funds

| Land Use Planning Mechanism | New Infrastructure Priority Ranking | | | Percent of All Priority Towns | Percent of All West-of-Hudson Towns |
| | Towns 1–7 (N=7) | Towns 8–22 (N=15) | All Priority Towns (N=22) | | |
	–number–				
Comprehensive Plan	4	3	7	32	42
Subdivision Regulations	6	13	19	86	87
Site Plan Review	4	8	12	54	57
Planning Board	6	13	19	86	95
Zoning Regulations	1	4	5	23	40

Source: New York State Legislative Commission on Rural Resources (1994).

these studies will help the public determine where growth in the community should be targeted or how future growth can be integrated with the economic development being proposed for the region (see below).

Planning for the Phosphorus Offset Pilot Program. The phosphorus offset pilot program allows three new WWTPs to be built in phosphorus-restricted basins west of the Hudson River (see Chapter 8 for a detailed review of the program). Unlike the two previous planning efforts described above, comprehensive planning required under the phosphorus offset pilot program is to take place at the county level rather than the town, village, or hamlet level. The West-of-Hudson county in which a new or existing WWTP is located must develop a comprehensive strategy to identify (1) existing economic resources, (2) water quality problems, (3) potential remedies for such problems, and (4) potential strategies and recommendations of economic development initiatives that could be undertaken to sustain the local economy while protecting the water supply.

Currently, the Cannonsville basin in Delaware County is the only West-of-Hudson basin subject to phosphorus restrictions and thus eligible for the offset program. To participate in this program, and for other purposes, Delaware County has developed a draft "Comprehensive Strategy for Phosphorus Reduction." The goal of the strategy is "to manage phosphorus (and associated pollutants) so that economic development is not impaired, and the economic and social future of Delaware County is sustained, based on scientifically sound research and a monitoring program" (NYS WRI, 1998). The County is working with small businesses, farms, industry, and residents to help meet the watershed regulations and still maintain some economic growth. Interestingly, the comprehensive strategy has investigated phosphorus trading (as embodied in the offset program), and found that there is not much interest in this approach among the county leadership (NYS WRI, 1998).

Delaware County's comprehensive planning is challenging for a variety of reasons. Reaching its goal of managing phosphorus requires that all phosphorus sources be identified. Although it is generally known that phosphorus emanates from WWTPs, OSTDS, municipal stormwater systems, other stormwater systems, agricultural lands, and forests, as of yet there is no specific information on sources within the watershed (i.e., particular farms or community's stormwater). The models that have been used to calculate phosphorus loading for different sources (GWLF, Reckhow model) often produce conflicting predictions (K. Potter, NYS WRI, personal communication, 1999). As a result, the county is not certain how much soluble phosphorus must be eliminated to bring the Cannonsville Reservoir into compliance. This impedes planning for and implementation of best management practices. Because of these complications, the county is considering adopting a phosphorus loading strategy for its lands within the Cannonsville Reservoir basin, rather than developing strategies based on individual source controls.

Economic Development Study

As mandated by the MOA, a comprehensive study of community and economic goals, opportunities, and strategies in the Catskill/Delaware watershed has been conducted by a consultant to the CWC. This study should assist the watershed communities in achieving their economic, social, and environmental goals in a manner consistent with New York City's water quality objectives and watershed regulations. The strategy developed during the study is intended to support and broaden the existing regional economic base, develop a process for long-range planning, and analyze infrastructure needs through development of an "Action Plan" (CWC, 1997). The study is also providing guidance to the Catskill Fund for the Future (see Table 7-3), a $59.7 million fund that will support projects that stimulate regional economic growth consistent with the stewardship of environmental assets.

Phase 1. The first phase of the Economic Development Study is a baseline economic analysis and community assessment that describes the status of and barriers to economic development in the watershed (HR&A, 1998a). Five counties within the watershed experienced a 7-percent decline in overall employment during the 1990s. These counties lag behind New York State in the proportion of the population over 25 years of age with a high school diploma and the proportion graduating from college. From 1990 to 1996, growth in the five watershed counties (1.85 %)[1] surpassed the growth of New York State (1.08 %). There is a limited supply of developable land in the region because much of the remaining land is on slopes too steep for development on any scale, and publicly owned lands are generally required to be preserved in an undeveloped state. There are no coordinated regional marketing efforts and little partnership between public and private agencies. The banking system in the watershed is characterized by historically low interest rates, a large amount of liquidity, and an adequate supply of capital.

Because of the watershed's beauty, its abundance of clean air and water, and its proximity to a huge population base, the study identified economic opportunities such as tourism in the hamlets and villages, arts, the construction of second homes, and telecommuting. In fact, the predominant form of projected development is low-density residential housing outside of villages and hamlets. Specialty manufacturing and artisans are identified as a source of quality jobs. Agriculture is perceived as currently healthy and as able to capitalize on the proximity of major markets and on growing demand for organic produce. Mining for bluestone and the harvesting of high-quality northern hardwoods are also identified as

---

[1] This growth rate is similar to the rate determined for the same period of time by the NRC committee: 1.5 percent from 1990 to 1996. The higher rate of the Economic Development Study results from the use of countywide population figures.

providing a foundation for local industries. The health care industry, one of the leading employers of the five-county region, is expected to continue as the residents in the area age.

Phase 2. The second report contains an assessment of the market sectors and an analysis of the program issues (HR&A, 1998b). Eight market sectors are identified as being vital to the Catskill/Delaware watershed economy:

1. Hamlets and villages
2. Tourism and the arts
3. Skiing
4. Specialty manufacturing and service
5. Adjacent centers of employment
6. Agriculture
7. Forestry and bluestone quarrying
8. Other resource needs

The Phase 2 report recognizes that no single program will solve all the watershed's economic woes and states that any strategy for expanding and strengthening the region's economy must be sensitive to the special characteristics of the economic base.

Phase 3. The Catskill Fund for the Future is the focus of the most recent phase of this study (HR&A, 1998c). The purpose of this fund is to finance business projects that can enhance job opportunities, incomes, and property values in an environmentally sound manner. The third report describes the practical steps necessary to make the fund "an engine for growth in the region, rather than an addition to the substantial inventory of existing academic studies on the Watershed" (HR&A, 1998c). The study states that the fund is not intended to be a substitute for private capital, a public works program, a giveaway, or a political tool. Rather, it must be a self-sustaining capital institution, providing a diverse range of financial products.

The study suggests that the Catskill Fund for the Future should fund four areas to develop a strong and environmentally sensitive economy:

1. Hamlets and villages, revitalizing and expanding for retail, specialty services, and tourism.
2. Tourism and the arts, drawing on existing and on new niche-based accommodations and attractions, including skiing, ecotourism, main street programs, and historic preservation.
3. Manufacturing and businesses, focusing on specialty manufacturing, artisans, and business services.
4. Agriculture, forestry, and mining, strengthening the traditional rural

economy and working through agricultural cooperatives, the Watershed Agricultural Council and Watershed Forestry Program, loggers and foresters, and bluestone miners.

The study recommends that some of the CFF funds go toward the development of Whole Hamlet Programs. These strategic plans would identify the inherent strengths of the hamlets and produce an action plan for capitalizing on those strengths. The plans would include elements such as infrastructure needs, size of the market, appearance and safety, existing land use patterns, access, circulation and parking, events, promotion and marketing, and historic preservation. Projects and programs presented in a Whole Hamlet Plan, after being certified as satisfying environmental requirements, would be fast-tracked within the CWC's loan process.

The success of the Catskill Fund for the Future is dependent upon public outreach and education provided by the CWC. As the primary advocate for maintaining high water quality in the Catskill/Delaware watershed, the CWC will have to devote significant staff time and effort to explaining the MOA and its provisions to interested stakeholders.

Role of the Catskill Watershed Corporation

The limited resources available to rural communities often hamper land use planning in sparsely populated watersheds, making it difficult to integrate competing development and watershed protection needs. When townships have different planning capacities, the result is often fragmented land use planning, uneven infrastructure, and other development within watersheds. Cooperation between rural West-of-Hudson communities can greatly enhance the current planning efforts, and the CWC is the appropriate agency to foster this cooperation.

The CWC was created by the MOA "to establish a working partnership between the City and the West-of-Hudson communities" (MOA, Article IV, Paragraph 117). This locally based and locally administered organization is starting to address the problems of limited financial and professional planning resources common to the West-of-Hudson communities.

The CWC administers a range of watershed programs, including infrastructure construction and improvement, economic development, and public information and education (see Box 7-3). The CWC's centralized administration of these programs provides an institutional basis for more integrated watershed protection and development. In addition to CWC programs, the New York State Department of State (NYS DOS) operates a $500,000 Master Planning and Zoning Awards Program to assist towns and villages in the development of environmentally sound master plans, zoning laws and standards, and capital investment plans among other activities (CWC, 1999). The NYS DOS also assists residents of the watershed (both east and west of the Hudson River) in determining what permits

BOX 7-3
CWC Watershed Protection and Development Programs

The CWC is a nonprofit organization made up of a specified number of representatives from West-of-Hudson communities, members appointed by the State governor, and one New York City employee appointed by the mayor. Over the next 15 years, the CWC will preside over the expenditure of $240 million in New York City-provided funds for infrastructure, economic development, conservation, education, and operations. One of the largest expenditures, $59.7 million over 15 years, is earmarked for the Catskill Fund for the Future, a program that supports responsible, environmentally sensitive economic development. In addition, funds are to be expended for infrastructure improvements that could be conducive to environmental protection as well as economic development, e.g., construction of new WWTPs or community septic systems as well as the formation of septic districts. Finally, more than $80 million is available for a variety of infrastructure improvements and educational and conservation efforts that are likely to have positive water quality impacts.

The CWC has established a number of standing and ad hoc advisory committees for specific areas of concern in the watershed. These committees integrate varied interests and experts into the process of planning and administering watershed development. For example, the Tax Advisory Committee is made up of town and tax attorneys, appraisers and evaluation consultants who advise the CWC's Land Committee regarding tax rules. The Sporting Advisory Committee is comprised of representatives of sporting and recreational organizations. This committee reviewed 85 parcels targeted for acquisition by New York City, recommending recreational uses for the properties.

Septic Rehabilitation and Replacement. This program concentrates on repairing septic systems identified as failing throughout the Catskill and Delaware watershed areas. A total of 673 systems have been repaired or replaced at a cost of over $4 million. Another 575 systems are either under review, design, or construction. Following high demand for this program, the CWC Board of Directors voted in December 1998 to restrict participation in the program to homeowners who received Notices of Violation of Failure from NYC DEP prior to January 1, 1999. In order to evaluate the operation of the septic program, in May 1998, the CWC established a Technical Working Group to study the program's impact on water quality and assess the likelihood of failure of septic systems throughout the West-of-Hudson watershed. Effective July 1, 1999, the CWC is focusing on priority areas such as the 60-day travel-time zones.

continued

BOX 7-3 Continued

Sand and Salt Storage. The CWC will fund and supervise the construction of 31 municipal sand and salt storage facilities throughout the watersheds as provided for in the MOA. Through October 1999, nine facilities have been completed.

Stormwater Controls. In June 1998, the CWC adopted rules for the West-of-Hudson Future Stormwater Controls Program, which provides funds for BMPs to limit and control runoff of stormwater and pollutants from impervious surfaces. Through September 1999, CWC has paid $199,000 to fund construction projects for stormwater control plans and erosion management devices required under NYC DEP regulation of new construction. In 1999, the CWC's Wastewater Committee developed rules for the Stormwater Retrofit Program to address existing stormwater problems throughout the watersheds.

Economic Development Study. The CWC appointed a Steering committee to work with the consulting firm hired to complete the economic development study called for in the MOA. The CWC Board of Directors adopted a plan for the expenditures from the $59.7 million Catskill Fund for the Future in July 1999.

Business Loans. The CWC has loaned ten businesses a total of $1,545,000 to assist in the expansion and improvement. The CWC Board of Directors made these loans as an interim measure while awaiting results of the economic development study.

Tourism Grants. The CWC allocated $250,000 to tourism agencies throughout the watershed. A fourth of these funds were invested in regional tourism promotion efforts. Funds were also used for promotional materials such as advertising, travel guides, and signage.

Environmental Financial Incentive Awards. To date, the CWC has made one grant for $27,000 to a children's clothing manufacturing firm for the installation of equipment to recycle up to 12,000 gallons of wastewater per day, eliminating chlorine and neutralizing dye that had been discharged to a City-run WWTP.

Watershed Education. The CWC awarded $100,000 to 13 schools and nonprofit organizations. The targeted audience in this first round of grants was school-age children and their teachers in West-of-Hudson communi-

continued

BOX 7-3 Continued

ties and New York City. A second round of grants worth $200,000 was made in May 1999.

Watershed Regional Museum. For the design and fabrication of museum exhibits for the Watershed Museum, CWC allocated $1 million.

Commemorative Project. The CWC established an advisory committee to explore the possibility of erecting informational kiosks and historic markers at six West-of-Hudson reservoirs. The project is intended to commemorate the communities that were moved or disincorporated in order to build the reservoirs and to explain New York City's extensive water supply system.

---

are required for various types of regulated activities. This program is based at the CWC office and has established a toll-free number that residents can call to obtain information. The CWC and NYS DOS also cooperate with a network of New York State and federal agencies to provide specialized regulatory and compliance-related information to watershed residents, villages, and towns.

Analysis of Planning Efforts

Assessing the quality of comprehensive plans and various planning efforts can be a substantial challenge. Professional planners can often differentiate high quality from low quality, but they are hard-pressed to explicitly define the key characteristics of plan quality (Berke and French, 1994). Given the diversity of opinion about what a plan is and how it is to be used, a standardized evaluation of the plans outlined in the MOA is not possible. Because these plans will be implemented after this report is published, their effect on the communities in the watershed region can only be estimated. However, some constructive comments on the plans being developed by the watershed communities are made, based on professional judgment informed by years of experience in different settings. Considerations regarding plan implementation are also discussed.

Croton Plans

The Croton Plans developed by Westchester and Putnam counties are taking quite different approaches. The Westchester County planning effort is a comprehensive, fully integrated process. It considers the complexity inherent in implementing future recommendations by looking at problems from both a political perspective (using town boundaries) as well as from a water resource perspective (using subwatershed boundaries). This creative approach should be emulated by other counties. Hopefully, the result of this process will be land use regulations with budgetary implications that are well understood by all stakeholders.

The Putnam County planning process appears to address the full range of management and regulatory tools and investment opportunities relating to maintenance and improvement of water quality. While the data compilation and analysis are largely the responsibility of the county, and NYC DEP has some ability to present its analyses and perspective, the towns are required to develop the management, regulatory, and investment proposals. This division of responsibilities for plan formulation and evaluation is awkward and is unlikely to produce the most productive array of actions. This is because towns, which are responsible for plan development, will not have access to all of the data compiled at the county level, which in turn may result in conflicting county and town goals. Implementation of economic development plans is most successful when local citizens have broad access to information. This problem may be partially alleviated because the Croton Plan requires the agreement of all parties. Although both the county and NYC DEP have veto power in principle, there will likely be pressure to defer to town proposals.

A significant issue is found under work item 3 in Phase II of the Putnam County Statement of Work: "To economize on limited assets, priority may be given to those investments which help maintain community character and meet community needs, while helping to achieve the water quality goals. In addition, to the extent available, [NYC] DEP's estimates of the future water quality of each basin...may be considered so that priority may be given to these measures which will achieve the greatest relative benefit to the basin." This footnote is a clear affirmation that the real emphasis of the Putnam County Croton Plan is on meeting community needs however they may be defined, with water quality impacts as a constraining factor. Such an approach may be the only politically feasible one, but it makes achieving reservoir quality objectives more problematic—particularly given the distribution of responsibilities in the Putnam County Scope of Work.

West-of-Hudson Plans

As noted earlier, there are several challenges that comprehensive planning and land use planning must overcome in the Catskill/Delaware watershed. First, many of the towns are very small—approximately 80 percent have a population of less than 2,000, with the population being spread over a large area fragmented by mountain ridges. Villages, generally comprised of small clusters of houses, are even smaller. Thus, the staff resources of local governments are correspondingly small. Most towns do not have a professional planner on staff, but rely on volunteer citizen planning and zoning boards or hire consultants to develop the required plans. Comprehensive plans and/or zoning ordinances are available in approximately half of the watershed towns (Table 7-4). Although in most rural areas this level of planning is appropriate, the emphasis on planning found in the MOA in effect mandates a change in the status quo. Fortunately, with the slow rate of growth projected by the economic development study, communities should have time to develop their plans. There are three important issues in West-of-Hudson planning considered below: fragmentation of planning efforts, citizen participation, and plan implementation. Some of these problems could be overcome through the use of a regional planning agency. The CWC could be such an organization, because it plays a critical role in economic development and it administers a wide range of programs that can favorably influence community development.

Fragmentation. The planning efforts under way in the West-of-Hudson communities are fragmented. None of the plans required by the MOA is truly "comprehensive" in that it considers housing, transportation, public utilities, environmental conditions, and other relevant factors. Rather, the RFPs issued for new wastewater treatment facilities and sewer extensions reveal a largely engineering focus. There are calculations of growth, assumed from historic growth patterns, and designs of wastewater systems to handle the projected growth. But there is no clear integration of planning efforts needed to prevent duplication, inadvertent omission of important issues, or inconsistent assumptions. For instance, it is not clear that economic planning being done at the county and regional levels is taking into consideration the housing that will be required of those taking advantage of new economic opportunities, as well as the additional wastewater that will accompany the new housing.

The institutional arrangement of counties, town, and villages within the watershed region creates additional complexities in achieving an integrated cohesive plan for development (see Box 7-2). A town represents an entire geographic region rather than just a cluster of residential and commercial areas, necessitating town governments to take agricultural and silvicultural practices into account along with urban services. Many small-town governments are not equipped to manage these additional issues. Villages are free to create their own plans, though they must file them with the county. Counties are left to integrate village and town plans into comprehensive plans or to mediate the differences between plans of neighboring areas. Fortunately, the central role of the CWC and the planning assistance offered by the NYS DOS will be useful in overcoming fragmentation of plans across watershed regions.

Economic development is being advanced primarily through the Catskill Fund for the Future, as administered by the CWC. It is not clear how this process is being coordinated with the other required planning efforts. This is due, in part, to that fact that it is very difficult to know where some types of economic development, such as specialized manufacturing, will occur. However, integration of these efforts would be beneficial. Economic revitalization of the villages and their promotion for tourism should mesh with the planning for wastewater treatment services, and the consultants hired to do such planning should make sure the efforts are integrated.

Citizen Participation. Citizen participation is a key component in planning efforts, even those deemed to be highly technical. L. Robert Neville, in discussing his experience working with stakeholders to protect the natural resources of Sterling Forest in New York and New Jersey, found that:

> "Effective watershed management at the local level is far more dependent on social and economic variables than on watershed science. Full community support is needed in order to create the necessary changes in land use law and policies that will institutionalize the protection of natural systems and processes during development and ensure their continued health through comprehensive natural resource management." (Neville, 1999)

Watershed planning has come to be thought of as an iterative process, in which citizen and stakeholder input enters the decision-making process on a recurring basis: "Long term, effective watershed management requires 25 percent science and technology and 75 percent human psychology and sociology" (Bowers, 1999). There are three forces that currently prohibit society from achieving watershed stewardship: (1) fragmentation of knowledge about the system, (2) unwillingness to change present behavior and practices, and (3) a belief that human ingenuity and technology can solve problems of resource depletion or scarcity (Cairns and Pratt, 1992). Watershed stewardship is better served by convincing people to adopt new ways of behaving that will sustain high water quality than it is by involving citizens for short, sporadic exercises. Simply put, current planning efforts in the Catskill/Delaware region, as outlined in the MOA, do not require a level of citizen participation sufficient to overcome these forces.

Plan Implementation. The last step in the planning process is implementation. It involves zoning, site reviews, subdivision ordinances, and regulations. A common assumption is that once a planning document and its implementing regulations are adopted, they will be followed. However, this is not always the case. When outside consultants prepare comprehensive plans, as opposed to community members, there is a greater tendency for these plans to not be implemented. This occurs because such plans often are developed without significant local input and oversight. Clearly, the commitment needed to implement comprehensive plans and their associated regulations must come from involved and informed citizens.

Recent research on the success of implementing environmental plans and their regulations shows a widespread lack of compliance (Brower and Ballenger, 1991; Burby and Patterson, 1993; Promersberger, 1984). For example, Johnston and McCartney (1991) have reported widespread violations of the mitigation requirements of the California Environmental Quality Act. In the Catskill/ Delaware watershed, where New York City is depending on the implementation of comprehensive plans and watershed regulations, ensuring compliance will be a challenge. This is especially problematic when the cost of complying with regulations outweighs any direct benefits to the developer. In such cases, deterrence may be necessary. Compliance with specifications and standards can be improved with greater use of deterrence, specifically frequent inspections to detect violations (Burby and Patterson, 1993).

Successful plan implementation also depends on the degree to which regulators can implement a cooperative approach. This means allowing adequate time for regulators at the office to review plans, for review in the field using pre-construction conferences, and for negotiation to resolve disagreements. Burby et al. (1998) found "effective enforcement is a function of multiple, interrelated agency activities and capabilities. Effective enforcement is most likely to occur in an agency employing a facilitative enforcement philosophy with (1) an adequate number of technically competent staff, (2) strong proactive leadership, (3) adequate legal support, and (4) a consistently strong effort to check building and development plans, inspect building and development sites, and provide technical assistance." The CWC could foster successful plan implementation by providing workshops for contractors and builders as part of the ongoing education programs and by providing the necessary legal support.

Economic Development

As described in Chapter 1, there are fundamental tensions between land management to promote economic development and land management to sustain or enhance water quality. The Economic Development Study recognizes this challenge and attempts to promote environmentally sensitive economic development. The study is comprehensive, exhaustive, and well-thought-out. It takes into account the problems of limited land for building, the narrow road system found throughout much of the region, and the established farming, mining, and forest products industries. The three volumes of work (HR&A, 1998a–c) maintain that the region will experience only slow growth, allowing time for needed adjustments.

There was a general lack of discussion on chemical and waste management needs of future residential, commercial, or industrial development. Undoubtedly, some of the suggested commercial development funded by the Catskill Fund for the Future will generate hazardous materials that will require proper disposal. However, no special requirements are outlined for handling or storing the materials to prevent or capture spills. Many organic solvents have the potential to volatilize and eventually pollute nearby waterways. Other pathways of contaminant migration are also significant in this region, as most of the developable land is prone to flooding (HR&A, 1998c). Specialty industrial development may require current water and wastewater facilities to install new technology or more equipment. This is an issue that requires considerable further attention from the CWC in choosing projects for the Catskill Fund for the Future.

Conclusions and Recommendations for Land Use Planning

Although it is too early to evaluate the net result of planning efforts, several general recommendations are offered. Perhaps the most important strength is that most programs are locally operated and locally driven and hence are better able to respond to the needs of watershed communities. It is important to emphasize that the effectiveness of watershed management will be enhanced by comprehensive planning.

1. The CWC should review local and county plans for the watershed areas to make sure they are compatible and are working in a coordinated and integrated fashion. The fragmented planning efforts called for in the MOA for the Catskill/Delaware watershed may produce less-than-optimal results. One solution would be to create comprehensive plans for those watershed towns that do not have them and to review and update existing plans for neighboring communities to be in accord with the newly created plans. At the next level, counties should attempt to integrate the plans for the towns and villages within its jurisdiction, and the CWC should review the county plans for compatibility and workability.

2. The mission of the CWC should be expanded so that it can play a greater role in strengthening local planning efforts of smaller West-of-Hudson communities. The CWC can provide for meaningful citizen involvement, can provide technical assistance to support local implementation and enforcement, and can provide a template for local watershed plans.

3. The CWC and the New York State Department of State should continue to assist citizen planning boards in plan-making and decision-making. The CWC should foster involvement by local communities in planning efforts undertaken by consultants. The CWC and the NYS DOS should continue to offer citizen planning boards technical assistance in determining the long-term impact of proposed plans.

4. The CWC should use external review of funding requests for grants. About 20 percent of the Catskill Fund for the Future is for grants to help communities and others. To ensure an open and fair process, external review of funding requests should be established.

5. Because of the constraints enumerated by the Economic Development Study, future growth in the Catskill/Delaware watershed region is likely to be slow. There are a limited number of areas in the watershed where growth can occur. Many of these areas are also targeted by the Land Acquisition Program in order to protect water quality. The growth areas identified by the study are in keeping with the spirit of the MOA.

6. The CWC and NYC DEP should identify social and economic factors that should be monitored to help determine whether the Watershed Partnership and Protection Programs are effective. As mentioned in Chapter 4, although they are frequently overlooked, social and economic factors are often key to the success of watershed management. Some parameters are currently being monitored as part of the MOA, including acreage of land acquired and economic trends in the watershed (via the Economic Development Study). Other parameters that should be considered for monitoring in the New York City watersheds include population growth in the watershed (both permanent and seasonal population), types and sizes of new development, and awareness levels of watershed citizens with regard to the MOA and watershed management. In particular, the latter metric could help direct educational resources to less-well-informed watershed residents, thereby increasing public participation in the Watershed Partnership and Protection Programs.

REFERENCES

Berke, P. R., and S. P. French. 1994. The influence of state planning mandates on local plan quality. Journal of Planning Education and Research 13(4):237–250.

Bowers, J. L. 1999. Watershed management: It's not just a job, it's a way of life. Water Resources Impact 1(1):11–13.

Brower, D. J., and G. G. Ballenger. 1991. Permit Compliance Assessment. Prepared for the Division of Coastal Management, North Carolina Department of Environment, Health and Natural Resources. Chapel Hill, NC: Center for Urban and Regional Studies, University of North Carolina.

Budrock, H. 1997. Summary Guide to the Terms of the Watershed Agreement. Arkville, NY: The Catskill Center for Conservation and Development, Inc. in collaboration with the Catskill Watershed Corporation.

Burby, R. J., P. J. May, and R. Patterson. 1998. Improving compliance with regulations: Choices and outcomes for local government. Journal of the American Planning Association 64(3):324–334.

Burby, R. J., and R. Patterson. 1993. Improving compliance with state environmental regulations. Journal of Policy Analysis and Management 12(4):753–772.

Cairns, J. J., and J. R. Pratt. 1992. Restoring ecosystem health and integrity during a human population increase to ten billion. Journal of Aquatic Ecosystem Health 1:59–68.

Catskill Watershed Corporation (CWC). 1997. Economic Development Consultant, Request for Proposal. Margaretville, NY, Catskill Watershed Corporation, October 28, 1997. Margaretville, NY: Catskill Watershed Corporation.

CWC. 1999. Watershed Advocate 1(1). Margaretville, NY: Catskill Watershed Corporation.

Hamilton, Rabinovitz and Altschuler, Inc. 1998a. West of the Hudson Economic Development Study for the Catskill Watershed Corporation Report 1: Baseline Economic Analysis and Community Assessment. Prepared by Hamilton, Rabinovitz and Alschuler, Inc., Alle King, Rosen and Fleming, Fairweather Consulting, The Saratoga Associates, and Shepstone Management. June 22, 1998.

Hamilton, Rabinovitz and Altschuler, Inc. 1998b. West of the Hudson Economic Development Study for the Catskill Watershed Corporation Report 2: Market Sector Assessment & Program Issues Analysis. Prepared by Hamilton, Rabinovitz, and Alschuler, Inc., Alle King, Rosen and Fleming, Fairweather Consulting, The Saratoga Associates, and Shepstone Management. September 25, 1998.

Hamilton, Rabinovitz and Altschuler, Inc. 1998c. West of the Hudson Economic Development Study for the Catskill Watershed Corporation Preliminary Draft Report: A Blueprint for the Catskill Fund for the Future. Prepared by Hamilton, Rabinovitz and Alschuler, Inc., Alle King, Rosen and Fleming, Fairweather Consulting, The Saratoga Group, Shepstone Management, and Sno. Engineering. December 4, 1998.

Johnston, R. A., and W. S. McCartney. 1991. Local government implementation of mitigation requirements under the California Environmental Quality Act. Environmental Impact Assessment Review 11:53-67.

Marx, R., and E. A. Goldstein. 1999. Under Attack: New York's Kensico and West Branch Reservoirs Confront Intensified Development. New York, NY: National Resources Defense Council.

Miele, J. A. 1999. Memo to Eric Goldstein of the NRDC regarding the report Under Attack: New York's Kensico and West Branch Reservoirs Confront Intensified Development. March 4, 1999.

Neville, L. R. 1999. Effective watershed management at the community level: What it takes to make it happen. Water Resources Impact 19(1):14–15.

New York City Department of Environmental Protection (NYC DEP). 1997. The New York City Land Acquisition and Stewardship program. Valhalla, NY: NYC DEP.

NYC DEP. 1998. Calibrate and Verify GWLF models for remaining Catskill/Delaware Reservoirs. Valhalla, NY: NYC DEP.

NYC DEP. 1999a. Land Acquisition and Stewardship Program Status Report. July 27, 1999. Valhalla, NY: NYC DEP.

NYC DEP. 1999b. Preliminary Report on Recreational Use of New York City Water Supply Lands. Shokan, NY: NYC DEP.

New York State Legislative Commission on Rural Resources. 1994. Local Planning and Zoning Survey: New York State Cities, Town, and Villages. Albany: New York State Legislature.

New York State Water Resources Institute (NYS WRI). 1998. Phase I Delaware County Comprehensive Strategy for Phosphorous Reductions. 1st Complete Draft. Ithaca, New York, December 1998.

Promersberger, B. 1984. Implementation of stormwater detention policies in the Denver Metropolitan Area. Flood Hazard News 14(1):10–1.

Putnam County. 1997. Scope for Work for Croton Plan, September 18, 1997. New York: Putnam County.

Virginia Department of Environmental Quality. 1998. Assessment of Virginia's Waters, 305b report to EPA for water years 1996 and 1997 Virginia Department of Environmental Quality. Richmond, VA: Virginia Department of Environmental Quality.

Westchester County. 1998a. The Croton Plan. Draft Work Plan for Development of the Comprehensive Croton System Water Quality Protection Plan in Westchester County. White Plains, NY: Westchester County, May 1998.

Westchester County. 1998b. Work Plan for Development of the Comprehensive Croton System Water Quality Protection Plan in Westchester County. White Plains , NY:Westchester County.

8

Phosphorus Management Policies, Antidegradation, and Other Management Approaches

This chapter discusses four policy tools that New York City is using, or could be using, within its watershed management strategy. Although somewhat unrelated to one another, these tools represent various ways in which protection strategies (both structural and nonstructural) can be implemented. In addition, they each highlight the difficulty of assessing and mitigating nonpoint source pollution in comparison to point source pollution. The Total Maximum Daily Load (TMDL) Program, mandated for impaired waters by the Clean Water Act (CWA), is a powerful tool for determining the relative contributions of point and nonpoint source pollution to the water supply reservoirs. The Phosphorus Offset Pilot Program, or "trading" program, was created to allow new point sources of pollution without increasing the overall level of pollution within a subwatershed. Antidegradation refers to a state and federal policy that is intended to prevent the lowering of water quality within water supply reservoirs (and other waterbodies). Finally, treatment processes beyond chlorine disinfection are being explored for use in the New York City water supply system.

TOTAL MAXIMUM DAILY LOAD PROGRAM

Section 303(d) of the CWA requires states to identify waters that do not meet the goal of "fishable, swimmable water quality" and to develop TMDLs for them. The TMDL process involves identifying the chemical(s) of concern that are causing impairment, defining a water quality standard for that chemical (if a federal or state standard does not already exist), and determining the allowable loading (TMDL) of that chemical such that the standard is not exceeded. If the current pollutant loading to a waterbody exceeds the TMDL, then the state must identify point and nonpoint sources of pollution and suggest ways to decrease loading from these sources.

Participation of all 19 New York City reservoirs in the TMDL program is required under the City's filtration avoidance determination. The New York City Department of Environmental Protection (NYC DEP) has been developing TMDLs for each reservoir in three phases; the TMDLs generated at each phase supersede the previous values and reflect more site-specific data, better modeling efforts, and improved implementation methods for meeting TMDLs. Because the overall goal of the TMDL program is to identify specific polluting areas and land uses and concentrate point and nonpoint source pollution controls to those specific areas, the TMDL calculations are viewed by NYC DEP as a planning exercise that will guide the implementation of best management practices (BMPs) through-out the New York City watershed (K. Kane, NYC DEP, personal communication, 1998). This fact places tremendous importance on the adequacy of the methods and models used in the TMDL program.

NYC DEP has decided to base its TMDL calculations on total phosphorus as the compound of concern. In doing so, it has assumed that phosphorus is the limiting nutrient for plant primary production and that it correlates best with disinfection byproduct (DBP) formation, nuisance algae and eutrophication, hypolimnetic anoxia, and taste and odor problems. Using phosphorus as the target of the TMDL program presents several challenges because it is somewhat ambiguous and controversial as an indicator of drinking water quality. Because there is no national maximum contaminant level (MCL) for phosphorus, NYC DEP has had to interpret qualitative state laws regarding phosphorus. The state currently endorses a guidance value for in-reservoir total phosphorus of 20 µg/L measured in the epilimnion during the growing season (NYS DEC, 1993). This value, based on aesthetic effects for primary and secondary contact recreation, was used for Phase I TMDLs. For Phase II, however, NYC DEP has interpreted the state's narrative standard for phosphorus, which states that, for all classes, "there shall be none in amounts that result in the growth of weeds, algae, and slimes that will impair the waters for their best uses." (NYCRR 701-703). Accordingly, a phosphorus concentration of 15 µg/L has been recommended by NYC DEP and used for Phase II TMDLs (NYC DEP, 1999a).

TMDL Methodology

The TMDL program in New York City was developed in phases in order to generate results quickly with currently available information and also to update TMDLs on a regular basis as more data and sophisticated models become available. The first two phases have been completed; Phase I values have been adopted by the New York State Department of Environmental Conservation (NYS DEC) and approved by the Environmental Protection Agency (EPA), and Phase

II values await approval in September 1999. Phase III is not expected to be complete until 2002.

Phase I Methodology

Step 1: Calculating the TMDL. The first step of the TMDL process determines the status of the basin with respect to phosphorus concentration. Phosphorus concentration data were collected from different points in each reservoir and at different depths, both biweekly and monthly. During Phase I, data from the growing seasons (May 1 through October 31) of 1990–1994 were collected, and the median value for each year was calculated.

Median values of phosphorus concentration were converted into annual phosphorus loads using the Vollenweider equation (see Equation 8-1 of Box 8-1). The Vollenweider equation is a simple, steady-state model of chemical flux through a completely mixed waterbody (Vollenweider, 1976). It was originally developed with data collected from Canadian lakes. The derivation of the Vollenweider equation and its use to convert phosphorus concentrations to phosphorus loads is discussed in Box 8-1.

It should be noted that the use of the Vollenweider equation to estimate loadings to the water supply reservoirs is unusual. In most cases, the Vollenweider equation is utilized to estimate the total phosphorus concentration in the reservoirs based on measured or estimated loadings from tributaries and other inputs. Agreement between the measured and calculated phosphorus concentrations indicates appropriate application of the model. In NYC DEP's application of the Vollenweider equation, measured total phosphorus concentration in the reservoirs is utilized to "back-calculate" the loading to each reservoir (and a comparison is made with an independent method, the Reckhow equation, to estimate loadings). This use of the Vollenweider equation requires less information to be gathered because monitoring of storm events from tributary inputs to derive accurate estimates of nutrient loadings is not needed.

The estimated phosphorus loads (L) from each of the five years between 1990 and 1994 were arithmetically averaged to determine one value for the "current load." The state guidance value for phosphorus, 20 μg/L, was also converted into an annual load (the "critical load") using the Vollenweider equation. The TMDL for a reservoir corresponds to this critical load. However, given the uncertainties inherent in the process, a margin of safety (10 percent for the New York City reservoirs) is usually subtracted from the critical load to give the "available load." It is this available load to which the current load is compared to determine if a waterbody is exceeding its TMDL. If the current load is below the available load, then the basin is not exceeding its TMDL. It the current load is greater than the available load, then the amount in exceedance must be reduced in the watershed by reducing the contribution of phosphorus from either point or nonpoint sources.

BOX 8-1
The Vollenweider Equation

The Vollenweider equation was derived as follows: a simple mass balance for a constituent like total phosphorus in a complete-mix basin (in which a fraction of material is settling at an apparent settling velocity of v_s) results in the following steady-state equation:

$$P = L / (q_s + v_s) \qquad (8\text{-}1)$$

where P = the total P concentration throughout the reservoir, mg/L
  L = the total P loading to the reservoir from all sources, g/m^2/day
  q_s = the overflow rate of the reservoir (flow rate/surface area), m/day
  v_s = the apparent mean settling velocity of total P, m/day

This mass balance equation can be written in an equivalent form as:

$$P = L \tau_w (1 - R) / H \qquad (8\text{-}2)$$

where τ_w = the mean hydraulic detention time of the reservoir, days
  R = the fraction of total P retained within the reservoir system (dimensionless)
  H = the mean depth of the reservoir (volume/surface area)

The form of the equation that the NYC DEP utilized varies from Equation 8-2 because it makes use of correlations to account for the fraction of total phosphorus that is not retained in the reservoir. The fraction of total phosphorus that passes through the reservoir is inversely proportional to the square root of the mean hydraulic residence time (Larsen et al., 1976). This empirical relationship, which has been validated on hundreds of lakes, was found to be accurate for the New York City reservoirs (Janus, 1989). Therefore, the final equation that is used by the NYC DEP is:

$$P = L\tau_w / \left[H \left(1 + \sqrt{\tau_w} \right) \right] \qquad (8\text{-}3)$$

Step 2: Modeling All Sources of Pollution. The second step in the Phase I TMDL process is to establish the relative contribution of phosphorus from different sources throughout the basin. In New York City, this step used data from 1993 only. To determine the contribution from wastewater treatment plants (WWTPs), effluent samples were collected on a regular basis and a flow-weighted mean phosphorus concentration was determined. This concentration was then multiplied by the average flow from the WWTP and the number of days in operation during 1993 to determine the annual phosphorus loading from WWTPs. Upstream reservoirs are also considered to be point sources for downstream reservoirs. To determine their contribution, the tunnels connecting the reservoirs were sampled for phosphorus at a point just prior to the receiving reservoir.

Nonpoint sources pose greater difficulties because few methods have been developed to measure their contributions. NYC DEP used the Reckhow model to predict the contributions of agricultural land, forest, urban land, atmospheric deposition, and septic tanks (Reckhow et al., 1980). As shown in Box 8-2, each contribution was calculated by multiplying the area of land devoted to that land use by an export coefficient, which was derived from the literature. For septic systems, only those sited within 100 ft of lakes, reservoirs, and watercourses were assumed to contribute loadings to the reservoirs. Because current state regulations prohibit septic systems within 100 ft of a waterbody, the septic system contribution is made up primarily of preexisting and noncomplying systems.

The Reckhow equation sums up all the point and nonpoint sources and produces an annual phosphorus loading in kg/year, which can be converted into phosphorus concentration, again using the Vollenweider equation. This concentration was compared to the measured annual median phosphorus concentration. If the difference between those values was less than 20 percent, then the Reckhow model was assumed to be an adequate model for that basin.

Although this second step may appear to be strictly academic and not required during the TMDL process, NYC DEP was obliged (because of the filtration avoidance determination and agreements with NYS DEC) to measure point and nonpoint phosphorus loading using the Reckhow model, even for those basins that were not exceeding their TMDLs.

Step 3: Determining Wasteload Allocations and Load Allocations. The last step of the Phase I TMDL process was only done if a basin exceeded its TMDL (as it did in Cannonsville Reservoir). In this step, the waste load allocations (WLA) from point sources and load allocations (LA) from nonpoint sources are determined to help optimize management strategies. NYC DEP determined the Cannonsville WLA by assuming that all WWTPs would meet the effluent phosphorus concentration goals of the Memorandum of Agreement (MOA) as a result of the upgrades. These effluent phosphorus concentrations were then multiplied by the State Pollution Discharge Elimination System (SPDES) permit-

BOX 8-2
The Reckhow Model

The Reckhow model is a lumped-parameter model that determines the contribution of different land uses to overall pollutant loads. Data on land use in the New York City watersheds are stored in a Geographic Information System (GIS) database derived from Landsat images taken at 30-m resolution. Four land use categories are distinguished in the Reckhow model: urban, forest, agriculture, and water, with the agricultural land use being sometimes subdivided into corn/alfalfa, bare soil, and grass/shrubs. For Phase I, the export coefficients for the Reckhow model were derived mainly from the literature. Slight variations in the Phase I coefficients were used to reflect known differences between the Croton, Catskill, and Delaware watersheds. In Phase II, only the Croton watershed TMDLs utilized the Reckhow model. Export coefficients for this phase were revised to include some site-specific data and information. Coefficients used for Phases I and II are shown below along with the governing equations for the Reckhow model.

$$L = EC_{ag}A_{ag} + EC_fA_f + EC_uA_u + EC_aA_s + Septic + PSI \qquad (8-4)$$

$$Septic = EC_s \text{ (#people per house)(# houses) } (1 - SR) \qquad (8-5)$$

where L = the total P loading to the reservoir from all sources, kg/year
EC_{ag} = export coefficient for agricultural land (kg/ha/yr)
EC_f= export coefficient for forest land (kg/ha/yr)
EC_u= export coefficient for urban land (kg/ha/yr)
EC_a= export coefficient for atmospheric input (kg/ha/yr)
EC_s= export coefficient for septic systems (kg/capita/yr)
A_{ag} = area of agricultural land (ha)
A_f = area of forest land (ha)
A_u = area of urban land (ha)
A_s = area of lake surface (ha)
PSI = point source input (kg/yr)
Septic = septic system input (kg/yr)
SR = soil retention coefficient.

Export Coefficients (kg/ha/yr):	Croton Phase I	Catskill	Delaware	Croton Phase II
Urban	0.70	0.70	0.70	0.90
Forest	0.05	0.07	0.05	0.05
Atmospheric	0.53	0.53	0.53	0.10
Agriculture:	0.3			0.3
Corn/alfalfa		2.00	2.00	
Bare soil		0.30	0.30	
Grass shrubs		0.15	0.20	

ted flows to obtain the WLA. The LA was obtained by subtracting the WLA from the available load.

The main purpose of calculating a WLA is to set a permanent upper bound on phosphorus loading that can be incorporated into the SPDES permit of a WWTP. The LA is sometimes used to target best management practices to the appropriate nonpoint sources.

Phase II Methodology

Phase II has a more complex methodology than Phase I, although the three steps of the process are analogous (NYC DEP, 1999b). Details are only provided where differences between Phases I and II exist.

Step 1. In Phase II, the geometric mean of the reservoir phosphorus concentration data was used rather than the median. The geometric mean was thought to be a more accurate representation of the phosphorus concentration data, which follow a lognormal frequency distribution (K. Kane, NYC DEP, personal communication, 1998). Values that were below the detection limit (2–5 µg/L) were set to half the detection limit. Unlike Phase I, phosphorus concentration data from 1992 to 1996 were used in Phase II. The geometric mean from each of these five years was arithmetically averaged to determine the current load.

The margin of safety in Phase II was changed slightly to account for large variations in phosphorus data from year to year (K. Kane, NYC DEP, personal communication, 1998). The margin of safety can vary between 10 percent and 20 percent, depending on interannual variability in phosphorus concentrations (NYC DEP, 1999b).

Finally, the Vollenweider equation was altered slightly for Phase II to accommodate its coupling to the Generalized Watershed Loading Function (GWLF) in Step 2. Unlike the Reckhow model, which does not vary over time, the GWLF can simulate large, storm-related pollutant loadings. These loadings may deliver particulate phosphorus that settles to the bottom of the reservoirs and does not affect midlake phosphorus concentration. As originally formulated, the Vollenweider equation does not take into account this fraction of input phosphorus lost to the sediments. Thus, an additional term representing this fraction was added to the Vollenweider model, resulting in Equation 8-6 (NYC DEP, 1999b).

$$P = \left(1 - f_s\right)L\tau_w / \left[H\left(1 + \sqrt{\tau_w}\right)\right] \qquad (8\text{-}6)$$

where P = phosphorus concentration (mg/L)
   L = the total P loading to the reservoir from all sources, g/m^2/day
   τ_w = the mean hydraulic detention time of the reservoir, days
   H = the mean depth of the reservoir (volume/surface area)
   f_s = the fraction of input phosphorus that is positionally unavailable

Step 2. Determining the contributions of point and nonpoint sources to phosphorus loading has changed dramatically from Phase I. For reservoirs in the Croton watershed, the nested Reckhow model was used to divide the region into multiple subbasins. The nested Reckhow was adopted because NYC DEP felt that the Phase I approach was not accounting for phosphorus retained in nine large waterbodies upstream of the 13 reservoirs. In Phase II, the waterbodies in these subbasins were assumed to retain 50 percent of the phosphorus and were then considered to be point sources to the downstream waterbodies. A 50 percent phosphorus retention rate was used for all of the subbasins in the absence of data on the water budget, phosphorus concentrations, and residence times for these subbasins. Previous studies on reservoirs with residence times greater than six months show that phosphorus retention tends to plateau around 60 percent (K. Kane, NYC DEP, personal communication, 1998). NYC DEP choose 50 percent because although the East-of-Hudson subbasins have longer residence times than the main East-of-Hudson reservoirs, they are also more eutrophic and may be creating phosphorus near the bottom waters as a result of algal decay.

As shown in Box 8-2, the Phase II Reckhow model was amended to include some locally measured export coefficients, and data from four years were used rather than data from one year. Another major change for Phase II is that the criterion for determining whether the Reckhow model is a good fit to a basin has been increased from 20 percent to 50 percent in some cases. The 20 percent acceptance criterion was thought to be too restrictive, especially in basins where the phosphorus concentration is very low.

In the Catskill/Delaware watershed, the Reckhow model was replaced in favor of the GWLF, which can predict the temporal and spatial variability in phosphorus loading. The GWLF is a numerical model that simulates hydrology, nonpoint source runoff of pollutants, and point source inputs. It is an advancement over the Reckhow approach because it utilizes daily time intervals to generate monthly, seasonal, or annual loadings to reservoirs. Also, it provides separate estimates of groundwater inputs, which are especially important in systems with thin soils such as the Catskill/Delaware reservoirs [in fact, the model was originally developed for the Catskill Region by Haith et al. (1983, 1992) and Haith and Shoemaker (1987) at Cornell University]. Figure 8-1 shows the three submodels that make up the GWLF, and a more detailed description of the model is given in Box 8-3. The model has been calibrated and validated using stream flow data from the Catskill/Delaware reservoirs.

Although the data requirements for the GWLF are extensive, literature values are often used for some of the needed parameters. NYC DEP monitoring program provides required data on precipitation, air temperature, land use, soil type, topography, and point source phosphorus loads. There are 11 land use categories in the GWLF as compared to six in the Reckhow model. Data of other model parameters, including soil erodability and pollutant concentrations in ground-

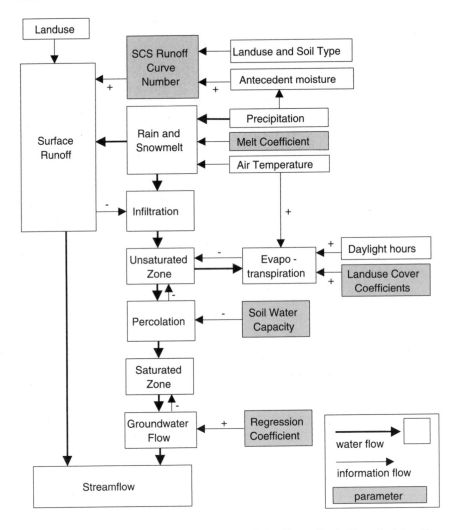

FIGURE 8-1 (A) Water balance submodel of the Generalized Watershed Loading Function.

water and surface runoff, are not currently available and have been derived from the literature up to the present time.

Step 3. During the Phase I TMDL process, the WLA and LA were calculated only for those basins exceeding their TMDLs. However, during Phase II they were calculated for all basins, regardless of their TMDL status. The methods for calculating the WLA and LA are unchanged from Phase I.

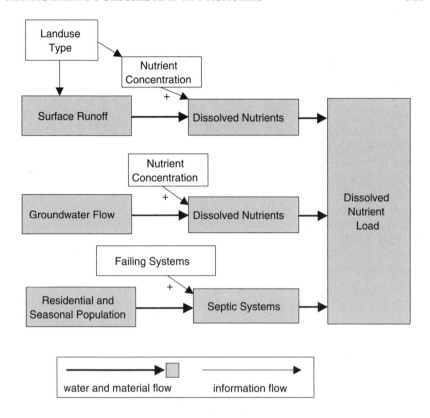

FIGURE 8-1 (B) Dissolved nutrient submodel of the Generalized Watershed Loading Function.

Phase III Methodology

Although Phase III TMDLs are not due for several years, NYC DEP has been developing more sophisticated models for this process. It is anticipated that the Vollenweider equation will be discarded in favor of a more complex water quality model that has both hydrothermal and eutrophication components. Such a model is currently being designed for the Cannonsville watershed (Cannonsville model), and similar efforts for the other Catskill/Delaware reservoirs are under way. A primary goal of Phase III is to link the GWLF to such a water quality model in order to predict long-term changes in water quality as a result of management practices in the watershed. More information on the hydrothermal and eutrophication models that make up the Cannonsville Model can be found in Auer et al. (1998), Doerr et al. (1998), Owens et al. (1998), and numerous NYC

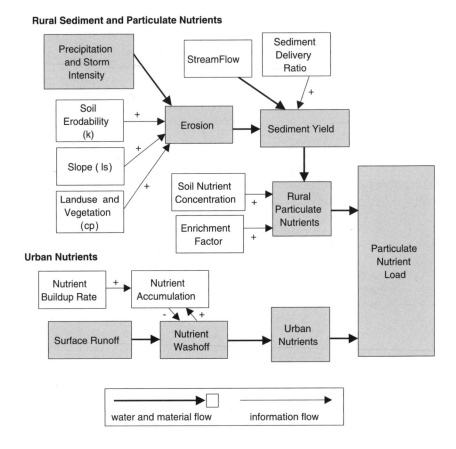

FIGURE 8-1 (C) Particulate nutrient submodel of the Generalized Watershed Loading Function.

DEP publications. As presently formulated, these water quality models focus specifically on nitrogen, phosphorus, dissolved oxygen, and sediment. Water quality and terrestrial models for simulating fate and transport of microbial pathogens and precursors of disinfection byproducts are not currently under development.

Phosphorus-Restricted Basins

NYC DEP has developed a water quality measure for its reservoirs that is similar in calculation to the TMDL and is used for similar management purposes.

BOX 8-3
Generalized Watershed Loading Function

The GWLF is a terrestrial runoff model than can predict contributions of point and nonpoint sources to overall pollutant loading. As shown in Figure 8-1, the GWLF is divided into three submodels that are combined to predict overall pollutant loadings to waterbodies.

In the water balance submodel, stream flow is divided into surface runoff and groundwater contributions. Precipitation, evapotranspiration, infiltration, percolation, snowmelt, and other processes contribute to overall stream flow. This submodel requires data on rainfall, temperature, land use, and soil characteristics. Rather than measuring all of the parameters, NYC DEP has used known data on land cover and soils (from the current GIS databases) combined with parameters derived from literature research to determine the contributions of groundwater and surface runoff to overall flow. The hydrology submodel was calibrated and validated using measured stream flow data from eight drainage basins west of the Hudson River.

The dissolved nutrient submodel generates loadings of dissolved nitrogen and phosphorus in groundwater and surface water and from septics systems. To use this model, NYC DEP has combined GIS data on land cover and septic system numbers with literature values of nutrient concentrations in different types of nonpoint source runoff.

The final submodel is the particulate nutrient submodel, which generates sediment and particulate phosphorus loadings from erosion and surface runoff. This submodel relies on empirical runoff and erosion relationships (e.g., the Universal Soil Loss Equation (USLE) developed by the USDA Agricultural Research Service). Data on land cover and sediment yield were used by the NYC DEP in combination with literature values for soil erodability, slope, vegetation, and other factors to generate loadings.

For all three submodels, the same land cover database used for the Reckhow model was used in the GWLF. Both of the nutrient submodels were calibrated and validated using actual stream loadings measured at one location in the Cannonsville watershed.

There are some important limitations of the GWLF that should be noted and effectively overcome as part of the Phase III TMDL calculations. First, it is not clear that the septic system component of the dissolved nutrient loading submodel accounts for effects of soil type and depth to groundwater. Nitrate concentrations may be much lower in anaerobic soils with high organic content than in sandy soils where most studies of nitrate plumes from septic systems have been conducted.

continued

BOX 8-3 Continued

Second, the SCS Curve Number and USLE do not necessarily yield accurate, absolute values of streamflow, erosion, or sediment delivery. Both methods can be orders of magnitudes off for certain events and soil-vegetation complexes because neither was developed for the forested, natural vegetated watersheds with thin soils that are characteristic of the Catskill/Delaware watershed. Using SCS curve numbers to characterize runoff potential can misrepresent true runoff processes in the interest of computational simplicity (D. Booth, University of Washington, personal communication, 1999). The USLE is a poor estimator of upland sediment generation in any watershed save those with predominantly agricultural land uses (for which the USLE was originally developed). Watersheds that generate much of the sediment load to streams via stream-channel erosion will not be represented by this approach. (It should be noted that the USLE has been adapted for forested conditions and has been shown to work well under such conditions. The committee assumes that this adaptation was used in the GWLF for forested parts of the watershed.) Further site-specific testing and validation of both nutrient submodels in each subwatershed are of paramount importance before the GWLF is used in Phase III TMDL calculations.

Source: NYC DEP (1998a).

"Phosphorus-restricted basins" are those which also exceed the state guidance value of 20 µg/L phosphorus. However, there are some fundamental differences between TMDL and phosphorus-restriction calculations. Phosphorus restriction is mentioned here (1) because of its similarity to TMDLs, (2) because it is an important part of the phosphorus offset pilot program to be discussed later, and (3) because it is used in conjunction with the controversial 60-day travel time (see Chapter 11).

The Watershed Rules and Regulations prohibit new or expanded WWTPs with surface discharges from being located within phosphorus-restricted basins. Thus, phosphorus restriction is a measure of the health of a reservoir. Phosphorus-restricted basins are determined by measuring total phosphorus concentration at all reservoir depths from May to October (NYC DEP, 1997a). It should be noted that because of the seasonality of the data used, the phosphorus restriction analysis as formulated yields the lowest concentrations (or "best picture") of the year. This is because utilization of phosphorus by microbes is highest during the summer and because there are relatively low influents to the reservoirs during summertime.

Measured phosphorus concentrations are expressed as a yearly geometric mean, and the geometric means are then averaged over the five most recent years to determine exceedance. In general, if a reservoir's five-year running average exceeds 20 µg/L for more than two years in a row, it is designated as phosphorus-restricted.

There are significant differences between the concepts of phosphorus TMDLs and phosphorus-restricted basins, although both were designed to protect water quality and to mitigate problems associated with eutrophication, algal blooms, and low dissolved oxygen in the hypolimnion. These differences are summarized in Table 8-1. In addition to the distinctions listed in Table 8-1, it should be noted that the concepts differ in spatial scale and extent. The calculation of phosphorus restriction uses data from a single waterbody, not an entire watershed. Thus, it is really not a "basin concept" like the TMDL process, which considers loading from point and nonpoint sources in the entire watershed.

Phase I and II TMDL Results

For the most part, the reservoirs in the Catskill/Delaware watershed are not exceeding their Phase I or Phase II TMDLs, nor are they phosphorus-restricted. Reservoirs within the Croton watershed, which has undergone rapid development

TABLE 8-1 TMDL Basin Concept vs. Phosphorus-Restricted Reservoir Concept

	TMDL	Phosphorus Restriction
Governing Agency	NYS DEC and EPA	NYC DEP
Requirements	Multiple data sets/models required	Phosphorus concentration data
Regulatory Implications	Determines compliance with the Clean Water Act and affects SPDES Permits	Affects siting of WWTPs in the New York City water supply watersheds only
Exceedance Test	Phosphorus load corresponding to 15 µg/L exceeded	Phosphorus concentration of 20 µg/L exceeded
Consequence of Exceeding the TMDL or Phosphorus Standard	Reduction in SPDES-permitted phosphorus loads and nonpoint source loads	No new WWTPs can be constructed in the basin; extra requirements for SPPPs.
Margin of Safety	90% of load is allocated	No margin of safety
Updating	Infrequent revisions	Annual revision

during the last 20 years, are considerably less healthy. In general, the basins that are exceeding their Phase II TMDLs are also phosphorus-restricted. Table 8-2 lists the Phase II TMDLs and phosphorus-restricted status of all 19 reservoirs in both watersheds.

TABLE 8-2 Phase II TMDLs and Phosphorus-Restriction Status

Reservoir	Phase II TMDL kg/yr	Available Load kg/yr	Current Load kg/yr	WLA[a] kg/yr	LA kg/yr	Average [P], 1992–1996 µg/L[b]
West of Hudson						
Ashokan East	19,542	17,588	16,484	4	17,584	13.4
Ashokan West	45,399	40,859	32,833	264	40,595	15.4
Cannonsville[c]	40,237	35,207	52,368	1,059	34,148	**21.1**
Neversink	16,914	15,223	6,863	0	15,223	6.4
Pepacton	59,375	53,437	37,327	388	53,049	9.7
Rondout	41,413	37,272	23,476	125	37,147	8.7
Schoharie	22,321	20,089	19,864	789	19,300	21.3
East of Hudson						
Kensico	28,276	25,448	16,926	0	25,448	9.9
Amawalk[c]	997	897	1,318	390	507	**21.3**
Bog Brook[c]	281	253	321	28	225	18.7
Boyds Corner[c]	725	652	687	0	652	14.6
Cross River	1,007	881	717	108	773	12.9
Croton Falls[c]	3,565	3,030	5,010	615	2,415	**28.7**
Diverting[c]	2,098	1,794	3,844	232	1,562	**30.0**
East Branch[c]	2,116	1,851	3,462	449	1,402	**26.4**
Middle Branch[c]	712	612	1,020	173	439	**24.3**
Muscoot[c]	7,048	6,343	11,560	1,405	4,938	**24.9**
New Croton	9,731	8,758	11,189	209	8,549	17.6
Titicus[c]	869	739	1,124	0	739	**22.8**
West Branch	12,760	11,484	8,662	28	11,456	13.0

Note: The Phase II TMDL column lists the TMDL value calculated from the modified Vollenweider model based on the 15-µg/L phosphorus standard. The Available Load column shows the available load, which ranges from 80 to 90 percent of the calculated TMDL. The Current Load column represents the current load derived from phosphorus concentration data collected in each reservoir. The current load must be less than the available load for a reservoir to be in compliance with its TMDL. The WLA and LA columns show calculated waste load allocations (WLA) and load allocations (LA), respectively. The final column shows the 1992–1996 calculation of average phosphorus concentration in each reservoir. Significant figures in the columns differ because of the different units used.

[a]The WLA was calculated by assuming that upgrades mandated by the Watershed Rules and Regulations would be in effect and that the SPDES-permitted flow would be used.

[b]Bolded values indicate phosphorus-restricted basins based on two consecutive years of exceedance.

[c]Basin exceeds its TMDL, which is based on 15-µg/L. Nonpoint sources will have to be reduced to meet the TMDL.

Source: NYC DEP (1997a, 1999c–t).

All of the basins that were exceeding their Phase I TMDLs are also exceeding their Phase II TMDLs. However, four basins in the Croton watershed that did not exceed Phase I TMDLs exceed Phase II TMDLs. This difference is attributable to the lower phosphorus standard of 15 µg/L and the additional years of phosphorus concentration data that were used during Phase II. During the Phase I TMDL process, there were only three basins (Diverting, East Branch, and Muscoot) that required a reduction in nonpoint loadings to meet the required TMDL. The other basins in exceedance were able to meet TMDLs by upgrading their WWTPs, as required by the Watershed Rules and Regulations. However, in Phase II, all of the basins that exceed their TMDLs must reduce phosphorus loadings from nonpoint sources in order to comply. As a result, meeting Phase II TMDLs in these basins will be considerably more difficult and challenging than implied from Phase I calculations. In the New York City watersheds, implementation of point source controls is under way in some basins and is planned for all others. Implementation of nonpoint source controls in Cannonsville and the nine Croton subwatersheds as a result of the TMDL calculations is not readily apparent. An important measure of success for the New York City TMDL program (or any TMDL program) is the degree of implementation of protection strategies to help the reservoirs meet their Phase II TMDLs.

Replacing the Reckhow model with the nested Reckhow model and with the GWLF had a significant impact on predicted nonpoint source loadings. In particular, the GWLF consistently predicted greater nonpoint loadings because of its ability to simulate storm events and the associated high loadings of particulate pollutants. Because the GWLF spatially and temporally simulates the inflows of both particulate and dissolved nutrient loads from a large number of land covers, it can be an important tool in targeting management practices for control of nonpoint source pollution.

Analysis of TMDL Program

Methods

Both the Phase I and II TMDL methodologies have received criticism, of varying degrees of validity. These criticisms and others identified by the committee are discussed below.

Needed Data Input for Models. The Vollenweider, Reckhow, and GWLF models are well based, highly functional, and predictive, *provided data inputs are accurate*. As stressed in Chapter 6, more event-based stream monitoring will improve the accuracy of these models, which are totally dependent upon influent discharge-based loadings and residency times within the reservoirs. The GWLF model is particularly data-intensive, requiring numerous parameter inputs that are currently in minimal supply, are not available, or are assumed from published values from other ecosystems that roughly approximate conditions of the reservoir catchment under consideration (NYC DEP, 1998a). Many of the key elements needed, such as estimations of snowmelt, soil chemistry data, and pollutant concentrations in surface runoff and groundwater, are lacking and can be feasibly collected. Real data for these and other parameters are needed at fine resolution scales (<30 m) within each of the reservoir drainage basins, particularly within half a kilometer of the boundaries of the reservoirs and major tributaries. At the very least, such measurements could be made in representative subwatersheds and generalized with other, less-intensive measurements to the rest of the system. Recent discussions on model data development, particularly using the Landsat imagery data of very high resolution (1 m), are a step in the correct direction (NYC DEP, 1998a).

Model Appropriateness. Critics of the TMDL methodology often point to the Vollenweider model as a significant weakness. The model was first constructed using lakes with residence times of 1–10 years. Thus, the steady-state assumption that is part of the Vollenweider may not be valid for the New York City reservoirs (with residence times of six months to one year). Unfortunately, it is not clear that a better model is currently available. Because the Vollenweider will probably not be used during Phase III (in favor of a time-variable water quality model), its use in Phases I and II is not perceived as a problem by this committee.

NYC DEP's use of the Vollenweider equation was somewhat unorthodox and was less than optimal. By using phosphorus concentration data to back-calculate phosphorus loadings, NYC DEP was forced to estimate an arbitrary net phosphorus retention (50 percent) in the large lakes of the Croton system and use an empirical relationship to determine phosphorus retention in the reservoirs. Simple input–output budgets (measurements of reservoir phosphorus loadings and mass phosphorus exports) could have been used to arrive at a specific net phosphorus retention in each reservoir, which would have been substantially more defensible for the calculation of TMDLs. In addition, NYC DEP's procedure required that two models be used to calibrate one another (Vollenweider and Reckhow) rather than using measurements of input loadings and lake concentrations to calibrate and validate the Vollenweider model. The committee presumes that this was because of a lack of resources required to measure all the inputs and outputs to the 19 reservoirs.

Margin of Safety. The margin of safety that differentiates the available load from the critical load has been judged by some as too low (Izeman and Marx, 1996 a,b, 1998; Novotny, 1996). There can be many purposes of the margin of safety. NYC DEP has stated that the margin of safety should account for yearly variations in hydrology and the reservoir response to the phosphorus load (NYC DEP, 1996). The agency also believes that the margin of safety should take variability in phosphorus concentrations into account, because it specifically altered the margin of safety between Phases I and II to do this.

If NYC DEP wants to determine an appropriate margin of safety given its selection of the Vollenweider model, it should conduct a formal uncertainty analysis. Equation 8-3 or 8-6 could be used in a Monte Carlo simulation to determine how errors in the inputs (total phosphorus concentrations in the reservoirs, mean depth, and mean hydraulic detention time) propagate through the equation and result in uncertainty in the estimated loadings (Reckhow and Chapra, 1983; Schnoor, 1996). In order to do this, frequency distributions with time must be constructed for (1) the measured phosphorus concentrations in the lake, (2) the mean depth (volume/surface area), and (3) the mean hydraulic residence time (volume/discharge).

As many as 10,000 realizations (simulations of the model equation) should be run while sampling the frequency distribution of each parameter using a random number generator in the Monte Carlo approach. The result is a distribution of loading values with its own frequency distribution that can be sorted to yield the median and quartile loading values. Such a distribution provides an estimate of uncertainty, which can then be characterized as a margin of safety. For example, the coefficient of variation (the standard deviation divided by the mean) can be utilized as the margin of safety.

Seasonal Variations. It has been charged that because they are averages of several years of data, the TMDL calculations do not take seasonal variations into account and are not made on a "daily" basis (Izeman and Marx, 1996a,b, 1998). Critics feel that this methodology may not capture important trends in phosphorus loading. NYC DEP has argued that because eutrophication parameters are not acutely toxic, variations in phosphorus concentration (occasionally including high concentrations) will not adversely affect reservoir water quality as long as average phosphorus concentrations are maintained at low levels (NYC DEP, 1999a), an assessment with which the committee agrees. In addition, given NYC DEP's unorthodox use of the Vollenweider model to calculate loads from concentration data, there is little utility in calculating daily loads. A meaningful calculation of daily loads would require daily sampling of all major tributaries and groundwater inputs to each reservoir, activities that are beyond the scope of NYC DEP's current monitoring program and, while desirable, are probably unnecessary for Phase I and II TMDLs.

There are, however, other important seasonal considerations that should have been taken into account. NYC DEP calculates the TMDL over the growing period because this is the time period during which NYS DEC requires that phosphorus concentrations be below the state standard. This requirement is based on the assumption that phosphorus triggers algal growth (and eutrophication) predominantly during summer months because of increased heat and light. However, problems of increased algal growth can occur at all times of the year.

In addition, it is clear that phosphorus loadings from discharges of rivers are strongly coupled to precipitation events, which are not predictable and certainly not seasonal. As currently conducted, the TMDL program may be underestimating the annual phosphorus loading, a supposition that stems from a comparison of relatively low phosphorus concentration data and high chlorophyll *a* data (NYC DEP, 1993, 1999a). Thus, the committee does not support limiting the TMDL calculation to the "growing season" only.

Implementation

The most frequently cited criticism of TMDL programs across the country is the failure of states to implement pollution control measures following calculations of waste load and load allocations. There is little information regarding implementation of Phase I TMDLs in New York City (see NYC DEP, 1998b), and none regarding Phase II (expected after September 1999). The City, via New York State, is currently under legal pressure to commence implementation. Although ongoing WWTP upgrades in the Cannonsville watershed were sufficient to meet Phase I TMDLs for the Catskill/Delaware system, meeting Phase II TMDLs will likely require implementation of nonpoint source controls in several watersheds, including the Cannonsville and Croton watersheds.

In those instances where water quality models indicate that TMDLs can be met entirely by point source controls, such improvements should be rapidly planned and implemented. Where nonpoint source reductions are needed, NYC DEP must embark on a vigorous plan to understand how BMPs can be employed to reduce phosphorus loading to the reservoirs. This is a difficult task because most watershed management actions required by the MOA are not tied quantitatively to the load reductions necessary to improve water quality. That is, there is little understanding of which BMPs should be used, how many are needed, and where they should be located to obtain a desired phosphorus loading reduction or water quality objective (e.g., 15 µg/L). These frustrations are particularly apparent in Delaware County where it is unclear what specific quantitative source reductions are needed and how to link BMP performance to such reductions.

Public Health and Ecological Protection

Phosphorus Guidance Value. The 20-µg/L guidance value for phosphorus used during Phase I has been the target of much criticism because the value is based on aesthetic and recreational concerns rather than on the use of a waterbody for drinking water (NYS DEC, 1993a). There is nothing sacred about the guidance value of 20 µg/L total phosphorus. In 1934, Clair Sawyer originally proposed a criterion of 30 µg/L of orthophosphorus at spring turnover as a criterion to prevent algal blooms in the spring and summer seasons for New England Lakes (Sawyer, 1947). Vollenweider's research on European and North American lakes indicated that the boundary separating so-called eutrophic lakes from mesotrophic lakes was about 20 µg/L, and the boundary between mesotrophic lakes and oligotrophic lakes was about 10 µg/L (Dillon and Rigler, 1974). The committee feels that the 20-µg/L guidance value is not sufficiently conservative, because it would allow the reservoirs to exist under mesotrophic and mildly eutrophic conditions. For filtration avoidance, it is desirable to have water supply quality with better than meso-eutrophic conditions.

Recently, the NYC DEP assessed the 20-µg/L guidance value by collecting data on phosphorus and chlorophyll *a* concentrations in all 19 reservoirs and relating these data to use-impairment criteria like taste, odor, algae, and other incidences of eutrophication (NYC DEP, 1999a). In order to ensure high-quality drinking water to the City's distribution system, these criteria specified that cyanobacteria cannot be the dominant algal class, and that algal numbers cannot exceed 2,000 SAU/mL, in more than 25 percent of all samples. Analyses indicated that a phosphorus guidance value of 15 µg/L would be required to meet these criteria and maintain eutrophication at or below an acceptable level, and this value was used to calculate all Phase II TMDLs. For the many reasons discussed above, the committee enthusiastically supports this value.

Other Parameters. It would be desirable to extend the TMDL calculations on phosphorus to water quality parameters that more directly influence health effects and the protection of the New York City drinking water supply. Phosphorus controls aid in the protection of ecological health of the reservoirs, they prevent taste and odor problems, and they help control sources of DBP precursors. The committee also believes that phosphorus controls can help prevent anoxia in bottom waters, although this is not the primary goal of the TMDL program.

However, for a major drinking water supply such as New York City's, there are other public health concerns, including DBPs and pathogens, that may not be fully protected using phosphorus controls alone. The remainder of this section focuses on approaches that the NYC DEP might take to model trihalomethane formation potential (THMFP), haloacetic acid formation potential (HAAFP), and other parameters for protection of the water supply. These analyses are targeted at Phase III of the TMDL program and as such will complement the use of time-variable models.

Phase III TMDLs

In Phase III, the Vollenweider model should be discarded in favor of a modeling approach where event-based, time-variable reservoir loadings (Longabucco and Rafferty, 1998) are used as input to a fully dynamic reservoir model (Schnoor, 1996). If phosphorus continues to be used as the priority pollutant, both particulate and dissolved forms of phosphorus must be modeled. Individual input–output models (which could have benefited Phases I and II) should be constructed and validated using site-specific data in order to estimate the fraction of phosphorus retained in each reservoir and the fraction of phosphorus returned to the water column from the sediment by internal regeneration processes (scour, resuspension, and mineralization/diffusion). Thus, monitoring of phosphorus loading from all prominent sources and monitoring of in-reservoir phosphorus concentrations must occur for each Catskill/Delaware reservoir.

NYC DEP has moved in this direction with the development and adaptation of a hydrothermal model (Owens, 1998), a nutrient/phytoplankton model (Doerr et al., 1998), and the GWLF. As shown in Figure 8-2, the nutrient/phytoplankton model considers nitrogen, phosphorus, dissolved oxygen, phytoplankton, and zooplankton and their interrelations. It is a dynamic mechanistic model that can be coupled with the GWLF to simulate hydrology and nutrient loading from tributaries. To date, these two models have been calibrated and validated using data from the Cannonsville watershed and have been extended to the other basins. Gathering site-specific data to validate these models in each watershed will be a necessary part of the Phase III TMDL process. For example, data on phosphorus bioavailability and phosphorus cycling similar to data collected in the Cannonsville Reservoir (Auer et al., 1998) are needed for each reservoir.

The models currently under development by New York City do not incorporate DBPs or DBP precursors (organic carbon compounds), even though compliance with future EPA rules may require modeling of these compounds. NYC DEP should make substantial efforts during Phase III to develop TMDLs for trihalomethane and haloacetic acid formation potential (THMFP and HAAFP). In order to calculate TMDLs for these parameters, it will be necessary to use numerical water quality models capable of predicting THMFP, HAAFP, dissolved organic carbon (DOC), turbidity, and taste and odor. These are the water quality parameters that affect drinking water quality directly. Such models should be developed now so that they can be tested prior to use in decision-making on the New York City water supply in the next decade.

Figure 8-3 is a schematic that shows some of the interactions that may be modeled. Soluble reactive phosphate and particulate phosphorus must both be simulated as state variables because of the need to account for nonbioavailable fractions of phosphorus. Vertical profiles of dissolved oxygen will be needed in order to estimate the internal phosphorus regeneration from sediments under anoxic conditions. Data on chlorophyll *a* (or phytoplankton biomass) will be needed because chlorophyll *a* measures ecosystem impacts and authochthonous (internally generated) DBP precursors. Knowledge of vertical variations in phytoplankton would be helpful for understanding the potential impairment of uses by excessive algae growth. DOC should be modeled both for the quantity and quality of carbon compounds that react to form DBPs, including the relative fractions of allochthonous and autochthonous sources of DOC. Dissolved silica is an important nutrient for diatom growth, and information on nitrogen species

(a)

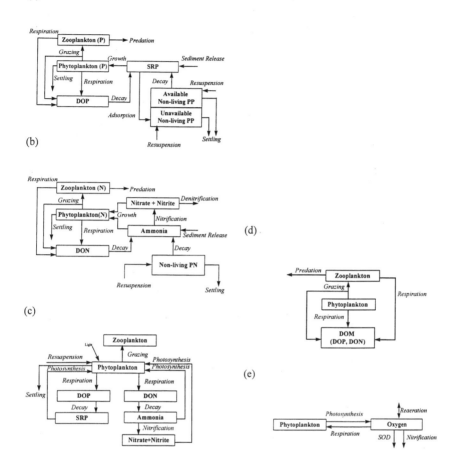

FIGURE 8-2 Conceptual submodels of the nutrient/phytoplankton model for Cannons-ville Reservoir: (a) phosphorus, (b) nitrogen, (c) phytoplankton, (d) zooplankton, and (e) oxygen. Source: Doerr et al. (1998). Reprinted, with permission, from Doerr et al., 1998. © 1998 by the North American Lake Management Society.

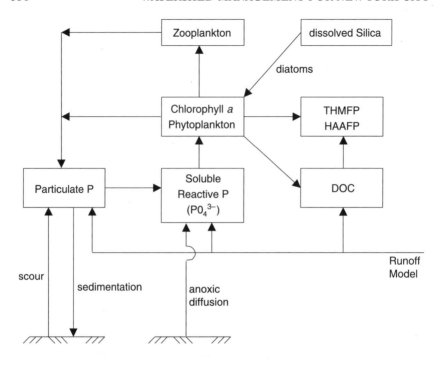

FIGURE 8-3 Modeling of THMFP, HAAFP, and other parameters.

may also be required to accurately simulate the growth and decay of phytoplankton standing crops. It remains to be seen whether taxonomic distinctions are necessary (green algae, diatoms, and cyanobacteria) to adequately simulate THMFP, HAAFP, and taste and odor thresholds.

Taste and odor are the most challenging parameters of all because little is known about the sources and fate of organic molecules that cause taste and odor. The redox status of sediments, bacteria assemblages such as *Actinomycetes* spp., and certain algal species are reported to influence taste and odor formation. Fortunately, there has not been a large problem with taste and odors in the New York City water supply to date, probably because turbidity and algae have been maintained at low levels.

It is not clear whether additional trophic levels, such as zooplankton grazing and fish predation, will be needed to simulate chlorophyll *a* concentrations accurately. In general, it is best to develop the simplest model that accurately simulates important drinking water quality variables. The required modeling of DBP precursors, DOC, color, and taste and odor goes beyond the current state of the art and will require development.

NYC DEP has made a good start on relating phosphorus concentrations and DOC concentrations to THM precursors in the Cannonsville Reservoir (NYC DEP, 1997b; Stepczuk, 1998a–c). The relationships between these parameters, however, need to move beyond simple correlations to deterministic models. Results to date indicate that it is a complicated problem; further research and development of these models is strongly encouraged.

Conclusions and Recommendations

1. In general, methods that were used for Phases I and II of the TMDL program were adequate. The Vollenweider model was the best choice for a water quality model given the limited data available and the time constraints placed on the program. It should be kept in mind that the phosphorus loading calculations are limited by the lack of real data used for the Reckhow and GWLF models.

2. Phase I and II TMDLs would have been improved if the following had been taken into consideration:

• The calculations should have used phosphorus concentration data for the entire year rather than for just the growing season.
• The modeling should have taken into account all septic systems, not only those septic systems within 100 ft of the reservoirs.
• Phosphorus retention should have been determined using input–output measurements rather than through selection of an arbitrary phosphorus retention value of 50 percent.
• The margin of safety should have been determined by conducting an uncertainty analysis to show how variability in phosphorus concentration, hydraulic detention time, and reservoir depth propagates through the Vollenweider model.

3. The new 15-µg/L phosphorus guidance value is appropriate for Phase II TMDLs. The Phase I goal of 20 µg/L was not adequately conservative for a drinking water supply, as it is based on ecological and aesthetic considerations. Conservatism in the choice of phosphorus standard is necessary because data for some of the New York City reservoirs (Cannonsville, Croton system) show that algal productivities, estimated from average and maximal summertime algal biomass as chlorophyll *a*, are in excess of recommended values for any drinking water system (NYC DEP, 1993, 1999a).

4. NYC DEP should place a high priority on implementing all necessary nonpoint source control measures to reach Phase II TMDLs. Currently, it is unclear what specific measures are being taken to reduce nonpoint source phosphorus loading to the Cannonsville Reservoir and several Croton reservoirs other than WWTP upgrades. Ongoing watershed management actions in the Catskill-Delaware watershed must be quantitatively linked to the source reductions necessary to improve water quality prior to successful implementation of the nonpoint source TMDLs.

5. In Phase III, time-variable analyses should be performed using dynamic reservoir models and the GWLF (or equivalent model) for inputs. This will require construction of input–output models for total phosphorus, measurements of dissolved and particulate phosphorus loadings and concentrations, and assessment of individual phosphorus retention coefficients for each reservoir. NYC DEP's efforts in this regard for the Cannonsville Reservoir are to be commended.

6. Data collection to support and validate the GWLF and other Phase III water quality models should be given a high priority. Currently, the use of the GWLF is limited by a lack of site-specific data on soil chemistry, pollutant concentrations in runoff and groundwater, and other factors. At the very least, such data should be collected in representative subwatersheds and should be generalized with other, less-intensive measurements to the rest of the system. The use of event-based monitoring data rather than fixed-frequency sampling, and increasing the spatial resolution of the collected data, would greatly improve model accuracy. Given the pervasiveness of the GWLF in many facets of the City's watershed management program, the time and money spent to improve the accuracy of this model through intensive data collection would be well justified.

7. NYC DEP should focus Phase III of the TMDL program on public health protection by developing models that can link phosphorus to DBP precursors and other relevant parameters. New York City should consider doing a TMDL calculation for THM and HAA formation potential by deriving models that can link phosphorus and other inputs to THMFP, HAAFP, algae, chlorophyll, and taste and odor.

8. The method for determining which reservoirs are phosphorus-restricted should use phosphorus concentration data for the entire year, not just for the growing season. In addition, the criterion for determining whether a basin is restricted should be set at 15 µg/L rather than at 20 µg/L. The phosphorus guidance value determined for the Phase II TMDL program is relevant to the calculation of phosphorus restriction, which is also used to express the degree of use impairment.

PHOSPHORUS OFFSET PILOT PROGRAM

The New York City MOA includes a five-year phosphorus offset pilot program that allows for the construction of new or expanded WWTPs in phosphorus-restricted basins, areas otherwise closed to further construction of WWTPs. The purpose of the offset program is to allow for some continued growth in these phosphorus-sensitive basins while preventing a net increase in phosphorus loading.

The phosphorus offset pilot program is similar to effluent trading programs for water and emissions trading programs for air, in which a discharger of pollution is allowed to increase its pollutant discharge if another party will concomitantly reduce its discharge. In general, these programs require that an overall net reduction of pollutant loading be achieved by the trade. Most programs specify a trading or offset ratio that indicates the amount of pollutant reduction that must occur to balance the increase in pollutant discharge. Offsets are provided by a variety of mechanisms that reduce pollutant loadings from other sources, either point or nonpoint.

The objective of this section is to assess the scientific and technical basis for the New York City phosphorus offset pilot program. The effectiveness of trading as an approach to pollution reduction is discussed, using lessons learned from elsewhere. The potential reliability of the program is evaluated by addressing the following issues:

- Are the offset ratios specified by the MOA scientifically sound?
- How is "surplus" phosphorus defined?
- Are the proposed offset mechanisms appropriate?
- Can the phosphorus offsets be effectively quantified?
- What are the expected net effects of offsets on reservoir water quality?
- How can offsets be integrated within the TMDL process?
- Will the proposed offset program be sufficiently cost-effective or include sufficient incentives to be successfully implemented?

Overview of Watershed-Based Trading

Since 1996, EPA has been promoting effluent trading as an innovative way to develop cost-effective, common-sense solutions for water quality problems in watersheds. Trading is an agreement between parties contributing to water quality problems on the same waterbody in which the allocation of pollutant-reduction responsibilities among the parties is altered (EPA, 1996).

Proponents of Effluent Trading

Effluent trading has been proposed as an alternative to command-and-control environmental regulation, which attempts to reduce emissions at all point sources.

Proponents of trading argue that mandated discharge reductions for WWTPs are not efficient mechanisms for reducing pollutant loads because they force dischargers to adopt identical practices for pollutant reduction and to bear identical shares of the pollution-control burden regardless of their relative impacts (Jacobson et al., 1994). Under this system, every discharger is required to reduce pollutant loading to a certain level, which is likely to be more costly for some discharges than for others. In contrast to command-and-control regulations, effluent trading attempts to control overall pollution in a given area (such as a watershed) by assuming that particular pollutants disperse evenly across the area. Some have argued that these two systems of control can achieve the same result but that trading is more flexible and achieves pollution control more efficiently than existing command-and-control policies (Hoag and Hughes-Popp, 1997; Jacobson et al., 1994).

Opponents of Effluent Trading

Effluent trading has received substantial criticism, primarily because it assumes that a certain level of pollution will always exist and thus does not lead to the elimination of all emissions (Lapp, 1994). Likewise, Mann (1994) argues that pollutant trading has the undesirable effect of shifting the terms of debate from public health and industry responsibility to economics and finances. Many environmental groups feel that effluent trading is inconsistent with the goals of the CWA for attaining fishable/swimmable waters because it would allow pollutant loadings from some sources to increase (see EPA, 1996). The spatial distribution of the remaining pollutants also presents an environmental justice dilemma. For example, pollution rights from low-polluting sources in affluent areas might be traded to high-level polluters in poor areas. If residents of poor areas are unable to resist such transactions, they may be exposed to greater pollution than their more affluent counterparts. These and other benefits and criticisms of effluent trading are considered below, with specific reference to the phosphorus offset pilot program mandated by the MOA.

Trading Frameworks and Their Implementation

EPA Guidance

EPA has developed a framework for effluent trading that allows for five types of trades: (1) point source/point source trading, (2) intraplant trading, (3) pretreatment trading, (4) point source/nonpoint source trading, and (5) nonpoint source/nonpoint source trading (EPA, 1996). Trading programs developed by the states must meet CWA water quality requirements by ensuring a number of important conditions. Trading partners must meet applicable technology-based treatment requirements. Thus, no participant is allowed to operate treatment processes at suboptimal levels and engage trading partners to compensate for their lack of efficiency. Trades must be consistent with federal and state water quality standards throughout a watershed, and they should be used only for certain appropriate types of pollutants. A TMDL or similar process should provide the framework for developing trades, because of the ease with which pollutant loadings are accounted for under such programs. Trades must be supported by regulatory and enforcement mechanisms, which can be a challenge when nonpoint source pollutant reduction is involved. Performance monitoring is needed to track the success of trading programs and the effectiveness of BMPs. In order to maintain desired water quality, trading areas should align with waterbody segments or watersheds. And finally, any trade must include stakeholder involvement and public participation.

Existing Applications of Trading

Pollution trading was first implemented for air emissions control under the Clean Air Act, and it has been widely utilized in that arena. Efforts to use pollution trading to protect water quality are more recent and are relatively uncommon. A few water pollution trading programs have existed since the 1980s and early 1990s (e.g., Fox River, Wisconsin; Cherry Creek Basin, Colorado), the most notable being the Tar-Pamlico program in North Carolina (Harding, 1993; Hoag and Hughes-Popp, 1997; Jacobson et al., 1994; Stephenson, 1994). Typically, the trades are initiated by WWTPs as a means of reducing overall phosphorus and nitrogen loads in the watershed. In effect, the WWTPs invest in off-site measures to reduce pollutant loads. These investments are expected to remove at least as much of the pollutant in question as innovations that the WWTP might make to reduce point source emissions, but at a lower cost to the WWTP. Boxes 8-4 and 8-5 present case studies of two effluent trades being implemented in the United States. Because most programs have only recently been developed, in general it is too early to judge their success.

The New York City Phosphorus Offset Pilot Program

The phosphorus offset pilot program for the New York City watershed is one of only a few programs in the northeastern United States. On Long Island, a trading program to improve marine water quality is in its initial stages (M. Tedesco, EPA, personal communication, 1997). A New Jersey program, in which copper is discharged from small industrial facilities into WWTPs, is reported as the first trade in the country between indirect dischargers (i.e., facilities that discharge effluent to a WWTP) (C. Tunis, EPA, personal communication, 1997). Box 8-5 describes a trading agreement that has recently been approved in Massachusetts.

Under the terms of the MOA, construction of six new WWTPs within phos-

BOX 8-4
Effluent Trading in Minnesota

The Minnesota Pollution Control Agency has identified four critical elements that constitute an effective effluent trading program:

- *Cost-effectiveness.* Trades must be cost-effective or provide sufficient incentives for trading to occur. Trading costs may include capital and operation/maintenance costs of trading mechanisms, transaction costs, and time associated with the complexities of involving multiple parties in the trade.
- *Equivalence.* Proposed trades must be equivalent, or sufficiently similar in physical attributes, to substitute for one another. Physical conditions that must be considered in determining equivalency include timing of discharges, spatial differences, and chemical differences (e.g., dissolved versus particulate forms).
- *Additionality.* The condition of additionality requires that load reductions credited to a source in a trade would not have occurred in the absence of the trading program. That is, reductions are "surplus" or beyond those already required by other regulatory programs.
- *Accountability.* Conditions of the trade must be met over time. For example, BMPs proposed to achieve trading reductions must actually be implemented and maintained at the proposed levels of effectiveness over time. The degree of monitoring associated with a trading program will depend on the degree of accountability desired.

These conditions have recently been put to the test as part of the first trading permit ever issued in Minnesota. The Rahr Malting Co. plant, which produces barley malt, discharges wastes into the Minnesota River under a National Pollutant Discharge Elimination System (NPDES) permit. The river, certain sections of which are highly degraded, must meet a Total Maximum Daily Load of Carbonaceous Biological Oxygen Demand (CBOD) of 53,400 lbs/day for the section just downstream of the Rahr plant. After approval of the TMDL in 1988, many upstream discharges, including Rahr, were required to reduce loading of CBOD by as much as 40 percent. At the same time, Rahr Malting Co. requested a new permit that would allow additional wastes to be discharged into a lower section of the river.

In designing an effluent trading scheme that would satisfy the chemical characteristics of the Rahr plant's treated wastewater and the Minnesota River, it was assumed that the plant's wastewater would provide nutrients (such as phosphorus) to the river that would be utilized by microorganisms and eventually would result in higher CBOD levels. Positive correlations between phosphorus, chlorophyll, and CBOD were dem-

BOX 8-4 Continued

onstrated for the Minnesota River to support the trade. The conversion of nutrients from wastewater into CBOD was assumed to be more rapid in stagnant areas of the river because organic material derived from decaying microorganisms would have a chance to accumulate.

The actual trade appears to be a 1:1 trade using phosphorus as the parameter that will be measured both by the point source (Rahr) and participating nonpoint sources. The trade assumes that a one-pound reduction of phosphorus could yield an 8- to 17-pound reduction in CBOD, depending of which sections of the river are involved. Nonpoint sources that have been targeted for participation include soil erosion BMPs, livestock exclusion, rotational grazing, set-asides of highly erodible land, and wetland treatment systems. However, trading partners have yet to be confirmed and approved by the Minnesota Pollution Control Agency. There is no indication of how difficult it will be to attract trading partners and monitor appropriate nonpoint source control measures.

Source: Adapted from Anderson et al. (1997).

BOX 8-5
Effluent Trading in Massachusetts

An effluent trading program has recently been approved in Massachusetts involving an office complex that discharges treated sewage into the Sudbury River. As with the New York City phosphorus offset pilot program, this trade requires a 3:1 offset for all new discharges of phosphorus, utilizing load reductions from other sources. The trading partners in this case are the office complex (Congress Group Ventures) and dozens of neighboring properties that house failing septic systems. The septic systems will be connected to the new sewerage service rather than have the office building install additional, expensive wastewater controls at its plant.

The trade, which was recently approved by EPA, the Massachusetts Department of Environmental Protection, Congress Group Ventures, and the town of Wayland, will be implemented over a two-year period. Although heavily contaminated by a variety of pollutants, the Sudbury River is most noteworthy for its bacterial pollution.

Source: Clean Water Report (1998).

phorus-restricted basins is allowed as part of the five-year pilot program (NYC DEP, 1997c). These plants must provide a 3:1 phosphorus offset through the reduction of existing point or nonpoint source discharges. In other words, every kilogram of additional phosphorus loading allowed from new WWTPs and accompanying nonpoint source discharges from associated new development must be offset by three kilograms of phosphorus reduction from elsewhere within the phosphorus-restricted basin.

The pilot program allows for the construction of up to three new surface-discharging WWTPs within the Croton watershed (totaling no more than 0.15 mgd) and up to three new or expanded surface-discharging WWTPs (0.10 mgd total) in the Catskill/Delaware watershed. Proposed WWTP discharges must lie outside the 60-day travel-time boundary and must be in basins that have committed to comprehensive water quality protection planning (such as the Croton Plan). If the pilot program successfully achieves water quality and regulatory goals, it will serve as the basis for establishing a permanent phosphorus offset program in the future.

A similar but separate program has been established to allow existing surface-discharging WWTPs in phosphorus-restricted basins to expand, given a 2:1 offset. This program does not place a limit on the number of participating WWTPs (NYC DEP, 1997d). In all cases, if it can be shown that a new or expanding WWTP can safely discharge its effluent to the subsurface, then it is not eligible for participation in either program.

Program Basis

The New York City phosphorus offset pilot program includes two of the trading mechanisms allowed by EPA: (1) point source/point source trading, in which an existing point source undertakes more stringent reductions than required and trades those credits to the new point source, and (2) point source/nonpoint source trading, in which the new point source arranges for stricter pollutant control from nonpoint sources. The responsible party for the new WWTP must identify appropriate offsets for the new phosphorus discharge. As described below, four important criteria must be satisfied by the proposed phosphorus offsets (NYC DEP, 1997c).

Surplus. Proposed offsets must be demonstrated as "surplus" before they can be included within the pilot program. NYC DEP guidance on the program has defined surplus as those phosphorus reductions that are "not otherwise required by federal, state, or local law" (NYC DEP, 1997c). Baseline, or minimum, requirements for phosphorus reductions that cannot count as surplus include (1) upgrades of existing WWTPs mandated by the MOA and (2) implementation of stormwater pollution prevention plans (SPPPs) for new developments larger than five acres and for new SPDES-permitted facilities. Reductions achieved by retrofitting developed areas with effective stormwater quality controls can qualify as surplus. In addition, any phosphorus reductions achieved under the Catskill Fund for the Future and the Stormwater Retrofit Program can qualify as surplus offsets.

Quantifiable. An offset is quantifiable if a reasonable basis exists for calculating and verifying the amount of reduction in phosphorus. Models can be used to estimate phosphorus reductions, but these reductions must be verified through routine monitoring. Some methods for calculating quantifiable phosphorus reductions are provided in program guidance (NYC DEP, 1997c).

Permanent. An offset is permanent if the phosphorus reduction is ongoing and of unlimited duration. Adequate maintenance and routine inspections are required to ensure the permanence of offsets. A contingency plan must identify alternative measures to be taken if the existing offset mechanisms fail.

Enforceable. The offset must be incorporated into a legally valid and binding agreement to qualify as enforceable. At a minimum, the offsets must be incorporated into the SPDES permit for the new WWTP.

Offset Mechanisms

The following text describes offset mechanisms found in NYC DEP guidance. Program applicants are free to suggest additional mechanisms, which must be approved by NYC DEP.

Stormwater Best Management Practice Retrofits. Best management practices (BMPs) can be added to reduce phosphorus loadings associated with stormwater runoff from existing sites, particularly those that were initially developed without BMPs. Stormwater retrofits are structures, such as ponds and wetlands, that remove urban pollutants through sedimentation, adsorption, and biological methods (Claytor, 1996). When properly located, designed, constructed, and maintained, stormwater retrofits can be an effective element of a phosphorus-trading program.

Land Reclamation. Lands that generate significant phosphorus loads can be altered to a condition that reduces overall phosphorus generated on the site. An example of land reclamation would be to remove impervious surfaces and replace them with a vegetated landscape that would export less phosphorus.

Reductions of Phosphorus Discharge from Existing WWTPs. Phosphorus from existing WWTPs can be reduced in several ways. Plant retrofits can provide an offset if they reduce phosphorus discharges below the levels to be achieved by City-funded upgrades of existing WWTPs (i.e., below 0.2 mg/L). Diversion of effluent flows from existing WWTPs outside of the City's water supply watershed can also provide offsets. Phosphorus reductions can be achieved by flow reductions of existing WWTPs, including complete elimination through facility closure. Finally, conversion of surface discharges from existing WWTPs to subsurface discharges, through a subsurface absorption field, can provide an offset. All WWTP flow reductions must be reflected in revised SPDES permits.

Removal of Septic Systems. Improperly functioning septic systems that contribute measurable amounts of phosphorus are generally subject to enforcement proceedings from a local health authority. For this reason, repair or replacement of such systems does not qualify as an offset. However, NYC DEP will allow the complete removal of improperly functioning and irreparable septic systems to qualify as an offset (Warne, 1999). In this case, the offset is determined by multiplying the volume flux (volume/day) of discharge from the septic system (based on the number of residents) by the phosphorus concentration in septic effluent (as determined by the New York State Department of Health).

Wetland Restoration. Restoration of degraded wetlands to reestablish stormwater runoff treatment functions can accomplish phosphorus reductions that qualify as surplus.

Offset Calculation Methodology

WWTP Phosphorus-Load Increase. The first step in calculating the appropriate offset is to determine the phosphorus load from the new WWTP. This load is calculated as the product of the maximum SPDES-permitted flow and the effluent limits for new WWTPs, or the flow increase and new effluent limits for WWTP expansions.

Nonpoint Source Phosphorus-Load Change. Associated changes in nonpoint source phosphorus loading that result from WWTP and all associated construction are then calculated. The incremental change in nonpoint source loadings is the difference between the phosphorus load generated under existing or predevelopment conditions and the phosphorus load generated under postdevelopment conditions. The guidance suggests the Simple Method (Schueler, 1987), the P8 Urban Catchment Model (EPA, 1997), or the Stormwater Management Model (EPA, 1997) to calculate nonpoint pollutant loads from development activities. The use of generalized runoff coefficients and pollutant loading coefficients developed from data collected in the Nationwide Urban Runoff Program (Schueler, 1987) is also recommended. If controls on nonpoint source runoff from the new development actually achieve a net *decrease* in phosphorus loading over predevelopment conditions, this decrease can be applied as an offset.

Net Change in Phosphorus Load. The phosphorus loads attributed to the new WWTP and the associated nonpoint sources are combined. If nonpoint sources are projected to decrease, the nonpoint portion of the equation is set equal to zero, not less than zero.

Offset Requirement. To determine the offset requirement, the net increase in the phosphorus load from the new WWTP and associated nonpoint sources is multiplied by 3.

Offset Mechanism Reductions. The final step of the process is to calculate the phosphorus reductions to be achieved through the proposed offset mechanisms. The calculation of offset reductions for modifications of existing WWTPs is straightforward: the reduction in flow is multiplied by the effluent limit. For nonpoint source discharges, the guidance allows for "any reliable method to predict their removal rate" and provides a few examples (NYC DEP, 1997c). For stormwater BMP retrofits, the reduction in the phosphorus load is a function of the predevelopment loading rate and estimated BMP removal rates. For land reclamation and wetland restoration mechanisms, the reductions are calculated as the difference in loads generated by the land in its restored form versus the load generated in its current form.

Existing Applications for the Phosphorus Offset Pilot Program

As of April 10, 1999, three applications are being considered by NYC DEP for the pilot program. Each of these applications has undergone extensive revision in order to comply with the requirements of the program. The applications concern developments in Putnam County, within the Croton system, where development pressures are greatest. Although no applications from the Catskill/Delaware watershed have been received, it is expected that the town of Delhi in the Cannonsville basin may choose to participate in the near future.

NYC DEP has developed criteria for evaluating all applications (NYC DEP, 1998c). First, the proposed plan must have technical merit. The project must be adequately described, and the models used to calculate offsets must be valid. This includes all calculations of pre- and postdevelopment phosphorus loading. In addition, technical merit must be found in the monitoring plan, the quality assurance/quality control (QA/QC) plan, the contingency plan, and the inspection and maintenance plan. The other major criterion for each application is that the plan can be implemented. Finally, the application must help further the goals of the phosphorus offset pilot program. NYC DEP is particularly interested in projects in which credible efforts are made to monitor the offset mechanisms and in which water quality changes will result from the construction of the WWTP and from the offset mechanisms. Box 8-6 describes one application currently under consideration by NYC DEP. For all the projects, the new source of phosphorus is a WWTP designed to treat wastewater from proposed new development. The phosphorus offset mechanisms include a variety of stormwater BMPs (e.g., wet ponds, compost filters, extended detention basins) and street sweeping. All the proposed mechanisms are claimed to achieve reductions in phosphorus loading that are at least three times the increase in loading attributable to the new WWTP.

The three applications appear to meet the eligibility requirements of the pilot program, and most have gained NYC DEP approval. However, the success of the proposed projects will depend almost entirely on the effectiveness and long-term reliability of the offset mechanisms. The analysis below suggests that many offset mechanisms will be difficult to monitor and that some cannot achieve the pollutant reductions claimed in the applications.

Analysis of the New York City Program

The committee has reviewed the phosphorus offset pilot program and provides recommended improvements to address concerns in several areas.

Ensuring that Reductions are Surplus

One of the most critical aspects of the offset program is ensuring that proposed offset mechanisms are actually surplus reductions. In order to make such assurances, baseline or minimum requirements for WWTP upgrades and SPPPs must be in place and operating effectively before offsets are identified and accepted. NYC DEP may need to develop criteria that will help determine whether baseline requirements are currently in place and operational, as none are specified in the current guidance materials.

Detailed recommendations for designing and implementing SPPPs are presented in Chapter 9, and should be considered for the phosphorus offset pilot program. In general, baseline stormwater practices, as required by SPPPs, must be permanent and well maintained. When reviewing program applications, NYC DEP must ensure that the reductions are not overstated because of insufficiently conservative assumptions. Because estimates of pre- and postdevelopment stormwater runoff loads are complex and rely on many assumptions, a careful review is required to ensure that proposed offsets are accurately quantified and that they qualify as surplus. The committee recommends that NYC DEP provide further explanatory language and more specific criteria to better define surplus. If possible, this definition should be flexible and receptive to changes and improvements in technology.

BOX 8-6
Applicant for the Phosphorus Offset Pilot Program:
Terravest Phase 3 - Highlands

This application was initially rejected because it proposed that the off-set mechanism be subsurface discharge. This was controversial because in order for a WWTP to participate in the program, it has to be shown that subsurface discharge is not possible. NYC DEP has subsequently decided that subsurface discharge cannot be used as an offset mecha-nism, and the applicant has chosen a new offset mechanism.

New Phosphorus Load. After its initial application was turned down, the applicant proposed a second, smaller WWTP, the Highlands WWTP, to serve a retail center of 384,000 square feet. The anticipated phos-phorus loading from the WWTP is 10.95 lbs/yr. The operator of the plant is confident that with a recycling system in place, the WWTP's phos-phorus load can be reduced to 3.7 lbs/yr. This is because recycling will cut the effluent flow volume by two-thirds (from 36,000 gpd to 12,000 gpd) while maintaining the low phosphorus concentration of 0.1 mg/L. A SPDES permit for 0.1 mg/L has been requested. (The site had previously obtained a SPDES permit for 0.2 mg/L phosphorus.)

Offset Mechanisms. Predevelopment phosphorus loadings were calculated to be 45.1 lbs/yr, while postdevelopment loadings are 11.96 lbs/yr. These reductions in phosphorus loading are based on a series of stormwater BMPs. Two detention basins are expected to achieve 60 percent removal of phosphorus, while one extended detention basin is expected to achieve 40 percent phosphorus removal. Four additional water quality basins are also included and are credited with 60 percent phosphorus removal. Because the predevelopment loading is greater than postdevelopment loading (by 33.1 lbs/yr), this difference can be used as credit toward the required offset.

Required Offset. Assuming that recycling is used for the new WWTP, the required offset will be $3.7 \times 3 = 11.1$ lbs/yr. If recycling is not used, the required offset will be 32.85 lbs/yr (10.95 $\times$ 3). The phosphorus load reductions achieved with the stormwater control practices (33.1 lbs/yr) should be able to satisfy either offset requirement.

Contingency Plan. The contingency plan for the project lists eight options, three of which NYC DEP thinks are most promising: (1) chemi-cally treating any malfunctioning detention basin until corrective measures restore the basin's functional capabilities, (2) creating another stormwater basin for short- and long-term use, and (3) pumping the efflu-ent from stormwater basins back through a series of basins. The appli-cant must expand on these options before the project can commence.

Comprehensive Plan and Town Approval. Putnam County is in the process of preparing the required Croton Plan, and the town of Southeast has signed a letter approving the project.

Appropriateness of Offset Mechanisms

NYC DEP guidance describes several offset mechanisms that can be used to achieve reductions in phosphorus loading. The committee has reviewed each mechanism and gives comments below, particularly for those mechanisms that require substantial improvement before they are able to provide the reliable, long-term protection needed for the program.

Stormwater BMP Retrofits. Stormwater retrofits can be an effective element of an overall watershed strategy to reduce phosphorus loads in stormwater runoff generated by existing urban development. Given the applications received by NYC DEP to date, they are the most popular mechanism chosen for achieving a phosphorus offset. However, the guidance document for the program (NYC DEP, 1997c) needs to be greatly strengthened in several areas to ensure that stormwater retrofits are effective.

1. The guidance appears to permit the use of any urban nonpoint source practice as an eligible stormwater retrofit, such as might be found in the "Urban/ Stormwater Runoff Management Practices Catalogue for Nonpoint Source Pollution Prevention and Water Quality Protection" (NYS DEC, 1996). In fact, only four of the 43 urban nonpoint source practices that are summarized in the NYS DEC catalogue appear to meet the quantifiable and/or permanent removal criteria of the pilot program. Sufficient research is presently available only to quantify the expected phosphorus removal capability of ponds, wetlands, sand filters, and swales (Brown and Schueler, 1997). The phosphorus offset pilot program should restrict eligible stormwater retrofit BMPs to these four groups until further independent research indicates that other practices have quantifiable phosphorus removal capability. In practice, most stormwater retrofitting employs stormwater ponds and wetlands that can cost-effectively treat large catchment areas.

2. The table of expected phosphorus removal rates provided in Appendix F of the guidance document for the phosphorus offset pilot program (NYC DEP, 1997c) is outdated and has been superseded by more recent data. Updated stormwater BMP removal rates are provided in MDE (1998).

3. The sizing of stormwater retrofit BMPs is not explicitly addressed in the offset program. The guidance document bases the phosphorus credit solely on the presumed pollutant removal capability of the stormwater retrofit that is ultimately designed. It does not specifically require that the retrofit have an adequate storage or treatment volume to actually accomplish the desired removal. The computational methodology needs to be revised to ensure that the stormwater retrofit has a minimum stormwater treatment volume to accomplish the desired degree of pollutant removal.

4. Program applications have assumed that BMP removal rates are constant for BMPs placed in series. (That is, if a BMP has a 40 percent removal rate, effluent pollutant concentrations will be decreased by 40 through each BMP

used.) However, removal efficiences of individual BMPs generally decline when placed in series, based on the handful of performance research studies that have examined the issue (Gain, 1996; McCann and Olson, 1994; Oberts, 1997; Urbonas, 1994). Removal efficiencies vary in accordance with the changing composition of stormwater as it passes through multiple BMPs. For example, the first BMP may accomplish 50 percent removal of sediment and sediment-associated particles. But because larger particles are more effectively removed, subsequent BMPs will be treating stormwater enriched with finer particles, and removal efficiency will drop below 50 percent. At some point, the incremental removal is negligible, and the pollutant concentration from the final BMP reaches an irreducible concentration, which represents the maximum treatment limit for gravity-driven practices (Schueler, 1996). Upper bounds on the amount of pollutant removal from BMPs in series vary depending on the specific pollutant, but none approach 100 percent.

5. Current stormwater treatment technology cannot reduce pollution loads to below predevelopment levels. In most cases, the asserted pollutant removal shown in stormwater offset applications is a result of computational methods that have no real basis in engineering or science (e.g., using BMPs in series, use of curve numbers rather than runoff coefficients, and over-sizing). Also, the committee is unaware of any field study that has actually documented that stormwater BMPs (or groups of BMPs) were actually able to reduce phosphorus loads to predevelopment levels for forest or meadow conditions. In a modeling study, Caraco et al. (1998) found that predevelopment nutrient loadings could not be achieved through any combination of better site design and stormwater BMPs.

6. Although the Simple Method (Schueler, 1987) used to compute pre- and postdevelopment pollutant loads is a general model that has been widely used across the country, it is important to utilize regional data for phosphorus event mean concentrations (EMCs) and background loads. The guidance relies heavily on stormwater EMCs developed from the mid-Atlantic Region and also employs the annual background phosphorus load of 0.5 lbs/ac/year that was derived from a mix of rural land from the Chesapeake Bay region. Stormwater monitoring data have been collected to derive more accurate stormwater EMCs for the Catskill/ Delaware region. Derivation of a regional background load should also be a priority, since a higher or lower regional background load will have a profound influence on the offset calculations.

7. The current guidance requires that applicants consider several physical feasibility factors, but it does not address several feasibility factors that are unique to stormwater retrofitting. Examples include locational factors (e.g., are retrofits allowed within watershed setbacks? Within jurisdictional wetlands?), public acceptance factors (acceptance by adjacent landowners, potential habitat/ restoration benefits, long-term maintenance capability), and watershed significance (minimum size or load reduced per retrofit). Implementation of stormwater retrofits in other regions of the country has generally been done in a watershed

context to ensure that retrofits not only meet phosphorus reduction targets, but also meet other restoration and community objectives. In addition, nearly all stormwater retrofits to date have been constructed on public lands and therefore require a much greater level of public involvement, agency coordination, and environmental permitting than is normally needed to construct BMPs to serve a new development (Claytor, 1996).

8. Language in the current guidance should be added to ensure regular inspection of stormwater BMPs and long-term (beyond the first few years) maintenance of these facilities, even if they are designed correctly. The magnitude of institutional commitment, in terms of funding and trained staff, needed to achieve effective pollutant reduction will be high.

Land Reclamation. Land reclamation is a preferred offset mechanism because of its long-term nature, because of the ability to effectively quantify load reductions from this mechanism, and because of the direct water quality benefits. Land reclamation should be given high priority as an offset mechanism.

Reductions of Phosphorus Discharge from Existing WWTPs. This mechanism holds considerable potential because available treatment technologies can significantly reduce phosphorus concentrations. The following caveats should be considered:

1. The diversion of phosphorus effluent to other watersheds must be carefully evaluated to ensure that the applicant is not transferring phosphorus problems to another area.

2. Reductions in flow at existing facilities must also be accompanied by overall reductions in phosphorus loadings (e.g., no increases in phosphorus concentrations) in order to be protective.

3. The transfer of discharges from the surface to subsurface should not be allowed as an offset since groundwater may also be a source of phosphorus to down-gradient surface waters.

Repair/Replacement of Improperly Functioning Septic Systems. As stated in NYC DEP guidance material, repair or replacement of improperly functioning septic systems is not allowed as an offset mechanism because these units are subject to enforcement proceedings by the local health authority. However, complete removal of failing septic systems is an option.

In the committee's opinion, the complete removal of failing septic systems is not an acceptable offset mechanism for the program. First, it goes against the idea that baseline treatment requirements should be met prior to identifying offsets. Septic systems should first be rehabilitated to use best available control technology (BACT), and suggestions for doing so are given in Chapter 11. Second, complete removal of septic systems may have the unintended consequence of converting the watershed population to more expensive central sewerage systems. Such shifts in infrastructure use often precede increases in development and population. Finally, the incremental phosphorus load reduction to be gained from individual septic systems is small. The phosphorus offset pilot program should focus on more effective and substantial offset mechanisms.

Wetland Restoration. The microbiota and higher aquatic plants of wetlands and littoral areas can function effectively to sequester phosphorus from runoff if they are designed and maintained specifically for that purpose. Thus, wetlands used effectively as stormwater BMPs are *constructed*, rather than *restored*. Constructed wetlands generally require large land areas and special design enhancements to achieve significant phosphorus removal (EPA, 1993a; Knight et al., 1995; Reed et al., 1995). Although not stated in NYC DEP guidance material, constructed wetlands are an acceptable offset mechanism for the pilot program (Warne, 1999), which the committee supports.

Restored wetlands are generally less amenable to specific design criteria and are usually ineffective in long-term net phosphorus removal. They may even increase phosphorus loadings under certain hydrologic conditions. Their effectiveness fluctuates seasonally, with more phosphorus removal being observed during the warmer growing season. Restoration of wetlands to treat stormwater could have associated regulatory and permitting requirements, as they are likely to be considered waters of the United States, which would require CWA certification and a Section 404 permit for any modifications. Finally, it is important to note that using natural or restored wetlands to treat stormwater runoff can result in long-term adverse effects on natural wetland functions (Azous et al., 1997; Kadlec and Knight, 1996; Richter and Azous, 1995).

Quantification of Offsets and Assessing the Impact of the Program on Water Quality

For all the offset mechanisms allowed under the program, there must be sufficient techniques to monitor their performance and demonstrate compliance. The collection and analysis of performance monitoring data provide the only sure method for assessing the impact of the program on long-term reservoir water quality. In particular, monitoring should be designed to take the cumulative impacts of various offsets and of development within the watersheds into account.

The establishment of a reliable, long-term monitoring program is probably the most challenging aspect of the New York City pilot phosphorus offset program. NYC DEP allows for the application of models to estimate phosphorus removals, but these estimates must be confirmed with actual data. A brief overview and analysis of monitoring requirements is presented below for stormwater runoff controls (stormwater BMP retrofits, land reclamation, and wetland restoration) and wastewater controls (WWTPs and septic systems).

Stormwater Runoff Controls. Phosphorus load reductions associated with stormwater runoff controls are a function of two factors: (1) storm flow volume and (2) removal effectiveness or percent reduction in concentration. For BMP facilities, data on phosphorus concentrations in the inflow and outflow should be collected during and shortly after storm events using paired flow-compositing automated samplers. Performance monitoring for phosphorus removal should be conducted for a minimum of ten storm events per year to adequately characterize the range of storm conditions. In addition, flow data should be collected over the entire year to estimate the total storm flow and baseflow volume. BMP effectiveness may change over time, depending on storm size, level of BMP maintenance, age of facility, influent concentrations, and conditions in the contributing watershed. To confirm continued removal effectiveness, it is recommended that estimates of load reduction be reviewed periodically (i.e., every 5–7 years).

The above recommendations on stormwater BMP monitoring also hold for wetlands and land reclamation with the following caveats. Because wetlands are not effective in reducing phosphorus during base flow conditions, it may be appropriate to incorporate only storm flow volumes in estimates of annual load reductions for wetlands. To estimate phosphorus removal benefits associated with land reclamation, information on event mean concentrations and runoff coefficients before and after reclamation is required.

Point Source Controls. Documenting pollutant load reductions from point sources such as WWTPs is relatively easy, and such documentation should be captured during regular SPDES monitoring. Septic systems are much more difficult to monitor. For this reason (and for those reasons mentioned previously), septic systems should not be used to provide offsets.

Offset Ratios

A trading ratio specifies how many units of pollutant reduction a source must achieve to receive credit for one unit of load reduction (EPA, 1996). A trading ratio of 1:1 indicates an equal exchange between sources. Typically, trading ratios exceed 1:1 to provide a margin of safety in the event that traded reductions are less effective than expected. Higher trading ratios can also reflect a net reduction strategy, although this is not a stated goal of the New York City pilot phosphorus offset program.

Establishing a scientific basis for trading ratios is an important component of any credible trading program. Unfortunately, detailed guidance for developing ratios is extremely limited to date. The 3:1 and 2:1 ratios found in the New York City program were chosen after negotiations among government agencies (S. Amron, NYC DEP, personal communication, 1998; D. Warne, NYC DEP, personal communication, 1999). The ratios apply to all program applicants, regardless of the nature of the trade and the proposed offset mechanism. The following cursory evaluation of the New York City ratio is based on information available from trading programs in Colorado (see Box 8-7), Minnesota, and Massachusetts.

The New York City ratio of 3:1 for new WWTPs is among the largest ratios found among the four state programs. The Cherry Creek Basin program (Box 8-7), which more explicitly discusses the types of uncertainty that should be considered when setting trading ratios, has developed ratios between 1.5:1 and 3:1 (Paulson, 1997). The ratio being used in Minnesota seems to be a 1:1 ratio, while that in Massachusetts is 3:1. Thus, at first glance, it appears that the New York City ratio may be sufficiently protective. However, because the scientific basis of the trading ratio is not stated, its protectiveness cannot be adequately evaluated by such a simple comparison.

The Cherry Creek Basin program justifies its trading ratio as providing a safety margin against several relevant uncertainties, including variability in phosphorus loading and BMP performance, uncertainties associated with laboratory analysis and data evaluation, and institutional uncertainties. All these uncertainties exist in the New York City phosphorus offset pilot program as well. In addition, there are several specific scientific uncertainties in the New York City watershed that must be taken into account. These issues, which are not mentioned in the program's guidance material, suggest that the 3:1 ratio may not be sufficiently protective of water quality.

First, the New York City program does not make a distinction between the forms of phosphorus that are used to acquire offsets. Generally, phosphorus discharged from WWTPs is in a soluble, reactive form. Phosphorus removed by many of the proposed offset mechanisms, particularly stormwater runoff controls, is predominantly particulate in form. A shift in the existing balance of phosphorus toward a larger soluble fraction could have significant implications for downstream reservoirs, such as higher concentrations in the water column and enhanced algal growth. This could in turn aggravate eutrophication problems in these already phosphorus-restricted basins. Monitoring of the soluble fraction of total phosphorus concentrations, both in reservoirs and tributary streams, is recommended before and after proposed offsets to track this possible outcome.

The second consideration is spatial—the location of the proposed offset versus the location of the proposed WWTP discharge. Natural losses of phosphorus released to a stream would be expected to occur, at varying levels, with transport to a downstream reservoir. The closer a WWTP is located to a downstream reservoir, the less time for potential retention en route. Unfortunately, no distinctions in the trading ratio are made for new WWTPs built closer to, or further from, terminal reservoirs, nor are relative locations of the WWTP and the proposed offsets discussed. A related issue is that diffuse sources of phosphorus experience more opportunities for phosphorus retention than single-point sources of equivalent concentration. In order to demonstrate that point sources and nonpoint sources are spatially "equivalent," a proposed offset should be located

BOX 8-7
Trading Ratios in the Cherry Creek Basin, Colorado

Trading ratios in the Cherry Creek Basin program have been developed primarily to reflect uncertainty associated with specific phosphorus removal facilities. The ratios do not reflect a net reduction strategy. The Cherry Creek Basin trading ratios incorporate two aspects of uncertainty associated with the trading program—scientific and institutional. The ratios are facility-specific and range from about 1.5:1 to 3:1.

In the Cherry Creek Basin program, scientific uncertainty stems from the variability in phosphorus load reductions that occurs from one year to another as well as from uncertainty in the estimate of reductions. A variability factor accounts for annual variations in load reductions, and is a function of both flow volume and removal effectiveness (i.e., concentration reduction). The variability factor is based on the relationship between the 95th and 50th percentile estimates of load reduction to ensure that the load reductions could be expected to occur 95 percent of the time. In addition, a best professional judgment (BPJ) factor is applied to account for the degree of uncertainty in the estimate. For example, if data on a given facility were limited, it would have a higher BPJ factor than if the facility had extensive site-specific data. The ratio associated with scientific uncertainty for a given facility is the product of the variability and BPJ factors [e.g., scientific uncertainty = (variability factor)(BPJ factor)].

Institutional uncertainty is a function of the entity responsible for implementation of the trading program and the degree to which the trades are documented and enforceable. Institutional uncertainty will be low if the entities responsible for the trading program are permanent and well established. In the Cherry Creek Basin program, the ratio associated with institutional uncertainty was set equal to 1:1, meaning that no institutional instability could adversely influence the continued effectiveness of the trades over the long term.

For most programs, a number of factors can determine the level of institutional uncertainty. These include finances, staffing, and general administrative stability, all of which play important roles in the regular functioning of any program. For example, a trading program needs some capacity for monitoring, verification, and enforcement of trading arrangements. Institutional uncertainty is lowest if trading programs have clearly defined tasks, unambiguous assignment of responsibilities, qualified personnel, and regular funding linked to performance of program activities. Conversely, institutional uncertainty rises when no clear lines of responsibility and authority are identified, personnel are appointed informally, and funding decisions are politically motivated.

within proximity of a proposed WWTP discharge or at some point further downstream toward the reservoir.

Finally, there are temporal considerations that must be reflected in the trading ratio. It is likely that reductions in phosphorus as a result of the offset mechanisms may not coincide with WWTP discharges of phosphorus. Phosphorus associated with stormwater runoff occurs intermittently, with storm events. In contrast, wastewater discharges are more consistent. Also, there is considerable seasonal variability in the performance of BMPs (especially wetlands). For a downstream reservoir with a relatively long residence time (i.e., longer than one year), temporal differences in offsets are of less consequence. The average residence times of the New York City reservoirs suggest that temporal fluctuations may be significant.

Integration of Offset Program with TMDL Program

EPA has recognized that trades should occur in the context of a TMDL or SPDES permit to help meet water quality requirements more cost-effectively. As such, the New York City program requires all offsets to be reflected in revised SPDES permits for involved WWTPs. The TMDL program provides another opportunity for consolidating monitoring efforts and ensuring the protectiveness of the pilot phosphorus offset program. Phosphorus TMDLs are expressed in terms of allowable annual loads that are allocated among several components: point sources, nonpoint sources, future growth, and a safety factor. Phosphorus offsets, along with allowed increases in point source discharges, should be incorporated into the wasteload and load allocations and/or the control strategy of a TMDL.

Economic Considerations

The attractiveness of the phosphorus offset program is that it allows increases in some sources of pollution to be offset by reductions in other sources. The immediate intention is to allow increased urbanization within phosphorus-restricted basins by allowing growth in WWTP capacity, provided that phosphorus reductions are realized elsewhere within the basin. Because offset programs typically require a more-than-proportional reduction for an increase in discharge of the pollutant, the quality of the water in targeted waterbodies can be maintained while accommodating demands for economic development within the watershed.

It should be kept in mind that the New York City program is not a "market-based" trading program and will not necessarily result in the greatest pollutant reduction per dollar. The BMPs that are chosen as offsets are determined more by the applicant's familiarity with operating and monitoring them and the potential pollutant reduction abilities of the BMPs rather than their cost-effectiveness.

There is the possibility that offset programs may be inappropriately relied upon to achieve water quality goals. It is important that they do not become substitutes for effective regulatory and environmental management actions by local government, which may want to shift the costs of environmental cleanup to elements of the private sector as opposed to carrying out discharge reductions itself or requiring existing pollutant dischargers to change their practices.

A final economic consideration for the phosphorus offset pilot program is whether offset mechanisms can be identified. In the Cannonsville watershed (which is the only eligible West-of-Hudson watershed), interest in the program has been low because of an inability to identify surplus phosphorus and appropriate offset mechanisms and because of limited demand for additional WWTPs. Agricultural BMPs implemented as part of the Watershed Agricultural Program cannot be used as offsets, although BMPs constructed outside the program are allowed. In the absence of identified offsets, the trading program is likely to remain untested in the Catskill/Delaware watershed during its five-year timeframe. This would leave the program unprepared for future full-scale implementation.

Conclusions and Recommendations

The phosphorus offset pilot program as currently formulated contains significant weaknesses that prevent the committee from endorsing it fully. However, improvements in the existing program can be achieved by addressing a few key areas. These recommendations should be incorporated into the program before it is expanded to full scale.

1. Baseline minimum requirements for phosphorus reduction must be in place and operating effectively before additional reductions can be defined as surplus. This refers to such activities as all planned upgrades to WWTPs that will reduce phosphorus loadings and to phosphorus reduction mechanisms that are part of an SPPP, among others. NYC DEP should develop clear criteria to determine whether baseline requirements are in use and operational and to further define surplus reductions.

2. NYC DEP should update its guidance on the use of stormwater retrofits to achieve offsets, as the stormwater retrofits currently allowed under the program are unreliable and scientifically indefensible. Specifically,

• the program should restrict eligible stormwater retrofit BMPs to ponds, wetlands, sand filters, and swales until further independent research indicates that other practices have quantifiable phosphorus removal capability;

• the methodology for calculating the phosphorus load reduction must ensure that a stormwater retrofit uses the minimum stormwater treatment volume required to accomplish the desired degree of pollutant removal;

• the pilot program should not give credit for reducing postdevelopment phosphorus loads to below predevelopment levels; and

• the appendixes of the NYC DEP guidance document should be updated to provide more recent, region-specific estimates of pollutant loading coefficients and BMP removal effectiveness, to discuss the limitations in their application, and to discuss regular inspection and long-term maintenance of BMPs.

3. Two offset mechanisms—transfer of WWTP discharges from the surface to the subsurface and removal of improperly functioning septic systems—are not adequate and should be dropped from the offset program. Subsurface discharge of phosphorus to groundwater is not acceptable because groundwater may be a source of phosphorus to down-gradient surface waters. NYC DEP has recognized this problem and is likely to alter its guidance document accordingly. Removal of septic systems is inappropriate as an offset because monitoring of this mechanism is difficult and because the baseline requirement should include effectively operating septic systems.

4. NYC DEP should reevaluate the 3:1 and 2:1 ratios and develop a technical basis for the ratios that reflects the unique conditions associated with specific proposed offset mechanisms. The offset ratios of 3:1 and 2:1 currently have neither scientific basis nor explanatory justification. In addition to providing a safety margin for BMP performance variability, erroneous data collection, and institutional uncertainty, the offset ratio should reflect conditions present in the New York City watershed such as the spatial and temporal variability of offset mechanisms, the relative locations of the offset mechanisms and the WWTPs, and the different forms of phosphorus produced in the effluents of WWTPs and the offset mechanisms. Because the ratios do not explicitly take these issues into consideration, they are likely to be underprotective.

5. There is no evidence that the phosphorus offset pilot program will result in a net reduction in phosphorus loading to the water supply reservoirs, because the offset ratios do not currently incorporate an additional factor to provide for net reductions in phosphorus. If this becomes a goal of the phosphorus offset pilot program, the offset ratio should be made more conservative.

6. NYC DEP must develop performance monitoring to document offset mechanism effectiveness and overall net effects on downstream reservoirs. NYC DEP guidance material describes the elements of such a monitoring program, but puts the burden of monitoring entirely on the program applicant. Monitoring and maintenance of BMPs often suffer in the absence of an enforcement presence or when accountability is unclear. Because program applicants generally do not have the needed equipment and technical expertise, and are not responsible for overall watershed health, NYC DEP should assist and supplement the monitoring efforts of the program applicants.

7. Phosphorus offsets, along with allowed increases in any point source discharges, should be incorporated into the wasteload and load allocations that have been developed as part of the TMDL program.

8. NYC DEP should establish a comprehensive set of criteria to systematically evaluate the effectiveness of the phosphorus offset pilot program after five years. Criteria for program evaluation might include (1) reduction in phosphorus loadings directly resulting from the BMPs implemented, (2) relative effectiveness of the different BMPs implemented, (3) the adequacy of the 3:1 offset ratio in achieving net phosphorus reductions, and (4) technical adequacy of local planning to support the program.

ANTIDEGRADATION

Unlike the other three programs that form the core of this chapter, antidegradation is a watershed management policy that is not directly part of the MOA. As described in detail in Chapter 3, antidegradation is a federal regulation related to the CWA stating that waterbodies must not be allowed to degrade in quality. The implementation and enforcement of this policy at the state level is highly variable, with some states creating a separate antidegradation program, while others use existing environmental programs to comply with federal requirements. In New York State, there has been considerable interest in updating antidegradation policy to explicitly protect the water quality of the New York City reservoirs (Izeman, 1998). This section compares the antidegradation policy of New York to other states, assesses its compliance with federal guidelines, and makes recommendations regarding implementation and enforcement of antidegradation that will positively impact the New York City reservoirs.

Federal Antidegradation Policy

As set forth in federal regulations, antidegradation dictates that waterbodies cannot be allowed to sustain pollutant loadings that will prevent them from meeting their specific use classification and associated water quality criteria. It is considered one of the three points to the "Clean Water Act triangle," along with waterbody use classifications and water quality criteria (EPA, 1994). Although antidegradation has been used for many purposes and supports a wide variety of regulatory activities, its most important role is to describe the necessary steps that must be taken when additional pollutant loading is proposed that would eliminate part or all of a waterbody's assimilative capacity (R. Shippen, EPA, personal communication, 1998).

States are required to develop and adopt antidegradation policies that mirror the EPA policy. As described in Chapter 3, state policy must define Tier 1 waters (quality below fishable/swimmable), Tier 2 waters (fishable and swimmable), and Tier 3 waters (outstanding natural resources) and discuss whether and how their assimilative capacity can be used. When deciding whether to approve new discharges into their waterbodies, the states are required to conduct an "antidegradation review" that will weigh economic, social, and public concerns and suggest alternatives to the proposed activities.

State Antidegradation Policies

All states have submitted antidegradation policies to EPA, and a wide range of effort is apparent. Some states such as Pennsylvania have devised an elaborate antidegradation review process; others, including New York, have made minimal efforts to establish any oversight activities unique to an antidegradation program. One reason for these disparities is that detailed guidance from EPA on when and how to conduct antidegradation reviews has been lacking. Some states have developed more comprehensive policies in response to legal challenges from environmental advocacy organizations. Common concerns from environmental organizations are that state antidegradation policy (1) does not define a distinct antidegration review that goes beyond current regulatory processes, (2) does not assure the highest statutory and regulatory requirements for all new and existing point sources, and (3) does not assure that cost-effective and reasonable best management practices to control nonpoint sources will be used.

EPA recently contracted with consultants to compare the antidegradation policies of 26 states, including New York (Cadmus Group, 1998). The study considered how each state has interpreted the three tiers of the EPA guidance and related them to their existing use classifications and water quality criteria. This report (summarized in Box 8-8) illuminates very interesting and important strategies regarding state implementation that are discussed below.

The Cadmus study reveals important trends in state antidegradation policy. First, states have focused antidegradation almost exclusively on point sources because of the ease in accounting for their pollutant contributions. In addition, antidegradation is generally applied to new and expanding point sources rather than being applied retroactively to existing point sources. Very few activities that cause nonpoint source pollution have been subject to an antidegradation review.

Tier 2 waters were identified in Chapter 3 as being the most controversial because they have assimilative capacity. For this reason, federal regulations require discharges into tier 2 waters to be supported by considerations of economic and social benefits and proposed alternatives. Table 8-3 shows that the criteria for determining social and economic benefits of a discharge vary from state to state and are extremely vague. Another important subtlety regarding

BOX 8-8
Review of State Antidegradation Policies

Following a request from EPA, 26 state antidegradation policies were compared. The study evaluated how the states classify waters into tiers 1, 2, and 3 and how discharges into those waters are regulated via an antidegradation review.

Tier 1

In tier 1 waters, existing uses—and water quality to protect those uses—must be maintained and protected. These waters are usually not fishable/swimmable for any or all parameters. The states have taken one of two general approaches to tier 1 waters:

1. Do not specifically assign any waters to tier 1, and simply state that all waters are protected for their existing uses (20 states). Because most states do assign waters to tiers 2 and 3, this means that, by default, all other waters are in tier 1. Fishable/swimmable waters can be assigned to tier 1 if they have no assimilative capacity.

2. Specifically assign waters to tier 1 (6 states). This has been accomplished in one of two ways: the waterbody approach and the parameter approach. In the waterbody approach, the tier 1 classification is assigned to the entire body of water without considering the concentrations of individual parameters. Antidegradation is implemented by making sure that water quality criteria in these waters are not violated. For parameters of *higher quality than the applicable criterion,* these parameters can be degraded to the criterion without an antidegradation review. Thus, assimilative capacity is not protected.

In the parameter approach, certain water quality parameters are assigned to tier 1 and others to tier 2. Tier 2 parameters have assimilative capacity; the tier 1 parameters do not. Here, even when states have use classifications, they do not coincide directly with antidegradation tiers. Antidegradation is implemented by conducting an antidegradation review for those parameters that have assimilative capacity. For those that do not, antidegradation is implemented by making sure that water quality criteria for those parameters are not violated. Massachusetts is an example of a state that uses the parameter approach.

Tier 2

For fishable/swimmable waters, assimilative capacity cannot be used without an antidegradation review that assesses the economic and social impacts of the discharge. In addition, the state should consider whether alternatives to the discharge exist. Most states specifically assign waters to tier 2 using the same two approaches mentioned for tier 1. That is,

continued

under the waterbody approach, all parameters must meet or exceed the criteria for fishable/swimmable for a body of water to be assigned to tier 2. Under the parameter approach, if any parameter is above its criterion for fishable/swimmable, then its assimilative capacity is protected, even if the whole body of water is not fishable/swimmable because of other parameters.

Some states have introduced "extra" conditions that must be met for a water to be classified as tier 2. This appears to be an effort on the part of some states to avoid classifying waterbodies as tier 2 (which would necessitate time-consuming antidegradation reviews and place limits on development). For example, Colorado's definition of a tier 2 water is more stringent than simply having fishable/swimmable quality, thereby allowing 43 percent of Colorado waters to be tier 1. Florida has no tier 2 waters whatsoever.

Tier 2 waters are protected by conducting an antidegradation review that analyzes the social and economic benefits of a discharge that would use some of the assimilative capacity of a body of water. The criteria for determining the social and economic benefits of a discharge vary from state to state and are extremely vague. Of the 26 states included in the study, 21 have statements regarding economic and social benefits that provide some guidance (see Table 8-3).

To refine the antidegradation process, only those discharges that are judged to be "significant" undergo an antidegradation review. The states have developed three different approaches for deciding when a discharge is significant: (1) defined numerical values for the percent assimilative capacity used (11 states), (2) qualitative descriptions of the percent assimilative capacity used, or (3) there are no criteria and cases are judged individually.

Finally, some tier 2 waters receive protection beyond that of justifying social and economic benefits of a discharge. In North Carolina and Pennsylvania, discharges into tier 2 waters must be pretreated with best available technologies. Ohio, Oklahoma, and North Carolina have multiple tier 2 levels with increasingly strict requirements.

Tier 3

The federal antidegradation policy states that water quality shall be maintained in outstanding national resources. Many states have not used this terminology, but do have a category of water that corresponds to tier 3. Although EPA envisioned only one category of tier 3 waters where no discharges would ever be allowed, the states have developed four basic categories of protection for tier 3 waters:

continued

BOX 8-8 Continued

1. No discharges allowed.
2. Discharges allowed only if they meet some special condition and do not lower water quality.
3. No lowering of water quality allowed.
4. Lowering of water quality allowed in some circumstances.

In reviewing state antidegradation policies, EPA has generally not considered the fourth category to be tier 3. Rather, waters falling into that category are termed tier 2.5. Categories 2 and 3 are also somewhat controversial, but in most cases, EPA has not yet made a final determination as to whether they qualify as tier 3.

In determining whether a discharge will "lower" a water's quality, some states require that the discharge meet "background" pollutant concentrations that currently exist in the water, while others insist that only pure water can be discharged.

Stormwater is a type of discharge that has traditionally been exempt from antidegradation polices. Most states do not consider stormwater inputs into tier 3 waters to be a violation of antidegradation. However, in Massachusetts, stormwater discharges into tier 3 waters are prohibited. Connecticut only allows stormwater discharge to tier 3 waters after significant pretreatment has occurred. These actions likely limit development around tier 3 waters.

A final consideration for tier 3 waters is whether drawdown of these waters is lowering water quality. If so, the states may want to consider allowing discharges into tier 3 waters to maintain water quantity. Arizona is the first state that has addressed this issue.

Additional Complexities

There is an interesting tradeoff between water quality criteria and antidegradation. If water quality criteria are very strict, then there will likely be no assimilative capacity to deal with in an antidegradation review (as is the case in Virginia). This must be taken into consideration when evaluating individual state policies on antidegradation. In states with strict water quality criteria, antidegradation may prove to be of little value.

In a related matter, some of the states surveyed have a particular drinking water use classification that specifically prohibits discharges into these waters (e.g., Massachusetts and Connecticut). These use classifications have been highly effective in preventing new development around these waters and are more protective than a tier 2 antidegradation review would be.

Source: Adapted from The Cadmus Group (1998).

TABLE 8-3 State Economic and Social Benefits Analysis

Arizona	There are five categories of social benefit that a proposal must discuss, including improved community tax base and employment benefits.
California	Social or economic benefit is determined on a case-by-case basis. State guidance suggests that the community's baseline socioeconomic profile be compared to the projected profile that would exist after the discharge is in place.
Colorado	Submitted evidence is evaluated to determine whether a proposed discharger demonstrates important social and economic development for the area that would be affected by the discharge.
Connecticut	A discharger must demonstrate that it will produce overriding economic and social benefits to the state. Such evaluations rarely occur.
Delaware	Social or economic benefit is determined on a case-by-case basis.
Florida	The state does not apply a tier 2-equivalent test of economic or social benefits.
Maine	Examples of social or economic benefits include increased employment, increased production, improved tax base, correction of an environmental or public health problem, pollution prevention, and increased conservation of energy and natural resources.
Massachusetts	Examples of social or economic benefits include new production by a discharger that cannot be accommodated by existing treatment or permit limits, and increased loading to a publicly owned treatment works because of community growth that cannot be accommodated by existing treatment facilities.
Montana	The applicant must demonstrate that important economic or social development spurred by the activity outweighs the cost to society of allowing the proposed change in water quality. In determining whether a proposed activity is necessary, the department considers the economic, environmental, and technological feasibility.
Nebraska	A discharger must demonstrate that a proposed activity is as minimally polluting as reasonable and is beneficial to the surrounding community.
New Hampshire	A discharger must show a "preponderance of evidence" that the discharge provides a net social benefit. The benefits, in terms of new jobs or increased taxes, must be greater than increased infrastructure costs.
North Carolina	The state may request, from a local government affected by a discharge, documentation that the discharge is necessary for important economic and social development.
Ohio	Social or economic benefit is determined on a case-by-case basis by considering factors such as condition of the local economy, the number of potential jobs, expected tax revenues, and the projected overall impact on the community.
Oklahoma	Until recently, social or economic benefits were not justifiable reasons for allowing degradation. Little information on future evaluation procedures is available.
Pennsylvania	There is a social or economic benefits checklist. A net present value is assigned to both the proposed discharge and to one that causes no degradation. The net present value of the social or economic benefit must be greater than the associated cost for the discharge to be accepted.

continued

TABLE 8-3 Continued

Rhode Island	Few tier 2 social or economic benefit analyses have been performed.
Texas	The procedure does not refer to social or economic benefits.
Vermont	The project associated with a proposed discharge cannot produce substantial social or economic costs unless those costs are offset by equal or greater benefits of maintaining or improving water quality. The state must also find that the discharge is necessary to prevent substantial adverse economic and social impacts.
Virginia	There are five categories of social or economic benefits, including an increase in the number of jobs and an increase in tax revenues.
Wisconsin	There are seven categories of social or economic benefits, including increasing production level and avoiding employment reductions.
Wyoming	Formal evaluation of social or economic benefits through the Wyoming Continuing Planning Process has not yet taken place because new dischargers are encouraged to use zero-discharge techniques.

Source: The Cadmus Group, Inc. (1998). Reprinted, with permission from The Cadmus Group, 1998. ©1998 by The Cadmus Group.

discharges into tier 2 waters is that only those discharges that are judged to be "significant" undergo an antidegradation review. According to the Cadmus study, states have developed very different approaches for deciding when a discharge is significant. In some states, significance is defined as the use of a certain percentage of a water's assimilative capacity. In others, the judgment is entirely site-specific.

Finally, the study reveals that some states perceive antidegradation reviews as unnecessary and duplicative. For example, they argue that national and state permitting requirements imposed on point sources result in a level of protection as high as would be achieved under a tier 2 antidegradation review. The main drawback to this approach is that the concept of assimilative capacity may not be made apparent to interested stakeholders. Assimilative capacity of waters is not always considered during the NPDES permitting process, nor is the applicant necessarily required to seek less-polluting alternatives to the proposed activity, both of which are required by antidegradation policy.

Antidegradation in New York State

Antidegradation Policy

Although it is one of the states reviewed by the Cadmus Group study, New York is not specifically mentioned at any point in the report. This is likely to be because the New York State policy is a minimal and simple restatement of a portion of the federal regulation:

"It is recognized that certain waters of New York State possess an existing quality which is better than the standards assigned thereto. The quality of these waters will be maintained unless the following provisions have been demonstrated to the satisfaction of the Commissioner of Environmental Conservation: (1) that allowing lower water quality is necessary to accommodate significant economic or social development in the affected areas; and (2) that water quality will be adequate to meet the existing usage of a waterbody when allowing a lowering of water quality.

Where waters are meeting higher uses or attaining quality higher than the current classification, the Department will use the SEQR process to assure that potential adverse environmental impacts are adequately mitigated and higher attained uses are protected.

In addition, the highest statutory and regulatory requirements for all new point sources and cost effective and reasonable best management practices for nonpoint sources shall be achieved: and the intergovernmental coordination and public participation provisions of New York's continuing planning process will be satisfied.

Water that does not meet the standards assigned thereto will be improved to meet such. The water uses and the level of water quality necessary to protect such uses shall be maintained and protected." (NYS DEC, 1985)

New York does not define the State's waters as tier 1, 2, or 3 waters, although the classification system devised for measuring stream quality (i.e., AA, A, B, C, D) could be used to do so. For example, Class N and AA-S waters in New York are comparable to EPA's tier 3 waters (outstanding natural resources). Paralleling the federal mandate to protect tier 3 waters, Article 17, Title 17, of the New York Environmental Conservation Law prohibits discharges into waterbodies with those classifications. Classes A, A-S, AA, B, C, and D are all fishable/swimmable waters and could correspond to EPA tier 2 waters (although it might be useful to create multiple tier 2 categories to differentiate between the existing use classifications). There does not appear to be a New York State use classification equivalent to EPA's tier 1. Figure 8-4 shows the current use classifications of the New York City water supply reservoirs.

Implementation

New York relies on three existing regulatory programs to implement its antidegradation policy: (1) SPDES permitting, (2) reclassification of waters, and (3) the State Environmental Quality Review (SEQR) Act. There are also some state laws that contain protection clauses for certain named waters, waters classi-

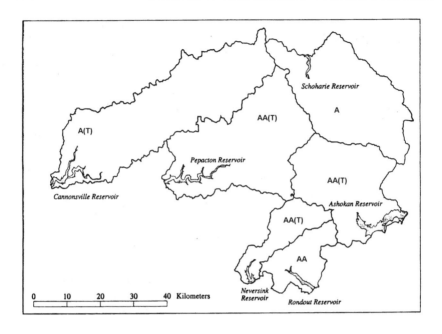

FIGURE 8-4 Use classifications of New York City reservoirs.

fied as AA-S or N, and waters in some agricultural and forest lands. However, none of the New York City reservoirs fall into these categories.

SPDES Permitting Process. As in many other states, the SPDES permitting process is used to help prevent degradation of waters by assessing the impact of nearby point source discharges. In New York, such discharges are not allowed to degrade water quality below the standards associated with a water's use classification. Depending on its complexity, the SPDES permitting process can be analogous to an antidegradation review for some types of discharges.

Reclassification of Waterbodies. Waterbodies can be reclassified sometime during the state's triennial water quality standards review process. An entire round of classifications generally takes ten years given the number of waters that must be assessed and the size of the NYS DEC staff. This process recently resulted in several waters moving from D to C and being afforded an extra level of protection from further degradation. Because reclassification does not deal directly with new discharges, its use as an antidegradation tool is limited to the designation of waters as tier 1, 2, or 3. It is not considered further in this analysis.

State Environmental Quality Review Act. The SEQR process, which is triggered when any State agency undertakes, approves, or funds actions with environmental consequences (such as permitting a WWTP), provides another level of antidegradation protection. SEQR requires that an environmental impact statement (EIS) be drawn up if it is determined that the activity will have a significant effect on the environment. SEQR is the only mechanism by which New York can implement its antidegradation policy for activities that cause nonpoint source pollution.

Analysis of New York State's Antidegradation Policy

NYS DEC currently argues that having a separate, distinct antidegradation review process is redundant because the SEQR process and the SPDES permitting process can accomplish the goals of maintaining high water quality (Cronin, 1997; N. G. Kaul, NYS DEC, personal communication, 1998). Most large-scale activities (such as business park construction) that produce both point and nonpoint source pollution will require review under SEQR. It is, however, possible that some small-scale activities will not trigger action under either regulatory program (N. G. Kaul, NYS DEC, personal communication, 1998).

Over the last 15 years, New York State has wavered on the issue of antidegradation. EPA, NYC DEP, and NYS DEC have irregularly noted that the State's policy and implementation procedures are deficient, although no action has been taken to correct the situation (EPA, 1988a,b, 1993b, 1994; NYS DEC, 1990, 1991). In fact, in 1993, a Proposed Antidegradation Review Process for New York State was abandoned (NYS DEC, 1993b). The only substantive action on antidegradation that has occurred in New York regards its participation in the drafting of an antidegradation policy for the Great Lakes Initiative (NYS DEC, 1997). However, this policy will only apply to watersheds feeding the Great Lakes.

In determining whether New York State's antidegradation policy is effective in fulfilling the intent of the CWA, many issues must be considered. These are discussed below, with particular reference to protecting water quality in the New York City drinking water reservoirs. When possible, the policies of other states are compared to that of New York for perspective.

Policy Document

As is evident from the previous discussion, New York's antidegradation policy is minimal in requirements, relying entirely on existing programs for implementation. This is in stark contrast to the lengthy procedures required elsewhere, particularly Ohio and Pennsylvania. Thus, outward appearances indicate that New York is less focused on antidegradation than are other states. This position is likely to translate into a less effective policy because of the lack of public interest and stakeholder participation in the process (see below).

Potential Applicability

Antidegradation is most useful when water quality criteria are less strict and allow waterbodies to have assimilative capacity, as discussed in Box 8-8. As shown in Table 3-7 for a limited number of parameters, New York's water quality criteria are similar to those of other eastern states and are slightly more strict than western states. Thus, there are likely to be fewer tier 2 waterbodies in New York that might benefit from antidegradation than there are in western states. However, without an accurate estimate of the state's assimilative capacity, this conclusion must remain entirely speculative.

Do Existing State Regulations Effectively Implement Antidegradation?

The SEQR and SPDES permitting processes are used to implement antidegradation in New York State. An analysis of how these programs function and what they achieve should be the best indication of the effectiveness of New York's antidegradation policy. The most important considerations include (1) whether these programs assess the economic and social benefits of proposed activities, (2) whether they consider all relevant alternatives to the proposed project, (3) how they determine the significance of proposed activities, and (4) whether they consider the assimilative capacity of the receiving waterbody through a quantitative assessment of background conditions and the pollutant loading from proposed discharges.

SEQR. The State Environmental Quality Review Act (Article 8 of the Environmental Conservation Law) is New York's most powerful tool for protecting water quality from adverse environmental impacts. It applies to all activities that require government funding, approval, or action, with some notable exceptions (such as agricultural activities). Because the SEQR process is described in Box 3-5, only those portions relevant to antidegradation are discussed here.

The SEQR process requires that the social and economic benefits of a proposed activity be discussed in a draft environmental impact statement (EIS). Minimal guidance for what relevant social and economic benefits are can be found in the SEQR Handbook (NYS DEC, 1992). This guidance describes benefits as satisfying a "need" or "perceived need" such as having a convenience store in the neighborhood or supplying water to an area that is being developed for residential housing. For the most part, the guidance is extremely limited regarding potential benefits, concentrating much more on what is *not* an acceptable social or economic benefit. For example, the profitability of the proposed project is not considered to be an appropriate benefit for mention in the draft EIS. Compared to the specific social and economic benefits that are considered by other states during an antidegradation review (Table 8-3), the SEQR guidance document is considerably less comprehensive.

The draft EIS must consider all relevant alternatives to the proposed action. SEQR guidance material is very explicit about the range and types of reasonable alternatives, discussing the use of alternative locations, alternative technologies, actions of a different scale, alternative project designs, alternative timing, and completely different substitute actions. Most importantly, all draft EISs must contain a consideration of the no-action alternative.

Determining the environmental significance of a proposed activity takes place early in the SEQR process. For those activities classified as Type I, the environmental assessment form is used to make an initial determination of significance. SEQR guidance stresses the importance of two factors: magnitude of the impact and importance of the impact in relation to its setting. The cumulative impact of multiple actions and long-term effects must be considered. Although a "substantial adverse change in existing water quality" is listed as an important criterion, SEQR does not define a particular percent change in assimilative capacity as significant, unlike 11 other states' antidegradation policies.

The use of the phrase "assimilative capacity" does not appear in the SEQR process, nor is there an explicit consideration of a waterbody's assimilative capacity. However, many of the provisions of SEQR imply that assimilative capacity is important, such as the determination of significance required in the environmental assessment form, and the description of the environmental setting and the consideration of the "no-action" alternative in the draft EIS. The guidance material associated with SEQR is general because it targets a wide variety of activities that go beyond those impacting water. As discussed below, the construction of new point sources, which almost always triggers review under SEQR, involves a more explicit consideration of assimilative capacity because of the additional oversight afforded by the SPDES permitting program.

SPDES Permitting Program. WWTPs that discharge to surface waters must obtain a SPDES permit. In addition, all WWTPs discharging greater than 1,000 gpd to the subsurface are required to have a SPDES permit. All applicants for a SPDES permit must undergo the initial steps of the SEQR process and submit an environmental assessment form. Thus, the SPDES permitting program and the SEQR review must be considered jointly when assessing New York State's antidegradation policy.

The SPDES permitting program does not require an explanation of the social and economic benefits of a proposed discharge above and beyond that required of a draft EIS. Thus, the SPDES permitting program satisfies this criterion of an antidegradation policy to the same extent as does SEQR. Similarly, because all applicants for a SPDES permit must fill out an EAF, the environmental significance of all proposed point source discharges must be assessed.

The SPDES permitting program includes an explicit call for considering the assimilative capacity of receiving waterbodies. That is, for those plants that discharge into surface waterbodies, the SPDES program may require that applicants conduct a "waste assimilative capacity analysis." This analysis is very similar to the mathematical calculations done under the TMDL program to determine waste load allocations (WLAs) (NYS DEC, 1998). The waste assimilative capacity analysis helps determine whether a proposed point source discharge will cause a waterbody to violate water quality criteria. If so, it is then used to develop alternative waste load scenarios that will achieve water quality standards. NYS DEC has developed multiple guidance documents for calculating waste assimilative capacity under a variety of conditions.

SPDES permits are required to be renewed by the state every five years. However, because of the large number of point sources holding SPDES permits in New York and the limited staff available for conducting reviews, NYS DEC has devised a scheme to prioritize permits for renewal. Under the Environmental Benefit Permit Strategy, NYS DEC ranks all permits by considering the size of the discharge, the present water quality of the receiving water, the nature of the discharged pollutants, and the time since the last review (NYS DEC, 1992). At the time of renewal, the waste assimilative capacity analysis is recalculated and best available technologies are assessed for possible inclusion.

Other Criticism

In addition to comments on the use of the SEQR process and the SPDES permitting program for implementing antidegradation, some other comments about the New York State antidegradation policy are warranted. First, as noted by others (Izeman, 1998), the New York antidegradation policy does not adequately address tier 3 waters. No procedures for assigning waters to tier 3, or for determining what discharges are allowed into tier 3 waters, are given. As demonstrated by the wide variation observed among the states, how tier 3 waterbodies are defined takes on great importance. Second, the antidegradation policy is not referenced in the State's water quality standards, nor is it a part of the standards, as required by federal regulations. This omission, however, is not particularly damaging and would be relatively simple to correct.

Conclusions and Recommendations

1. NYS DEC should define how tier 3 waterbodies are assigned, as the necessary criteria are currently not stated. This definition may have a significant impact on the New York City drinking water reservoirs if their use classification and water quality criteria are sufficiently high to allow them to qualify for tier 3 status.

2. The SPDES permitting program and the SEQR process are adequate tools for implementing antidegradation, but both would benefit from minor changes. For the most part, these regulations take into account the reasonable alternatives to a proposed action, the social and economic benefits of a proposed action, and the significance of potential environmental impacts. The latter two issues require additional attention, as recommended below.

3. NYS DEC should provide additional guidance on the types of economic and social benefits that should be part of draft environmental impact statements. Although limited information is available, the guidance material should be considerably more comprehensive.

4. NYS DEC should better define what a significant lowering of water quality is in a tier 2 waterbody. That is, it should set a quantitative criterion for altering the assimilative capacity of a waterbody. Other state antidegradation programs suggest that a 5 percent to 25 percent change in a water's assimilative capacity is significant.

5. An explicit consideration of a receiving water's assimilative capacity should be required as part of draft environmental impact statements. Consideration of assimilative capacity should be stated clearly to facilitate understanding by the public in written guidance documents, within draft EISs, and during public hearings. The stated purpose of antidegradation is for communities, regulators, and dischargers to consider the assimilative capacity of waterbodies. However, this language is not part of federal regulations and, as a consequence, most state antidegradation policies do not require an explicit consideration of assimilative capacity. Although such a consideration is an integral part of the SPDES permitting program, it is less obvious during the SEQR process. Because SEQR is the only avenue for regulating nonpoint sources that will impact water quality, this requirement for addressing assimilative capacity is critical if the SEQR process is to be relied upon for implementing New York's antidegradation policy. Further guidance from EPA on how to implement antidegradation policy for nonpoint source activities should be taken into consideration as soon as it is made available.

ADDITIONAL TREATMENT OPTIONS

Dual-Track Approach

When EPA gave New York City its second conditional waiver from filtration in 1993, it required (in addition to various watershed management activities) a series of studies on filtration of the Catskill and Delaware systems, leading up to completion of conceptual and draft preliminary designs of a filtration plant.

Activities also included a study on alternative disinfectants that might be used in the absence of filtration. These activities have been carried out in parallel with all the other filtration avoidance determination watershed management activities. In 2002, EPA will decide (using the results of these studies) whether filtration or other treatment processes are needed for the Catskill/Delaware water supply.

Simultaneously carrying out watershed management and planning for filtration, a relatively new policy concept, has been dubbed the "dual-track approach" by EPA (Krudner, 1997). This approach, as applied to New York City, was designed to ensure that no time is lost if filtration is later determined to be necessary. Although not completely analogous, the dual-track approach is similar to the multiple-barrier approach to producing high-quality drinking water introduced in Chapter 4, provided that both tracks continue into the foreseeable future.

Filtration Plant for the Catskill/Delaware Water Supply

Phase I Pilot Plant Study

To determine optimal conditions for building a filtration plant for the Catskill/Delaware system, NYC DEP has recently completed a two-year pilot study. Several different treatment trains were investigated, including conventional treatment, ozone/direct filtration, and ozone/dissolved air flotation (DAF)/filtration. All the treatment trains contain ozonation followed by filtration, but some have additional clarification processes such as sedimentation or flotation. Phase I consisted of several different sources of water being tested in small, mobile treatment units comprised of these various treatment processes. Treatment goals were to inactivate 99 percent of *Cryptosporidium* oocysts, 99.9 percent of *Giardia* cysts, and 99.99 percent of viruses (via the combined action of ozonation and filtration). Goals were also set for chemical and physical parameters such as turbidity, DBPs, inorganic compounds, and taste and odor.

In order to provide flexibility in the potential siting of a filtration plant, water from several locations was tested during the pilot studies: (1) Shaft 18, effluent from the Kensico Reservoir, (2) Millwood, upstream from Kensico in the Catskill aqueduct, and (3) Shaft 17, upstream from Kensico in the Delaware aqueduct. Tested waters had a turbidity around 1–3 NTU, an average TOC of 1–3 mg/L, and source water particle concentrations between 4,500 and 8,000/mL in the 2–30 μm range (Hazen and Sawyer, 1996). No significant differences existed among the chosen water sources, except during one winter season when the Catskill aqueduct source had higher turbidity.

At Shaft 18 and Shaft 17, all three treatment trains produced low levels of turbidity (< 0.1 NTU), particle counts (< 30/mL in the 5–15 micrometer range)[1],

---

[1] No information was provided for particle removal in the 2–30 μm size range.

and total trihalomethanes (TTHMs) and the sum of five haloacetic acids (HAA5s) (less than 0.040 mg/L and 0.030 mg/L, respectively). During the season in which the Catskill aqueduct experienced higher turbidity, only DAF/filtration was effective in meeting established water quality goals at the Millford location during the entire year. As a result of Phase I, conventional treatment was eliminated from further consideration because of its high cost and its inability to reliably lower turbidity in the Catskill aqueduct water.

Phase II Pilot Plant Study

The objective of Phase II of the pilot plant study was to further test and optimize filtration technologies. In addition, consultants were hired to determine how many filtration plants would be necessary and to find a proper location for the treatment plant(s). These activities were conducted simultaneously in order to meet EPA deliverables. First, seven configurations for a treatment facility were designed that included from one to three plants located at several different points along the Catskill and Delaware aqueducts. The water demand was projected to be 1,700 mgd. Second, sites within the watershed region that were suitable for a treatment plant were identified. Of 577 initial locations, most were eliminated by taking into consideration such issues as distance from the aqueducts, acreage of land available, and vacancy of the land. Twelve (12) sites that could not be eliminated from consideration were then combined with the seven configurations to produce 25 treatment schemes. A weighted matrix analysis of the schemes was conducted, taking costs, acceptability, site considerations, flexibility and reliability, implementation, and water quality into account. The preferred treatment scheme consisted of one filtration plant at the Eastview location, which had been identified previously by NYC DEP as a potential site.

Eastview is located downstream from Kensico Reservoir and upstream from Hillview Reservoir. Its location would allow water from both the Catskill and Delaware aqueducts to be diverted into the filtration plant. The advantages of the Eastview location are that the appropriate aqueduct connections are possible, the use of Kensico as a settling basin can be maximized, and minimal pumping would be necessary.

DAF/filtration and direct filtration were tested further using water derived from sources downstream of Kensico that would be similar to Eastview water. The treatment requirements for the combined ozonation/filtration processes were made more stringent: 99.9 percent inactivation of *Cryptosporidium*, 99.9 percent inactivation of *Giardia*, and 99.99 percent inactivation of viruses. It was also required that finished water turbidity be less than 0.10 NTU 95 percent of the time. Ozone doses were chosen that would achieve a 1.5-log removal of oocysts, as determined by a literature review. Filtration was expected to achieve a 1.5-log removal of cysts and oocysts, and this was tested by determining percent removals of surrogate particles of many size ranges during filtration. Overall treatment was optimized by varying the filtration rate (gallons per minute per square foot), the filter bed depth, the ozonation time and dose, the flocculation time, and the DAF rate (gpm/ft^2).

Direct filtration achieved the expected 1.5-log removal of particles in the size range of *Cryptosporidium* (2–30 μm) (Hazen and Sawyer, 1998). The ozone dose was varied between 1.4 and 3 mg/L. FeCl$_3$ was the primary coagulant, a cationic polymer was used (1.5 mg/L), and a filtration rate between 10 and 13 gpm/ft^2 was found to be best. In this test, 84 inches of 1.5-mm anthracite were used for filter media. DAF/filtration was also successful in achieving at least a 1.5-log removal of *Cryptosporidium*-sized (2–30 μm) particles. Because DAF removes natural organic matter prior to ozonation, the required ozone dose was somewhat lower (between 0.75 and 2 mg/L). A filtration rate between 12 and 16 gpm/ft^2 and 58 inches of 1.3-mm anthracite for filter media were found to be acceptable.

The engineering consultant has recommended, and NYC DEP has approved, one plant at the Eastview location using direct filtration (because of its lower cost) combined with pre-ozonation, as shown in Figure 8-5. The filtration rate would be 13 gpm/ft^2 (10 gpm/ft^2 in the winter), which would be unprecedented for the East Coast. Hillview Reservoir, the posttreatment storage facility, would need to be covered to protect the superior quality of the filtered water. Cost estimates for construction of the recommended plant range from $2–3 billion (Nickols, 1998) to $4 billion (J. Miele, NYC DEP, personal communication, 1999). A key to this low cost estimate is the use of Hillview as a storage reservoir so that the system does not have to be engineered to meet peak daily demands.

Disinfection Study

The filtration avoidance determination also includes a study of alternative disinfectants to assess their ability to render inactive *Cryptosporidium* oocysts in raw water from the Catskill/Delaware system. The motivation for such research was threefold. First, disinfection with chlorine is the primary chemical treatment received by Catskill/Delaware water, but there have been no previous attempts to

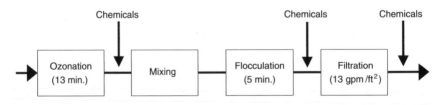

FIGURE 8-5 Proposed treatment train for direct filtration of the Catskill/Delaware water supply. Source: NYC DEP (1998d).

demonstrate that chlorine is the most effective disinfectant. Second, if regulations for TTHMs and HAA5 in finished water are tightened, then the current reliance on chlorine could take Catskill/Delaware water out of compliance with the Safe Drinking Water Act (SDWA). Third, future regulations for *Cryptosporidium* log inactivation in water supply systems are likely; a system relying solely on chlorine, in the absence of filtration, would find itself ill-prepared to comply.

The first phase of an ongoing study of alternative disinfectants was recently completed (NYC DEP, 1998e). A major assumption was that the New York City water supply would continue to be of high enough quality to qualify for filtration avoidance. The study evaluated ozone, chlorine dioxide, and chlorine for their disinfecting abilities. However, no actual experiments were performed in which the inactivation of either indicator organisms or pathogens in water samples was measured. Rather, the literature was used to develop target CT (disinfectant concentration multiplied by contact time) values that corresponded to (1) 3-log *Giardia* inactivation and 4-log virus inactivation (the EPA requirement for unfiltered systems), (2) 1-log inactivation of *Cryptosporidium*, and (3) 2-log inactivation of *Cryptosporidium*. These CT values are given in Table 8-4. The suitability of disinfectants was determined solely by comparing the achievable CT values (measured in actual water samples) to the estimated requirements for inactivation of pathogens (gathered from the literature).

TABLE 8-4 Target CT Values for the Disinfection Study Developed from Multiple Literature Sources. CT values are in (mg/L)(min).

Disinfectant	Season (°C)	CT		
		3-log *Giardia* or 4-log virus	1-log *Cryptosporidium*	2-log *Cryptosporidium*
Ozone	November 1997 (13°)	1.1	6.5	13
	March 1997 (7°)	1.9–1.5	8.5	16
	July 1998 (19°)	0.8	5	10
Chlorine	November 1997 (13°)	86–100	1.4×10^6	
	March 1997 (5–9°)	139–182		
	July 1998 (19°)	52–68	3,000–4,000	6,000–8,000
Chlorine Dioxide	November 1997 (13°)	28		
	March 1997 (5°)	33.4	60–80	120–160
	July 1998 (20°)	15	30	40

Source: NYC DEP (1998e).

In order to determine achievable CT values for New York City water, decay rates were measured for ozone, chlorine, chlorine dioxide, and disinfectant combinations in water from the Kensico Reservoir Shaft 18 location. Three sets of tests were conducted, using water collected in November 1997 and in March and July 1998. Chlorine decay and chlorine dioxide decay profiles were generated using a variety of initial doses in batch systems. From this, the integrated CT value was computed (using a first-order decay model fit to the data) and was compared to the target levels shown in Table 8-4. Ozone decay experiments were conducted in a continuous-flow sparged pilot plant comprising five columns in series (with ozone application to the first column only). To obtain varying times, the pilot plant was run at different flow rates, and samples were obtained at multiple intermediate times within the ozonation train. Tracer studies were also conducted to obtain the T_{10} value (time of exit of the fastest 10 percent of the influent). The consultants then used the T_{10} values at each sampling location along with the observed ozone residual at each point to estimate an apparent "first order" decay rate.

The conclusions of the study are that the "CT goals were satisfied for all three conditions established...with reasonable ozone and chlorine dioxide doses and contact times. The applied ozone doses and the contact times used to meet the high CT goal were 1.5 mg/L and 21 minutes in the cold water (5°), and 1.7 mg/L at 25 minutes and 1.9 mg/L at 9 minutes in the warm water (19°). The chlorine dioxide doses and contact times used to meet the high CT goal were 1.0 mg/L at 240 minutes and 1.2 mg/L at approximately 200 minutes in the cold water (5–9°) and 1.7 mg/L at 30 minutes in the warm water (19°). Chlorine disinfection and sequential disinfection (ozone followed by chlorine) were not effective in meeting the established CT goals" (NYC DEP, 1998e).

Analysis

Use of Literature Review Data. The primary literature used to estimate *Cryptosporidium* CT values consisted of conference papers by Finch and colleagues, and details of the CT computations were not provided. The use of conference proceedings for CT values is questionable, particularly because there is recent published information on oocyst inactivation. In addition, the limited information on oocyst inactivation used for this study (in which all data were obtained on buffered, demand-free water) is apparently being used to compute a "best estimate" CT value, with no recognition of the uncertainty surrounding this estimate. The CT values computed by EPA ranged from 1.5 to 3 mg/L × min because of varying safety factors. Furthermore, recent work on oocyst inactivation indicates a considerable variability in ozone inactivation efficiency for *Cryptosporidium* (Oppenheimer et al., 1999), which may result in a large safety factor for regulatory purposes. Hence, the CT values for *Cryptosporidium* used in this study may be not on a commensurate scale with the *Giardia* and viral CT values derived by EPA.

Decay Rate Experiments. The methodology used to analyze the ozone decay process is problematic. The use of a semilog plot to extract data from a reactor with backmixing is inappropriate, as it confounds the processes of hydrodynamic dispersion and reaction. Hence, the computed "CT" values achievable via this method are not entirely accurate. A more appropriate methodology to analyze the data should be employed (i.e., use of the full mathematical solution for a reactor with reaction and dispersion).

Use of Ozone. The committee questions the use of ozone for treating New York City's drinking water in the absence of other treatment processes. Although ozone is a powerful disinfectant for many microbial pathogens, it can react with dissolved organic matter to produce ketoacids, carboxylic acids, and aldehydes. The reaction of ozone with dissolved organic matter can also change largely refractory humic materials into biodegradable products that can support bacterial regrowth in the distribution system. (It should be noted that regrowth is much less apparent in ozonated systems that carry a distribution system residual provided by another disinfectant.) In addition, in the presence of bromide ion, ozone can lead to formation of bromate and bromine-substituted organic compounds. Because bromide levels were not measured in the disinfection study, the potential for bromate formation in the Catskill/Delaware water supply is currently unknown.

To avoid these problems, ozone should only be applied to water with the lowest possible organic content, and ozonation should be followed by a granular media step or secondary disinfection. (The results of using sequential ozone–chlorine disinfection are not encouraging, suggesting that other combinations of disinfectants should be investigated.) Granular media would provide filtration as well as a surface on which heterotrophic organisms can grow and degrade newly formed DBPs and other biodegradable matter. In the absence of these steps, ozonation is not recommended.

Conclusions and Recommendations

1. The dual-track approach allows New York City to focus the bulk of its resources on improvements in the watershed. This initial focus will help establish a strong source water protection program without diverting attention and resources toward the details of a filtration plant. The pollution prevention achieved through watershed protection reduces influent pollutant concentrations that would be treated via filtration. If the source water protection program is effective, the cost of filtration can be reduced.

2. The results of the filtration pilot plant study show that the present New York City water supply can be effectively treated by ozonation combined with coagulation/filtration. Treated water from direct filtration had a turbidity of less than 0.1 NTU, average particle (2–30 μm) counts ranging from 8 to 153/mL, and total trihalomethane and haloacetic acid formation potential of less than 0.040 mg/L and 0.030 mg/L, respectively. At least 3-log oocyst inactivation/removal is expected for the entire treatment train. These low effluent pollutant concentrations from a potential filtration plant are dependent on maintaining high source water quality via aggressive watershed management. The construction cost of such a treatment facility ranges from $265 to $400 per capita.

3. New York City should conduct studies on the actual inactivation of pathogens in its water under potential design conditions. In view of the potential effect of as yet unknown water quality factors on inactivation efficiency, and in view of the large potential investment that enhanced disinfection might require, it is not prudent to rely upon literature values for oocyst inactivation efficiency. These studies should be conducted using best available methodology for assessing cyst, oocyst, and virus viability and for susceptibility testing.

4. Additional studies to assess the potential of ozone as a treatment technique are required. Any consideration of ozonation should include measurements of bromide in the source waters to determine the potential for bromate formation. The literature to date suggests that ozone has the potential to increase biodegradable organic carbon in finished water and to foster regrowth of heterotrophic plate count organisms and possibly coliforms, although distribution system disinfectant residuals may counter this phenomenon. Without assurance that such regrowth would occur, it is imprudent to consider ozone as a sole treatment method.

5. A decision to construct a filtration plant should in no way deter New York City from pursuing an aggressive watershed management program. If a coagulation/filtration plant is put in place, it should be treating the best-quality source water possible. For that reason, high water quality in the Catskill/Delaware system must be maintained via aggressive implementation of the watershed management strategy.

REFERENCES

Anderson, W. P., J. A. Klang, and R. Peplin. 1997. Water pollutant trading: From policy to reality. Environmental Engineer October.

Auer, M. T., S. T. Bagley, D. A. Stern, and M. J. Babiera. 1998. A framework for modeling the fate and transport of *Giardia* and *Cryptosporidium* in surface waters. Journal of Lake and Reservoir Management 14(2-3):393–400.

Azous, A. L., L. E. Reinelt, and J. Burkey. 1997. Management of freshwater wetlands in central Puget Sound Basin. Chapter 12 In Zous, A. L., and R. R. Horner (eds.) Wetlands and Urbanization: Implications for the Future. King County, WA: Department of Natural Resources Water and Land Resources Division.

Brown, W., and T. Schueler. 1997. Pollutant Removal of Urban Stormwater Best Management Practices—A National Database. Ellicott City, MD: Center for Watershed Management, Chesapeake Research Consortium.

Caraco, D., J. Zielinski, and R. Claytor. 1998. Nutrient Loading from Conventional and Innovative Site Development. Ellicott City, MD: Cheseapeake Research Consortium, Center for Watershed Protection.

Claytor. 1996. Stormwater Retrofits–A tool for watershed management. Technical Note 48 Watershed Protection Techniques 1(4):188–192.

Clean Water Report. 1998. Nation's first pollutant trading water permit approved by EPA Region I. Clean Water Report 36(23):224.

Cronin, R. 1997. NYS DEC. Presentation given at the first meeting of the NRC Committee, September 25–26, 1997, New York City, NY.

Dillon, P. J., and F. H. Rigler. 1974. Limnology and Oceanography 19:767.

Doerr, S. M., E. M Owens, R. K. Gelda, M. T. Aurer, and S. W. Effler. 1998. Development and testing of a nutrient-phytoplankton model for Cannonsville Reservoir. Journal of Lake and Reservoir Management 14(2-3):301–321.

Environmental Protection Agency (EPA). 1988a. Memorandum on State Antidegradation Program Review to R. L. Caspe, Region II Director of Water Management Division from M. G. Portho, Director of the Office of Water Regulations and Standards. September 12.

EPA. 1988b. Memorandum on EPA's Review of New York's Toxic Control Program to D. Barolo, Director of Division of Water, NYS DEC from R. L. Caspe, Region II Director of Water Management Division. September 30.

EPA. 1993a. Subsurface Flow Constructed Wetlands for Wastewater Treatment: Design Manual Number 74. EPA 832-R-93-008. Washington, DC: Office of Water.

EPA. 1993b. Memorandum on State Antidegradation Implementation Methods to Regional Water Quality Standards Coordinator from B. Shippen, Water Quality Standards Branch. May 26.

EPA. 1994. Water Quality Standards Handbook. Washington, DC: EPA.

EPA. 1996. Draft Framework for Watershed-based Trading. EPA 800-R-96-001. Office of Water. Washington, DC: EPA.

EPA. 1997. Compendium of Tools for Watershed Assessment and TMDL Development. EPA-841-B-97-006. Office of Water. Washington, DC: EPA.

Gain, W. 1996. The Effects of Low Flow Path Modification on Water Quality Constituent Retention in an Urban Stormwater Pond and Wetland System. USGS Water Resources Investigations Report 95-4297. Orlando, FL: USGS.

Haith, D. A., and L. L. Shoemaker. 1987. Generalized watershed loading functions for stream flow nutrients. Water Resources Bulletin 23(3):471–478.

Haith, D. A., and L. L. Shoemaker, R. L. Doneker, and L. D. Delwiche. 1983. Non-Point Source Control of Phosphorus–A Watershed Evaluation, Volume 4, Modeling of Streamflow, Sediment Yield, and Nutrient Export from the West Branch Delaware River Watershed, U.S. Environmental Protection Agency, ORD, EPA-P002-137-02.

Haith, D. A., R. Mandel, and R. S. Wu. 1992. GWLF Generalized Watershed Loading Functions, Version 2.0 User's Manual. Ithaca, NY:Department of Agricultural & Biological Engineering, Cornell University.

Harding, D. 1993. TMDL Case Study: Tar-Pamlico Basin, North Carolina, Washington, DC: EPA.

Hazen and Sawyer. 1996. Phase I Pilot Testing Program Preliminary Summary of Results. June 18, 1996.

Hazen and Sawyer. 1998. Task 4: Phase II Pilot Testing Report. EPA Submittal. Corona, NY: NYC DEP.

Hoag, D. L., and J. S. Hughes-Popp. 1997. The Theory and Practice of Pollution Credit Trading in Water Quality Management. Review of Agricultural Economics, 19(Fall/Winter): 252–262.

Izeman, M. A. 1998. NRDC. Memorandum to the National Research Council dated May 13, 1998.

Izeman, M. A., and R. Marx. 1996a. Comments of the Natural Resources Defense Council, Inc. on the New York State Department of Environmental Conservation's Proposed Phase I Total Maximum Daily Loads for Phosphorus in the Croton River System. November 29, 1996.

Izeman, M. A., and R. Marx. 1996b. Comments of the Natural Resources Defense Council, Inc. on the New York State Department of Environmental Conservation's Proposed Phase I Total Maximum Daily Loads for Phosphorus in the Catskill and Delaware Systems. December 27, 1996.

Izeman, M. A., and R. Marx. 1998. Comments of the Natural Resources Defense Council, Inc. on the New York City Department of Environmental Protection's Draft Methodology for Calculating Phase II Total Maximum Daily Loads (TMDLs) of Phosphorus for New York City Drinking Water Reservoirs.

Jacobson, E. M., D. L. Hoag, and L. E. Danielson. 1994. Theory and Practice of Pollution Credit Trading in Water Quality Management. The Colorado Agricultural Experiment Station and the North Carolina Agricultural Research Service.

Janus, L. 1989. The relationshop of nutrient residence times to the trophic condition of lakes. Ph.D. dissertation, McMaster University, Ontario.

Kadlec, R. H., and R. L. Knight. 1996. Treatment Wetlands. Boca Raton, FL: Lewis Publishers.

Knight, R. L., R. Randall, M. Girst, J. A. Tess, M. Wilhelm, and R. H. Kadlec. 1995. Arizona Guidance Manual for Constructed Wetlands for Water Quality Improvement. Arizona Department of Environmental Quality.

Krudner, M. 1997. EPA Region 2. Presentation given at the first meeting of the NRC Committee, September 25-26, 1997, New York City, NY.

Lapp, D. 1994. Managing market mania. Environmental Action Winter:17-21.

Larsen, D. P., J. VanSickle, K. W. Maleug, and P. D. Smith. 1976. Water Research 13:1259–1268.

Longabucco, P., and M. R. Rafferty. 1998. Analysis of Material Loading to Cannonsville Reservoir: Advantages of Event-Based Sampling. Journal of Lake and Reservoir Management 14(2-3): 197–212.

Mann, E. 1994. Trading delusions. Environmental Action Winter:22–24.

Maryland Department of Environment (MDE). 1998. Stormwater Design Manual. Volume 1 and 2. Baltimore, MD: Water Management Administration.

McCann, K., and L. Olson. 1994. Pollutant removal efficiencies of the Greenwood Urban Wetland Treatment System. Journal of Lake and Reservoir Management 9(2):97–106.

New York City Department of Environmental Protection (NYC DEP). 1993. Implications of Phosphorus Loading for Water Quality in NYC Reservoirs. Corona, NY: NYC DEP.

NYC DEP. 1996. Methodology for Calculating Phase II TMDLs of Phosphorus for NYC Drinking Water Reservoirs. December 31, 1996. Valhalla, NY: NYC DEP.

NYC DEP. 1997a. Methodology for Determining Phosphorus Restricted Basins. July 1997. Valhalla, NY: NYC DEP.

NYC DEP. 1997b. The Relationship between Phosphorus Loading, THM Precursors, and the Current 20 µg/L TP Guidance Value. Valhalla, NY: NYC DEP.

NYC DEP. 1997c. Guidance for Phosphorus Offset Pilot Programs. April 1997. Valhalla, NY: NYC DEP.

NYC DEP. 1997d. Guidance for Preparing a Variance Application for a WWTP Expansion Involving a Phosphorus Offset Plan. May 1997. Valhalla, NY: NYC DEP.

NYC DEP. 1998a. Calibrate and verify GWLF models for remaining Catskill/Delaware Reservoirs. January 27, 1998. Valhalla, NY: NYC DEP.

NYC DEP. 1998b. NYC DEP Review of the Phase I TMDL Implementation. June 29, 1998. Valhalla, NY: NYC DEP.

NYC DEP. 1998c. Report on Applications to Participate in the Pilot Phosphorus Offset Program. January 1998. Valhalla, NY: NYC DEP.

NYC DEP. 1998d. Capital Project No. WM-30 Catskill and Delaware Water Treatment: Treatment Process Selection to Commence Preliminary Design.

NYC DEP. 1998e. Disinfection Study for the Catskill and Delaware Water Supply Systems. October 30, 1998. Valhalla, NY: NYC DEP.

NYC DEP. 1999a. Development of a water quality guidance value for Phase II Total Maximum Daily Loads (TMDLs) in the New York City Reservoirs. Valhalla, NY: NYC DEP.

NYC DEP. 1999b. Methodology for Calculating Phase II Total Maximum Daily Loads (TMDLs) of Phosphorus for New York City Drinking Water Reservoirs. March 1999. Valhalla, NY: NYC DEP.

NYC DEP. 1999c. Proposed Phase II Phosphorus TMDL Calculations for Ashokan Reservoir. Valhalla, NY: NYC DEP.

NYC DEP. 1999d. Proposed Phase II Phosphorus TMDL Calculations for Cannonsville Reservoir. Valhalla, NY: NYC DEP.

NYC DEP. 1999e. Proposed Phase II Phosphorus TMDL Calculations for Neversink Reservoir. Valhalla, NY: NYC DEP.

NYC DEP. 1999f. Proposed Phase II Phosphorus TMDL Calculations for Pepacton Reservoir. Valhalla, NY: NYC DEP.

NYC DEP. 1999g. Proposed Phase II Phosphorus TMDL Calculations for Rondout Reservoir. Valhalla, NY: NYC DEP.

NYC DEP. 1999h. Proposed Phase II Phosphorus TMDL Calculations for Schoharie Reservoir. Valhalla, NY: NYC DEP.

NYC DEP. 1999i. Proposed Phase II Phosphorus TMDL Calculations for Kensico Reservoir. Valhalla, NY: NYC DEP.

NYC DEP. 1999j. Proposed Phase II Phosphorus TMDL Calculations for West Branch Reservoir. Valhalla, NY: NYC DEP.

NYC DEP. 1999k. Proposed Phase II Phosphorus TMDL Calculations for Amawalk Reservoir. Valhalla, NY: NYC DEP.

NYC DEP. 1999l. Proposed Phase II Phosphorus TMDL Calculations for Bog Brook Reservoir. Valhalla, NY: NYC DEP.

NYC DEP. 1999m. Proposed Phase II Phosphorus TMDL Calculations for Boyds Corner Reservoir. Valhalla, NY: NYC DEP.

NYC DEP. 1999n. Proposed Phase II Phosphorus TMDL Calculations for Cross River Reservoir. Valhalla, NY: NYC DEP.

NYC DEP. 1999o. Proposed Phase II Phosphorus TMDL Calculations for Croton Falls Reservoir. Valhalla, NY: NYC DEP.

NYC DEP. 1999p. Proposed Phase II Phosphorus TMDL Calculations for Diverting Reservoir. Valhalla, NY: NYC DEP.

NYC DEP. 1999q. Proposed Phase II Phosphorus TMDL Calculations for East Branch Reservoir. Valhalla, NY: NYC DEP.

NYC DEP. 1999r. Proposed Phase II Phosphorus TMDL Calculations for Middle Branch Reservoir. Valhalla, NY: NYC DEP.

NYC DEP. 1999s. Proposed Phase II Phosphorus TMDL Calculations for Muscoot Reservoir. Valhalla, NY: NYC DEP.

NYC DEP. 1999t. Proposed Phase II Phosphorus TMDL Calculations for New Croton Reservoir. Valhalla, NY: NYC DEP.

NYC DEP. 1999u. Proposed Phase II Phosphorus TMDL Calculations for Titicus Reservoir. Valhalla, NY: NYC DEP.

New York State Department of Environmental Conservation (NYS DEC). 1985. Water Quality Antidegradation Policy. Albany, N.Y.: Division of Water.

NYS DEC. 1990. Director Pagano on Antidegradation. NYSDEC Water Bulletin August 1990.

NYS DEC. 1991. Final Combined Supplemental Environmental and Regulatory Impact Statement for Revision of Water Quality Regulations: Surface Waters and Groundwaters. 6 NYCRR Parts 700-705.

NYS DEC. 1992. SEQR Handbook. Albany, N.Y.: Division of Regulatory Affairs.

NYS DEC. 1993a. New York State Fact Sheet: Ambient Water Quality Value for Protection of Recreational Uses. Albany, N.Y.: Bureau of Technical Services and Research.

NYS DEC. 1993b. Proposed Antidegradation Review Process for New York State: Community Assessment Form. Albany, N.Y.: Division of Water.

NYS DEC. 1996. Urban/Stormwater Runoff Management Practices Catalogue for Nonpoint Source Pollution Prevention and Water Quality Protection in New York State. November, 1996.

NYS DEC. 1997. Implementation of the NYSDEC Antidegradation Policy—Great Lakes Basin (Supplement to Antidegradation Policy dated September 9, 1985). Division of Water Technical and Operational Guidance Series 1.3.9. June 30.

NYS DEC. 1998. Total Maximum Daily Loads and Water Quality—Based Effluent Limits. Divison of Water Technical and Operations Guidance Series 1.3.1. February 26.

Nickols, D. 1998. Hazen and Sawyer. Presentation given at the Fourth Meeting of the NRC Committee. August 31–September 1, 1998, Washington, DC.

Novotny, V. 1996. Evaluation and Comments on Proposed Phase I TMDL Calculations for Catskill-Delaware Reservoir. Mequon, W.I.: Aqua Nova International, Ltd.

Oberts, H. 1997. Lake McCarrons Wetland Treatment System: Phase III Study Report. St. Paul, MN: Metropolitan Council Environmental Services.

Oppenheimer, J. A., E. M. Aieta, R. R. Trussell, J. G. Jacangelo, and I. N. Najm. 1999. Evaluation of *Cryptosporidium* Inactivation in Natural Waters. Denver, CO: AWWA Research Foundation. Draft.

Owens, E. M. 1998. Development and testing of one-dimensional hydrothermal models of Cannonsville Reservoir. Journal of Lake and Reservoir Management 14(2-3):172–185.

Paulson, C. L. 1997. Testimony on Cherry Creek Basin Water Quality Authority before the Water Quality Control Commission of the State of Colorado.

Reckhow, K. H., M. N. Beaulac, and J. T. Simpson. 1980. Modeling phosphorus loading and lake response under uncertainty: A manual and compilation of export coefficients. Vol. 440/5-80-011. Washington, DC: EPA.

Reckhow, K. H., and Chapra, S.C. 1983. Engineering Approaches for Lake Management, Vol. 1— Data Analysis and Empirical Modeling, Boston, MA: Butterworth Publishers,

Reed, S. C., E. J. Middlebrooks, and R. W. Crites. 1995. Natural Systems for Waste Management and Treatment, McGraw Hill, 2nd Edition.

Richter, K. O., and A. L. Azous. 1995. Amphibian occurrence and wetland characteristics in the Puget Sound Basin. Wetlands 15(3):305–312.

Sawyer, C. N. 1947. J. New England Water Works Association 61:109.

Schnoor, J. L. 1996. Environmental Modeling—Fate and Transport of Pollutants in Water, Air and Soil. New York, NY: John Wiley and Sons.

Schueler, T. 1987. Controlling Urban Runoff—A Manual for Planning and Designing Urban Best Management. Washington, DC: Metropolitan Washington Council of Governments.

Schueler, T. 1996. Irreducible pollutant concentrations discharged from urban best management practices. Watershed Protection Techniques 2(2):369–372.

Stepczuk, C. L., A. B. Martin, P. Longabucco, J. A. Bloomfield, and S. W. Effler. 1998a. Allochthonous contributions of THM precursors in a eutrophic reservoir. Journal of Lake and Reservoir Management 14(2–3):344–355.

Stepczuk, C. L., A. B. Martin, S. W. Effler, J. A. Bloomfield, and M. T. Auer. 1998b. Spatial and temporal patterns of THM precursors in a eutrophic reservoir. Journal of Lake and Reservoir Management 14(2–3): 356–366.

Stepczuk, C. L., E. M. Owens, S. W. Effler, J. A. Bloomfield, and M. T. Auer. 1998c. A modeling analysis of THM precursors for a eutrophic reservoir. Journal of Lake and Reservoir Management 14(2-3): 367–378.

Stephenson, K. 1994. Effluent Allowance Trading: A New Approach to Watershed Management. Blacksburg, VA: Virginia Water Resources Research Center.

The Cadmus Group, Inc. 1998. Technical Analysis of State and Regional Antidegradation Procedures. Final Draft. Prepared for the U.S. Environmental Protection Agency under Contract 68-C4-0051. Waltham, MA: The Cadmus Group, Inc.

Urbonas, B. 1994. Performance of a pond/wetland system in Colorado. Technical Note 17. Watershed Protection Techniques 1(2):68–71.

Vollenweider, R. A. 1976. Advances in defining critical loading levels for phosphorus in lake eutrophication. Mem. Ist. Ital. Idrobiol. 33:53–83.

Warne, D. 1999. NYC DEP. Memorandum to the National Research Council dated April 1999.

9

Nonpoint Source Pollution
Management Practices

Nonpoint source (NPS) pollution is widely dispersed in the environment and is associated with a variety of human activities. These activities produce pollutants such as nutrients, toxic substances, sediment, and microorganisms that may be delivered to nearby waterbodies following rainfall or directly via atmospheric deposition. Under pristine conditions, land generally has an enormous capacity to remove pollutants from rainwater. For example, research in the Catskills has shown that undisturbed forests can remove as much as 90 percent of the nitrogen from rainwater before it can reach nearby streams (Lovett et al., 1999). However, activities that produce NPS pollution also cause changes in vegetative cover, disturbance of soil, or alteration of the path and rate of water flow. These physical changes may prevent the land from naturally removing pollutants in stormwater. Thus, there are two interacting effects of NPS activities: (1) production of a pollutant and (2) alteration of the land surface in a way that increases pollutant loading to receiving waters. The goals of NPS pollution best management practices (BMPs) are to maintain or restore the ability of the land to remove pollutants and to limit production of the pollutant.

The intent of this chapter is to evaluate the BMPs that are being used to control NPS pollution in the Catskill/Delaware watershed. Because most BMPs are implemented within the framework of an NPS pollution control program, this chapter reviews and critiques a variety of such programs (1) the Watershed Agricultural Program, (2) the Watershed Forestry Program, and (3) the Stormwater Pollution Prevention Plans. Conclusions and recommendations are made for the BMPs and for the NPS pollution control programs in general.

Table 9-1 lists potential nonpoint sources, priority pollutants, and some of the qualitative and quantitative criteria that can be used to rate NPS pollution

TABLE 9-1 Nonpoint Sources, Priority Pollutants, and Potential Criteria for Evaluating NPS Pollution in the New York City Watersheds (Not all table entries are considered in this report)

Nonpoint Sources	Priority Pollutants	Evaluation Criteria

• Agriculture
• Urban stormwater
• Construction/
  roads
• Forestry
• On-site sewage
  treatment and
  disposal systems
• Atmospheric
  deposition
	• Phosphorus
	• Organic carbon
	  compounds
	• Turbidity/TSS
	• *Cryptosporidium*
	• *Giardia*
	• Fecal coliforms
		• Total land area covered under the MOA
		• Export coefficient
		• Best management practices used
		• NYC DEP performance monitoring data of
		  BMPs
		• Are BMPs implemented using best available
		  technology?
		• National performance/general effectiveness
		• Does BMP efficiency depend on regular
		  maintenance?
		• Are there appropriate institutions to ensure
		  full-scale implementation?

control programs. This review is not comprehensive; for example, atmospheric deposition is not specifically addressed. This is because nitrogen is the primary pollutant associated with atmospheric deposition, and nitrogen loading is not a particular concern in the watershed region. In addition, atmospheric deposition of pollutants onto the land surface might be treated by the BMPs designed to treat other sources of nonpoint pollution.

NONPOINT SOURCE PROGRAMS

NPS pollution is a problem that is becoming increasingly important in the Catskill/ Delaware watershed, as evidenced by at least 30 different programs developed to reduce its impact. In fact, the New York City Department of Environmental Protection (NYC DEP) identifies almost all the regulations in the Memorandum of Agreement (MOA) as dealing with NPS pollution (NYC DEP, 1998a). Because these programs are not integrated into one overall program, there can be some confusion when trying to evaluate the methods being used to manage NPS pollution in the New York City watersheds. Sources for NPS pollution programs that affect the New York City water supply include (1) the Watershed Rules and Regulations of the MOA, (2) the May 1997 Filtration Avoidance Determination (FAD), (3) the Watershed Protection and Partnership Programs of the MOA, and (4) various State programs.

Watershed Rules and Regulations

The Watershed Rules and Regulations contain most of the NPS pollution programs conducted by NYC DEP. These programs include three basic mechanisms for controlling NPS pollution (NYC DEP, 1998a): (1) strict performance standards applied to activities that produce NPS pollution, (2) a review and approval process for activities that produce NPS pollution, and (3) prohibition of certain activities in a "setback" region between the activity and nearby waterbodies.

Performance standards exist for wastewater treatment plants (WWTPs) that discharge to the subsurface but not for on-site sewage treatment and disposal systems (OSTDS), stormwater BMPs, agricultural BMPs, and forestry BMPs. Although it is technically feasible to monitor performance standards for all these activities, it can be difficult and expensive because of the diffuse and episodic nature of pollutant transport.

The Stormwater Pollution Prevention Plan (SPPP) is a good example of a review and approval process used to control NPS pollution. All construction activities affecting more than five acres must prepare an SPPP and receive approval before the project commences. Finally, setbacks have been designated for a range of activities, including OSTDS construction, hazardous materials storage, the construction of impervious surfaces, siting of landfills, and residential pesticide application. These setback distances, which vary depending on the activity and the type of waterbody nearby, are evaluated in detail in Chapter 10.

Filtration Avoidance Determination and the Watershed Protection and Partnership Programs

The FAD and the Watershed Protection and Partnership Programs indirectly manage NPS pollution. The FAD requires that Total Maximum Daily Loads (TMDLs) be calculated for the reservoirs and that a phosphorus offset pilot program be developed, both of which require implementation of NPS pollution BMPs. The Watershed Agricultural Program is included under the FAD, although agricultural BMPs are not specified. Finally, the Watershed Protection and Partnership Programs include the Watershed Forestry Program and the Stream Management Program, which discuss control of NPS pollution.

New York State NPS Pollution Programs

NYC DEP participates in a number of State activities relating to NPS pollution, including the New York State Nonpoint Source Coordinating Committee. NYS DEC is the primary state agency for controlling NPS pollution as part of its enforcement of the Clean Water Act (CWA). In many cases, the Watershed

Rules and Regulations mirror provisions in State regulations, and NYC DEP oversight provides additional protection.

AGRICULTURE IN THE CATSKILL/DELAWARE WATERSHED

Agriculture, the predominant industry in the Catskill/Delaware watershed, is concentrated primarily in the Cannonsville and Pepacton watersheds. As shown in Figure 9-1, farms are relatively evenly distributed across the Delaware watershed, with many found close to major tributaries such as the West Branch of the Delaware River. Farms become more sparse in the eastern portion of the watershed because of more mountainous terrain and relatively poor soils.

Ninety (90) percent of the 351 farming operations in the Catskill/Delaware watershed are dairy farms, most of which have between 50 and 200 animals.[1] Most farms in the region also support crop production and contain significant tracts of forest. Although dairy cows are the predominant animal at farms participating in the Watershed Agricultural Program, a wide variety of other animals can be found (Table 9-2).

Watershed Agricultural Program

The Watershed Agricultural Program (WAP) is a voluntary program intended to standardize and improve environmental practices among watershed farmers. Because of the WAP, all agricultural activities in the Catskill/Delaware watershed are exempt from MOA regulations such as setback distances, discharge permits, and rules regarding pesticide application. To date, many in the farming community and most of those concerned about the quality of New York City drinking water have been ardent advocates of the program. This broad support is testimony to the strong affinity for agriculture as an important economic endeavor in the Catskill/Delaware watershed region.

The WAP is administered by the Watershed Agricultural Council (WAC), a grassroots organization composed of farm, agribusiness, and environmental leaders. Since 1993, the primary role of the WAC has been to review and approve changes being made on individual farms to improve the water quality of nearby receiving waters (both surface and subsurface).

One of the unique features of the WAP is its strong connection to basic research being conducted by Cornell University on the sources and transport of pollutants from agricultural practices. With funding usually supplied by NYC DEP, scientists at Cornell have studied the hydrology, phosphorus transport, parasitology, and economics of the Catskill/Delaware watershed. Cornell

---

[1]"351" refers only to those farms having a gross annual salary of at least $10,000, making them eligible for the Watershed Agricultural Program.

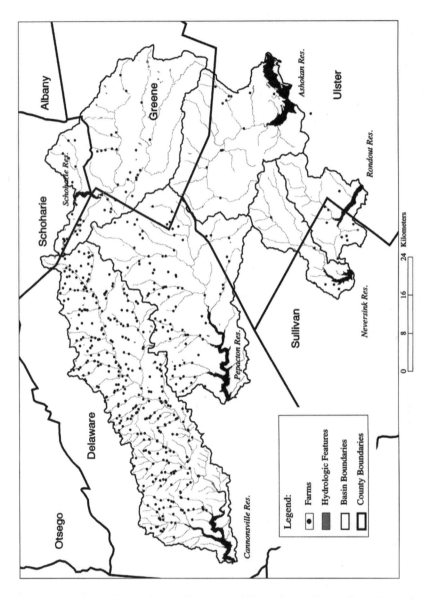

FIGURE 9-1 Farm locations in the Catskill/Delaware watershed. Courtesy of the NYC DEP.

TABLE 9-2 Livestock Types and Numbers in the Catskill/ Delaware Watershed

Livestock Type	Number
Mature Dairy	12,636
Dairy Heifers	8,758
Chickens	2,655
Beef Cattle	1,566
Veal Calves	790
Sheep	569
Horses	565
Deer	375
Pheasant	250
Other	284

Source: NYC DEP (1997a).

researchers have also developed a mix of very useful process-based and empirical models to use as planning tools or to use in support of the planning process. Thus, the WAP has been in a position to apply research findings to actual farm practices at an early stage in the process. The WAP has largely focussed on understanding the role of agriculture in generating pathogens and phosphorus and on application of BMPs for pathogen and phosphorus control. Two activities that may change the overall focus of the program are manure export and the Conservation Reserve Enhancement Program.

Whole Farm Plans

Whole Farm Plans are comprehensive strategies for controlling potential sources of pollution at individual farms. For those farmers that participate in the WAP, a Whole Farm Plan, which addresses their specific problems and needs, is developed by a planning team consisting of the farmer and representatives from local Soil and Water Conservation Districts, the U.S. Department of Agriculture (USDA) Natural Resources Conservation Service (NRCS), and the Cornell Cooperative Extension. The contributions of each organization to the Whole Farm Plans are described in Table 9-3, and the 11-step Whole Farm Plan process is outlined in Table 9-4. To date, 199 Whole Farm Plans have been completed and approved by the WAC (NYC DEP, 1999a).

The main purpose of the Whole Farm Plan is to develop and implement BMPs that address pressing environmental concerns while being compatible with the farmer's mission and objectives. Specific BMPs are evaluated and approved

TABLE 9-3 Organizations Involved in Whole Farm Planning

Organization/Party	Role
Farmer	The focus of the Watershed Agricultural Program. Participation is voluntary.
NYC Department of Environmental Protection	A source of major funding and technical and administrative support to the WAP. One NYC DEP staff member is a member of the WAC.
NYS Department of Environmental Conservation	Ex officio member of the WAC, providing technical expertise on nonpoint source pollution control measures.
USDA Natural Resources Conservation Service	A source of technical and scientific expertise for the WAP. Its role in the watershed predates the WAP.
Soil and Water Conservation Districts	Grassroots organizations created by individual counties to supply technical expertise to farmers on conserving soil/water.
Cornell Cooperative Extension	Technical and managerial expertise to farmers as part of New York State's land grant university, Cornell University.
Cornell University	Original research focusing on water resources and pollution prevention in the New York City watershed.

Source: WAP (1997).

TABLE 9-4 Steps in the Whole Farm Plan Process

Step	Description
1	Identify farm mission, objectives, business plan, and resources, both short-term and long-term.
2	Inventory and analyze water, soil, air, plant, and animal resource information.
3	Determine the priority water quality (and other) issues for the farm.
4	Identify practices (BMPs) to address the priority water quality (and other) issues.
5	Evaluate the effects of these practices on water quality (and other) issues from Step 3.
6	Identify adequate alternatives that satisfy the WAP's water quality criteria.
7	Quantify the economic and management effects of the alternative practices.
8	Select and integrate the practices to be included in the Whole Farm Plan. Submit the plan to the Soil and Water Conservation District and the WAC for approval.
9	Develop tactical plans to ensure successful implementation of the approved Whole Farm Plan.
10	Implement the Whole Farm Plan.
11	Assist, monitor, and evaluate implementation of the Whole Farm Plan and evaluate progress toward addressing the priority issues.

Source: NYC DEP (1997a).

by the planning team prior to their implementation. Once a plan is approved, the planning team develops a strategy for implementing the plan and measuring its success. Monitoring for changes in water quality at this stage is an ideal approach to measuring the impact of farm BMPs. However, in some cases the necessary monitoring techniques do not exist, or the staff or technical equipment is not available to conduct the monitoring. Thus, extensive monitoring of soil and water quality does not take place on a regular basis at all farms participating in the program. Planning teams ensure maintenance of the BMPs via field inspection and try to keep abreast of improvements in technology that should be considered for use.

Focus of Agricultural Best Management Practices

The BMPs that have been chosen for the farms in the Catskill/Delaware watershed are a direct outgrowth of the priority pollutants and the specific environmental problems present. They can be classified as one of three general types: (1) barriers to pollutant transport (source control), (2) landscape barriers, and (3) stream corridor barriers. As discussed in Chapter 5, livestock generates pathogens, underscoring the importance of managing manure on farms (NYS WRI, 1997). Nutrients, especially phosphorus, are also priority pollutants in the watershed. Typical phosphorus inputs occur in animal feed and fertilizers; typical outputs include animal waste, plant material, and fertilizer that does not penetrate the soil. Finally, pesticides are potential pollutants in the watershed, and the MOA does not regulate agricultural uses of these compounds.

Common BMPs used in the watershed include barnyard management, improved manure storage, and the separation of calves from cows. A complete list of agricultural BMPs used in the WAP is given below (WAP, 1997):

- Barnyard water management system
- Conservation cropping sequence
- Cover and green manure crop
- Diversion
- Fencing
- Filter strip
- Grassed waterway or outlet
- Obstruction removal
- Pasture and hayland planting
- Pipeline
- Planned grazing system
- Access road
- Pathogen management
- Spring development
- Trough or tank

- Stock trails and walkway
- Stripcropping—contour
- Structure for water control
- Nutrient management plan (manure spreading based on soil phosphorus recommendations)
- Pesticide management
- Subsurface drain
- Underground outlet
- Waste utilization (manure management)
- Windbreak

Measuring Success

Typically, the measures of success for water quality BMPs in agricultural settings have been related to the numbers of practices installed and the numbers of farms participating, an approach that was initially taken by the WAP. Several success metrics for the WAP are included in New York City's waiver from filtration: (1) the number of participating farms, (2) the number of Whole Farms Plans that are developed and approved, (3) the number of Whole Farm Plan implementations, and (4) the number of plans for which an annual evaluation has been completed. These measures were seen as important because of the contentious interactions between watershed farmers and City and State regulators.

The WAP has not been prioritized by targeting specific farms with known pollution problems. Rather, farmers have been allowed (after observing the prototype of ten pilot farms) to join the program on a voluntary basis, regardless of the extent of pollution at their farm. These criteria have resulted in a large number of participating farms (317, over 90 percent of all eligible watershed farms).

The most important measures of program success are monitoring data and other evidence that demonstrate a positive impact of new farm practices on water quality. This approach requires performance monitoring of BMPs, determining pollutant loadings emanating from these BMPs, and modeling the resultant water quality in nearby receiving waters. Unfortunately, linking the performance of BMPs with nearby water quality conditions is *much* more difficult to accomplish than determining the number of farms in the program or the number of plans implemented.

To use this latter approach, more information is needed on (1) the accumulation of pathogens and nutrients in source areas (as suggested in Chapter 6), (2) pathogen and nutrient wash-off from source areas during rainfall (i.e., pollutant loading in overland flow), and (3) transport of pathogens and nutrients through the shallow subsurface prior to reaching waterbodies (i.e., pollutant loading in subsurface flow). As suggested in a recent independent review of the WAP, this approach relies on understanding the baseline pollutant conditions in the watershed and the receiving waters (CTIC, 1998).

Monitoring of farm conditions is currently concentrated at only two locations–the Shaw Road control area and Robertson Farm—in the hope that data gathered at these locations will be applicable to the entire Catskill/Delaware watershed. This monitoring is based on a modified paired watershed approach to understanding the integrated effects of Whole Farm Plan implementation rather than being based on monitoring of individual BMPs. Preliminary data gathered since 1996 indicate that there are significant differences between the loading rates of different pollutants at the farm and at the control area (WAP, 1997). However, additional research must be conducted before these data can be used.

Analysis of the Watershed Agricultural Program

The whole farm planning approach taken by the WAP represents a significant advance compared to the standard conservation planning done by USDA-NRCS and associated agencies. Because of the level of resources available to the WAP and the early recognition that new scientific information was needed to accomplish the Whole Farm Plans, the WAP has been instrumental in attempting to integrate scientific information and farm plans for controlling nonpoint source pollution.

Although the accomplishments of the WAP and the whole farm planning projects are substantial, gaps in knowledge, implementation, and monitoring underscore the difficulty of determining the level of ecological and environmental sustainability provided at the watershed scale by a suite of agricultural BMPs. As discussed below, specific areas that must be addressed to improve the whole farm planning process include the transport and loading of phosphorus (including issues of scale, modeling, and monitoring) and the transport of microbial pathogens.

Transport and Loading of Phosphorus

There have been numerous efforts over the last 20 years to study nutrient loadings (particularly phosphorus) from agricultural lands to the Cannonsville Reservoir via the West Branch of the Delaware River (WBDR), other tributaries, and subsurface flow. As described in Box 9-1, these efforts have revealed a wealth of information about the relative sources of phosphorus, average phosphorus concentrations in different types of runoff, and the efficacy of certain agricultural BMPs for reducing pollutant loading from some farm sources. Despite these efforts, there are still no scientifically based nutrient load reductions developed for the Cannonsville watershed. The following section discusses the importance of such requirements and the information needed to develop them.

BOX 9-1
The Model Implementation Program of the 1980s

Phosphorus has been of considerable interest in the New York City water supply watersheds, especially in the West Branch of the Delaware River–Cannonsville watershed, for at least two decades. The first national demonstration of agricultural BMPs as a means of water pollution control was undertaken by USDA and EPA in 1977 as part of the Model Implementation Program (MIP). One of the MIP sites was the West Branch of the Delaware River (WBDR). MIP focused on control of dissolved phosphorus from animal wastes (particularly barnyard runoff), manure-spreading schedules, and conservation practices including waste storage, stripcropping, conservation tillage, conservation cropping, cover cropping, critical area planting, woodland improvement and harvest, and tree planting.

As part of the MIP, barnyards on 275 farms were prioritized for treatment based on their distance from defined watercourses. Out of 154 high-priority barnyards, 91 were treated with BMPs. The practices applied included diversions, open drains, water-control structures, roof gutters, grading, fencing, livestock exclusion from streams, and pavement of heavily used areas. Treatment of some cropland with phosphorus and with erosion-control measures was accomplished, and an educational and advisory program for silviculture was established.

Cornell University and NYS DEC subsequently studied the effects of the MIP in the Cannonsville watershed (Brown et al., 1984). As a result, a great deal was known about the potential sources of phosphorus, the appropriate analysis methods for phosphorus, and phosphorus transport mechanisms in agricultural watersheds. The MIP first focused on changing the size and management of dairy barnyards. Detailed monitoring data showed that barnyards can be treated to achieve a high level of phosphorus control (50–90 percent load reduction). Diversion of water flow from areas above the barnyard was found to be a critical factor in controlling runoff from barnyards. However, even though dairy barnyards were a significant source of phosphorus, the studies concluded that the

---

Developing Phosphorus Reduction Goals. Two components are needed to assess the impact of farm activities on nearby water quality: (1) a water quality model for the Cannonsville Reservoir that predicts long-term phosphorus concentrations in relation to varying inputs and (2) phosphorus loadings from adjacent agricultural areas, considering both surface and subsurface contribu-

barnyard control would achieve less than a 5 percent reduction in overall dissolved phosphorus loading (Brown et al., 1984). MIP-related studies also modeled the effects of manure-spreading schedules, which indicated that manure scheduling could lead to an average reduction of total phosphorus loading of 35 percent.

A watershed-based model was used to show that direct runoff (both surface and subsurface), WWTPs, and baseflow accounted for 38, 23 and 33 percent, respectively, of dissolved phosphorus loadings and 70, 10, and 16 percent, respectively, of total phosphorus loadings to the Cannonsville Reservoir (Brown et al., 1984). The 30-month record developed for this study showed that more than 85 percent of dissolved phosphorus, total phosphorus, and sediment loadings occurred during periods of snowmelt and rainfall in January to March. The model illuminated the most relevant parameters for assessment of BMPs. Average concentration of dissolved phosphorus in direct runoff (both surface and subsurface) estimated from load vs. runoff regressions at the small watershed scale was found to be the best parameter for gauging the effectiveness of a phosphorus control program. Interestingly, a substantial portion of the dissolved phosphorus budget (39 percent) for the entire WBDR was due to groundwater inputs to streamflow, considered at the time "to be largely uncontrollable" (Brown et al., 1984).

Based on manure scheduling (Robillard and Walter, 1984b) and a watershed-scale model of the WBDR (Haith et al., 1984), Brown et al. (1984) found that the spreading of manure during winter was a major source of dissolved phosphorus in surface runoff in the WBDR. Another major phosphorus source identified in this work was flooded cropland along the WBDR itself. Recommendations were that (1) efforts should focus on cropland and manure-management practices in the watershed, (2) field-scale investigations were needed to understand phosphorus losses from upland farming in the watershed, and (3) barring changes in land use or agricultural practices, a concentration of 0.23 mg/L could be used to estimate dissolved phosphorus in direct runoff. This estimate for dissolved phosphorus in surface runoff was used in some of the modeling studies undertaken in 1997 (NYC DEP, 1997a).

tions. Several issues regarding reservoir water quality modeling must be addressed. First, as suggested by others (CTIC, 1998), monitoring and modeling must take both dissolved and particulate phosphorus into account. The Cannonsville Reservoir regularly exceeds the water quality standard for total phosphorus (20 µg/L), although concentrations of dissolved phosphorus entering the reservoir have decreased from the 1980s to the 1990s by 52 percent (CTIC, 1998).

Auer et al. (1998) reports that most of the phosphorus (95 percent to 97 percent) available to algae in the Cannonsville Reservoir comes from soluble phosphorus inputs from the WBDR. These results should be used to guide the development of a comprehensive model of nutrient/phytoplankton dynamics in the reservoir. This would allow the development of a load management strategy as part of the WAP and TMDL process.

The WAP and New York City should consider using the existing information developed from both the phosphorus cycling models and the TMDL process to develop specific phosphorus load reduction targets for the Cannonsville watershed. Scientifically based load reduction targets would (1) provide an endpoint to be reached in the planning and implementation process of the WAP, (2) be a quantifiable measure of success, and (3) provide feedback to the process of BMP application during whole farm planning. If loads were unequally distributed among subwatersheds, the monitoring of subwatersheds could be related to specific load reduction goals and the effectiveness of specific agricultural BMPs.

Monitoring to Document Success. The second goal mentioned above was to determine phosphorus loadings from adjacent agricultural areas, considering both surface and subsurface contributions. This task requires that individual BMPs (or systems of BMPs) be monitored for their effectiveness in reducing pollutant loadings. Because of technical limitations, such an approach has not traditionally been taken in farm management. Rather, it is more common to ensure that BMPs use Best Available Control Technology (BACT) and simply assume certain BMP removal rates. As suggested by others (CTIC, 1998), new techniques will be needed in order to *demonstrate* that agricultural BMPs can meet load reduction goals.

Although a large number of BMPs have already been installed and Whole Farm Plans have been implemented on 198 farms, there is very little known about the net result. Monitoring data are available for the Robertson Farm and Shaw Road sites for a two-year period prior to BMP installation at Robertson Farm (Pre 1 and Pre 2) and for the first year of postimplementation (Longabucco et al., 1999). At both locations, stream discharge and precipitation are monitored continuously, and concentrations of pollutants in stream discharge are regularly monitored, including three forms of phosphorus, three forms of nitrogen, organic carbon, suspended solids, *Giardia,* and *Cryptosporidium.* As expected, loading rates of most pollutants are higher at the farm (WAP, 1997; Longabucco et al., 1999). NYC DEP has found that most of the pollution on Robertson Farm emanates from a small, heavily used area. Practices implemented on Robertson Farm include manure storage, spreading schedules based on hydrologic sensitivity, tile drainage of wet fields, and diversion of milkhouse waste discharge away from streams.

The use of small, paired watersheds such as the Robertson Farm versus Shaw Road comparison is a good prototype for the WAP and NYC DEP to follow. According to the first year of monitoring results after the implementation period, the Whole Farm Plan has reduced certain loadings, especially soluble reactive phosphorus loading. However, as noted by the authors (Longabucco et al., 1999), one year of postimplementation monitoring data is insufficient to make conclusions about the success or failure of the Robertson Farm Plan. It should be noted that monitoring was suspended from May 1995 until fall 1996 during the implementation of BMPs. Future monitoring efforts should include the implementation period because the first year of postimplementation data show increases in total suspended solids and particulate phosphorus, attributed to construction activities. Understanding the impact of these construction activities should be part of evaluating Whole Farm Plan implementation.

Although this type of paired-watershed monitoring activity is expensive, more of it is essential for both documentation of the effects of Whole Farm Plans and for determining whether load reduction goals are being met. The paired-watershed monitoring will not yield information on individual BMPs, but it provides the best approach to understanding the aggregate effects of Whole Farm Plans. More watershed sites should be established, with the goal being to compare loads from reference areas and farms before and after implementation of Whole Farm Plans. Such data can later be used to test models of loadings from farm-scale watersheds. The primary determinants of research cost are the number of samples analyzed and the numbers and types of constituents. Flow proportional sampling on more sites would yield more useful information with a lower marginal cost.

The WAP should consider contracting with appropriate agencies to set up a monitoring system with the following attributes: (1) a series of representative reference (forested) watersheds that will provide comparisons for all agricultural watersheds, (2) preimplementation, implementation, and postimplementation monitoring for a representative set of agricultural watersheds, (3) selection of farm-scale watersheds that represent the variety both of the type and the intensity of BMPs being used (these may or may not be just one farm, as long as the use of BMPs can be quantified), and (4) use of flow proportional sampling, instead of daily sampling, with the basic sample interval being weekly. Thus, the sample load per sampling station can be reduced to 50–100 per year, rather that 350–365 per year. Although the monitoring system would involve a substantial commitment of resources, without it there will be no way to quantify the water quality effects of Whole Farm Plans.

Integration of Modeling and Monitoring. The overall goal of modeling is to estimate the effects of Whole Farm Plans on nonpoint source pollution. Models should be clear and explicit in the approaches taken to simulate the effects of BMPs, the generation of nutrient budgets for farms, and the simulation of watershed-scale results. They must also be tested and validated using monitoring data taken at the field, farm, and watershed scale. Unfortunately, a major limitation to current efforts is the lack of monitoring data from specific BMPs that can be used for model development.

The needed integration of monitoring and modeling can be accomplished using the expanded paired-watershed approach described above. Another approach is to establish a farm on which individual BMPs can be demonstrated and monitored, as was done under the Model Implementation Program (MIP) (see Box 9-1). Although a single demonstration farm could not provide monitoring data for all BMPs, particular high-priority BMPs (as identified by modeling) could be targeted. For example, preliminary modeling efforts indicate that changes in manure-spreading schedules based on manure storage may provide the largest decrease in phosphorus transport (NYC DEP, 1997a). Testing these model simulations might be a priority for a demonstration farm. Although the effectiveness of particular BMP types may change from farm to farm because of site-specific conditions, performance data for specific BMPs would be very useful to the WAP in validating models and formulating future Whole Farm Plans.

Modeling has already played a prominent role in the preliminary evaluation of the farm-scale impacts of WAP. For example, NYC DEP has recently modeled the effects of source barriers and landscape barriers on the transport of pollutants, primarily phosphorus, and has compared model results to field observations at a demonstration farm (NYC DEP, 1997a). The source barriers involved changes in the size and structure of the barnyard, while the landscape barriers included changes in the schedule of manure applications to fields. Modeling results showed that installing barnyard gutters and pads and decreasing the size of the barnyard decreased barnyard runoff by 47 percent and phosphorus export by approximately 30 percent. In addition, variable manure-spreading schedules resulted in 2 percent to 50 percent reduction in phosphorus export. Field observations of barnyard runoff at the demonstration farm, however, did not always corroborate model results. This shows why better integration of modeling and monitoring is needed. For example, monitoring how changes in barnyard runoff routing affect filter areas and adjacent fields could inform the modeling effort.

This type of model-based evaluation must be reinforced by efforts to generate phosphorus load reduction goals, as recommended above. The NYC DEP study describes a useful approach to estimating phosphorus load reduction. However, these estimates are not particularly useful unless there is an overall phosphorus load reduction goal at the watershed scale.

Future Use of Phosphorus Control BMPs. Several years of WAP operation highlight opportunities for prioritizing and coordinating phosphorus control BMPs. First, although not historically done, manure application rates should be based on phosphorus, which is a higher priority pollutant in the Catskill/Delaware watershed and reservoirs than nitrogen. Second, manure storage is one structural measure that most analyses show should be emphasized. In fact, Robillard and Walter (1984b) indicates that changes in manure-spreading schedules may allow a maximum 35 percent phosphorus load reduction if they are combined with effective manure-storage facilities. There is contradictory evidence about the efficacy of barnyard treatment. Monitoring studies (Robillard and Walter, 1984a) and model estimates (NYC DEP, 1997a) indicate significant nutrient load reductions from barnyard treatments, while estimates of nutrient runoff from watershed studies indicated little effect on overall watershed loading from barnyard treatments (Brown et al., 1984). What is clear from earlier studies is that phosphorus loadings from barnyards are primarily dependent on the *amount of runoff* rather than on the *amount of manure phosphorus* in contact with the runoff. This implies that marginal changes in manure phosphorus concentrations would have little or no effect on phosphorus transport from barnyards.

Transport of Pathogens

Other than nutrient reductions, the main goal of the WAP is to establish the risk of viable/infective *Cryptosporidium* and *Giardia* to the New York City water supply and reduce pathogen loadings from agricultural areas. Transport of pathogens is a relatively new concern in agricultural watersheds. Hence, there is much less known about the transport and fate of pathogens. The pathogen-related research undertaken by Cornell to support the WAP was, by necessity, oriented toward developing methods for (oo)cyst detection, determining (oo)cyst viability, and determining the source of (oo)cysts. Although a clear picture has begun to emerge concerning sources, risk factors associated with these sources, and factors affecting the viability of parasites in the environment, little is known about the actual transport of parasites from farms or the importance of agricultural areas as a source of parasites at the reservoir watershed scale (NYS WRI, 1997). A recent modeling study by Walker and Stedinger (1999) suggests that agricultural loading of pathogens to reservoirs is largely dependent on the ability of BMPs to enhance degradation of pathogens before they enter nearby streams.

Current Research Efforts. NYC DEP currently conducts a monitoring program for *Giardia* and *Cryptosporidium* at over 50 sites throughout the water supply watershed, including the Robertson Farm and the Shaw Road sites. Preliminary analysis of these data indicates sites influenced by agriculture have a lower incidence of detection of *Giardia* and *Cryptosporidium* than do urban watersheds or WWTP effluent, although the differences are not large (NYC DEP, 1998b). Undisturbed watersheds have the lowest incidence of both protozoans. More specific studies have investigated farm animal sources of these pathogens, as described in Chapter 5. Because calves were identified as a major source of parasites, procedures were recommended to segregate calves from cows to prevent cross-infection. At a few farms, greenhouses or separate shelters have been installed to house calves, improve ventilation and bedding conditions, and reduce cow-to-calf contact (NYC DEP, 1997a). Other research has focused on the prevalence and incidence of protozoan infection in entire dairy herds and on management practices that contribute to development and perpetuation of protozoans in dairy herds (NYS WRI, 1997).

Cornell research has also made contributions to understanding the viability of *Cryptosporidium* oocysts in various farm environments. Laboratory studies show that free ammonia in manure causes significant inactivation of oocysts. The viability of oocysts in calf manure piles was reduced by more than 90 percent in 100 days during late fall and early winter, clearly suggesting that calf manure should be stored prior to spreading (NYS WRI, 1997). Other studies indicate contact with soil material is necessary for more rapid inactivation of oocysts during freeze/thaw cycles. The implications of this research are that winter spreading of manure may require bare soil without snow cover, limiting the time periods when winter spreading can occur.

Future Needs. Because agricultural areas are a significant source of pathogens, monitoring, controlling, and preventing pathogen loading from these areas should be a primary concern of the WAP. Effective management of pathogens requires that (1) sources of (oo)cysts from agricultural areas be identified and quantified and (2) the effectiveness of BMPs in removing pathogens be quantitatively assessed. Although substantial progress has been made in understanding the factors controlling parasite incidence on farms, the state of knowledge does not allow quantitative assessment of changes in pathogen loading to the water supply associated with improved agricultural practices. Thus, the WAP's current approach to parasite risk reduction is to control manure storage in order to maximize exposure of pathogens to detrimental environmental factors and minimize the potential of pathogen transport in runoff, and to maintain a healthy calf-rearing environment (NYC DEP, 1997a).

In support of the first goal, it would be ideal to develop a source term for cysts and oocysts from various types of agricultural practices, especially calf-rearing areas. To accomplish this, the WAP should continue supporting basic research on the transport and fate of (oo)cysts and associated modeling efforts such as the model developed in Walker and Stedinger (1999). The second goal has received limited attention, and as a result much more is known about phosphorus removal by BMPs than about pathogen removal by BMPs. Monitoring of individual BMPs should focus heavily on pathogens or appropriate microbial indicators. This information is critical to making informed management decisions about which BMPs to use at specific farms.

New Directions for the Watershed Agricultural Program

Conservation Reserve Enhancement Program. The Conservation Reserve

Enhancement Program (CREP) involves "retiring" miles of riparian land from agricultural use and will allow the WAP to effectively manage croplands inundated by floods. Based on field observations (Brown et al., 1984), annual or periodic inundation of manured fields may be a significant problem in manure phosphorus transport. Much of the most productive cropland is located in the active floodplain of the WBDR and has historically received high manure application rates (Brown et al., 1984).

Prioritization of lands within CREP can be difficult, as it depends on voluntary participation. The establishment of riparian forest buffers could lead to changes in land use that last beyond the 10- to 15-year life of a rental agreement. Unless new information is available indicating otherwise, CREP should focus on lands that are frequently inundated. (It should be noted that flow patterns and natural/enhanced levees may cause some of the most frequently inundated land to be away from streams.) Consideration should be given to obtaining permanent easements on these lands to allow them to be maintained in riparian forest buffers after expiration of the CREP contracts.

In evaluating the vegetation on CREP land, the potential for grass filters to fail because of either inundation from floods or inundation by overland flow from barnyard diversions or tile drains should be considered. Grass filters are known to fail under conditions of concentrated flow (Dillaha et al., 1989). These failures have apparently been observed in filter areas installed for barnyard runoff (NYC DEP, 1997a).

Another consideration is whether CREP can be used as a vehicle for achieving voluntary limits on direct livestock access to streams. Although a complete prohibition on direct access may be desirable to maximize water quality (CTIC, 1998), it is unlikely to be achieved. In lieu of a mandatory exclusion and fencing program, incentives within CREP and other cost-share measures such as alternative water supplies and fencing can be employed to enhance compliance.

Manure Export. Manure export is being considered as an alternative in the Catskill/Delaware watershed because of an excess of usable manure. Based on both watershed and field-scale mass balance considerations and on manure-spreading schedule changes, manure export would likely have a large effect on phosphorus transport (NYC DEP, 1997a). The effects of manure export on pathogen loading seem less certain because of the cost of collecting and transporting calf manure separately.

Manure export has several important implications for watershed management. Most importantly, manure should be exported to locations (1) where it is economically feasible given a certain sustainable level of subsidy, (2) where the manure is needed as an agronomic resource, and (3) where there will not be environmental quality problems caused by manure imports. At first analysis, the manure-producing areas of the Catskill/Delaware watershed do not seem to be near areas in need of manure imports.[2] Potential agronomic problems in the Catskill/Delaware watershed also warrant consideration. Farmers will need to keep an adequate amount of manure to provide the phosphorus needs of their crops, especially on low-phosphorus fields. If manure resources are exported, farmers may decide to provide phosphorus by adding inorganic fertilizer, which would be an additional (and hence, undesirable) source of phosphorus in agricultural runoff. In short, manure export plans must be carefully considered.

Science Program. The WAP's connection to basic science research has provided significant new information on the sources of *Cryptosporidium* oocysts and the relation between herd management and oocyst production. In addition, a great deal of knowledge on factors affecting oocyst survival and tracking has been gained. The field- and farm-scale modeling and management tools developed by Cornell University researchers represent significant advances. In order to assess whether agriculture will be able to meet the water-quality goals required of an unfiltered drinking water supply, more research will be necessary. The collaboration between the WAP and basic research should continue until reliable predictions of the following can be made: (1) reductions in pathogen loading rates for specific BMPs applied to dairy farms, (2) reductions in phosphorus loading rates for specific BMPs applied to dairy and other farms, and (3) reductions in watershed-scale loading rates for pathogens and phosphorus for systems of BMPs and Whole Farm Plans.

In order to accomplish these tasks, the science program should explicitly consider the relationship between system function and the management tools under development. An example is the Field Risk Table, a tool used by planning teams to prioritize areas where manure should and should not be spread (NYS WRI, 1997). There are some indications that the Field Risk Table does not take into account existing knowledge about the production of nonpoint source pollutants in association with manure spreading. For instance, it does not include snow cover, even though manure application on top of or in snow would seem to be a significant problem because of increased pollutant transport during melt or rain-on-snow events and limited inactivation of oocysts not in contact with soil. In addition, the Field Risk Table only addresses flooding in a very general way, and it does not address effects of tile drains. Appropriate refinements should be made in future research conducted in support of the WAP.

---

[2]Export immediately north or west would impact the upper Susquehanna, which forms the headwaters of the Chesapeake Bay watershed. Areas directly east are still within the New York City watersheds. Export south down the Delaware River Valley may be possible, but it is not known if there are agronomic needs for imported manure.

Conclusions and Recommendations for the
Watershed Agricultural Program

Although agriculture is exempt from many provisions of the MOA, given its role as a pollution source and its predominance as a land use, the committee chose to examine the performance and principles involved in the WAP. Although this report does not assess whether the funds administered by the Watershed Agricultural Council are sufficient for accomplishing the intended outcomes, it is worth noting that the success of the WAP and related programs is directly related to the level and constancy of financial support.

1. A scientifically based phosphorus load reduction goal that will achieve the desired phosphorus concentration in the water supply reservoirs should be developed. This process should take into account in-reservoir generation of phosphorus and phosphorus cycling and different forms of phosphorus. This load reduction can then be apportioned between the phosphorus outputs from individual subwatersheds.

2. Additional farm-scale monitoring should be implemented to determine the ability of Whole Farm Plans to reduce nonpoint source pollution. Current analyses of Whole Farm Plans are limited to two sites within the watershed. NYC DEP and the WAP should establish additional demonstration farms for evaluating Whole Farm Plans and systems of BMPs. The monitoring program that would be necessary may be costly and will require the participation of additional scientists and technicians.

3. Modeling and monitoring efforts should be better integrated. The WAP has access to multiple models with implications for management practices such as manure spreading and barnyard treatment. Additional models should be developed to take streambank protection BMPs into account. In relation to all models, more focused monitoring of individual BMPs is required to test and validate these models (for those BMP types that have not yet been tested). In particular, monitoring to determine the effectiveness of BMPs in removing pathogens (or microbial indicators) is greatly needed.

4. Although the WAP has met the current milestones required for the FAD, these metrics do not establish the net impact of agriculture on water quality. The WAP, in conjunction with outside research teams, should determine appropriate monitoring and modeling tools available for the establishment of metrics that relate to actual water quality improvement.

5. New York City must develop a greater understanding of factors controlling pathogen transport in the watersheds to determine the relative influence of agriculture as a source of pathogens as compared to other land

uses. The goal of such efforts should be to develop source terms for *Giardia* cysts and *Cryptosporidium* oocysts from various types of agricultural practices, especially calf-rearing. NYC DEP could benefit by assimilating work done on pathogen inactivation under variable environmental conditions and by making use of watershed models developed for nonpathogens (e.g., nutrients).

6. The WAP and NYC DEP should continue to support science for the whole farm planning process. Funding gaps for this research has halted several relevant lines of work that could hold considerable benefits to the WAP. It is important that scientific findings continue to be integrated into the evolution of management plans.

7. Lands enrolled in the Conservation Reserve Enhancement Program should be prioritized based (1) on frequency of inundation, (2) on vegetation type, and (3) on whether the landowner will voluntarily exclude livestock from riparian zones. If prioritization is not possible, rental and cost-share incentives should be increased to retire frequently flooded farmland into riparian forest buffers and to exclude livestock from streams.

8. The WAP should conduct a comprehensive assessment of the relative merits and feasibility of manure storage versus manure export. All aspects of both scenarios should be considered. If it is determined that manure export is the only viable way to maintain agriculture in the watershed, the WAP should consider the implications of this for sustainability of animal-based agriculture within the Catskill/Delaware watershed.

9. Implementation of Whole Farm Plans and long-term maintenance of best management practices must be addressed. Experience in other localities has shown that Whole Farm Plans are often not fully implemented, long-term maintenance of BMPs is difficult to achieve and, as a result, BMP effectiveness decreases with time (T. Dillaha, Virginia Polytechnic and State University, personal communication, 1999). Research and field experience has underscored the importance of active BMP maintenance in sustaining water quality. Program managers should consider what lessons could be learned from the maintenance of the Model Implementation Program (Box 9-1) when evaluating BMPs implemented under Whole Farm Plans. Program success will likely be greater if the WAC, NYC DEP, or some other appropriate watershed authority assumes responsibility for such maintenance.

WATERSHED FORESTRY PROGRAM

Forests are the most desirable land cover for the protection of water supply watersheds in temperate ecoregions. For this reason, when water supply watersheds are owned by the public or by a water utility, they are often converted to or maintained as forest preserve. In the New York City watersheds, much of the land area is privately owned (74 percent in the Catskill/Delaware region). In such areas as these, forestry (silviculture) is an important land use that can benefit both the private landowner and the water supply reservoirs.

Compared to other land uses, sustainable forest management leads to small-scale, infrequent, distributed environmental change while generating a wide array of goods and services. Therefore, it holds considerable promise for balancing resource use and watershed protection. However, in order for forestry to be the preferred land use in any region, common-sense principles must consistently guide management actions. Watershed managers for New York City, Boston, Providence, New Haven, Hartford, and scores of smaller communities and their rural neighbors throughout the Northeast are currently exploring the tradeoffs between active forest management, preservation, and other land uses.

The reckless disruption of forest ecosystems characteristic of late nineteenth and early twentieth centuries inspired the conservation movements in the United States and led directly to the establishment of the forestry profession in North America. Outdated images of massive clearcuts extending from ridgeline to ridgeline, forest fires roaring out of cutover[3] into nearby towns, streams choked with sediment, and barren landscapes still influence the values and attitudes of an increasingly urban public. However, such scenes should not define what is possible, practical, and desirable for modern forestry. Box 9-2 lists commonly accepted principles of forestry sustainability; they are compatible with the goals and objectives of watershed management programs.

Program Description

The Watershed Forestry Program (WFP) was established in 1997 to improve the economic viability of forest land ownership and the forest products industry in ways compatible with water quality protection and sustainable forest management. Patterned after and affiliated with the Watershed Agricultural Program, the WFP was formed following the deliberations of a Watershed Forest Ad Hoc Task Force. The Task Force—comprised of foresters, local landowners, loggers, local and regional forest products industry representatives, representatives of nonprofit groups, and New York City and State officials—synthesized information about forests and forestry in the watershed region, identified problems and opportunities, and developed five overarching position statements for the WFP (Table 9-5).

The ideas, goals, and objectives set forth by the WFP correspond closely

---

[3]Cutover is a term used in the late 1800s and early 1900s to describe large areas that were logged without regard for the condition of the site and other forest values, such as water quality, wildlife, recreation, and aesthetics.

BOX 9-2
Northern Forest Lands Council
Principles of Sustainability

• Maintenance of soil productivity.
• Conservation of water quality, wetlands, and riparian zones.
• Maintenance of a healthy balance of forest age classes.
• Continuous flow of timber, pulpwood, and other forest products.
• Improvement of the overall quality of the timber resource as a foundation for more value-added opportunities.
• Addressing scenic quality by limiting adverse aesthetic impacts of forest harvesting.
• Conservation and enhancement of habitats that support a full range of native flora and fauna.
• Protection of unique or fragile areas.
• Continuation of opportunities for traditional recreation.

Source: Northern Forest Lands Council (1994).

with other recent approaches, including the Massachusetts Metropolitan District Commission's management of the Quabbin Forest (Barten et al., 1998; MDC, 1995), the USDA Forest Service-sponsored Stewardship Incentive Program, and other contemporary examples (Bentley and Langbein, 1996; NRC, 1990, 1998). Although they are more comprehensive and sophisticated, most of the Task Force's findings and recommendations echo more general turn-of-the-century calls for conservation of forest resources. In fact, enabling legislation to acquire and manage U.S. national forests (now 220 million acres) was justified solely on the basis of watershed protection and restoration, not timber supply or wildlife conservation (Dana and Fairfax, 1980; Ellefson, 1992). Foresters and policymakers did, however, recognize the ecological and economic necessity of balancing water yield and quality, timber harvesting, wildlife populations, and recreational use.

Program Accomplishments

For a variety of reasons, the Catskills region has lagged behind many other areas of the Northeast and New England in the development and implementation of private lands programs. However, the WFP has served as a catalyst for changes

TABLE 9-5 Position Statements of the Watershed Forestry *Ad Hoc* Task Force and Corresponding Policy Recommendations[a]

Position Statements	Policy Recommendations
1. Well-managed forests provide the most beneficial land cover for water quality protection.	• Educate the public about the linkages between forests, forestry, water quality, and rural economies. • Use conservation easements that allow for traditional uses and maintain undeveloped land.
2. Existing forest management activities are a negligible nonpoint source of pollution; however more extensive use of Best Management Practices (BMPs) will further reduce sediment and nutrient loading from forest management activities.	• Expand the logger training and certification program. • Develop a user-friendly BMP field manual tailored to watershed conditions. • Expand forest management outreach and plan development with landowners. • Conduct a watershedwide, posttimber harvest survey to assess the effectiveness of New York State's timber-harvesting guidelines. • Develop regulatory and economic incentives to improve BMP compliance and on-the-ground performance.
3. High property taxes discourage stewardship of private forest land.	• Reform New York State forest tax law (RPTL 480-a) to a current-use strategy equivalent to that of neighboring states with reimbursement to local governments. • Develop alternative or supplemental funding sources for local governments and school districts (in addition to local property tax revenue). • Establish a system of incentives to encourage owners of smaller parcels (<50 acres) to maintain their holdings.
4. Retention and growth of primary and secondary forest products manufacturing are essential to a healthy forest-based economy, forest land retention, natural resource protection, and sound forest conservation and management.	• Provide technical and financial assistance to forest landowners interested in long-term management to ensure a continuous supply of high-quality timber for local manufacturing. • Foster an improved business climate to develop and sustain local forest-based industry. • Promote the inclusion of forestry and the forest products industry in economic development studies.
5. Existing public forest lands should provide a model for sound resource management that complements private stewardship.	• NYS DEC and NYC DEP forest lands should, where appropriate, serve as examples of sustainable forest management.

[a]Position statements are direct quotations from the original source while policy recommendations are paraphrased.
Source: WFAHTF (1996).

in the practice of forestry and in long-term prospects of forest landowners in the region. Now it is frequently cited as a model for other areas of the United States (Jannik, 1998).

Although the WFP is new, progress toward objectives has been rapid and substantial. Work on most policy recommendations is now under way. Several BMP (focused on sediment control) and logger training workshops have been held in the region, and temporary bridges for stream crossings are available as a no-cost loan to area loggers (Kittredge and Woodall, 1997; Kittredge et al., 1997). A pilot project using geotextiles to stabilize forest roads in sensitive areas is under way. Postharvest field assessment of BMP effectiveness by independent scientists is in progress on more than 60 sites throughout the Catskill and Delaware watersheds.

Thirty-four (34) professional foresters have qualified as "approved contractors" to prepare and implement forest management plans on private lands. Administrative procedures for forest management plan specifications, cost sharing, and NYS DEC review have been developed and are being implemented. To date, 78 landowners and a combined area of 21,000 acres have been enrolled in the program. Figure 9-2 shows that the model forest demonstration sites and forest management plans are evenly distributed across the Catskill/Delaware watershed.

Education and outreach efforts have included (1) watershed forest tours for members of New York City-based environmental groups, (2) informational meetings with members of local, New York City, and State governments, and (3) active collaboration with NYC DEP and NYS DEC foresters. In addition, the WFP has established four demonstration forests with the assistance of the New York State College of Environmental Science and Forestry, local scientists and foresters, and the USDA Forest Service.

Enhancement of local and regional markets for wood products is under way. Forest land owners and loggers who are harvesting low-value species (e.g., red maple and beech) and low-quality oak logs during thinning and stand-improvement operations are being connected with pallet lumber and fuelwood buyers in the New York City metropolitan area. For high-value species and logs (oak, white pine, black cherry, sugar maple, yellow birch), connections are being made with local and regional mills and manufacturing facilities. A recent study of the community and economic benefits of the Massachusetts MDC Forestry program on the 55,000-acre Quabbin Reservoir watershed indicates what is possible for the WFP (NWF, 1998). At least 110 jobs in local communities, with a total annual payroll of $4 million, are directly related to the wood harvested from the Quabbin forest. In 1995, for example, the MDC forest management added an estimated total of $34 million to the state's economy and supported 290 jobs. Between 1992 and 1996, the Quabbin forest produced 20 million board-ft of sawtimber, 16,000 cords of fuelwood, and 22,500 tons of pulpwood from an area of approximately 40,000 acres, with a total estimated economic benefit of $120 million.

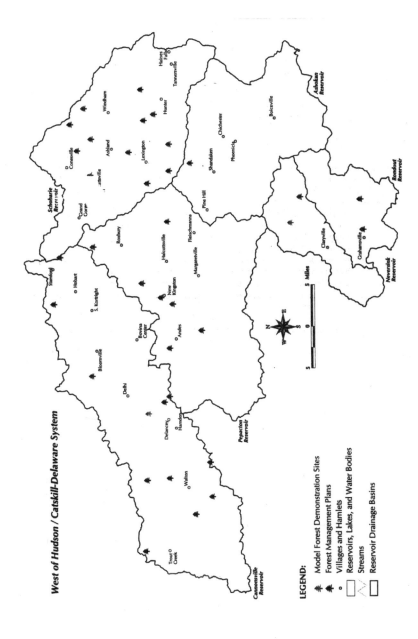

FIGURE 9-2 Watershed Forestry Program activities as of December 1998. Source: NYC DEP (1999b).

Program Analysis

Absolute measures of program performance are not available because of the young age of the WFP. It is anticipated that monitoring criteria and tools for forestry BMPs will be the most significant future need. As such, principles for developing and monitoring forestry BMPs are discussed below. In addition, important economic and ecological considerations for the sustainability of forestry in the Catskills are discussed.

Forestry BMPs

The basic principles guiding the design and implementation of forestry BMPs center on key hydrologic processes and ecosystem attributes, with the primary aim of preventing negative impacts. These principles include the following:

- Full integration of soil and water conservation into planning and operations.
- Protection of the litter layer and soil surface.
- Avoidance or dissipation of overland flow.
- Avoidance of hydrologic connections to wetlands, streams, and lakes.
- Adjustment of forest operations to weather and terrain.
- Combining of biological and physical controls.

These principles and practices proactively avoid or treat causes rather than reacting only to symptoms or unacceptable conditions. The reasons for many practices are self-evident (e.g., avoiding soil compaction, erosion, and sediment transport), yet only rarely are they assembled and implemented as a system.

A system of BMPs should encompass the full operational cycle of forest management activities from planning, developing access, and harvesting to renewal and restoration (Table 9-6). In recognition of their potential impact, most BMPs focus on roads, road-stream crossings, and skidding operations to minimize adverse long-term impacts and on riparian buffers and forest regeneration to ameliorate short-term changes.

Although a complete list of forestry BMPs being used in the Catskill/ Delaware watershed is not yet available (the WAC is developing such a manual to be released in 1999), guidelines for forestry BMPs have been published by NYS DEC. Its catalogue of silvicultural management practices (NYS DEC, 1993) includes planned harvest operations, riparian buffer protection, planned watercourse crossings, planned access routes, sediment barriers, vegetation establishment, and hazardous materials management. These guidelines, developed by a multiorganization task force, correspond closely with the system outlined in Table 9-6. The addition of on-the-ground performance evaluations (monitoring and research) would be appropriate and necessary in public water supply watersheds.

TABLE 9-6 System of Forestry Best Management Practices

Nonstructural Measures	Planning, site reconnaissance, scheduling, walking the road centerline and harvest units, contracts (performance standards and bonds), field supervision
Road Design, Construction, and Maintenance	Placement and alignment, minimizing grades and width, stabilization and restoration, scheduled inspection and maintenance
Stream and Wetland Crossings	Hydraulic structures, approach, and stabilization measures designed to protect road, stream channel, wetlands, and aquatic resources
Landings	Minimal size, distributed drainage, woody debris returned to adjacent forest, controlled equipment maintenance and fueling
Skid Trails	Equipment specifications consider ground pressure and total weight, skidder bridges or brush mats to cross fragile or wet areas, bumper trees to protect residual stand
Felling Operations	Marking (to guide or limit equipment access), operator training and supervision, directional felling and pre-bunching
Riparian Buffers	Zone 1 – shade, cover, food web functions, and bank stability Zone 2 – nonpoint source pollutant assimilation Zone 3 – stormflow control (as needed)
Site Stabilization and Restoration	Restore secondary roads, limit access in cooperation with landowners and local communities
Monitoring and Research	To verify the effectiveness of BMPs, improve operational procedures, develop better techniques, reduce scientific uncertainty

Source: Reprinted, with permission, from Barten, 1998.

Nonindustrial Private Forest Owners

People buy forest land for many reasons: privacy, recreational opportunities, a convenient and reliable source of fuelwood, periodic revenue from timber harvesting, direct use of lumber (after processing at a local sawmill or with a portable mill) for construction or woodworking, or some combination of the above (Leak et al., 1997; NRC, 1998; Rickenbach et al., 1998). Turnover or subdivision of smaller parcels (<50 acres), often occurring every 5–15 years, can be the bane of long-term forest and watershed management programs. Because trees grow slowly, this is usually not enough time for the multiple benefits of forest stewardship to accrue. The WFP and similar initiatives around the United States are designed to counteract frequent turnover and substandard management to ensure source water protection.

Forest Land Ownership and Taxation

A major impediment to the sustainability of forestry in the Catskill/Delaware region is the imbalance between taxes and expected forestry revenues. Throughout the United States, the most frequently cited reason for sale or subdivision is property tax burden (NRC, 1998). In the Catskill region, average annual property taxes on private forest lands are about $14 per acre while average annual revenue from forest product sales (fuelwood and sawtimber) is about $6 per acre (WFAHTF, 1996). Short-term financial pressures obviously discourage long-term investments in silvicultural techniques to improve the value and condition of the forest, such as stand improvement, low-intensity thinning, or uneven-aged management (Davis and Johnson, 1987; Smith et al., 1997).

The USDA Forest Service Stewardship Incentive Program, local conservation easements and current-use tax assessments, and education and outreach are expressly designed to counteract financial pressures and land turnover. They are having a positive effect in most eastern states (Campbell and Kittredge, 1996; Jannik, 1998; NRC, 1998). Conservation easements on forest land in the Catskill/Delaware watershed are likely to become more common because the area has been declared part of the Forest Legacy Program of the USDA Forest Service (NYC DEP, 1999b). In the absence of these types of programs, the gap between forestry expenses and revenues widens. This compels some landowners to cut only high-value trees such as black cherry, northern red oak, yellow birch, and sugar maple simply to maintain ownership of the larger parcel. In the worst case, when high-value trees are continually removed ("high-grading"), biological diversity, tree health, and vigor are steadily reduced. Resistance and resilience to insects, diseases, and natural disturbances such as severe storms, as well as the ameliorative effect of forests in headwater areas, are gradually diminished.

Forest Health

In addition to economic pressures and land tenure issues, the ecological legacy of exploitative logging during the 1800s (clearcutting hemlock for tanbark and subsequent hardwood regeneration for charcoal and fuelwood) described in an earlier chapter is a degraded, predominantly even-aged forest. Exotic insects like gypsy moth can seriously damage oak stands. A woolly adelgid, accidentally introduced from Japan, threatens the few remnant stands of eastern hemlock. Dutch elm disease and chestnut blight have virtually eliminated their host tree species from northeastern forests. More recently, atmospheric deposition has caused widespread mortality of high-elevation red spruce and balsam fir (Lovett et al., 1998; Lovett and Lindberg, 1993; Lovett and Kinsman, 1990). In sum, there are substantial economic and ecological challenges that highlight the need for forest conservation and stewardship in the Catskills region.

Conclusions and Recommendations for the Watershed Forestry Program

1. The Watershed Forestry Program should continue its comprehensive efforts to promote sustainable forest management in the region. Outreach, training, monitoring, and research efforts are self-reinforcing when they are advanced as a balanced program. As with the WAP, success in the WFP will depend largely on the active implementation and maintenance of forestry BMPs over the long term.

2. A spatially referenced database of forest lands, owners, management objectives, and activities should be developed to track on-the-ground progress and performance of forestry practices. The Geographic Information System (GIS) should be used to spatially link (1) databases with landowner information, forest resources inventory (at the stand level), management objectives, and silvicultural prescriptions and treatments, (2) watershed characteristics (e.g., topography, soils), and (3) current and future subwatershed conditions. Its use would enable watershed managers to evaluate changes (positive, negative, or non-detectable) in the quantity, quality, and timing of streamflow as influenced by the WFP and other land or resource uses.

3. Water quality monitoring should be implemented on the model forests and in conjunction with NYC DEP and NYS DEC networks on other tributaries. The WFP is encouraged to merge landowner, management plan, and field assessment data with water quality data to evaluate program performance. It is likely that reference watersheds could be used for WFP *and* WAP monitoring. The GIS/forest management database suggested above (recommendation 2) would provide data and information about WFP activities upstream of NYC DEP and NYS DEP monitoring stations. When coupled with a paired watershed and/or "above and below" sampling, valuable data and information about the cumulative impact (positive or negative) of the WFP and WAP would be available.

4. The Watershed Forestry Program should foster the movement toward third-party green certification of forest management (Forsyth et al., 1999; Hayward and Vertinsky, 1999; Mater et al., 1999). The Vermont Family Forests landowner cooperative is an excellent prototype for the Catskill/Delaware region. Resource manager and public land certification in New York, Pennsylvania, Massachusetts, and other New England states can serve as a model for consulting foresters.

5. It is recommended that New York State consider tax policy changes that would promote sustained management of private forest land including forests on relatively small parcels that are suitable for development. Because of the benefits of such programs extend beyond the local political boundaries, impacts on town revenues should be evaluated, and mitigation payments to local governments by the State should be made as an integral part of the program.

STORMWATER POLLUTION PREVENTION PLANS

Urban stormwater is the final type of NPS pollution considered in this chapter. Most types of new, large-scale development in the New York City watershed region are required to submit a Stormwater Pollution Prevention Plan (SPPP) for controlling the quantity and quality of stormwater runoff generated by new impervious cover (MOA, Appendix X, Section 18-39). SPPPs specify best management practices that will prevent erosion and sedimentation during construction and any increase in the rate of pollutant loading in stormwater after construction. These plans must include a quantitative analysis demonstrating that runoff quantity and quality from postconstruction conditions will be less than or equal to that of preconstruction conditions. Whether or not an SPPP or some other type of stormwater plan is developed depends on a number of factors, including the proximity of the project to nearby waterbodies. For detailed information on the multiple types of stormwater plans and the activities that require them, see NYC DEP, 1997b.

Although they have existed since 1993 as part of the NYS DEC's General Permit for stormwater discharges, SPPPs have recently received considerable attention because of their inclusion in the MOA for a variety of activities. Prior to the MOA, fewer activities required the drafting of an SPPP, and the regulatory oversight was spread among multiple agencies. It is not surprising, then, that SPPPs have spawned a great deal of confusion among engineers, developers, and local and State agencies about how SPPPs should be interpreted and implemented, since most of these organizations had no prior experience with stormwater quality control and/or stormwater BMP design.

Stormwater Pollution Prevention Plan Contents

An SPPP must include a description of the proposed construction activities. An estimate of pre- and postdevelopment runoff is required, considering both the quantity and quality of stormwater. Pollutants of concern vary, but often include biological oxygen demand, phosphorus, nitrogen, total suspended solids, organic matter, and bacteria. Measures that might be undertaken to reduce runoff rates and pollutant loading from stormwater are then presented. These measures are committed to a Stormwater Management Plan, which describes the specific BMPs that will be used to ensure that the postdevelopment runoff rates will not exceed predevelopment runoff rates for the 2-year, 10-year, and 100-year 24-hour storms. To prevent pollutant loadings, the Stormwater Management Plan must control the "first flush"—the first half inch of runoff from the 1-year, 24-hour storm event. However, there are no numeric standards requiring a certain amount of pollution to be removed by stormwater BMPs. Erosion and Sediment Control Plans are also part of an SPPP. These contain a complete description of the BMPs that will be used to control erosion during each phase of construction. The methods, criteria, and documentation for preparing an SPPP are contained in a series of guidance and permit documents (NYC DEP, 1997b; NYS DEC, 1996).

Performance of Stormwater BMPs

Throughout the SPPP guidance document, it is clear that the goal is to prevent postdevelopment loadings of pollutants from exceeding predevelopment levels (NYC DEP, 1997b). The requirement is stated by the following: "Regulations require that pre- and postconstruction runoff characteristics not be substantially altered." In order for this to be achieved, the SPPPs rely on an underlying premise that current engineering technologies (i.e., stormwater BMPs) are capable of reducing postdevelopment pollutant loadings to predevelopment levels. Unfortunately, there is little basis for confidence that the current generation of urban stormwater BMPs can reduce pollutant loads to levels that approach predevelopment conditions. Table 9-7 provides a summary of reported nutrient removal rates for stormwater BMPs.

Phosphorus Removal

Although their removal rates are variable, most BMP groups have median annual removal rates in the 30 percent to 50 percent range for both soluble and total phosphorus (Table 9-7). Dry extended detention ponds and open channels

TABLE 9-7 Median Removal Rates for Selected Groups of Stormwater Practices

BMP Groups	n	TSS	Total P	Sol P	Total N	Nitrate	Carbon
			Median Removal Rate, %				
Wet Ponds	36	67	48	52	31	24	41
Stormwater Wetlands	35	78	51	39	21	67	28
Sand Filters[a]	11	87	51	-31	44	(-13)	66
Channels	9	0	(-14)	(-15)	0	2	18
Water Quality Swales[b]	9	81	29	34	ND	38	67

Notes: n is a number of performance monitoring studies. The actual number for a given parameter is likely to be slightly less. Sol P is soluble phosphorus, measured as orthophosphate, soluble reactive phosphorus, or biologically available phosphorus. Carbon is a measure of organic carbon (BOD, COD, or TOC). ND = not determined.

[a]Excluding vertical sand filters and vegetated filter strips, but including organic filters.

[b]Includes biofilters, wet swales and dry swales.

Source: Brown and Schueler (1997). Reprinted, with permission, from Center for Watershed Protection, 1997. ©1997 by Center for Watershed Protection.

showed low or no ability to remove either total or dissolved phosphorus. Interestingly, several BMP groups—wetlands, water quality swales, and sand filters— exhibit very wide variation in phosphorus removal, suggesting internal nutrient cycling can be an important factor in determining BMP effectiveness. Some BMPs, such as sand filters, actually increase soluble phosphorus concentrations via desorption, dissolution, or extraction of phosphorus into the aqueous phase.

These removal rates are average annual load reductions, and the removal rates do not account for diminished removal related to poor design or construction, age, or lack of maintenance. It is also important to remember that trapping of phosphorus within a stormwater BMP is only a temporary form of removal; ultimate removal is dependent on the cleanout, removal, and safe disposal of trapped sediments through periodic maintenance. For stormwater wetlands, continued phosphorus removal may require periodic replacement of wetland media as adsorption sites diminish over time (Oberts, 1997).

The moderate phosphorus removal of stormwater BMPs needs to be balanced against the sharp rise in phosphorus loads produced by new development. The effect of stormwater BMPs on phosphorus load as a function of impervious cover is shown in Figure 9-3. At a low density of development (5 percent to 25 percent site impervious cover), the reduction in phosphorus load by stormwater BMPs keeps pace with the increased load produced by impervious cover. After that point, however, stormwater BMPs can no longer achieve predevelopment phosphorus loads.

Bacterial Removal

To date, studies evaluating the performance of stormwater BMPs in removing microbial pathogens have focused on bacteria. Urban stormwater BMPs must be extremely efficient if they are to produce stormwater effluents that meet the 200-CFU/100 mL standard for fecal coliforms at a site. Assuming a national mean storm inflow fecal coliform concentration of 15,000/100 mL (see Table 5-6), a 99 percent removal rate is needed to meet the standard. The limited research conducted to date indicates that current BMPs cannot meet this standard on a reliable basis. Only 24 BMP performance-monitoring studies have measured the input and output of fecal coliform bacteria from stormwater BMPs during storm events. These data, collected for fecal coliform, fecal streptococcus, and *E. coli,* are summarized in Table 9-8.

For stormwater ponds, the mean fecal coliform removal efficiency was about 65 percent (range was from –5 percent to 99 percent). The mean removal efficiency calculated for sand filters was lower (about 50 percent), but these practices had a wider range in reported removal (–68 percent to 97 percent). It should be noted that most sand filter performance data have been collected in warm seasons, and most sites were in Texas—conditions unlike those in the Catskill/ Delaware watershed. Grass swales and biofilters were found to have no ability to

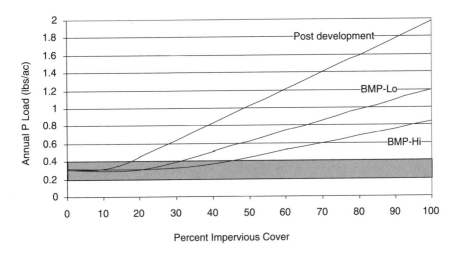

FIGURE 9-3 Relationships between impervious cover, phosphorus loads, and storm-water BMP performance in a typical watershed. The grey band indicates typical "background" phosphorus loads from undeveloped watersheds. The BMP-Hi line illustrates the effect of reducing phosphorus loads using BMPs with an average long-term removal rate of 60 percent. The BMP-Lo indicates a 40 percent removal rate. It should be noted that actual curves in individual watersheds may be different. For example, the contribution of new septic systems that accompany development to overall phosphorus loading is not represented by these curves. Source: Schueler (1996). Reprinted, with permission, from Center for Watershed Protection, 1996. © 1996 by Center for Watershed Protection.

TABLE 9-8 Comparison of Mean Bacterial Removal Rates Achieved by Different Stormwater BMP Groups

Bacterial Indicator	Bacterial Removal Rate, %		
	Ponds	Sand Filters	Swales
Fecal Coliform	65% (n =9)	51% (n=9)	−58% (n=5)
Fecal Streptococci	73% (n =4)	58% (n=7)	ND
E. coli	51% (n=2)	ND	ND

ND=not determined
Source: Schueler (1999). Reprinted, with permission, from Center for Watershed Protection, 1999.
© 1999 by Center for Watershed Protection.

remove fecal coliform bacteria, with zero or negative removal reported in four of the five studies. Pet wastes and *in situ* multiplication of bacteria were cited as the primary reason for the poor performance. No performance monitoring data are available to assess the capability of infiltration or wetland BMPs to remove coliforms. The limited data on fecal streptococcus and *E. coli* removal by stormwater BMPs are generally comparable to the more abundant fecal coliform data, suggesting that fecal coliform is a sufficient indicator of these organisms. There are no current monitoring data on *Giardia, Cryptosporidium,* or *Salmonella* removal by stormwater BMPs. Based on the mediocre effectiveness of BMP removal mechanisms for fecal coliforms, and the survivability of cysts and oocysts in sediments, the committee expects that it will be difficult to reliably remove the protozoan pathogens from urban runoff using traditional stormwater BMPs.

In summary, current BMP technology is not capable of removing fecal coliform bacteria to meet the 200-CFU/100 mL standard in stormwater discharges, assuming a national average bacterial concentration in stormwater influent. If no net increase in bacterial concentrations in postdevelopment runoff is desired, it may be necessary to install stormwater BMPs at both current development projects as well as at older, neighboring development sites. As written, the SPPPs call for a level of BMP performance that simply cannot be met with current stormwater techniques at most highly developed sites. As discussed in detail in Chapter 8, the use of multiple BMPs in series at individual sites cannot reduce postdevelopment loadings below predevelopment levels.

Acreage Requirements for Stormwater Quality Controls

With some exceptions, the Watershed Rules and Regulations exempt development projects of less than five acres in size from the stormwater pollution prevention plan requirements[4]. Many kinds of small-scale industrial and commercial development are thus allowed to produce phosphorus and bacteria loads without treatment. Most states and localities that currently regulate stormwater discharges have a much lower threshold for stormwater requirements (usually less than one acre) (Watershed Management Institute, 1997). Although it is true that even very low-acreage thresholds (30,000 ft[2]) have been found to allow as much as 25 percent of stormwater to pass through untreated (Booth and Jackson, 1997), the efficacy of stormwater management will be markedly improved by lowering the current threshold from five acres to one acre. A one-acre threshold

---

[4]SPPPs are also required, regardless of acreage, for construction of a subdivision; construction of an industrial, commercial, multifamily, or municipal project where more than 40,000 sq. ft. of impervious surface will be created; a landclearing or grading project involving two or more acres that are partially located on slopes greater than 15 percent or within setback distances from waterbodies; construction of an impervious surface in a village, hamlet, village extension, or area zoned for commercial or industrial uses West-of-Hudson; or construction of an impervious surface within a East-of-Hudson designated Main Street area.

should be the basis for further refinement, using the GIS database to identify the most appropriate long-term threshold given land development patterns in the watershed.

Sizing Criteria for Designing Stormwater BMPs

The SPPPs contain three different sizing criteria that must be considered when designing stormwater treatment facilities. The chosen BMPs must treat the greater of (1) the first half inch of runoff from impervious areas of the site or (2) the runoff produced by the one-year, 24-hour storm event (approximately 2.5–3.0 inches of rainfall). In phosphorus- and coliform-restricted basins, the BMPs must treat the runoff produced by the two-year, 24-hour storm event (approximately 3–4 inches of rainfall). The latter two sizing criteria are among the largest sizing criteria for stormwater runoff in the United States. The regulations, however, provide no guidance for designing BMPs that can fulfill these requirements, either in terms of the hydrologic models that should be employed or standardized parameters. Consequently, design engineers and state and local regulatory agencies are in frequent conflict as to how SPPPs should be interpreted and applied. The derivation of more effective sizing criteria should become a high priority for NYC DEP.

It should be noted that the larger stormwater treatment volumes do not necessarily lead to proportionately greater levels of pollutant removal. For example, a BMP designed to capture runoff from the one-year storm is able to treat 90 percent of the annual stormwater runoff volume each year (MDE, 1999). A BMP designed to capture runoff from the two-year storm is able to treat only 95 percent to 97 percent of the annual runoff volume produced each year, even though it is four times larger in size (and cost). BMP research has shown that treatment volume alone is not a reliable predictor of pollutant removal performance. Other design variables, such as internal geometry, pretreatment, conveyance, and multiple treatment pathways, are very important in determining pollutant removal. Yet the SPPP requirements offer minimal guidance on these other important design parameters. The lack of a stormwater design and engineering manual and of performance criteria for individual BMPs for the New York City watersheds is a major impediment to achieving higher and more consistent pollutant removal. Other states such as Maryland have recently produced detailed and useful manuals to assist engineers in designing and building more effective BMPs (MDE, 1999).

Need for Program Support

The Watershed Rules and Regulations of the MOA have introduced stormwater control technologies into a region of the country that had little or no prior experience with stormwater management. The regulations emphasize a permit-driven approach rather than a performance-based approach. That is, an SPPP relies strongly on quantitative (and highly theoretical) calculations, rather than on performance monitoring or strict requirements for BMP size and treatment efficiency. The SPPP program does not currently have the basic support needed to foster success in the areas of training, engineering manuals, performance monitoring, review staffing, program financing and demonstration projects, maintenance requirements, design methods, BMP feasibility guidance, construction and maintenance inspection criteria, or BMP specifications. As the Watershed Management Institute (1997) notes, strong program support in these areas has been the critical ingredient in effective implementation of stormwater requirements in other localities. It will be critical to the success of stormwater management in the New York City watersheds as well.

Incentives to Reduce Impervious Cover

The SPPP approach relies heavily on the use of structural stormwater practices such as ponds, wetlands, filters, and infiltration. Although these practices are an essential component of an effective stormwater quality strategy, they need to be combined with site design practices that reduce the amount or impact of impervious cover created by land development. Better site design techniques (narrower streets, open-space subdivisions, smaller parking lots, and on-lot bioretention) are being advocated by many stormwater agencies (Arendt, 1997; BASMAA, 1997; CWP, 1998; MDE, 1999). Recent modeling work has indicated that widespread application of better site design techniques can provide stormwater pollutant reduction equivalent to that achieved by structural stormwater practices (Caraco et al., 1998). When better site design techniques and structural practices are combined, nutrient loadings are projected to decline to levels 30 percent to 50 percent lower than what can be achieved using conventional subdivision designs.

The Watershed Rules and Regulations and the SPPP requirements provide no incentives for developments that employ better site design techniques. Recently, the state of Maryland provided a series of stormwater quality credits for developments that use better site design (MDE, 1999). These credits could be adapted for developments in the New York City watersheds.

Conclusions and Recommendations

1. For phosphorus, current stormwater best management practices are only moderately effective. In almost all cases, they cannot reduce post-development loading to predevelopment levels.

Most current practices show some ability to remove bacteria but not enough to meet current water quality standards. Swales are capable of no net removal. Because urban areas are a source of *Cryptosporidium* oocysts (see Chapter 5), this deficiency is particularly notable.

2. Stormwater Pollution Prevention Plans should be required for activities greater than one acre, rather than for those greater than five acres. The five-acre measure was likely derived from the fact that activities that affect less than five acres are generally not required to obtain an NPDES permit for stormwater. However, most communities have recognized that one acre is a more appropriate lower limit. Lowering the size requirement is important because much of the new development in the Catskill/Delaware watershed may be on a small scale.

3. NYC DEP should develop guidance material for designing stormwater BMPs that can meet the one-year, 24-hour storm event and the two-year, 24-hour storm event.

4. NYC DEP should embrace a performance-based approach to stormwater management rather than the permit-based approach embodied by the current SPPPs. Among other things, guidance material for such a new approach should include information on performance monitoring of stormwater BMPs for a variety of pollutants, including *Cryptosporidium,* and on long-term maintenance of stormwater BMPs.

5. The Stormwater Pollution Prevention Plans should encourage the use of nonstructural BMPs that limit the amount and the adverse effects of impervious surfaces. Excellent examples of good site design practices using such BMPs are found in Maryland.

REFERENCES

Arendt, R. 1997. Designing Open Space Subdivisions. Media, PA: National Lands Trust and American Planning Association.

Auer, M. T., K. A. Tomasoski, M. J. Babiera, M. L. Needham, S. W. Effler, E. M. Owens, and J. M. Hansen. 1998. Phosphorus bioavailability and P-cycling in Cannonsville Reservoir. Journal of Lake and Reservoir Management 14(2-3):278–289.

Barten, P. K. 1998. Conservation of soil, water, and aquatic resources of the NorSask Forest. Prepared for Mistik Management, Ltd., Meadow Lake, Saskatchewan.

Barten, P. K., T. Kyker-Snowman, P. J. Lyons, T. Mahlstedt, R. O'Connor, and B. A. Spencer. 1998. Managing a watershed protection forest. Journal of Forestry 96(8):10–15.

Bay Area Stormwater Management Agencies Association (BASMAA). 1997. Start at the Source: Residential Site Planning and Design Guidance Manual for Stormwater Protection. Oakland, CA: BASMAA.

Bentley, W., and W. Langbein. 1996. Seventh American Forest Congress: Final Report. Yale University, School of Forestry and Environmental Studies, New Haven, CT.

Booth, D. B., and C. R. Jackson. 1997. Urbanization of aquatic systems—degradation thresholds, stormwater detention, and limits of mitigation. Journal of American Water Resources Association 33(5):1077–1090.

Brown, M. P., M. R. Rafferty, P. B. Robillard, M. F. Walter, D. A. Haith, and L. R. Shuvler. 1984. Nonpoint Source Control of Phosphorus: A Watershed Evaluation. Albany, NY: NYS DEC Bureau of Water Research.

Brown, W., and T. Schueler. 1997. Pollutant Removal of Urban Stormwater Best Management Practices—A National Database. Ellicott City, MD: Center for Watershed Management, Chesapeake Research Consortium.

Campbell, S. M., and D. B. Kittredge. 1996. Application of an ecosystem-based approach to management on multiple NIPF ownerships: A pilot project. Journal of Forestry 94(2):24–29.

Caraco, D., J. Zieluski, and R. Claytor. 1998. Nutrient Loading from Conventional and Innovative Development Sites. Ellicott City, MD: Chesapeake Research Consortium, Center for Watershed Protection.

Center for Watershed Protection (CWP). 1998. Better Site Design—A Manual for Changing Development Rules in a Community. Site Planning Roundtable. Ellicott City, MD: Center for Watershed Protection.

Conservation Technology Information Center (CTIC). 1998. Review of the Watershed Agricultural Program. West Lafayette, IN: CTIC.

Dana, S. T., and S. K. Fairfax. 1980. Forest and Range Policy. Second Edition. New York, NY: McGraw Hill.

Davis, L S., and K. N. Johnson. 1987. Forest Management. Second Edition. New York, NY: McGraw Hill.

Dillaha, T. A., J. H. Sherrard, and D. Lee. 1989. Long-term effectiveness of vegetative filter strips. Water Environment and Society 1:419–421.

Ellefson, P. V. 1992. Forest Resources Policy: Process, Participants, and Programs. New York, NY: McGraw Hill.

Forsyth, K., D. Haley, and R. Kozak. 1999. Will consumers pay more for certified wood products? Journal of Forestry 97(2):18–22.

Haith, D. A., L. L. Shoemaker, R. L. Doneker, and L. D. Delwiche. 1984. Modeling of streamflow, sediment yield, and nutrient export from the West Branch Delaware River Watershed. Volume 4 in Brown, M., et al. (eds.) Nonpoint Source Control of Phosphorus: A Watershed Evaluation. Albany, NY: NYS DEC Bureau of Water Research.

Hayward, J., and I. Vertinsky. 1999. High expectations, unexpected benefits: What managers and owners think of certification. Journal of Forestry 97(2):13–17.

Jannik, P. 1998. Keynote address for Intl. Conference on the Science of Managing Forests to Sustain Water Resources. 9–11 November 1998, Sturbridge, MA.

Kittredge, D. B., and C. Woodall. 1997. Massachusetts loggers rate portable skidder bridges. The Northern Logger 46(4):26–27, 36.

Kittredge, D. B., C. Woodall, and A. M. Kittredge. 1997. Skidder bridge fact sheet. Amherst, MA: University of Massachusetts Extension.

Leak, W. B., M. Yamasaki, D. B. Kittredge, N. I. Lamson, and M. L. Smith. 1997. Applied ecosystem management on nonindustrial forestland. USDA Forest Service GTR NE-239. Radnor, PA: USDA Forest Service.

Longabucco, P., M. Rafferty, and J. L. Ojpersberger. 1999. Preliminary Analysis of the First Year of Sampling Data Following BMP Implementation at the Robertson Farm. Albany, NY: NYS DEC Bureau of Watershed Management.

Lovett, G. M., and J. D. Kinsman. 1990. Atmospheric pollutant deposition to high-elevation ecosystems. Atmospheric Environment 24A:239–264.

Lovett, G. M., and S. E. Lindberg. 1993. Atmospheric deposition and canopy interactions of nitrogen in forests. Canadian Journal of Forest Research 23:1603–1616.

Lovett, G. M., A. W. Thompson, J. B. Anderson, and J. J. Bowser. 1998. Elevational patterns of sulfur deposition at a site in the Catskill Mountains, NY. Atmospheric Environment (in press).

Lovett, G. M., Weathers, K. C. and W. V. Sobczak. 1999. Nitrogen saturation and retention in forested watersheds of the Catskill Mountains, NY. Ecological Applications (in press).

Maryland Department of Environment (MDE). 1999. Stormwater Design Manual. Volume I. Baltimore, MD: Maryland Department of Environment.

Mater, C. M., V. A. Sample, J. R. Grace, and G. A. Rose. 1999. Third party, performance-based certification: What public forestland managers should know. Journal of Forestry 97(2):6–12.

MDC (Metropolitan District Commission). 1995. Quabbin watershed: MDC land management plan 1995-2004. Boston, MA: Division of Watershed Management.

National Research Council (NRC). 1990. Forestry Research: A Mandate for Change. Washington, DC: National Academy Press.

NRC. 1998. Forested Landscapes in Perspective: Prospects and Opportunities for Sustainable Management for America's Non-Federal Forests. Washington, DC: National Academy Press.

National Wildlife Federation (NWF). 1998. Quality not quantity: Community and economic benefits from MDC forestry at the Massachusetts Quabbin Reservoir watershed. Montpelier, VT: Northeast Natural Resource Center.

New York City Department of Environmental Protection (NYC DEP). 1997a. Watershed Agricultural Program Preliminary Evaluation. December 1997. Valhalla, NY: NYC DEP.

NYC DEP. 1997b. Applicants Guide to Stormwater Pollution Prevention Plans and Crossing, Piping, or Diversion Permits. May 1997. Valhalla, NY: NYC DEP.

NYC DEP. 1998a. Quarterly Report on the Status of Implementing Projects Designed to Reduce Nonpoint Source Pollution. January 1, 1998–March 1, 1998. Valhalla, NY: NYC DEP.

NYC DEP. 1998b. DEP Pathogen Studies of *Giardia* spp., *Cryptosporidium* spp., and Enteric Viruses. Semi-Annual Report. Valhalla, NY: NYC DEP.

NYC DEP. 1999a. Watershed Agricultural Program Status Report. April 30, 1999. Valhalla, NY: NYC DEP.

NYC DEP. 1999b. Filtration Avoidance Annual Report for the period January 1 through December 31, 1998. Corona, NY: NYC DEP.

New York State Department of Environmental Conservation (NYS DEC). 1993. Silviculture Management Practices Catalogue for Nonpoint Source Pollution Prevention and Water Quality Protection in New York State. Albany, NY: NYS DEC.

NYS DEC. 1996. Stormwater Guidance Documents.

New York State Water Resources Institute (NYS WRI). 1997. Science for Whole Farm Planning: Cornell University Phase II Twelfth Quarter and Completion Report. Ithaca, NY: NYS WRI.

Northern Forest Lands Council. 1994. Finding Common Ground. Conserving the Northern Forest. Concord, NH: Northern Forest Lands Council.

Oberts, H. 1997. Lake McCavrons Wetland Treatment System—Phase III Study Report. Metropolitan Council Environmental Services. St. Paul, MN: Metropolitan Council Environmental Services.

Rickenbach, M. G., D. B. Kittredge, D. Dennis, and T. Stevens. 1998. Ecosystem management: Capturing the concept of woodland owners. Journal of Forestry 96(4):18–24.

Robillard, P. D., and M. F. Walter. 1984a. Phosphorus losses from dairy barnyard areas. Volume 1 in Brown, M. et al. (eds.) Nonpoint Source Control of Phosphorus: A Watershed Evaluation. Albany, N.Y.: NYS DEC Bureau of Water Research.

Robillard, P. D., and M. F. Walter, 1984b. Development of manure spreading schedules to decrease phosphorus loading to streams. Volume 2 in Brown, M. et al. (eds.) Nonpoint Source Control of Phosphorus: A Watershed Evaluation. Albany, N.Y.: NYS DEC Bureau of Water Research.

Schueler, T. 1995. The importance of imperviousness. Watershed Protection Techniques 1(3):100–112.

Schueler, T. 1999. Microbes in urban watersheds—Concentrations, sources, and pathways. Watershed Protection Technique 3(1):554–565.

Smith, D. M., Larson, B. C., M. J. Kelty, and P. M. Ashton. 1997. The Practice of Silviculture. 9th Edition. New York, NY: John Wiley & Sons.

Watershed Agricultural Program (WAP). 1997. Pollution Prevention through Effective Agricultural Management. Walton, NY: Watershed Agricultural Program.

Walker, F. R., and J. R. Stedinger. 1999. Fate and transport model of *Cryptosporidium.* Journal of Environmental Engineering April 1999:325–333.

Watershed Forest Ad Hoc Task Force (WFAHTF). 1996. Policy recommendations for the watersheds of the New York City's water supply. Ithaca, NY: NYS Water Resources Institute, Cornell University.

Watershed Management Institute. 1997. Institutional Aspects of Urban Runoff Management—A Guide for Program Development and Implementation. Washington, DC: Office of Water, EPA.

10

Setbacks and Buffer Zones

Regulations governing the use of private land within a specified distance of a watercourse, lake, wetland, or tidal shoreline have been in effect in many states and localities since the early 1960s. Such "setbacks" or "buffer strips" serve diverse purposes, for example, protection of surface waters from pollution, protection of structures from flooding or erosion, and preservation of riparian habitat and shoreline amenities. One of the most prevalent features of the Memorandum of Agreement (MOA) Watershed Rules and Regulations is the use of setback distances for separating waterbodies from potentially polluting activities. Depending on the activity, 25–1,000 ft of land must separate the activity from nearby waterbodies. Greater distances are required for setbacks around reservoirs, reservoir stems, and controlled lakes than for those around wetlands and watercourses, which encompasses all perennial streams and in some cases intermittent streams.

Although the use of setbacks is quite common in watershed regulations across the country, little research has been done regarding the effectiveness of setbacks per se in preventing contamination of waterbodies from nonpoint source pollution. Rather, research has focused on the use of *buffer zones* for nonpoint source pollutant removal. Buffer zones are natural or managed areas used to protect an ecosystem or critical area from adjacent land uses or sources of pollution. They are an increasingly used best management practice (BMP) for many activities. Effective buffers along rivers, reservoirs, and lakes (riparian buffers) either retain or transform nonpoint source pollutants or produce a more favorable environment for aquatic ecosystem processes.

Setbacks, in contrast to buffer zones, are simply prescribed distances between pollutant sources and a resource or aquatic ecosystem that needs protection. Only if a setback is subject to management or natural preservation can it be considered a "buffer" that reliably insulates ecosystems and resources from nonpoint source pollution. Because of the lack of information regarding unmanaged setbacks, this review focuses on management of buffer zones for achieving pollutant removal. In the absence of management, it is virtually impossible to predict what effect the setback distances in the MOA will have on the water quality of the New York City watershed. However, if the management practices reviewed and recommended in the following sections are used, then the setbacks may approach the pollutant-removal capabilities predicted for buffer zones.

The next section enumerates and explains key functions and characteristics of riparian buffer zones. It should be noted that waterbodies have a substantial effect on the characteristics of the surrounding buffer zones. That is, depending on whether they border wetlands, reservoirs, or streams, buffer zones will function differently. These differences are discussed when appropriate. Another important consideration is that buffer zones may not be permanent pollutant sinks, but rather may act as temporary storage areas that can be both sources and sinks of pollution. This is especially true for sediment and phosphorus, for which no degradation processes exist in the buffer (nitrogen can be removed via denitrification). Factors that enhance the long-term storage potential of riparian buffer zones, such as harvesting of vegetation, are important in evaluating their long-term effectiveness.

STRUCTURE AND FUNCTION OF RIPARIAN BUFFER ZONES

Riparian buffer zones refer to lands directly adjacent to waterbodies such as lakes, reservoirs, rivers, streams, and wetlands. These land areas have a significant impact on controlling nonpoint source pollution and on the associated water quality in nearby waterbodies. As a result, they are widely used in water resource protection programs and are the topic of intense investigation, especially in agriculture and forestry. Unfortunately, as noted in a recent symposium on buffer zones, policy-driven initiatives that have accelerated the debate on buffer zones have, at the same time, stretched scientifically based management to the limits of knowledge on this issue (Haycock et al., 1997). This is the case in the New York City watersheds and most other regions of the country.

Hydrology

Evaluating the effectiveness of riparian buffers to remove diffuse pollution from runoff requires a basic understanding of their hydrologic structure and function. Because of their proximity to waterbodies, riparian buffers are sometimes flooded by stream overflow. Riparian buffer zones are also strongly influenced by water from upslope areas, which is generally divided into three categories: (1) overland or surface flow, (2) shallow subsurface flow, and (3) groundwater flow (Figure 10-1).

Overland flow across buffer zones can occur via two pathways. *Infiltration-excess overland flow* is generated when rainfall intensity or snowmelt rate exceeds the rate at which water moves through the soil surface (the infiltration process). Infiltration-excess overland flow typically occurs when the soil surface is frozen, is compacted, or is otherwise unable to transmit water to the root zone. Extreme rainfall events may deliver water rapidly enough to generate infiltration-excess across a wide range of soil types and watershed locations. As in most predominantly forested areas, this mechanism of overland flow is rare in the Catskill/Delaware region.

Saturation-excess overland flow occurs when soil water storage capacity is exceeded by precipitation volume combined with lateral inflow from upslope areas. When total inflow exceeds total outflow, saturation from below is the obligate result. Once the zone of saturation reaches the soil surface, any new input (rain or snowmelt) is immediately converted to overland flow. As shown in Figure 10-1, saturation-excess overland flow typically occurs at the transitions from the uplands to the riparian zone. Saturation-excess overland flow is usually less damaging to water quality than is infiltration-excess overland flow and, though still uncommon, it is more likely in the Catskill/Delaware region.

Both infiltration-excess and saturation-excess overland flow occur during rain or snowmelt events and constitute the bulk of *stormflow.* Because this water

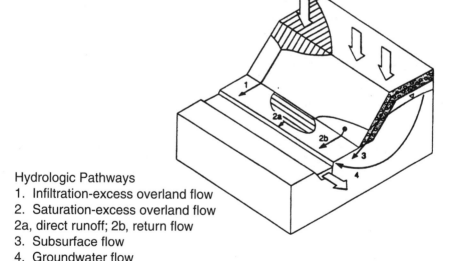

Hydrologic Pathways
1. Infiltration-excess overland flow
2. Saturation-excess overland flow
2a, direct runoff; 2b, return flow
3. Subsurface flow
4. Groundwater flow

FIGURE 10-1 Hydrologic pathways surrounding riparian buffer zones. Source: Burt (1997).

travels overland, it can accumulate high levels of particulate matter such as sediment, bacteria, and particulate-phase phosphorus. Depending on its velocity and on the soil water status of the riparian zone, overland flow can either infiltrate into buffer zones (generally desirable) or flow across buffer zones and discharge directly into neighboring waterbodies.

Shallow subsurface flow travels laterally through the root zone below the land surface (see Figure 10-1). It may be caused by an abrupt decrease in soil permeability or simply because a shallow soil is underlain by slowly permeable or impermeable bedrock. The latter is the most common case in the Catskill/ Delaware region. The response of subsurface flow to storm events is more attenuated than that of overland flow, although increases in subsurface flow do generally occur during and for a period of time following precipitation or snow-melt. Filtering and biogeochemical transformations in the soil limit shallow subsurface transport of suspended particulate matter. By contrast, the concentration of dissolved solids may increase in proportion to residence (travel) time (Burt, 1997).

Groundwater flow occurs when vertical flow extends beyond the root zone into lower strata. This may occur in deeper unconsolidated material (e.g., glacial or lacustrine sands and gravels) and/or through bedrock fractures. Travel time increases in proportion to the length of the flow path and in relation to hydraulic limitations imposed by the media. Hence, groundwater can have high dissolved solids but transports little, if any, suspended solids. Shallow subsurface flow and groundwater flow combine to generate *baseflow,* the water entering streams, wetlands, lakes, and reservoirs during dry periods. Shallow subsurface flow is much more likely to interact with riparian buffer zones than groundwater flow because it passes laterally through the root zone. In some instances, shallow subsurface flow in upstream areas can become saturation-excess overland flow by the time a buffer zone is reached (exfiltration or seepage).

Pollutant Removal and Other Functions

The structure and function of riparian buffer zones are determined by (1) the soil, vegetation, and hydrologic characteristics of the buffer and (2) the interactions with upslope and downslope water. For management purposes and for conceptualization of the various functions, the U.S. Department of Agriculture (USDA) guidelines suggest riparian buffers can be divided into three zones, each of which has certain physical characteristics and pollutant-removing abilities (Figure 10-2) (NRCS, 1995; Welsch, 1991). Zone 1 is the area immediately adjacent to the waterbody; Zone 2 is an intermediate zone upslope from Zone 1 where most active woody BMPs are used; and Zone 3 is the vegetated areas upslope from Zone 2. Although this conceptualization has not been universally adopted, it is particularly useful in this report for describing how riparian buffer zone functioning varies with distance from nearby waters.

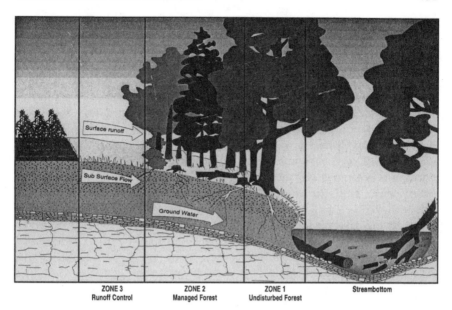

FIGURE 10-2 USDA's three-zone schematic of a riparian buffer zone. Source: EPA (1995).

Zone 3

The pollutant-removal abilities of riparian buffers are maximized when overland flow infiltrates into buffer zones rather than discharging directly into adjacent waterbodies. For this reason, the most important function of Zone 3 is to alter the hydraulic properties of rainfall runoff such that the overland flow interacting with Zones 2 and 1 is *sheet flow* rather than *channelized flow*. Depending on the characteristics of nearby land, overland flow reaching Zone 3 may be predominantly channelized flow. By design, when channelized flow reaches Zone 3, it is usually converted to sheet flow and subsurface flow by the hydraulically rough surface and the enhanced infiltration of the buffer zone.

Coarse sediment removal from stormwater is also predominantly accomplished in Zone 3, although it can also occur in Zones 2 and 1. When the hydraulic properties of stormwater change from channelized flow to sheet flow, infiltration of the water is enhanced. Sediment and other materials entrained in overland flow (such as particulate phosphorus) are deposited on the surface as water infiltrates into the soil. If Zone 3 land is properly managed, sediment removal can reach 80 percent (Sheridan et al., 1999).

Zone 2

Zone 2 is designed to remove, sequester, and transform nonpoint sources of pollution in stormwater (Lowrance et al., 1997; Welsch, 1991). Nutrients such as nitrogen and phosphorus, microbes, and sediment can all be altered during passage through a Zone 2 riparian buffer. Zone 2 can also reduce pesticide transport (Lowrance et al., 1997) and may trap other pollutants, including metals and hydrocarbons. In general, the ability of a Zone 2 riparian buffer to remove pollutants depends on (1) whether sheet flow has been established in Zone 3, (2) the type of vegetation present, and (3) the length of the buffer zone. The first criterion is determined by the condition of the Zone 3 riparian buffer upslope from Zone 2. The other two criteria are characteristics of Zone 2, some of which can be altered or managed to maximize pollutant removal. A literature review found later in this chapter discusses the extent of pollutant removal that can be achieved when rainfall runoff travels through riparian buffers. This review focuses on the pollutants of greatest concern in the New York City watersheds, including phosphorus, microbial pathogens, and sediment.

Zone 1

Zone 1 is the area of the riparian buffer closest to the waterbody. In forest ecosystems, it is characterized by a canopy of trees and shrubs that provide shade to near-shore areas of lakes, larger streams, and rivers during a portion of each day. The cumulative effects of the canopy on the energy balance can have a substantial (10–15°C) effect on water temperature. Because dissolved oxygen concentration is inversely proportional to water temperature, increases in temperature caused by the removal of riparian vegetation can impose chronic or acute stress on invertebrate and fish populations. Riparian vegetation has the greatest influence along headwater streams where vegetation can cover the entire width of the stream. The microclimate effect decreases downstream as the width of the stream, river, or lake increases relative to the height of the riparian vegetation.

Leaves, needles, and wood supply energy—as carbon—to headwater streams. Like microclimate effects, the relative importance of carbon inputs from riparian vegetation decreases as the receiving water becomes larger. However, the inflow of dissolved and particulate carbon from headwater areas remains an important supplement to *in situ* primary production by algae and other aquatic plants in rivers and lakes. The contribution of vegetated riparian zones to the total dissolved carbon load at the point of water withdrawal is an important issue in watershed management. Because of the concern over the role of dissolved organic carbon in producing disinfection byproducts, there are possible drawbacks to increasing the dissolved organic carbon levels in streams.

In addition to being a persistent source of carbon, woody debris ranging from small twigs to branches, boles, and entire trees is a critical structural feature for stream ecosystems. As woody debris lodges and jams along streambanks, it forms a matrix that captures leaves and other small organic matter as they drift downstream. The interlocking roots of riparian vegetation anchor streambank, floodplain, and lakeshore soils and substantially increase their resistance to erosion and slumping.

Finally, the zone nearest the waterbody can be responsible for unique aquatic habitats. Woody debris and leaf packs at the land/water interface increase the variation of flow velocity in headwater streams. Quiet water and eddies behind leaf packs and larger debris jams lead to the formation of alternating pool and riffle sections in headwater streams. Diverse hydraulic conditions provide a continuum of spawning, rearing, feeding, hiding, and overwinter habitat for fish, amphibians, and invertebrates.

All Zones

In all zones of a riparian buffer, vegetation helps to reduce soil erosion. All zones are also capable of increasing the thickness of the unsaturated zone through evapotranspiration of water from the soil profile. As a consequence, available storage for rain, snowmelt, and upland inflow is maximized.

Riparian buffer zones exhibit soil physical and hydraulic properties that further enhance the ability of the land to attenuate stormwater. Decomposition of vegetation in the riparian zone leads to reduced bulk density, thereby increasing porosity (storage capacity), infiltration capacity (rate of water movement into the soil), and permeability (rate of water movement through the soil). Water retention characteristics also are enhanced by the addition of organic matter. The growth, senescence, and death of roots, along with the actions of invertebrates and small mammals, produce a complex system of macropores that augment the permeability of the soil matrix. Collectively, these soil properties maximize the likelihood that rain, snowmelt, or overland flow from adjacent uplands will pass beneath the soil surface and travel as subsurface flow through the riparian zone. In addition, small-scale variations in slope, woody debris, herbaceous plants, and leaf litter on the forest floor present additional barriers to overland flow.

ACTIVE MANAGEMENT OF BUFFER ZONES

The most important management practice for influencing functioning of buffer zones is to stabilize the hydraulic properties of stormwater so that channelized flow does not reach nearby streams. Channelized flow can form very quickly during rainfall. In urban areas, stormwater concentrates into channelized flow within as few as 75 ft of its source (Schueler, 1996; Whipple, 1993). Given the typical land uses found on the East Coast, only about 10 percent to 15 percent of a watershed area produces sheet flow during precipitation

(Schueler, 1995). The remaining runoff is usually delivered to streams in open channels or storm drains, the flow from which can be extremely difficult, if not impossible, to dechannelize.

Hence, converting channelized flow to sheet flow or to multiple smaller channels is a critical aspect of buffer zone management. Most regulations involving setbacks and buffer zones have been written and enacted with no consideration of this important issue. Converting channelized flow to sheet flow may require the installation of a structural BMP in Zone 3. For example, at sites with significant overland flow parallel to the buffer, water bars should be constructed perpendicular to the buffer at 45- to 90-ft intervals to intercept runoff and force it to flow through the buffer before it can concentrate further. Low berms or vertical barriers, known as level-lip spreaders, have been used successfully to spread concentrated flow before entering a forest buffer (Franklin et al., 1992). Buffers should not be used for field roadways because vehicles and farm equipment will damage the buffers and may cause concentrated flows (Dillaha and Inamdar, 1997). Specific suggestions for dechannelizing urban stormwater are given later.

Buffers may accumulate significant amounts of sediment and nutrients over time. To promote vegetative growth and sediment trapping, herbaceous vegetated buffers should be mowed and the residue should be harvested two or three times a year (Dillaha et al., 1989a,b). Mowing and harvesting will increase vegetation density at ground level, reduce sediment transport, and remove nutrients from the system. Herbaceous vegetated buffers that have accumulated excessive sediment should be plowed out, disked and graded if necessary, and reseeded in order to reestablish shallow sheet flow conditions. Although natural herbaceous buffer zones are rare in the Catskills, those created during active management of setbacks should be harvested. The primary management for Zone 1 is to reestablish and maintain native woody vegetation.

Although it is known that vegetation type can greatly influence buffer zone functioning, field data are not available for most types of buffers. There are numerous aspects of vegetation management for which more information is needed. For instance, the rooting depth will influence nutrient uptake from shallow or deeper groundwater, and more must be known about the differences among root systems of various types of vegetation. Different types of vegetation also have different management requirements, with woody vegetation providing a natural longer-term sink for nutrients than does herbaceous vegetation. Some general conclusions can be drawn. First, riparian forest buffers require native woody vegetation near the waterbody. States make determinations as to the appropriateness of different species, with native hardwoods required in most states of the eastern United States. Second, in areas experiencing runoff high in sediment, herbaceous vegetation is recommended between a forest buffer and the runoff source because a well-managed grass buffer can be more effective at trapping sediment and associated contaminants. Combinations of vegetation may prove most effective at removing a range of pollutants. For example, an outer grassed strip followed by an inner forested strip has been suggested for complete sediment, phosphorus, and nitrogen removal (Correll, 1991; Osborne and Kovacic, 1993). In all cases, the hardiness of riparian vegetation will determine how well it accomplishes pollutant removal and other functions. The New York City Department of Environmental Protection (NYC DEP) has recently noted that high densities of white-tailed deer may be preventing regrowth of forests around Kensico Reservoir (NYC DEP, 1997), which should be considered when determining vegetation requirements for buffer zones.

The preceding discussion applies only to those nonpoint source pollutants found in stormwater. Active management of buffer zones will have no effect on atmospheric deposition of pollutants directly over the surface of waterbodies, nor can it control in-stream increases in dissolved organic carbon derived from terrestrial vegetation. Additional suggestions for the active management of buffer zones are given below in relation to specific activities that produce nonpoint source pollution.

Agriculture

In general, agricultural land uses tend to increase surface runoff and decrease infiltration and groundwater recharge in comparison to perennial vegetation such as forest or grassland. Grazing animals can cause compaction of soils, especially under wet soil conditions. Tillage may increase subsurface compaction and lead to crust formation at the soil surface. The severity of these effects depends on soil properties and climate. In some watersheds, increased surface runoff, often combined with ditches and drainage enhancements, can change a groundwater-flow-dominated system to a surface-runoff-dominated system (Schultz et al., 1994). Increases in surface runoff cause increases in the stormflow/baseflow ratio and in the amount of sediment and chemical transport.

Many of the effects of agriculture on hydrologic and transport processes can be mitigated through the use of properly managed buffer zones. The USDA launched a National Conservation Buffer Initiative in 1997 to increase the adoption of conservation buffers and the integration of conservation buffers into farm plans. Conservation buffers include many practices designed to impede and retain surface flows and pollutants such as vegetated filter strips, contour filter strips, and riparian forest buffers. The general guidance given above on establishing and maintaining buffer zones is largely derived from studies in agricultural areas and is of primary importance. Hydrologic enhancement (conversion of channelized flow to sheet flow) can be accomplished through grading of soils, removal of berms or channels, and creation of shallow overland flow paths. Vegetation establishment may involve fertilizer and lime application, seeding, or other planting. Active management may also involve harvesting of vegetation to remove nutrients.

Other important aspects of buffer zone management on agricultural lands include restrictions on grazing and pesticide application. Grazing of riparian buffer systems, including riparian forest buffers and filter strips, is generally not allowed under programs such as the USDA Conservation Reserve Program. Riparian buffers should be combined with practices such as fencing and alternative water supplies to exclude domestic animals from the entire buffer zone. Certain pesticides have setback restrictions from watercourses for storage, mixing, and application that are part of the label restrictions issued for the chemical by the Environmental Protection Agency (EPA). For example, metolachlor, a common herbicide used in corn production, cannot be mixed or stored within 50 ft of lakes, streams, and intermittent streams. The herbicide atrazine may not be mixed or loaded within 50 ft of an intermittent or perennial stream, it may not be applied within 66 ft of where field runoff exits a field and enters a stream, and it may not be applied within 200 ft of natural or impounded lakes or reservoirs. Caution should be used when applying herbicides to adjacent fields to avoid damage to buffer zone vegetation.

Finally, it should be noted that artificial subsurface drainage (tile drains) may short-circuit the functioning of riparian buffers in agricultural settings. Although drain lines are not supposed to enter streams directly, they sometimes do, providing a direct conduit for pollutant movement to streams. Allowing tile drain water to flow through a spreading device before entering a riparian buffer is desirable.

Forestry

As noted in earlier chapters, the majority of nonpoint source pollution (primarily sediment) from timber-harvesting operations emanates from the road and skid trail network needed to remove sawtimber, pulpwood, or fuelwood from the forest. Overland flow is generated when the litter layer is scrapped away *and* the soil is compacted. This disturbance is usually limited to about ten percent of the harvest unit. The remainder of the site retains high infiltration capacity, with shallow subsurface flow as the predominant mechanism of streamflow generation. Therefore, it is usually unnecessary (and impractical) to construct stormwater control devices (e.g., level-lip spreaders) at the transition between the harvest unit and the riparian forest buffer. By contrast, a large proportion of agricultural fields or urban areas can generate overland flow and associated nonpoint source pollution because of changes in soil surface conditions.

Riparian forest buffers are subject to special operating restrictions, often specified by state forest practice acts, to minimize undesirable changes in site conditions. The most important restriction is the prohibition of direct access by heavy equipment. Selective harvesting of trees within Zones 2 and 3 can and should occur. However, logs can only be winched on a steel cable to a machine (skidder, specially equipped farm tractor, or small 4WD tractor) located outside of the buffer or removed by a mechanical harvester with a hydraulic boom. The latter is capable of reaching up to 40 ft into the buffer. Restricting access by heavy equipment virtually eliminates the soil disturbance and compaction responsible for generating and conveying nonpoint source pollution. Except for the restriction on equipment access, the transition between the harvest unit and the riparian forest buffer should, by design, be gradual and indistinct.

Basic silvicultural practices to maintain or enhance the health, vigor, and growth rate of trees should be implemented in the riparian forest buffer. Typically, trees are marked and removed if they have been overtopped by their neighbors (low thinning), damaged by storms or careless logging (stand improvement), or severely damaged by insects and/or diseases (sanitation cut). Stand treatments reallocate the productive capacity of the site to larger, more vigorous trees, and they naturally regenerate the forest by fostering the establishment and growth of seedlings. Although tree planting is still used by forest products companies in parts of the United States and Canada to establish fast-growing, even-aged stands for pulpwood and sawtimber production, it is a costly and unnecessary practice when natural regeneration can be assured (as in the Catskills, where abandoned land will revert to forest after three or four years without mowing). Furthermore, one of the objectives of silviculture is to control the seed/sprout source and site conditions to ensure a mixed-species, uneven-aged stand will result from a series of carefully planned and implemented harvests.

Foresters and landowners should deliberately retain long-lived, commercially valuable species (e.g., northern red oak or yellow birch) and diversify the vertical structure, spatial arrangement, and species composition of the residual stand. An actively growing diverse forest maximizes resistance to and resilience from widespread natural (e.g., hurricanes, insect and disease outbreaks) and anthropogenic (e.g., atmospheric deposition) disturbance that may threaten source water quality (Barten et al., 1998; MDC, 1995).

Stormwater

As mentioned previously, stormwater runoff from impervious surfaces can become channelized quickly if not immediately, partly because of storm drains and pipes. In many cases, no amount of management will allow riparian buffers (in the absence of other BMPs) to convert this channelized flow to sheet flow. Thus, "active management" of urban stream buffers must include maintenance of physical structures *in addition to* the buffer zone. Because the ability of urban stream buffers to remove many pollutants has not been tested in the field, the following design suggestions are based solely on engineering theory. If buffer zones are ever elevated to the status of other highly engineered stormwater BMPs (by accumulating the necessary field data), design improvements to help achieve sheet flow will become apparent.

An urban stream buffer is ideally comprised of three zones: a stormwater depression area that leads to a grassed filter strip that in turn leads to a forested buffer. The stormwater depression should be designed to capture and store stormwater during most storm events. Stormwater detained by the depression can then be spread across a grass filter designed for sheet flow conditions, which in turn discharges into a wider forested buffer. The outer boundary of an urban stream buffer must be carefully engineered in order to satisfy these demanding hydrologic and hydraulic conditions. In particular, simple structures are needed to store, split, and spread surface runoff within a stormwater depression area at the boundary. Although past efforts to engineer urban stream buffers were plagued with hydraulic failures and maintenance problems, recent experience with similar bioretention areas has been much more positive (Claytor and Schueler, 1996). Consequently, it may be useful to consider elements of bioretention design for the outer boundary of an urban stream buffer.

Wetlands

Natural wetlands extend along all bodies of water to varying extents. In some cases, the saturated hydrosoils characteristic of wetlands may extend only a few feet; in other cases, wetlands can cover a number of acres beyond the watercourse or reservoir boundary. Because natural wetlands sequester certain nutrients, there may be a tendency to reduce the size of terrestrial buffers in the vicinity of natural wetlands. However, at certain times the hydrology of natural wetlands may be quite channelized, which can overwhelm and diminish the excellent sequestering of pollutants observed at low flows. Hence, it is essential that terrestrial buffers be maintained maximally in the vicinity of wetlands and that the waterbody boundary be delineated at the upgradient boundary of wetlands. Buffer management for wetlands should be no different than for reservoirs and streams.

SETBACKS IN THE CATSKILL/DELAWARE WATERSHED

Numerous setbacks are specified in the MOA regulations. The activities for which setback distances are proscribed are not all-inclusive. For example, agriculture, a contributor of nonpoint source pollution, is specifically excluded from setbacks. Table 10-1 is an inventory of the setback distances prescribed in the Watershed Rules and Regulations. Note that reservoir stems are defined to be the major tributaries within 500 ft of a reservoir.

In order to assess how effective setback distances in the Catskill/Delaware watershed are likely to be in protecting water quality, information on soil type, land use, and other factors is necessary. Presently, conditions on land within potential setbacks are not well known. New York City owns a substantial amount of land immediately surrounding each water supply reservoir, for which land cover is indicated in Table 10-2. Because the construction of residences is prohibited on these City-owned lands, and because there are few structures, this

TABLE 10-1 Inventory of Setback Distances in the MOA for Different Activities

Regulated Activity	Watercourse, Wetland	Reservoir, Reservoir Stem, or Controlled Lake
Storage of hazardous substances (new tanks at an existing facility or a new facility altogether)[a]	100	500
New aboveground and underground petroleum storage facilities with NYC DEP registration[b]	100	500
New home heating oil tanks installed underground	100	500
New aboveground and underground petroleum storage tanks >185 gallons without NYC DEP registration[b]	25	300
Subsurface discharge from a wastewater treatment plant (WWTP)	100	500
Absorption field from new septic tanks	100	300
Raised septic systems	250 (100)[c]	500 (300)[c]
Impervious surfaces (basic guidelines only)	100 (50)[d]	300
New impervious surfaces at individual residences[e]	100	300
Siting or expansion of a solid-waste landfill or junkyard	250	1,000
Pesticide application without approval from NYC DEP[f]	250	1,000

[a]For storage facilities between 100 and 250 ft from a watercourse or wetland, additional forms and reports required.

[b]Expansion of pet. storage at existing facilities allowed within setbacks if business can prove it would otherwise fail.

[c]Setbacks in parentheses allowed if size or location of the land makes it impossible to abide by the current setbacks.

[d]Setbacks in parentheses allowed for intermittent streams. Many impervious surfaces can be built within the setbacks provided that a Stormwater Pollution Prevention Plan is drafted.

[e]Applies to surfaces constructed after October 16, 1995. If planned before October 16, 1995, impervious surfaces can be built within the 100-ft setback provided that an Individual Residential Stormwater Permit is drafted.

[f]Within the setbacks, NYC DEP approval must be gained annually. It should be noted that this setback has *not* yet been approved as part of the Watershed Rules and Regulations.

land is mostly forested. However, it is not necessarily representative of land use on setbacks throughout the watershed (D. Warne, NYC DEP, personal communication, 1998). On privately owned lands, there may be buildings on the banks of watercourses.

NYC DEP's Geographic Information System (GIS) was used to approximate land use on privately owned land that might fall within a 100-ft setback. As

TABLE 10-2 Land Cover in the New York City-Owned Land Surrounding
Reservoirs in the Catskill/Delaware Watershed

Cover Type	Acres	Percent of Total
Deciduous Forest	19,898	61
Evergreen Forest	6,430	20
Mixed Forest	1,146	3.5
Grass/Shrub	462	1.4
Grass	4,587	14
Bare Soil	37	0.1
Impervious Surface	156	0.5

Data courtesy of the NYC DEP.

shown in Figure 10-3, land use varies substantially among the six reservoirs west
of the Hudson River. Land use within setbacks surrounding reservoirs and reservoir stems only (additional data were not made available) for the six basins is as
follows: (1) Ashokan is predominantly deciduous forest, (2) Cannonsville is
predominantly grass, (3) Neversink has both deciduous and coniferous forest,
(4) Pepacton has multiple land uses of approximately equal acreage, (5) Rondout
is predominantly coniferous forest, and (6) Schoharie is split between deciduous
forest and grass.

Land slope is another critical factor to take into consideration when evaluating the setbacks for their water protection capabilities. In fact, some suggest that
lands with slopes greater than 15 percent should not be considered when assigning setback distances (Nieswand et al., 1990). Around reservoirs and reservoir
stems in all six West-of-Hudson basins, slopes are predominantly between 0 and
6 percent (Figure 10-4). However, median and maximum land slopes are expected
to increase higher in the watersheds. First- and second-order tributaries occur in
steep, mountainous terrain throughout much of the Catskills region. Streams are
incised into narrow (<100 ft) valleys with adjacent slopes that routinely exceed
15 percent. Finally, the soils within the setback areas of the reservoirs and
reservoir stems are primarily 8,000- to 10,000-year old alluvial material.

EFFECTIVENESS OF THE MOA SETBACKS

Several approaches can be used to evaluate the effectiveness of the setback
distances prescribed in the MOA, some of which were used in the 1993 environmental impact statement (EIS) prepared for the Watershed Rules and Regulations. The most straightforward approach is to monitor setback influents and
effluents to demonstrate removal of pollutants. To our knowledge, this has not

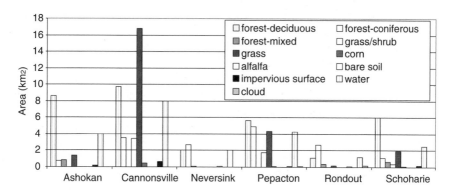

FIGURE 10-3 Land use in the 100 feet surrounding West-of-Hudson reservoirs and reservoir stems only. Courtesy of the NYC DEP.

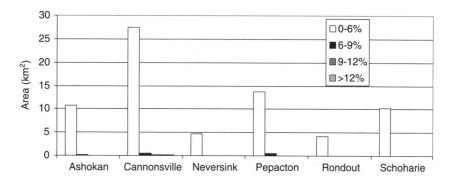

FIGURE 10-4 Land slope in the 100 feet surrounding West-of-Hudson reservoirs and reservoir stems only. Courtesy of the NYC DEP.

yet been accomplished anywhere in the New York City watersheds, primarily because of the newness of the Watershed Rules and Regulations and because it is not a stated goal of the NYC DEP's monitoring program. Monitoring activities are under way for specific projects involving the fate and transport of microbes through soil and groundwater (the Septic Siting Study). However, as of March 1999, no results are available.

In the absence of monitoring data, all other attempts to determine setback effectiveness are necessarily indirect. Experience from other localities that have imposed setback distances around waterbodies can provide an initial assessment of the adequacy of New York City's setback distances, although the site-specific nature of setback effectiveness limits this type of evaluation. The discussion below considers setback distances in several states. A second alternative is to search the literature for field research conducted elsewhere. The removal efficiencies of other setbacks may have implications for setback effectiveness in the New York City watersheds if land conditions are similar and the pollutant loadings upstream of the setbacks are similar.

In lieu of literature and actual field data, a model can be used to simulate the functioning of setbacks. Several models, with varying levels of complexity, are available to examine this issue (Nieswand et al., 1990; Phillips, 1989b). The more detailed models require input data on pollutant concentrations in runoff entering setbacks; they generate pollutant concentrations leaving setbacks. If combined with models of reservoir water quality, setback models can potentially be used to determine the net effect of setbacks on reservoir health. However, no such data on pollutant concentrations in runoff for the Catskill/Delaware watershed are available. Additional data requirements for the more complex models effectively prevent their use. Thus, a simplified model of subsurface hydrology using slope and soil type data from the Catskill/Delaware watershed is used to draw some general conclusions about how effective a 100-ft setback may be (see below). This approach is similar to that found in the 1993 EIS, and comparisons are drawn between the results of each effort.

Setbacks Across the United States

The use of setbacks in watershed regulations throughout the United States is becoming increasingly common. Setbacks are either of fixed width, or they vary depending on such factors as slope, vegetation, and stream size. Fixed-width setbacks usually range between 50 and 200 ft (Robbins et al., 1994).

Some setback/buffer provisions simply require a permit for certain activities on private land within the stated regulatory area. This gives the permitting authority the opportunity to impose specific conditions on the way construction is designed in order to minimize impacts on the aquatic resource. Others entirely prohibit most alterations of the setback area. Some provisions take a middle position, requiring a permit for any land use change while forbidding certain uses such as underground storage tanks within the regulatory setback. The constitutionality of imposing setbacks on private land is considered in Box 10-1.

Massachusetts

Massachusetts has pioneered the use of setbacks bordering various aquatic resources. The Massachusetts Wetlands Protection Act (MGLA, Ch. 131, sec. 40) requires local permits for most activities within specified distances of aquatic resources. Prior to 1996, the law generally permitted work to proceed within

BOX 10-1
Constitutionality of Setbacks and Buffers

Public restrictions on the use of portions of private land are familiar and longstanding. Since the early days of land use zoning in the 1920s, residential construction has been subject to minimum street setbacks in the interest of providing front yards. Typically such space, while not literally accessible to the general public, is open and unfenced and is landscaped to provide mutual visual amenity and sense of spaciousness. Minimum side and rear yards are also commonly required in traditional zoning ordinances. The use of limited restrictions along waterbodies, wetlands, and streams to protect water quality and avoid flood damage is another common use of public setback regulations, as in the case of the New York City MOA.

Setback and buffer regulations, like other public land use regulations, typically do not involve compensation to the private landowner. As such, they sometimes are challenged as "takings" of the value of the land affected without compensation, in violation of the Fifth Amendment of the U.S. Constitution, which states, "Nor shall private property be taken for public use without just compensation." It has long been assumed that land may be "taken" by unfair or excessive regulation even if title to the property is not removed from the owner. Justice Oliver Wendell Holmes stated in the famous 1922 decision in Pennsylvania Coal Co. versus Mahon (260 U.S. 395), "The general rule at least is, that while property may be regulated to a certain extent, if regulation goes too far, it will be recognized as a taking." Since that time, courts have struggled to define the limits of this regulatory power (i.e., how far is "too far") (Platt, 1996).

In resolving takings challenges, courts examine two primary factors: (1) do they reasonably serve a legitimate public purpose and (2) does the owner retain a reasonable (not necessarily optimal) economic return from the use of the total parcel of land. Two particular problems may lead to judicial overturn of setback regulations: (1) public access is allowed to the area covered by the setback for recreation or other purposes without compensation to the owner or (2) an owner's entire parcel is rendered valueless by the restriction. When a restriction is held to be an invalid taking, the court may order the public authority to compensate the owner, to remove the restriction, or both.

The most famous property rights case of the 1990s that involved a regulatory setback was Lucas versus South Carolina Coastal Council (112 S.Ct. 2886, 1992). The Lucas case arose from a takings challenge by the owner of two undeveloped lots on a coastal inlet in South Carolina. Because earlier developed adjoining lots had experienced severe erosion problems, the South Carolina Coastal Council denied Lucas approval

continued

BOX 10-1 Continued

to develop his lots under authority of the 1988 South Carolina Beachfront Management Act (Platt, 1992).

The trial court ordered the state to pay Lucas $1.2 million in compensation for an invalid taking. In a 3-2 vote, the South Carolina Supreme Court (404 S.E.2d 895, 1991) reversed the trial court, holding the permit denial to be a reasonable restriction on the right of a private owner to build in an unsafe location. The state court was reversed by the U.S. Supreme Court in a 6-3 decision, which held that where a regulation "denies all economically beneficial or productive use of land" (112 S.Ct., at 2893), it is a "categorical" taking equivalent to a physical invasion of the property by governmental action. In dissent, Justice Harry Blackmun wrote, "[S]ince no individual has a right to use his property so as to create a nuisance or otherwise harm others, the State has not 'taken' anything of value when it asserts its power to enjoin the nuisance-like activity. . . . It would make no sense under this theory to suggest that an owner has a constitutionally protected right to harm others, if only he makes the proper showing of economic loss" (112 S.Ct., at 2912).

The Lucas decision may be read narrowly to apply only to cases where regulation prevents an entire parcel of land from being developed, not just a portion of it. This may be viewed as a healthy counterbalance to the tendency of some regulatory agencies to act harshly toward private owners. On the other hand, it may have had a chilling effect on the ability of public agencies to impose and enforce necessary restrictions in areas of sensitive natural resources, such as along coasts and rivers.

In another taking issue case, Dolan versus City of Tigard (114 S.Ct. 2309, 1994), the U.S. Supreme Court cast further light on the bounds of public regulations along watercourses. The plaintiff, in applying to enlarge her hardware store, was required by the defendant to donate a portion of her land within a 100-year floodplain to the city, along with a 15-foot strip adjoining the floodplain to be used as part of the city's pedestrian and bikeway system. These exactions amounted to about 15 percent of her total land area. Upon a "takings" challenge by the owner, the city was upheld by the Oregon Supreme Court, which was in turn reversed by the U.S. Supreme Court. However, the majority opinion in this 5-4 decision was sympathetic with the need for land use planning, while holding the public access requirement to be too demanding: "Undoubtedly the prevention of flooding along Fanno Creek and the reduction of traffic congestion in the central business district qualify as the type of legitimate public purposes we have upheld. . . . It seems equally obvious that a nexus exists between preventing flooding along Fanno Creek and limiting development within the creek's 100-year floodplain" (114 S.Ct., at 2318). In dissent, Justice Stevens argued that "in our changing world, one thing is certain: uncertainty will characterize predictions about the impact of new urban developments on the risks of floods, earthquakes, traffic congestion, or environmental harms. When there is doubt concerning the magnitude of those impacts, the public interest in averting them must outweigh the private interest of the commercial entrepreneur" (114 S.Ct., at 2329). With a shift of one justice, this could have been the majority view and thus the "law of the land" (Platt, 1994).

In conclusion, the ability of public authorities to impose setback and buffer restrictions on private property owners without violating the Fifth Amendment depends on the following factors:

1. The objective of the restrictions must be clearly related to a valid public purpose (e.g., protection of public health by ensuring the purity of public drinking water supplies).

2. The restriction should be no wider than necessary to accomplish the stated purpose.

3. The achievement of secondary benefits (e.g., flood hazard reduction and protection of riparian habitat) is valid as long as public access to private property is not promoted without compensation to the owner (the Dolan problem).

4. The economic impact on the entire parcel of land should be considered, namely all use of the setback/buffer area may be prohibited if the parcel is large enough to support reasonable economic use elsewhere. If the entire parcel lies within the setback/buffer (the Lucas problem), the owner must be allowed some limited economic use of that area.

5. A provision for variance should be included to provide relief in cases of unnecessary hardship caused by literal enforcement of the restriction.

regulated areas subject to permits, land use zoning, and other applicable regulations. In 1996, the Act was amended to impose tighter limits on new construction within "riverfront areas," defined to extend 200 ft from a river's mean annual high-water line or 25 ft in certain urban areas. For proposed land use activities in riverfront areas, there can be no significant impact on the natural resources associated with the riverfront area, placing a higher level of protection (through closer scrutiny) on proposed work in such areas.

Massachusetts has imposed an even higher level of restriction for setback areas within the watersheds of the Quabbin and Wachusett reservoirs. As part of its program to qualify for a filtration avoidance determination, the 1992 Watershed Protection Act (Mass. Laws of 1992, Ch. 36) was adopted, establishing a "primary buffer zone" extending 400 ft from either reservoir and 200 ft from any "tributary or surface water" within the source watersheds. Within the primary buffer, the act prohibits outright "any alteration, or the generation, storage, disposal, or discharge of pollutants." Secondary buffer zones were defined to include areas between 200 and 400 ft of tributaries or surface waters, floodplains, wetlands, and certain areas overlaying aquifers. Within secondary buffer zones, many activities—such as the outdoor storage of road salt, fertilizers, and manure and the operation of junkyards—are prohibited. Limits are imposed on impervious cover and on the density of residential construction in relation to septic systems.

North Carolina

The setbacks in Massachusetts are fixed-width rather than variable-width. Fixed-width setbacks are easier to administer than variable-width, but they may not be sufficiently flexible to protect all sensitive riparian areas. In North Carolina, watershed control programs generally advocate variable-width setbacks around surface water supplies. Beginning with a minimum width of 50 ft, the additional distance is generated by taking slope into account according to Equation 10-1 (a simple calculation if a GIS and digital elevation model are available):

$$\text{Setback Distance} = 50 \text{ ft} + (4 \times \%\text{slope}) \qquad (10\text{-}1)$$

Other watershed programs in North Carolina consider vegetation as well as slope when calculating setback distances. Depending on slope and on whether the land is forested or grassland, setbacks around University Lake in Orange County vary between 50 and 250 ft.

It should be noted that not all North Carolina setback requirements are variable-width. Fixed widths of 150 and 100 ft are required for perennial streams feeding Lake Michie and Falls Lake, respectively, which supply drinking water to Durham and Raleigh. In addition, the recently proposed Neuse River Rule delineates a 50-ft setback along every watercourse in the Neuse watershed, both perennial and intermittent. Land clearing for any purpose on the setback would be prohibited, regardless of whether the land is publicly or privately owned.

Other States

Setbacks in other eastern states provide further comparisons. Baltimore County, Maryland, requires from 75 to 150 ft around all streams, depending on slope and stream size. Several streams in Newport News, Virginia are protected by 100- and 200-ft setbacks for intermittent and perennial streams, respectively. The Chesapeake Bay Preservation Act of Virginia takes both land use and the existence of BMPs into account when determining appropriate distances. Setbacks around Bear Creek and Dog River in Douglas County, Georgia, measured from the centerline of the streams rather than from the stream edge, vary between 100 and 300 ft. The same measurement technique is used around Town Lake and Lake Austin in Austin, Texas, where setbacks range from 100 to 300 ft.

Analysis

Compared to the setbacks described above, those found in the New York City Watershed Rules and Regulations are of similar extent. They are fixed-width setbacks whose distance depends on land use (like the Chesapeake Bay Preservation Act) and the type of nearby waterbody. Setbacks found in the MOA do not take slope, soil type, vegetation, and other factors into account, nor do they vary for particular land uses. Although this means that the provisions may be relatively easy to enforce, they may also be less protective of nearby waterbodies, especially in areas with steep slopes and where vegetation is less capable of dissipating rainfall runoff and sequestering pollutants.

Literature Review of Pollutant Removal in Buffer Zones

A limited body of research quantifies the effectiveness of buffer zones in removing pollutants such as nitrogen, phosphorus, sediment, pesticides, and some microbes from storm and drainage water. This literature is summarized below, and additional material is presented regarding other pollutants such as viruses from septic systems and landfill leachate. Because these research results were gathered from buffer zones with either native or managed vegetation, they represent best-case scenarios for pollutant removal in setbacks. Several attempts have been made to generalize on the required buffer widths necessary to accomplish pollutant removal (Castelle et al., 1994; Fennessy and Cronk, 1997; Nieswand et al., 1990). Extreme care should be taken when using such generalizations, because they may be based on particular pollutants (such as nitrogen) that are subject to higher removal efficiencies than are other pollutants.

Phosphorus

Phosphorus is found in both particulate and dissolved forms, with the major-ity of phosphorus in surface runoff being particulate phosphorus. Large inputs of dissolved phosphorus may originate from certain sources, such as when fertilizers are broadcast on the soil surface without incorporation (Sharpley et al., 1992). Dissolved phosphorus, usually orthophosphate, is thought to be generally bioavailable.[1] Particulate phosphorus, on the other hand, is only partially bioavailable, depending on the specific form of phosphorus involved.

---

[1] Bioavailability refers to the physical state of the pollutant with respect to its uptake by micro-organisms and other potential human or ecological "receptors" of the pollutant.

Particulate phosphorus, which is almost always associated with sediment and soil particles, is removed when suspended particles (especially fine particles) settle out of overland flow in Zones 2 and 3. Dissolved phosphorus can be removed via adsorption to leaf litter and soil particles associated with the litter, but it may also be *produced* from the desorption, dissolution, and extraction of phosphorus into solution as runoff moves across the surface (Sharpley, 1985; Uusi-Kamppa et al., 1997). Thus, the removal of dissolved phosphorus from surface runoff in buffer zones appears to be less effective than the removal of particulate phosphorus.

Phosphorus removal from surface runoff in buffer strips has received limited study (Table 10-3). Reports of the percentage of phosphorus removal in buffer zones are highly variable, ranging from zero percent to 95 percent, depending on the buffer length and other factors. A literature review conducted by Castelle et al. (1994) claims that buffer widths from 30 to 300 ft are necessary to accomplish nutrient removal. In general, studies of phosphorus removal in buffer zones have indicated that removal is most effective in the first several feet of the buffer, that dissolved phosphorus removal was less evident than particulate phosphorus removal, and that infiltration and deposition of fine particles was necessary for effective phosphorus removal.

These findings suggest that even narrow buffers might be important for phosphorus removal. Perennial vegetation that sequesters phosphorus in biomass is important, but should be removed as harvested material. Also, although total load of phosphorus may be reduced by the buffer because of the deposition of phosphorus-laden sediment, the proportion of bioavailable phosphorus may increase because of the release of dissolved phosphorus from the vegetation comprising the buffer zone. For example, both grass buffers and native vegetation have been found to increase dissolved phosphorus output in a number of studies (Correll and Weller, 1989; Dillaha et al., 1988; Jordan et al., 1993; Uusi-Kamppa et al., 1997).

Phosphorus can originate from many of the activities and structures prohibited on setbacks in the Catskill/Delaware watershed, particularly impervious surfaces, septic systems, and wastewater treatment plant (WWTP) discharges. The 100- and 300-ft setbacks in the Catskill/Delaware watershed for septic systems and impervious surfaces, and the 100- and 500-ft setbacks for WWTPs with subsurface discharge, fall in the range of widths encompassed by the literature. However, the wide range of reported removal efficiencies makes it impossible to predict the adequacy of these setback distances. The only conclusion that can be drawn is that if managed or maintained as buffer zones, the MOA setbacks may approach the removal efficiencies found in Table 10-3. Regarding soluble phosphorus, several studies show that buffer zones can be sources of this pollutant. Thus, the setbacks should not be expected to reduce this pollutant in stormwater runoff entering receiving waters.

TABLE 10-3 Phosphorus Removal in Buffer Zones

Reference	Location	Buffer Vegetation	Buffer Widths (ft)	Tot-P retention (%)	Ortho-P retention (%)	Comment
Dillaha et al. (1989a)	Virginia	Grass	15 30	49–85 65–93	69–83 48–81	Less effective after initial events, simulated rainfall
Magette et al. (1989)	Maryland	Grass	15 30	41 53		Less effective after initial events, simulated rainfall
Syversen (1995)	Norway	Grass	16 33 49	45–56 56–85 73	2–77 0–88 10	Natural rainfall, slope of 12–17%, and strips with native grass
Uusi-Kamppa and Ylaranta (1996)	S. Finland	Grass	33	20–36	0–62	Natural rainfall, increase of ortho-P
Schwer and Clausen (1989)	Vermont	Grass	85	89	92	Wastewater inputs to "overloaded" buffer; greatest removal in growing season
Vought et al. (1994)	Sweden	Grass	26 52	– –	66 95	Greatest removal in first several feet
Peterjohn and Correll (1984)	Maryland	Native hardwood forest	164	81	–	Based on surface flow and groundwater input–output budgets
Lowrance et al. (1983)	Georgia	Native hardwood forest	66–131	23	–	Based on subsurface flow budgets

Bacteria

Most research on microbial removal by buffer zones has concentrated on fecal coliform bacteria as indicators of human and animal contamination of natural waters. This research has most frequently been conducted in locations with high bacterial loads (such as feedlots and manured areas) that are affected by overland flow. Bacterial contamination of agricultural waters often exceeds the primary contact standard of 200 fecal coliforms per 100 mL (Walker et al., 1990). Bacterial loss in runoff can be as high as 90 percent from soils with fresh, unincorporated manure applications (Crane et al., 1983). Heavy rain and rapid surface flow may keep fecal coliforms in suspension while denser soil particles are trapped, leading to preferential transport of bacteria compared to inorganic sediment. Because fecal coliform bacteria are 1–2 μm in diameter, they behave much like fine clay particles once they are entrained in overland flow (Coyne et al., 1995).

Coyne et al. (1995) found grass filter strips were much more effective at trapping soil in surface runoff than in trapping fecal coliforms. Approximately 99 percent of soil particles in surface runoff were removed in filter plots compared to 74 percent and 43 percent removal of fecal coliforms. Earlier studies also indicated that grass buffer strips were less effective for bacteria versus sediment trapping. Dickey and Vanderholm (1981) found grass buffers with short flow-lengths did not significantly reduce bacteria levels in surface runoff. Young et al. (1980) found that grass buffers reduced fecal coliforms about 70 percent in a 89-ft flow length. As expected, channelized flow further reduces the trap efficiency for fecal coliform bacteria (Coyne et al., 1995). Loading rate is also a very important factor; buffer strips recommended by the USDA for dairy barnyard runoff may be undersized, especially in cold regions. An overloaded filter 75 ft in length showed no significant filtration of either fecal coliform or fecal streptococcus bacteria in stormwater from a dairy barnyard (Schellinger and Clausen, 1992).

These studies all support the conclusion that bacteria in overland flow can move through buffer zones without significant reduction. In general, their removal is much less effective than that of sediment and particulate-phase phosphorus. The effectiveness of buffer zones in removing bacteria is reduced because bacteria are small compared to other particulate matter and have very slow settling rates. Other studies of bacterial transport through saturated porous media (shallow subsurface or groundwater flow) indicate that bacteria can travel up to 50 ft in as few as 8 days (Hagedorn et al., 1978).

Impervious surfaces, failing septic systems, and subsurface WWTP discharges can be significant sources of bacterial pathogens. Because bacterial loadings from these sources are likely to be different in concentration, duration, and mode of transport (i.e., overland flow vs. subsurface flow) than loadings from agricultural areas, it is not possible to say whether the associated 100-, 300-, and 500-ft setback distances will be protective of water quality with regard to such organisms. Active management of setbacks to improve bacterial removal should focus on increasing the residence time of runoff waters in buffer zones.

Other Microbial Pathogens

There are no published reports on removal of *Cryptosporidium* oocysts or

Giardia cysts in riparian buffer zones. Given their larger size, it is expected that they may be removed more efficiently than fecal coliforms. However, because they are more environmentally resistant than bacteria, whether buffer zones are effective sinks of these organisms depends heavily on prolonging their residence time in the buffer, which is related to the frequency of overland or channelized flow.

The transport and fate of viruses in porous media (groundwater) has been studied in relation to the siting of septic systems (Yates et al., 1986; Yates and Yates, 1989). The reports suggest that in order to reduce virus concentrations to acceptable levels, 50–300 ft of porous media are needed. Because the decay rate of viruses increases with longer residence times, greater than 300 ft of buffer zone may be needed to sufficiently reduce virus levels in overland flow. The lack of data on removal of nonbacterial microbes in buffer zones makes any evaluation of the MOA setbacks regarding these pollutants highly speculative.

Sediment

Soil erosion and sedimentation occur as a result of three separate but inter-related processes—soil detachment, sediment transport, and sediment deposition. Raindrop impact and stormwater can dislodge sediment particles when their shear forces exceed the critical sheer forces of the particles. Once detached, sediment particles are transported in overland flow until the total energy becomes less than the energy required for particle transport. When this occurs, deposition begins, with large-sized sediment depositing first. Thus, the sediment remaining in runoff has a higher proportion of fine particles compared to the original material.

Detailed studies of sediment retention in buffer systems, as well as field observations, have been used to draw conclusions about the ability of Zone 2 buffer zones to remove sediment (Dillaha and Inamdar, 1997; Dillaha et al., 1989a,b). First, high removal efficiencies have been observed for sediment traveling through grass buffers. These removal efficiencies are generally better than those observed for phosphorus and bacteria. For example, a 30-ft buffer was shown to remove 84 percent of suspended solids (Dillaha et al., 1989a) and 19- and 36-ft grass buffer strips were shown to remove 69 to 90 percent and 69 to 97 percent of suspended solids, respectively (Patty et al., 1995). The required buffer width for removing sediment has been proposed to range from 33 to 200 ft (Castelle et al., 1994). A second conclusion is that buffer zones of reasonable width are effective for sediment removal only if flow is shallow and uniform and if the buffers have not been previously inundated with sediment. Third, the effectiveness of herbaceous buffers for sediment removal appears to decrease with time as sediment accumulates in the buffer and encourages concentrated flow across the buffer. Finally, in some instances, herbaceous buffers may be ineffective for sediment removal if flow accumulates in channels, rills, or gullies before reaching the buffer zone.

The prohibited activity in the Catskill/Delaware watershed most likely to generate sediment is the construction of impervious surfaces. Because of the efficient sediment removal rates found in the literature for much shorter distances, the 100- and 300-ft setbacks required for impervious surfaces are likely to be sufficiently protective of pollution via sedimentation. However, as with phosphorus, prolonged sediment removal is dependent upon active management of the setback areas to maintain the vegetation and ensure sheet flow. Without such measures, sediment removal diminishes over time as buffers become saturated.

Pesticides

Because of the extensive use of buffer zones as an agricultural BMP, there have been numerous studies of pesticide removal in buffer zones. In general, managed buffer zones can be highly effective in reducing pesticide concentrations in agricultural drainage water. For example, Patty et al. (1997) found that 99 percent of isoproturon, 97 percent of diflufenican, from 72 to 100 percent of lindane, and from 44 to 100 percent of atrazine were removed in grassed buffer strips of 20, 39, and 59 ft under a range of soil and cropping conditions. In addition to these pesticides, grassed buffer strips are also efficient at reducing concentrations of 2,4-D, trifluralin, metolachlor, metribuzoin, and cyanazine in agricultural waters. The highly degradable structure of many pesticides and their rapid sorption onto organic matter and vegetation in grassed buffer strips contribute significantly to the effectiveness of these buffers (Patty et al., 1997). The 250- and 1,000-ft setback distances found in the MOA are expected to sufficiently protect nearby waterbodies from pesticide contamination. (It should be noted that these setbacks are not yet approved by the state for inclusion in the Watershed Rules and Regulations.)

Landfill Leachate

There have been no published studies on removal of pollutants from landfill leachate via buffer zones of any kind. However, by examining the typical constituents of landfill leachate, it may be possible to draw conclusions about the ability of buffer zones to remove some constituents. The chemicals originating from solid-waste landfills can be highly variable. Those landfills that contain municipal waste only are generally composed of high concentrations of total organic carbon (TOC), total dissolved solids, nitrogen compounds, and inorganic compounds (salts). Typical data on leachate components at solid-waste landfills are presented in Table 10-4.

Constituents of hazardous-waste landfills can vary greatly from municipal landfills and include chemicals such as methyl ethyl ketone, acetone, methyl isobutyl ketone, methylene chloride, phthalic acid, phenol, arsenic, and barium (Pavelka et al., 1994). Some landfills contain mixtures of hazardous waste and

TABLE 10-4 Typical Leachate Components of Solid-Waste Landfills, Both New (less than 2 years) and Mature (greater than 10 years)

Component	Typical concentration (mg/L)	
	New Landfills	Mature Landfills
pH (standard units)	6	6.6–7.5
Alkalinity as $CaCO_3$	3,000	200–1,000
Total hardness as $CaCO_3$	3,500	200–500
Total Suspended Solids	500	100–400
Biological Oxygen Demand	10,000	100–200
Chemical Oxygen Demand	18,000	100–500
Total Organic Carbon	6,000	80–160
Total phosphorus	30	5–10
Ortho phosphorus	20	4–8
Nitrate	25	5–10
Organic nitrogen	200	80–120
Ammonia nitrogen	200	20–40
Calcium	1,000	100–400
Magnesium	250	50–200
Potassium	300	50–400
Sodium	500	100–200
Chloride	500	100–400
Sulfate	300	20–50
Total Iron	60	20–200

Source: Reprinted, with permission, from Tchobanoglous et al., 1993. © by McGraw-Hill, Inc.

municipal waste. Because of the wide and unpredictable variability in landfill leachate composition, evaluating the 250- and 1,000-ft setbacks in the MOA is necessarily limited. Phosphorus and nitrogen compounds, for which buffer zone studies have been conducted, can be expected to be removed along the setback distances; however, it is impossible to draw conclusions about the ability of the setbacks to greatly reduce concentrations of other chemicals. The attenuation of dissolved landfill leachate components has been investigated in unconsolidated sandy/gravel groundwater aquifers (Christensen et al., 1994). Leachate components were detected within a 3,280-ft radius from the landfill. In addition, evidence from the Richardson landfill in the Cannonsville watershed suggests that landfill components dissolved in groundwater can migrate from their source to local drinking water wells (EPA, 1999). These studies suggest that the 250- and 1,000-ft setbacks are not sufficiently protective for many landfill leachate constituents.

Petroleum Products and Hazardous Substances

There are no published studies on removal of petroleum substances and hazardous substances via setbacks or buffer zones. However, the transport and fate of these compounds through groundwater aquifers and unsaturated media are well studied. A recent study from the Lawrence Livermore National Laboratories and the University of California that included 271 petroleum-release sites in California suggests that most petroleum compounds do not migrate beyond 260 ft of their source in groundwater aquifers (Rice et al., 1995). This is attributed to natural microbial processes that break down petroleum compounds (a process known as natural attenuation). However, many hazardous-waste compounds are not amenable to biodegradation. In addition, petroleum storage tanks may contain hazardous substances that behave quite differently than petroleum compounds. The more highly mobile of these hazardous compounds are expected (and have been shown) to extend well beyond 260 ft of their source in certain circumstances. For example, the same studies on petroleum sites in California found that the fuel oxygenate MTBE migrated beyond the 260-ft mark at multiple sites. One Navy installation in California currently contains a groundwater plume of dissolved MTBE greater than 1,000 ft in extent (Department of the Navy, 1998). This research implies that the 100- and 500-ft setbacks separating hazardous-waste storage, petroleum storage, and heating oil storage are not sufficiently protective of water quality in nearby reservoirs.

Expert Panel Recommendations

To gain further insight into the MOA setbacks, 12 experts on buffer zone structure and functioning in the United States and Europe were polled for their opinions (see the Preface). Each expert was asked to judge the adequacy of the 22 setback provisions found in Table 10-1; six responses were received. The opinions were consistent with the conclusions drawn from the literature review.

Almost all respondents rated the setback requirements for hazardous-waste storage, petroleum storage, heating oil storage, and solid-waste landfills and junkyards as inadequate, with one exception—the 1,000-ft setback for landfills and junkyards around reservoirs was thought to be adequate. Although each respondent did not provide details, one scientist noted that "the risks associated with accidental contamination from these activities are too great." Almost all respondents thought that the setback requirements for pesticide application were adequate, which also supports the conclusion derived from the literature review.

The setbacks for septic systems and impervious surfaces were judged to be adequate by almost all the respondents. In this case, comparison with the literature review is not possible because the respondents were not asked to consider individual pollutants. The literature review supports the adequacy of these setbacks for sediment removal, but it is inconclusive or negative with respect to phosphorus and microbial pathogens. In addition, evidence gathered as part of the septic siting study (see Box 11-1) suggests that 100 ft is not sufficient to retard movement of viruses or bacteria in the subsurface of the Catskill/Delaware region (M. Sobsey, University of North Carolina, personal communication, 1999). Finally, respondents' opinions were mixed regarding the adequacy of the setbacks for subsurface-discharging WWTPs. The larger 500-ft setback required around reservoirs, reservoir stems, and controlled lakes was viewed more favorably than the 100-ft setback required around watercourses and wetlands. The potentially significant risks associated with accidental contamination from WWTPs were responsible for this split decision.

General Comments

Perhaps the most important point made clear from the literature review and the survey of buffer zone experts is that determining setback or buffer zone distances should be done on a site-specific basis. Research on buffer zone effectiveness has revealed several important environmental parameters that should be taken into account when determining buffer zone width. Castelle et al. (1994) note that there are at least four criteria that should be used to size buffer zones: (1) resource functional value (e.g., removal of nonpoint source pollution or maintenance of aquatic habitat), (2) intensity of adjacent land use, (3) buffer characteristics, and (4) specific buffer functions required (such as nitrogen removal). Most buffers are sized taking only the first criterion into consideration. That is, buffers are given a fixed width based on one parameter, rather than having variable widths that optimize all four parameters.

Several buffer characteristics are critical to predicting its pollutant removal capabilities. The most frequently cited parameter is slope (Barling and Moore, 1994; Phillips, 1989a,b). In fact, some models of buffer zone efficiency take only slope into consideration (Barling and Moore, 1994; Nieswand et al., 1990). Researchers have even recommended that any land with a slope greater than 15 percent should be considered unacceptable as a buffer zone (Nieswand et al., 1990). Phillips (1989a,b) has shown that buffer zone roughness (which corresponds to the amount of vegetation present), soil hydraulic conductivity, and soil moisture can also significantly affect buffer removal efficiencies.

Quantitative Analysis of Setbacks

This final section examines the potential effectiveness of the 100-ft setback by estimating travel times of subsurface flow from available site-specific data. It revisits a similar analysis found within the 1993 EIS for the Watershed Rules and Regulations. The inferences drawn from our analysis and salient research in the mid-Atlantic and southeastern United States are tempered with a discussion of sources of scientific uncertainty.

The status of knowledge about the structure and function of riparian forest buffers is changing rapidly. At present, we have a well-rounded *qualitative* understanding of the multiple functions and importance of the riparian zone in relation to the quantity, quality, timing of water flow, and other ecosystem attributes. The effectiveness of riparian forest buffers has been quantified for some water quality constituents (e.g., nitrogen, phosphorus, pesticides, and sediment) on several experimental sites (Lowrance et al., 1997) but is currently unknown for many other contaminants of concern.

Analysis of Setbacks in the 1993 Environmental Impact Statement

In 1993, NYC DEP prepared an evaluation of setback regulations as part of the EIS for the proposed Watershed Rules and Regulations. Various modeling approaches, primarily Darcy's Law for subsurface flow and Manning's equation for surface flows, were used to estimate pollutant travel times or other measures given certain setback distances, slope, hydraulic conductivity, porosity, roughness, and rainfall intensity. Table 10-5 presents the travel-time results of the EIS analyses for each of the setback provisions considered. For most activities, the pollutants were assumed to behave like conservative tracers. In general, the EIS shows that requiring setbacks will increase travel times, and this increase is dependent on the values used for the parameters mentioned above. In particular, 250- to 1,000-ft subsurface flow paths (used for underground storage of hazardous wastes, septic systems, and junkyards) generate travel times on the order of days to years, while surface flow paths of the same length generate travel times on the order of hours. The travel-time information can be used to gauge how much time would be available for remedial measures if a pollutant discharge were to occur. It should be noted that there was no quantitative analysis of the effect of setback distances adjacent to new impervious surfaces.

For pesticides (nonagricultural uses) and on-site sewage treatment and disposal systems (OSTDS), the analysis was extended to estimate percent pollutant removal for variable setback distances or changes in pollutant loading, respectively. The analysis for OSTDS evaluated the effect of variable setback distances, slope, and hydraulic conductivity on pollutant removal in the subsurface. Using removal efficiencies from the literature[2], physical data from the Catskill/Delaware watershed, and professional judgment, the analysis showed that for properly functioning OSTDS, setbacks between 100 and 550 ft would provide nearly 100 percent removal of biological oxygen demand, total suspended solids, coliforms, *Giardia,* and viruses. Removal efficiencies of setbacks around malfunctioning systems (those with less than 2 ft of unsaturated soil beneath the soil

---

[2] It was assumed that *Giardia* and coliform bacteria would be filtered out of OSTDS effluent in a relatively short distance (<100 ft).

TABLE 10-5 Model Evaluation of Setback Distance

Prohibited Activity	Location	Equation	Variables	Values Used	Travel Times
Hazardous Substance and Waste Storage, Generation, and Disposal	Above-ground	Manning's Equation	Roughness coefficient 24-hour rainfall Slope Setback length	0.011, 0.15, 0.24, 0.4 0.5 and 2.75 in 0.01 and 0.25 ft/ft 250 to 1,000 ft	0.5 to 2 hours
	Under-ground	Darcy's Law	Hydraulic conductivity Porosity Hydraulic gradient Setback length	0.352, 2.17, 15.7 in/hr 0.3 6% and 25% 250 to 1,000 ft	10 to 1,000 days
Petroleum Products	Above-ground	Manning's Equation	Roughness coefficient 24-hour rainfall Slope Setback length	0.011, 0.15, 0.24, 0.4 0.5 or 2.75 in 0.01 or 0.25 ft/ft 750 ft	0.04 to 6 hours
Septic Systems	Under-ground	Darcy's Law	Hydraulic conductivity Porosity Hydraulic gradient Setback length	0.352 to 15.7 in/hr 0.3 8%, 15%, and 25% 100 to 550 ft	60 days
Landfills	Above-ground	Manning's Equation	Roughness coefficient 24-hour rainfall Slope Setback length	0.011, 0.15, 0.24, 0.4 0.5 or 2.75 in 0.01 or 0.15 ft/ft 250 and 1,000 ft	0.01 and 3 hours
Junkyards, Composting, Sludge, Transfer Stations	Under-ground	Darcy's Law	Hydraulic conductivity Porosity Hydraulic gradient Setback length	0.352, 2.17, 15.7 in/hr 0.3 1%, 8% 250 to 1,000 ft	18 to 40,000 days
Pesticides	Above-ground	PESTRUN[a] and trap efficiency[b]	Setback length	250 to 1,000 ft	Pollutant loading estimated

[a] PESTRUN generates pollutant loadings rather than travel times.
[b] The "trap efficiency" method of Wong and McCuen (1982) only applies to particulate-phase pollutants. Trap efficiency is similar to pollutant removal.
Source: NYC DEP (1993).

adsorption system) were slightly lower, especially under conditions of high slope and hydraulic conductivity.

Pesticide loading to water bodies was estimated for chemicals attached to sediments in surface runoff. Trap efficiencies were based on results reported in Wong and McCuen (1982), which provides a graphical representation of percent trapping of suspended solids based on buffer length, slope, and cover conditions. Pesticides were divided into groups based on their adsorption properties, and half-lives were estimated. In general, the setback distances of 250 and 1,000 ft showed large load reductions (77 percent to 85 percent) for strongly adsorbed pesticides (group 1—glyphosate, diquat, and methoxychlor).

Many of the results in the EIS are directly influenced by the assumptions and limitations of the analysis. For instance, in determining trap efficiencies, it was assumed that the setback area was managed as a buffer zone and that channelized flow did not occur. Trap efficiencies were assumed to be zero in slopes greater than 15 percent, and dissolved pesticides were not considered. Although the study concluded that only a small fraction of the pesticides applied outside a 1,000-ft limiting distance would reach a reservoir, this conclusion is primarily based on best professional judgment and is only addressed for certain chemicals and conditions. Similarly, assumptions on pollutant removal downslope from OSTDS largely controlled the results generated in that analysis. Changing setback distances, slope, and hydraulic conductivity had minimal effect on pollutant removal.

Committee Analysis

Because the analysis in the EIS generates travel times rather than pollutant removals for almost all categories of setbacks considered, its use in predicting the effectiveness of the New York City setbacks is limited. However, given present modeling capabilities, the EIS analysis is still a relatively current approach. The types of data that could be used to refine the EIS analysis (e.g., pollutant concentrations emanating from different land uses and rates for degradation processes within a setback) are not available for the Catskill/Delaware watershed. The analysis that follows has three goals: (1) to generate new results on travel time, (2) to present the results in a format that highlights the interaction of the two most important physical parameters, and (3) to discuss assumptions that will substantially influence the predicted travel times.

Method. Unlike the EIS analysis, this analysis does not consider multiple land uses and their associated pollutant travel times through a setback. Rather, Darcy's Law (Hanks and Ashcroft 1980; Hillel, 1980) (Equation 10-2) is used to generate a single estimate that can be generically applied to all land uses and pollutants emanating from those land uses. The results are expressed as time of travel through a 100-ft setback for multiple combinations of slope and saturated hydraulic conductivity.

$$v = (K/n)(dh/dl) \qquad (10\text{-}2)$$

where v = average linear velocity (length/time)
      K = hydraulic conductivity (length/time)
      n = porosity (dimensionless)
      dh/dl = hydraulic gradient (dimensionless)

In preparation for flux calculations using Darcy's Law, USDA soil surveys from the Catskill/Delaware region were reviewed to determine the observed range of saturated hydraulic conductivity (K_s) of soils—0.06 to >20 in/hr, or 0.04 to >12.2 m/day. Slopes ranging from 0 to 45 degrees were used to represent the potential range of hydraulic gradients (sin[slope angle]) for lateral flow through the riparian zone into adjacent streams. The flow velocity through the soil matrix was determined by dividing flux by total porosity (0.5 was used for this analysis). Dividing the riparian setback width (100 ft) by flow velocity (ft/day) yields the time required to travel 100 ft in relation to multiple combinations of hydraulic conductivity and gradient.

The contours (isochrones) in Figure 10-5 show estimated travel (residence) times for combinations of saturated hydraulic conductivity (1–12.5 m/day, or 1.6–20.0 in/hr) and slope angle (1–25 degrees, or ~2–50 percent). Larger values of conductivity or gradient generate travel times of only a few days. Locations where both slope and hydraulic conductivity are high are rare since high-conductivity sand and gravel deposits typically occur in valley bottoms with limited gradients. Conductivity and gradient values less than 1 m/day and 1 degree, respectively, yield estimated travel times of hundreds or thousands of days. In the Catskill and Delaware watersheds, low-conductivity soils or low-gradient sites are usually riparian wetlands. The travel times generated in the EIS falls within the range presented in Figure 10-5.

Not all combinations of slope and hydraulic conductivity occur simultaneously in the Catskill/Delaware watershed. A GIS could be used to map those areas of Figure 10-5 that represent conditions within the watershed. Data on terrain and soil layers could be used to generate spatial statistics for hydraulic conductivity and gradient in the riparian zone. The GIS also could be used to cross-tabulate soil and gradient data to characterize their association at the landscape scale, sometimes referred to as the drainage catena. For example, do clay and silt loam soils (low-conductivity) largely occur on gentle slopes while sandy loams and stony till soils (high-conductivity) are found on steeper slopes? A digital elevation model in the GIS could be used to quantify other key watershed characteristics such as contributing area, slope shape (concave, planar, convex), landform (divergent, planar, convergent), vegetative cover, and land use—all of

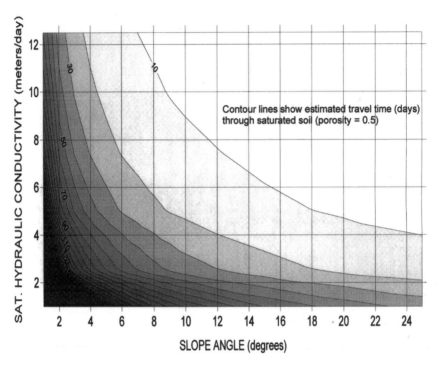

FIGURE 10-5 Estimated pollutant travel times for different combinations of soil hydraulic conductivity and slope.

which may influence the pathway and rate of water movement through the riparian zone (Bevin and Wood, 1983; O'Loughlin, 1986).

Assumptions, Field Conditions, Interpretation, and Inference. Several implicit and explicit assumptions influence the travel-time estimates. In essence, the data presented in Figure 10-5 are an idealized and simplified representation of field conditions. Saturated hydraulic conductivity data are derived from a laboratory determination on nominally undisturbed 3-inch (7.62-cm) diameter cores. Macropores (sometimes called preferential flow paths or soil pipes) formed by roots, organic matter, microbes, earthworms, and other invertebrates, and small mammals may increase *in situ* conductivity by several orders of magnitude (Mosley, 1982; Mullholland et al., 1990). Cobbles, gravel, and other coarse fractions (>2 mm) also may increase effective conductivity. Both sources of heterogeneity, which happen to be common in forest soils, *decrease* travel times for subsurface flow.

The analysis also assumes no overland flow in or through the riparian zone. Overland flow is uncommon but can occur when soil frost or compaction reduces infiltration capacity. Saturation from below can occur when lateral inflow from upslope areas, combined with rainfall or snowmelt, exceeds the outflow—usually where subsurface flow converges at the base of concave slope. Impervious surfaces adjacent to the riparian zone can collect and concentrate water in rills or ephemeral channels that simply cut across the riparian zone. (The MOA mandates NYC DEP approval of new development and of modifications to existing infrastructure to prevent this circumstance. This, however, presents a substantial challenge for permitting, enforcement, and long-term maintenance.) Whatever the cause(s), overland flow short-circuits the subsurface flow path and *decreases* travel time.

The analysis assumes complete saturation of the soil mantle. This is a conservative assumption for at least two reasons. First, flow through the saturated zone is considerably faster than flow through the unsaturated zone. Flow through the unsaturated zone is both vertical and horizontal, while saturated flow is predominantly horizontal. In addition, unsaturated hydraulic conductivities may be as much as six to ten orders of magnitude lower than saturated hydraulic conductivities. Second, saturated conditions are uncommon in the Catskill/ Delaware watershed. In this region, evaporation of soil water and transpiration by plants routinely produce unsaturated conditions. Rainfall is intercepted by and subsequently evaporated off leaf surfaces or enters the soil and is promptly sequestered by plants. During winter, precipitation is stored in a persistent snow-pack. Deprived of new inflow, continuous subsurface flow steadily reduces soil water content. Consequently, there are only brief periods of the year—in October or November after prolonged rains and March or April during snowmelt—when soils approach a saturated condition. Hence, the direct and indirect effects of vegetation may override the influence of climate, soil hydraulic properties, terrain features, and other physical attributes and have the net effect of *increasing* travel time.

Summary

The foregoing discussion highlights multiple sources of uncertainty, their potential interaction and effect upon the interpretation of Figure 10-5, and the efficacy of 100-ft setbacks. Until research in the Catskill/Delaware region, or transferable work on similar sites, quantifies the performance of riparian setbacks and buffers, conservative interpretation and common-sense application of well-established basic principles, salient research, and operational experience should guide riparian zone protection efforts. Specifically, the width of riparian setbacks should be increased as slope steepness and/or soil permeability yields insufficient residence time to effectively assimilate pollutants. Conversely, zoning variances and setback reductions should be granted when *in situ* data and information consistently demonstrate residence times and biological activity sufficient to meet in-stream water quality standards. Although 100-ft setbacks have become the default standard for watershed protection in the United States (Welsch, 1991), it is unlikely that a "one-size-fits-all" approach will be optimal in any particular system.

CONCLUSIONS AND RECOMMENDATIONS

1. The setbacks prescribed in the MOA are not buffer zones. As described in the MOA, the setbacks are distances between activities and waterbodies. Setback descriptions do not discuss the characteristics of the setback land that are known to influence pollutant removal in buffer zones, such as slope, hydraulic conductivity, soil moisture, vegetation or surface roughness, and flow rates. Only setback width is defined.

2. Active management for the setbacks is necessary to achieve the pollutant removal efficiencies attributed to buffer zones. NYC DEP should create incentives for managing setback areas to meet the following goals:
• Setbacks function most effectively as buffer zones if runoff entering the setbacks is sheet flow rather than channelized flow. Thus, best management practices such as filter strips, bioretention areas, and level spreaders should be installed upslope from buffer zones to create sheet flow in areas prone to significant concentrated flow. Converting channelized flow to sheet flow can be extremely difficult in areas where storm drains and pipes convey runoff to nearby bodies of water.
• Setbacks must be naturally regenerated or planted with the appropriate vegetation for retaining nutrients, sediment, and other pollutants. In most cases forested setbacks are the most effective buffers.
• If setbacks are managed as buffers, they should be managed as described in the USDA three-zone buffer specification (Figure 10-2), and consideration should be given to periodic vegetation harvesting in Zones 2 and 3.
• Compensation to private owners may be required to permit or conduct management activities in setback areas.

3. Setback distances in the New York City watersheds are similar to or greater than those found in other locations. This suggests that they will be as effective as setback requirements found elsewhere.

4. NYC DEP should set a slope threshold above which land cannot be included in setback considerations. The literature review suggests that areas with a slope of greater than 15 percent do not function as effective buffer zones. Fixed-width setbacks are desirable because of their easy demarcation and implementation compared to variable-width setbacks. However, fixed-width setbacks may pose unfair burdens on some landowners with primarily riparian properties, and they may be underprotective in areas with increased travel times (such as steep regions).

5. The existing state of knowledge on pollutant transport in buffer zones and the lack of site-specific information about pollutants that might be derived from activities behind a setback make a detailed quantitative analysis somewhat limited. Based on existing literature, an expert panel, and two analyses of travel times, the following setback requirements are judged most likely to be inadequate:
- 100 and 500 ft for hazardous wastes
- 100 and 500 ft for petroleum underground storage tanks
- 100 and 500 ft for heating oil
- 250 and 1,000 ft for landfills
- 100 ft for septic systems, impervious surfaces, and WWTPs because of potential breakthrough of microbial pathogens.

6. NYC DEP should undertake a program of field research to better justify and/or amend the current setbacks. In order to evaluate the effectiveness of particular setback distances, significantly more detailed data are needed on land use and pollutant transport through buffer areas to recipient streams and reservoirs. Performance monitoring of riparian buffer zones, in which shallow subsurface pollutant concentrations are measured above and below the buffer zone, is required. Integration of these data is possible by use of remote sensing and GIS techniques. Data of very fine resolution are needed prior to extensive application of predictive models. It should be noted that performance monitoring of buffer zones is much more difficult in the presence of channelized flow compared to sheet flow.

7. Because of the uncertainty associated with evaluating specific setback distances, the present setback distances should not be reduced for any activities or land uses to which they currently apply. NYC DEP should consider requiring setbacks for new agricultural activities.

8. Buffer zones should not be relied upon to provide the sole nonpoint source pollution control and are instead best used when integrated with appropriate source controls on pollutant releases. Although riparian buffers can ameliorate some nonpoint sources of pollution, they are most effective when used as part of an overall pollution control or conservation plan.

9. The MOA setbacks are likely to be found constitutional based on judicial precedents elsewhere, unless they preclude all economic use of a parcel of land.

REFERENCES

Barling, R. D., and I. D. Moore. 1994. Role of buffer strips in management of waterway pollution: A review. Environmental Management 18(4):543–558.

Barten, P. K., T. Kyker-Snowman, P. J. Lyons, T. Mahlstedt, R. O'Connor, and B. A. Spencer. 1998. Managing a watershed protection forest. Journal of Forestry 96(8):10–15.

Bevin, K., and E. F. Wood. 1983. Catchment geomorphology and the dynamics of runoff contributing areas. Journal of Hydrology 65:139–158.

Burt, T. P. 1997. The hydrological role of floodplains within the drainage basin system. Pp. 21–32 In Haycock, N., T. Burt, K. Goulding, and G. Pinay (eds.) Buffer Zones: Their Processes and Potential in Water Protection. Harpenden, UK: Quest Environmental.

Castelle, A. J., A. W. Johnson, and C. Conolly. 1994. Wetland and stream buffer size requirements—A review. J. Environ. Qual. 23:878–882.

Christensen, T. H., P. Kjeldsen, H. J. Albrechtsen, G. Heron, P. H. Nielsen. P. L. Bjerg, and P. E. Holm. 1994. Attentuation of landfill leachate pollutants in aquifers. Critical Reviews in Environmental Science and Technology 24(2):119–202.

Claytor, R., and T. Schueler. 1996. Design of Stormwater Filtering Systems. Ellicott City, MD: Chesapeake Research Consortium, Center for Watershed Protection.

Correll, D. L., and D. E. Weller. 1989. Factors limiting processes in freshwater wetlands: An agricultural primary stream riparian forest. Pp. 9–23 In Sharitz, R., and J. Gibbons (eds.) Freshwater Wetlands and Wildlife. Oak Ridge, TN: U.S. Department of Energy.

Correll, D. L. 1991. Human impact on the functioning of landscape boundaries. Pp. 90–109 In Holland, M. M., P. G. Risser, and R. J. Naiman (eds.) The Role of Landscape Boundaries in the Management and Restoration of Changing Environments. New York, NY: Chapman and Hall.

Coyne, M. S., R. A. Gilfillen, R. W. Rhodes, and R. L. Blevins. 1995. Soil and fecal coliform trapping by grass filter strip during simulated rain. J. Soil & Water Conserv. 50:405–408.

Crane, S. R., J. A. Moore, M. E. Gismer, and J. R. Miller. 1983. Bacterial pollution from agricultural sources: a review. Trans. ASAE 26:858–872.

Department of the Navy. 1998. Department of Navy Environmental Restoration Plan for Fiscal Years 1998–2002. Arlington, VA: Chief of Naval Operations.

Dickey, E. C., and D. H. Vanderholm. 1981. Vegetative filter treatment of livestock feedlot runoff. J. Environ. Qual. 10:279–284.

Dillaha, T. A., J. H. Sherrard, D. Lee, S. Mostaghimi, and V. O. Shanholtz. 1988. Evaluation of vegetative filter strips as a best management practice for feedlots. Journal of the Water Pollution Control Federation 60:1231–1238.

Dillaha, T. A., R. B. Renau, S. Mostaghimi, and D. Lee. 1989a. Vegetative filter strips for agricultural nonpoint source pollution control. Transactions American Society of Agricultural Engineers 32:513–519.

Dillaha, T. A., J. H. Sherrard, and D. Lee. 1989b. Long-term effectiveness of vegetative filter strips. Water Environment and Society 1:419–421.

Dillaha, T. A., and S. P. Inamdar. 1997. Buffer zones as sediment traps or sources. Pp. 33–42 In Haycock, N. et al. (eds.). Buffer Zones: Their Processes and Potential in Water Protection. Harpenden, UK: Quest Environmental.

Environmental Protection Agency (EPA). 1995. Water Quality Functions of Riparian Buffer Systems in the Chesapeake Bay Watershed. EPA 903-R-95-004. Washington, DC.

EPA. 1999. National Priority Site Fact Sheet. http://www.epa.gov/regimo2/superfnd/.

Fennessy, M. S., and J. K. Cronk. 1997. The effectiveness and restoration potential of riparian ecotones for the management of nonpoint source pollution, particularly nitrate. Critical Reviews in Environmental Science and Technology 27(4):285–317.

Franklin, E. C., J. D. Gregory, and M. D. Smolen. 1992. Enhancement of the Effectiveness of Forested Filter Zones by Dispersion of Agricultural Runoff. Report Number UNC-WRRI-92-270. Raleigh, NC: Water Resources Research Institute.

Hagedorn, C., D. T. Hansen, and G. H. Simonson. 1978. Survival and movement of fecal indicator bacteria in soil under conditions of saturated flow. J. Environ. Qual. 7(1):55–59.

Hanks, R. J., and G. L Ashcroft. 1980. Applied Soil Physics. New York, NY: Springer-Verlag.

Haycock, N. E., T. P. Burt, K. W. T Goulding, and G. Pinay. 1997. Introduction. Pp. 1–3 In Haycock, N., T. Burt, K. Goulding, and G. Pinay (eds.) Buffer Zones: Their Processes and Potential in Water Protection. Harpenden, U.K: Quest Environmental.

Hillel, D. 1980. Fundamentals of Soil Physics. New York, NY: Academic Press.

Jordan, T. E., D. L. Correll, and D. E. Weller. 1993. Nutrient interception by a riparian forest receiving inputs from adjacent cropland. J. Environ. Qual. 22:467–473.

Lowrance, R. R., R. L. Todd, and L. E. Asmussen. 1983. Waterborne nutrient budgets for the riparian zone of an agricultural watershed. Agric. Ecosys. Environ. 10:371–384.

Lowrance, R., L. S. Altier, J. D. Newbold, R. S. Schnabel, P. M. Groffman, J. M. Denver, D. L. Correll, J. W. Gilliam, and J. L. Robinson. 1997. Water quality functions of riparian forest buffers in Chesapeake Bay watersheds. Environmental Management 21(5):687–712.

Magette, W. L., R. B. Brinsfield, R. E. Palmer, and J. D. Wood. 1989. Nutrient and sediment removal by vegetated filter strips. Transactions American Society of Agricultural Engineers 32:663–667.

Metropolitan District Commission (MDC). 1995. Quabbin Watershed: MDC Land Management Plan 1995–2004. Boston, MA: MDC.

Mosley. M. P. 1982. Subsurface flow velocities through selected forest soils, South Island, New Zealand. Journal of Hydrology 55:65–92.

Mullholland, P. J., G. V. Wilson, and P. M. Jardine. 1990. Hydrogeochemical response of a forested watershed to storms: Effect of preferential flow along shallow and deep pathways. Water Resources Research 26:3021–3036.

National Resources Conservation Service (NRCS). 1995. Riparian Forest Buffer, 392. Seattle, WA: USDA NRCS Watershed Science Institute.

New York City Department of Environmental Protection (NYC DEP). 1993. Draft Generic Environmental Impact Statement for the Draft Watershed Regulations for the Protection from Contamination, Degradation, and Pollution of the New York City Water Supply and Its Sources. Corona, NY: NYC DEP.

NYC DEP. 1997. Water Quality Surveillance Monitoring. Valhalla, NY: NYC DEP.

Nieswand, G. H., R. M. Hordon, T. B. Shelton, B. B. Chavooshian, and S. Blarr. 1990. Buffer strips to protect water supply reservoirs: A model and recommendations. Water Resource Bulletin 26(6):959–965.

O'Loughlin, E. M. 1986. Prediction of surface saturation zones in natural catchments by topographic analysis. Water Resources Research 22:794–804.

Osborne, L L., and D. A. Kovacic. 1993. Riparian vegetated buffer strips in water quality restoration and stream management. Freshwater Biology 29:243–258.

Patty, L., J. J. Gril, B. Real, and C. Guyot. 1995. Grassed buffer strips to reduce herbicide concentration in runoff—Preliminary study in Western France. Symposium at the University of Warwick, Coventry, UK, April 3–5, 1995. BCDC Monograph #62: Pesticide Movement to Water (1995):397–406.

Patty, L., B. Real, and J. J. Gril. 1997. The use of grassed buffer strips to remove pesticides, nitrate, and soluble phosphorus compounds from runoff water. Pestic. Sci. 49:243–251.

Pavelka, C., R. C. Loehr, and B. Haikola. 1994. Hazardous waste landfill leachate characteristics. Waste Management 13(8): 573–580.

Peterjohn, W. T., and D. L. Correll. 1984. Nutrient dynamics in an agricultural watershed: Observation on the role of a riparian forest. Ecology 65:1466–1475.

Phillips, J. D. 1989a. Nonpoint source pollution control effectiveness of riparian forests along a coastal plain river. Journal of Hydrology 110:221–237.

Phillips, J. D. 1989b. An evaluation of the factors determining the effectiveness of water quality buffer zones. Journal of Hydrology 107:133–145.

Platt, R. H. 1992. An Eroding Base. 9 The Environmental Forum 10–11.

Platt, R. H. 1994. Parsing Dolan. Environment 36(8):4–5, 43.

Platt, R. H. 1996. Land Use and Society: Geography, Law, and Public Policy. Washington, DC: Island Press.

Rice, D. W., J. C. Michaelson, B. P. Dooher, D. H. MacQueen, S. J. Cullen, L. G. Everett, W. E. Kastenberg, R. D. Grose, and M. A. Marino. 1995. California's Leaking Underground Fuel Tank (LUFT) Historical Case Analyses. Lawrence Livermore National Laboratory and the University of California.

Robbins, R. W., Glicker, J., Bloern, D. M., Niss, B. M. 1994. Effective Watershed Management for Surface Water Supplies. AWWA Research Foundation and the AWWA.

Schellinger, G. R., and J. C. Clausen. 1992. Vegetative filter treatment of dairy barnyard runoff in cold regions. J. Environ. Qual. 21:40–45.

Schueler, T. 1995. The importance of imperviousness. Watershed Protection Techniques 1(3):100–112.

Schueler, T. 1996. The Architecture of Urban Stream Buffers. Watershed Protection Techniques 1(4):155–165.

Schultz, R. C., J. Colletti, W. Simpkins, M. Thompson, and C. Mize. 1994. Developing a multispecies riparian buffer strip agroforestry system. Pp. 203–225 In R. Lowrance, ed., Riparian Ecosystems in the Humid U.S.: Functions, Values and Management. Nat. Assoc. Conserv. Districts, WA.

Schwer, C. B., and J. C. Clausen. 1989. Vegetative filter treatment of dairy milkhouse wastewater. J. Environ. Qual. 18:446–451.

Sharpley, A. N. 1985. Depth of surface soil-runoff interaction as affected by rainfall, soil slope, and management. Soil Sci. Soc. Am. J. 49:1010–1015.

Sharpley, A. N., S. J. Smith, O. R. Jones, W. A. Berg, and G. A. Coleman. 1992. The transport of bioavailable phosphorus in agricultural runoff. J. Environ. Qual. 21:30–55.

Sheridan, J. M., R. Lowrance, and D. D. Bosch. 1999. Management effects on runoff and sediment transport in riparian forest buffers. Transactions of ASAE 42:55–64.

Syversen, N. 1995. Effect of vegetative filters on minimizing agricultural runoff in Southern Norway. Pp. 70–74 in Persson, R. (ed.) Agrohydrology and Nutrient Balances. Uppsala: Swedish University of Agricultural Science.

Tchobanoglous, G., H. Theisen, and S. Vigil. 1993. Integrated Solid Waste Management: Engineering Principles and Management Issues. New York, NY: McGraw-Hill, Inc..

Uusi-Kamppa, J., and T. Ylaranta. 1996. The use of vegetated buffer strips on controlling soil erosion and nutrient losses in Southern Finland. In Mulamoottil, G., B. G. Warner, and E. A. McBean (eds.) Wetlands: Environmental Gradient, Boundaries, and Buffers. Boca Raton: Lewis Publishers.

Uusi-Kamppa, J., E. Turtola, H. Hartikainen, and T. Ylanranta. 1997. The interaction of buffer zones and phosphorus runoff. Pp. 43–54 In Haycock, N. et al. (eds.) Buffer Zones: Their Processes and Potential in Water Proteciton. Harpenden, UK: Quest Environmental.

Vought, L. B.-M., J. Dahl., C. L. Pedersen, and J. O. Lacoursiere. 1994. Nutrient retention in riparian ecotones. Ambio 23:342–348.

Walker, S. E., S. Mostaghimi, T. A. Dillaha, and F. E. Woeste. 1990. Modeling animal waste management practices: Impacts on bacterial levels on runoff from agricultural land. Transactions American Society of Agricultural Engineers 33:807-817.

Welsch, D. J. 1991. Riparian forest buffers: Function and design for protection and enhancement of water resources. USDA Forest Service NA-PR-07-91.

Whipple, W. 1993. Buffer Zones around Water-Supply Reservoirs. J. Water Res. Plann. Manage. 119(4):495-499.

Wong and McCuen. 1982. The design of vegetable buffer strips for runoff and sediment control. A technical paper developed as part of the Maryland Coastal Zone Management Program. College Park, MD: University of Maryland.

Yates, M. V., and S. R. Yates. 1989. Septic tank setback distances: A way to minimize virus contamination of drinking water. Ground Water 27(2): 202–208.

Yates, M. V., S. R. Yates, A. W. Warrick, and C. P. Gerba. 1986. Use of geostatistics to predict virus decay rates for determination of septic tank setback distances. Applied and Environmental Microbiology 52(3):479–483.

Young, R. A., T. Huntrods, and W. Anderson. 1980. Effectiveness of vegetated buffer strips in controlling pollution from feedlot runoff. J. Environ. Qual. 9:483–487.

11

Wastewater Treatment

Treatment and disposal of wastewater in the Catskill/Delaware watershed is a major factor in determining the quality of New York City drinking water. This is because almost all wastewater from the region is discharged either directly into streams that feed the water supply reservoirs or into the subsurface where it can eventually migrate to the reservoirs. Over 30,000 on-site sewage treatment and disposal systems (OSTDS) and 41 centralized wastewater treatment plants (WWTPs) are the major sources of wastewater in the watershed. Although the quality and quantity of wastewater discharged from WWTPs is known from monitoring data, such information is not available for OSTDS. In addition, the aggregate impact of either type of discharge on an individual reservoir, or on the system of water supply reservoirs, is extremely difficult to estimate.

A qualitative approach was used to overcome these limitations and answer the following critical questions:

1. Can the watershed sustain new WWTPs and OSTDS without declines in water quality?

2. Are effluent standards adequate, and are proposed technologies for WWTPs and OSTDS appropriate?

3. Are the rules governing the locations of new WWTPs and OSTDS adequate?

These questions were answered by determining the effects of technology upgrades for WWTPs and OSTDS on overall pollutant loading to the Catskill/Delaware watersheds. Some of these technology upgrades are mandated by the Memorandum of Agreement (MOA), while others are not. For example, all surface-discharging WWTPs are to be fitted with microfiltration units or their equivalent.

It should be noted that this type of evaluation was conducted for WWTPs in the 1993 environmental impact statement (EIS) created for the Watershed Rules and Regulations (NYC DEP, 1993). However, upgrades to OSTDS were not considered in the 1993 EIS. Rather, that document evaluated the effect of variable OSTDS setback distances on pollutant loading to reservoirs. The similarities and differences between this analysis and that of the EIS are discussed throughout this chapter.

ANALYSIS OF WASTEWATER TREATMENT PLANTS AND ON-SITE SEWAGE TREATMENT AND DISPOSAL SYSTEMS

Impact Index

The approach used by the committee views each reservoir in the water supply system as a "black box" that receives effluent from OSTDS and WWTPs. The volume and pollutant concentrations of OSTDS and WWTP effluent that are currently entering the Catskill/Delaware reservoirs represent a baseline "impact index" of 100 percent. Improved effluent quality related to better treatment technology or regulatory strategies decreases the impact index to a percentage less than 100. Increased effluent volume related to population growth increases the impact index by a proportional percentage. The questions posed above are answered by estimating the impact index produced by various strategies or options. For the purpose of this study, the New York City water supply system was limited to those watersheds west of the Hudson River as well as those supplying the Kensico and West Branch reservoirs.

Parameters

Six parameters were selected to represent the quality of effluent from OSTDS and WWTPs: total phosphorus, total suspended solids (TSS), fecal coliforms, viruses, *Giardia* cysts, and *Cryptosporidium* oocysts. Each of these parameters is considered a priority pollutant by New York watershed planners as evidenced by their inclusion in the EIS prepared for the Watershed Rules and Regulations (NYC DEP, 1993).

The efficacy of removal of these six parameters varies between OSTDS and WWTPs (see Appendix D for more detail). Both processes are efficient in removing TSS. They are also effective in removing fecal coliforms because these bacteria die when exposed to the intolerable environmental conditions presented by either process. WWTPs are much more effective than passive OSTDS in removing dissolved phosphorus. Microorganisms in the WWTP treatment process use the phosphorus in the wastewater as a nutrient, and it is subsequently deposited in the residuals. Phosphorus may also be removed chemically by tertiary WWTP treatment processes. Dissolved phosphorus passes through the septic tank of a passive OSTDS and enters the drainfield, where it is either used by aerobic microorganisms in the soil or adsorbed to soil particles. In either case, the ability of the soil to retain phosphorus is eventually exhausted, after which phosphorus-rich water can enter the environment. WWTPs are also more effective than OSTDS in eliminating viruses and bacteria because the effluent of WWTPs is chlorinated. Finally, the requirement for microfiltration at WWTPs likely enhances their ability to remove *Giardia* cysts and *Cryptosporidium* oocysts in comparison to OSTDS.

Treatment Options

Treatment technologies that produce the highest-quality effluent, termed Best Available Control Technology (BACT), are required for WWTPs discharging to surface water under the terms of the MOA. BACT for WWTPs includes sand filtration, disinfection, phosphorus removal, and microfiltration, and all upgrades are scheduled for completion by 2002. Although not mandated in the New York City watersheds, BACT for OSTDS consists of aerobic treatment units (ATUs) that use mechanical devices such as pumps and impellers to create aerobic conditions within the septic tank. Support media may also be provided to increase the relative number of aerobic bacteria colonizing the tank, and a conventional drainfield is used after the aerobic unit. Dead bacteria, and associated phosphorus, are retained in the sludge layer in the tank, and this enhances the ability of ATUs to remove phosphorus. When these systems are properly maintained, the effluent quality of these systems is significantly better than that of passive septic tank and drainfield systems. Although there are other innovative aerobic OSTDS available (see Chapter 5), in the committee's opinion, only ATUs can be considered best available control technology because they have been field-tested and backed by third-party certification.

The existence of an effective regulatory strategy is an additional element that must be present for an OSTDS treatment process to be considered BACT. Failure rates and the period of time during which failed systems are a public health hazard can both be reduced by rigorous inspection and enforcement strategies. Typically, failures occur when solids in the tank become too deep and wash into the drainfield. These solids clog the perforated pipe and underlying biomat and restrict the amount of liquid that can percolate through the soil. This results in partially treated sewage surfacing and potentially flowing into surface waters. Such failures often back wastewater into the residence or limit the use of toilets and laundries, which serves as an incentive for the homeowner to have the OSTDS repaired. Regulatory strategies that rely on self-reporting result in a failure rate (defined as an inability of a system to accept wastewater) of about five percent during the year (Sherman, 1998). More stringent regulatory strategies, such as requiring inspection upon sale of the property, can decrease this failure rate.

In theory, a requirement for annual operating permits that mandate inspection of OSTDS each year can bring the annual failure rate to zero, and this strategy was assumed for the BACT calculations. Annual inspection must accompany the use of aerobic systems because these systems rely on mechanical devices that require electrical power and need preventative maintenance. In addition, aerobic systems produce a far greater volume of sludge than passive systems and require much more frequent pumping of sludge. Some states such as Florida already require annual operating permits for such systems, and these provide both revenue for inspections and a legal tool if enforcement is needed. Florida also requires aerobic systems to be under a maintenance contract to minimize the need for emergency repairs and to make sure maintenance is performed by qualified individuals with timely access to spare parts, rather than by homeowners. Maintenance is particularly important because ATUs function optimally when in continuous use and can experience disruption of function when shut down for extended periods of time. In fact, the addition of recirculating sand filters to ATUs is recommended for those systems operated on a seasonal basis to maintain a supply of aerobic bacteria. It should be noted that not all states have rigorous inspection and maintenance requirements, nor do they always enforce existing requirements, which can lead to significantly higher failure rates for ATUs (P. Miller, Virginia Department of Natural Resources, personal communication, 1999). A zero failure rate was assumed for this analysis in order to represent the greatest possible change that could accompany the use of BACT for OSTDS.

Effluent Pollutant Concentrations

Wastewater Treatment Plants. Determining the contribution of WWTPs to overall pollutant loading in a watershed is relatively straightforward. The daily discharge of each pollutant by a WWTP can be estimated with certainty from monitoring data required under the terms of its State Pollutant Discharge Elimination System (SPDES) permit. Performance testing data can be used to determine the decrease in effluent pollutant concentrations following installation of BACT. Quantitative increases in effluent pollutant concentrations arising from future growth can be estimated using the maximum discharge limits in the SPDES permit and/or estimates of growth and average per capita wastewater production.

On-Site Sewage Treatment and Disposal Systems. There is much less certainty regarding daily discharges of the priority pollutants from OSTDS. Removal efficiency of the six constituents by OSTDS is related to the type and age of the OSTDS, geological conditions, and loading rates. Therefore, it is necessary to make a number of assumptions in order to estimate the current daily discharge of each parameter from OSTDS. Improvements in effluent quality for OSTDS can be estimated by applying treatment efficiencies for ATUs and by assuming that a failure rate of zero is achieved through a rigorous enforcement strategy. Quantitative increases in effluent from OSTDS caused by population increases can also be estimated using projected rates of population growth and average per capita wastewater production. The assumptions used for both WWTPs and OSTDS and the calculated values derived from these assumptions are given in Appendix D.

Spatial Considerations. Data for effluent quality from both WWTPs and OSTDS do not consider changes between the discharge point and the receiving reservoir. The discharge point for WWTPs is typically a small watercourse, while effluent quality for OSTDS is usually measured 2 ft below the drainfield. In either case, travel of the effluent to the reservoir may result in qualitative improvements in the six parameters of concern. These changes have not been incorporated into these estimates because of a lack of studies on such changes and because of variations in locations and treatment techniques of individual OSTDS and WWTPs. Information from the Septic Siting Study currently being conducted in the watershed (see Box 11-1) may shed light on the fate and transport of

---

BOX 11-1
The Septic Siting Study

The Septic Siting Study is evaluating the transport and fate of indicator organisms at six subsurface locations throughout the New York City watersheds—two in the Croton watershed, two in the Catskill watershed, and two in the Delaware watershed. The experiments in this study involve spiking septic tank effluent with indicator organisms that would mimic the behavior of enteric viruses and protozoa. Multiple groundwater wells downstream of the drainfields are used to collect monthly samples, which are analyzed for the indicator organisms. Experiments are being conducted under both wet (spring) and dry (autumn) conditions. Each of the sites is using a septic system built or maintained according to the state health standards (75A). Preliminary work at the sites yielded information about soil characteristics and water table topography that was used to design the septic systems and optimize placement of the groundwater wells. Routine monitoring has been ongoing for 18 months, and results from the study are expected in December 1999. Information from the study could be used to refine this report's analysis of the impact of OSTDS.

pollutants beyond the drainfield. Our analysis of pollutant removal in setbacks (Chapter 10) and a similar exercise from the 1993 EIS (NYC DEP, 1993, Table VIII.F-12) indicate that subsurface travel of sewage effluent can significantly decrease concentrations of some pollutants, particularly TSS and microbial constituents. On the other hand, overland or surface flow is not expected to decrease effluent concentrations to the same extent (NYC DEP, 1993, Table VIII.F-14). Because almost all WWTPs in the watershed discharge to surface waters, they are less likely than OSTDS to receive the beneficial effects of subsurface travel on effluent pollutant concentration. Thus, we expect this analysis to slightly overestimate the contribution of OSTDS to the overall pollutant loading in those basins with WWTPs.

Impact Index vs. EIS Analysis. As noted above, the 1993 EIS contains an analysis of upgrades to WWTPs similar to the impact index presented here. The same information on current effluent pollutant concentrations and future concentrations following upgrades was used in both analyses. However, methods used to calculate effluent concentrations in 2010 are different. The EIS applied various methods for calculating the population that would be served by municipal plants, nonmunicipal plants and expanding plants. The impact index applied the growth rate observed for the period 1990–1996, developed in Chapter 2, to the period from 1990 to 2010. The annual growth rate of 0.25 percent for the West-of-Hudson region resulted in a total population increase of 5 percent between 1990 and 2010. The annual growth rate of 1.06 percent for the East-of-Hudson watersheds resulted in a total population increase of 21.2 percent between 1990 and 2010.

Regarding OSTDS, the EIS did not consider the effect of treatment process upgrades on current and future pollutant loading, which is the goal of the impact index. For that reason, a meaningful comparison cannot be made between the two analyses. However, a few issues are worthy of discussion. First, the EIS estimates that 14,600 new OSTDS will be constructed in the Catskill/Delaware watershed between 1990 and 2010. These additional units are taken into consideration in the impact index by estimating future growth (using the same growth rates quoted above). Second, the EIS supports failure rates for OSTDS of 2 percent to 2.5 percent, depending on the type of system. The impact index assumes that all systems without technology upgrades suffer a 5 percent failure rate, while those that install BACT have a zero percent failure rate. Finally, the EIS analysis implies that only those OSTDS residing within proscribed setback distances (between 100 and 500 ft) significantly contribute to pollutant loadings in the reservoirs. (Likewise, the Total Maximum Daily Load calculations for all basins only considered those OSTDS within 100 ft of waterbodies.) The impact index does not exclude any OSTDS based on its proximity to nearby waterbodies.

Results

The impact index for each reservoir in the West-of-Hudson watershed, as well as for the Kensico and West Branch watersheds, is presented graphically in Figures 11-1 through 11-4. These figures provide the total daily discharge of each of the six index parameters under baseline conditions, and they project changes if BACT is implemented and for population growth until the year 2010.

LONG-TERM WATER QUALITY CHANGES

From a simple mass-balance perspective, new sources of wastewater (e.g., population growth or new industry) in a watershed will degrade water quality. This can only be prevented by transporting the effluent to a discharge point outside the watershed or by using treatment technology that does not increase the level of contaminants above the baseline (i.e., zero-discharge). Neither of these methods is practicable in the New York City watersheds; therefore, some degradation of water quality will be associated with new WWTPs and OSTDS.

The magnitude and impact of a decline in water quality depend on the treatment efficiency of the new WWTP or OSTDS receiving the wastewater. First, the impact of new growth can be mitigated by rerouting wastewater flow from relatively inefficient OSTDS to more efficient WWTPs. A second technique is to upgrade existing WWTPs or OSTDS so that the combined impact of the newly upgraded plant and new growth is equal to or less than the baseline impact. This latter technique is currently being used for WWTPs because the MOA requires existing WWTPs to install BACT. The potential for BACT to greatly improve effluent quality is shown dramatically in Figures 11-1 through 11-4.

Role of Best Available Control Technology

Wastewater Treatment Plant Effluent

According to the impact index analysis, the largest reductions attributable to the use of BACT at WWTPs occurred with phosphorus, total suspended solids, and *Cryptosporidium* oocysts. Reductions for these three parameters exceeded 95 percent in the Cannonsville watershed and were also significant in the Ashokan, Pepacton, and Schoharie watersheds. There were also large decreases in *Giardia* cysts, coliforms, and viruses in all three of these watersheds. There was no reduction in contaminant levels for the six parameters in the Rondout watershed and little reduction in the West Branch watershed. This indicates that existing WWTPs in those watersheds are already at or near BACT. The Kensico and Neversink watersheds have no WWTPs and therefore do not benefit from any offsets brought about by BACT.

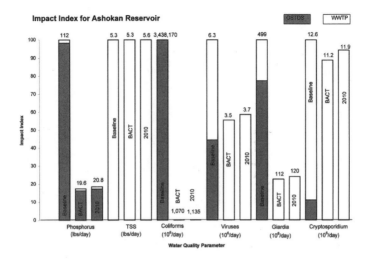

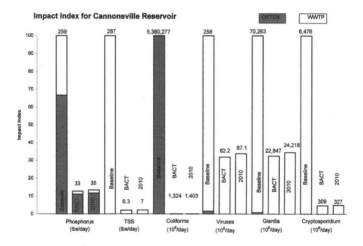

FIGURE 11-1 Impact index for Ashokan and Cannonsville reservoirs. Each of the six water quality parameters is represented by a cluster of three bars. The leftmost bar in each cluster represents loading of the parameter to the reservoir watershed under current conditions. The middle bar shows the loading of the parameter with BACT implemented. The rightmost bar represents the loading of the parameter with BACT in place and population growth to 2010. Each bar is divided into two parts: one representing the percentage contribution from WWTPs (unshaded) and the other the percentage contribution from OSTDS (shaded). Care should be taken when reading the graphs to note the absolute value of each parameter, given at the top of each bar. These values may be very low, although the height of the bar is high.

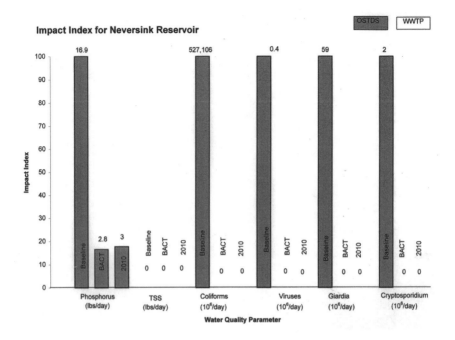

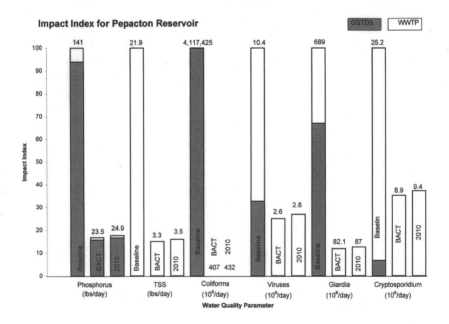

FIGURE 11-2 Impact index for Neversink and Pepacton reservoirs.

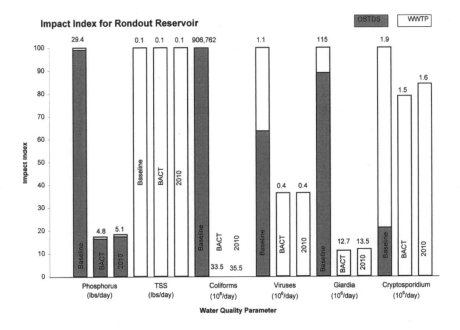

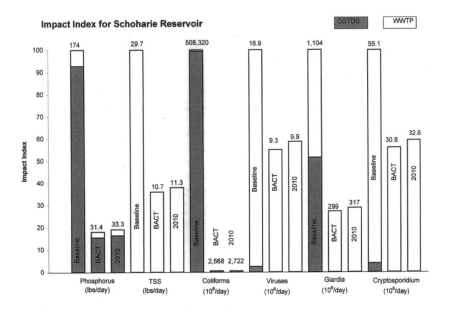

FIGURE 11-3 Impact index for Rondout and Schoharie reservoirs.

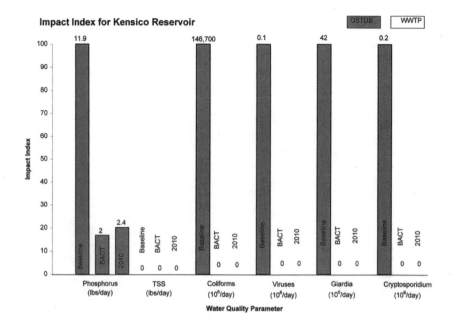

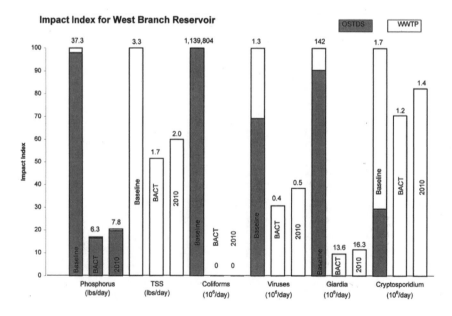

FIGURE 11-4 Impact index for Kensico and West Branch reservoirs.

On-Site Sewage Treatment and Disposal System Effluent

As shown in Figures 11-1 to 11-4, the use of aerobic treatment units as BACT for OSTDS significantly reduces effluent concentrations of the microbial parameters *Giardia, Cryptosporidium*, coliform bacteria, and viruses. BACT does not substantially change effluent concentrations of total suspended solids from OSTDS because any OSTDS that uses a drainfield already provides excellent removal of total suspended solids. BACT for OSTDS would reduce the loading of total phosphorus to receiving waters as long as residuals are properly managed.

BACT is not specifically required for OSTDS under the terms of the MOA, and there is no evidence that it is being used. Instead, apparent practice sets the minimum acceptable technology to be the passive septic tank and drainfield combination, although ATUs and experimental systems are allowed. This is unfortunate because passive septic tank and drainfield combinations have a limited life and provide inferior treatment of residential wastewater compared to properly functioning ATUs. Requiring existing and new OSTDS to achieve BACT would significantly reduce their contribution of *Giardia* cysts, viruses, coliforms, and *Cryptosporidium* oocysts in direct proportion to the number of OSTDS in the watershed. It should be noted that much of the estimated decrease in pathogen loading that would accompany the use of BACT can be attributed to improved enforcement and the associated elimination of surfacing sewage caused by OSTDS failure. Although there is an enforcement presence in the watershed, there is no requirement for an annual operating permit or inspection.

Effect of Population Growth

Population growth will eventually return the concentrations of the six contaminants emanating from WWTPs to pre-BACT levels. Annual population growth in the West-of-Hudson watersheds for the period from 1990 through 1996 was only 0.25 percent, while annual population growth in the East-of-Hudson region during that same period was 1.06 percent. The calculated increase in total population for the period 1990 to 2010 is 5 percent for the West-of-Hudson watersheds and 21.2 percent for the East-of-Hudson watersheds. Because WWTPs in the Rondout and West Branch watersheds are already at or near BACT, population growth will begin to degrade water quality to below pre-BACT conditions almost immediately in those watersheds. In the Ashokan, Cannonsville, Pepacton, and Schoharie watersheds, where implementation of BACT will significantly improve effluent quality, over 100 years will be required to reach previous contaminant levels. Applying these same growth estimates to OSTDS shows that the impact of each of the six parameters (except phosphorus) on receiving waters would be virtually eliminated if BACT were required for

OSTDS. For phosphorus, about 24 years beyond 2010 would be required before concentrations in OSTDS effluent reach pre-BACT levels.

TECHNOLOGY UPGRADES AND EFFLUENT STANDARDS

Technology is chosen with an emphasis on lowest cost, and applicable standards are typically met but not exceeded. Regulatory agencies attempt to influence this decision by requiring the use of BACT. BACT is a moving target that changes in response to technological advances. When upgrades to BACT are warranted, it is easier to retrofit BACT to a relatively few WWTPs than to thousands of OSTDS.

Wastewater Treatment Plants

Currently, BACT for new WWTPs discharging to surface waters is defined as sand filtration, disinfection, phosphorus removal, and microfiltration. The following effluent standards are also required, which the SPDES permits for these plants must be updated to reflect:

- Phosphorus: 1.0 mg/L for <50,000 gallons per day (gpd)
            0.5 mg/L for 50,000–500,000 gpd
            0.2 mg/L for >500,000 gpd
- *Giardia* and enteric viruses: 99.9% removal and/or inactivation.

These effluent standards for WWTPs, which are similar to effluent standards for drinking water treatment plants, are adequate and reflective of BACT. They should be viewed as upper limits that cannot be exceeded, even during large storm events.

At non-city-owned WWTPs, it is likely that the New York City Department of Environmental Protection (NYC DEP) will be substituting Continuous Backwash Upflow Dual Sand Filtration (CBUDSF) for microfiltration (Gratz, 1998). There is continuing controversy as to whether CBUDSF is functionally equivalent to microfiltration in meeting BACT for WWTPs (Marx and Izeman, 1998; NYC DEP, 1998a). In September 1996, NYC DEP presented oocyst removal data collected between 1994 and 1996 to the Environmental Protection Agency (EPA) indicating that the two methods were equivalent. EPA did not support this statement of equivalency, citing the fact that the research had been conducted at different labs, using different detection limits and testing protocols (Gratz, 1998). Thus, a new testing protocol was developed by NYC DEP with input from EPA and the New York State Department of Health (NYS DOH). A series of 12 tests using the new protocol as well as microfiltration was completed in October 1997, and a subsequent NYC DEP report concluded that both technologies "exceeded the requirements of the Surface Water Treatment Rule (SWTR) and proposed regulations governing log removals of *Giardia* cysts and *Cryptosporidium* oocysts for drinking water filtration plants" (NYC DEP, 1998a). Results from this report have been subject to criticism (Marx and Izeman, 1998).

In the committee's experience, there have been very few demonstrations of the CBUDSF technology. Dual-media filters are more frequently operated in down-flow mode and at drinking water treatment plants rather than at wastewater treatment plants. Indications from wastewater treatment plants using the down-flow technology are that the filters need to be cleaned regularly (beyond simple backwashing), a process that is not well understood and is often overlooked by plant managers (A. Amirtharajah, Georgia Tech, personal communication, 1999).

From a theoretical standpoint, no sand filtration system can be equivalent to properly operating microfiltration (G. Logsdon, Black and Veatch, personal communication, 1999; R. Trussell, Montgomery Watson, personal communication, 1999). However, based on the information available from NYC DEP, EPA, and independent expert scientists, it appears that the CBUDSF is an adequate substitute for microfiltration under relatively low oocyst loading rates and ideal operating conditions. Plants using this technology should be subject to rigorous long-term monitoring of particle counts in the 2- to 30-μm range, of turbidity, and occasionally of oocysts to verify that equivalency is maintained. This monitoring should determine the effectiveness of the filtration process and backwashing, and it should detect operational problems that may occur, such as clogging of the filter.

On-Site Sewage Treatment and Disposal Systems

Allowable treatment techniques for OSTDS in the watershed are contained in Appendix 75A of 10 NYCRR Part 75 (State of New York), which requires that raw sewage from a household sewer must be discharged to an absorption treatment system (i.e., a drainfield) and must undergo treatment prior to such discharge. The law allows only septic tanks or aerobic units for this treatment. These rules may be augmented by local codes, and NYC DEP has delegated regulatory responsibilities for OSTDS to the watershed counties. Section 75A.2(b) states that "local health departments may establish more stringent standards. Where such standards have been established, or approval by another agency is required, the more stringent standard shall apply."

Effluent standards have not been set for OSTDS, and standards relate only to the technology of the system. The current practice allows a choice between a passive septic tank or an ATU prior to the drainfield. Because of the higher cost of aerobic units, the passive septic tank and drainfield combination is invariably chosen. As previously mentioned, passive septic tank and drainfield combinations provide inferior treatment of residential wastewater compared to ATUs. Thus, the most rigorous standards for OSTDS treatment technology are not being applied.

The BACT strategy envisioned by the committee calls for the mandatory use of aerobic systems to maximize effluent quality entering the drainfield, combined with an annual operating permit and other regulatory methods to limit failures and minimize repair time. This strategy was the basis for reducing estimated annual failure rates of OSTDS from five percent to zero percent. Although it is possible to set effluent standards for OSTDS (a requirement for advanced wastewater treatment is now being applied to effluent from OSTDS in the Florida Keys), this strategy is not recommended for the New York City watershed because this technology has not been demonstrated in northern climes. Instead, it is more reasonable to forego sampling and rely on a technology (e.g., ATUs) that has been certified by a third party, such as the National Sanitation Foundation (NSF) or the American National Standards Institute, as being capable of reliably meeting certain treatment efficiencies.

The cost of rapidly achieving BACT for OSTDS in the New York City watersheds would be significant. There are an estimated 38,854 OSTDS in the Kensico, West Branch, and West-of-Hudson watersheds (NYC DEP, 1993). Purchase and installation of an NSF-approved aerobic system is estimated to cost $8,000 each (Ayres Associates, 1998), for a total capital cost of $311 million. Annual operating costs for electricity and maintenance are estimated to add another $300 per year for each household, resulting in an annual total of $11.7 million. The cost of increasing inspection staff and/or implementing a requirement for an annual operating permit have not been estimated. Although the initial capital costs are large, they are likely to be less than the cost of constructing sewers and new WWTPs to serve these homes. In addition, there are strategies that could allow these benefits to be phased in over time. One such strategy is to implement a rigorous enforcement effort to effectively detect and repair malfunctioning OSTDS. Those OSTDS found to need replacement and any new OSTDS required for population growth would use ATUs. In addition, older OSTDS could be retrofitted to ATUs at a pace that would result in BACT being achieved for OSTDS over a ten-year or longer period.

SITING ISSUES

On-Site Sewage Treatment and Disposal Systems

Suitable locations for OSTDS described in the NYS DOH publication "Individual Residential Wastewater Treatment Systems Design Handbook" appear to be adequate (NYS DOH, 1996). There must be 100 ft separating OSTDS from nearby waterbodies and wetlands. In general, these locations require proper separation from groundwater, suitable soils, and control of slope. New York requires a vertical separation of two feet between the bottom of an absorption-trench drainfield and groundwater. The purpose of this separation is to allow the wastewater sufficient time to travel through oxygenated, unsaturated soils that will support aerobic degradation. Soil suitability is related to the particle size of the soil, and either extreme may be problematic. Soils of very small particle size such as clays cannot accept wastewater rapidly, which leads to ponding of wastewater on the surface. Soils of overly large particle size such as sands will not detain wastewater for a sufficient period of time to allow aerobic degradation.

Separation from groundwater and soil suitability can be overcome by site preparation, such as mounding of the drainfield or the import of less permeable material for the drainfield. However, it is generally not possible to control slope over a large area. Thus, the New York State code does not allow OSTDS on slopes of greater than 15 percent. It should be emphasized that this requirement is based more on practicality than on science. The perforated piping that comprises the upper portion of the drainfield must be level or the wastewater will pond in the lowest area. In addition, because a minimum of 2 ft of aerobic soil is required beneath the distribution piping, and because the shoulders of the mound must be gently sloped to prevent erosion, the mounded area can become quite large.

It appears that waivers from the siting requirement for OSTDS have been authorized in the Kensico and West Branch watersheds (Fox, 1998) and that slopes of up to 20 percent may be used. This is unfortunate because regulatory codes represent minimum requirements. Their lessening may necessitate additional regulatory oversight to ensure conditions are maintained. Waivers are especially ill advised in areas that are environmentally sensitive or are of public health significance.

Wastewater Treatment Plants

The Watershed Rules and Regulations include a 60-day travel-time delineation that is used to regulate siting of new or expanding WWTPs in the Catskill/Delaware watershed. The regulations forbid the construction of new surface-discharging plants within the 60-day travel-time boundary and within phosphorus or coliform restricted basins. Outside the boundary, surface-discharging WWTPs are allowed if they discharge into watercourses but not into reservoirs, reservoir stems, wetlands, or controlled lakes. There are no 60-day travel-time restrictions for new or expanding subsurface-discharging plants.

The 60-day travel time refers to the time necessary for water to travel from its point of origin to the distribution system. This boundary is intended to protect the water supply reservoirs from pathogenic microorganisms originating in upstream watersheds. The choice of 60 days, which was based in part on die-off rates for *Giardia*, has been approved by NYS DOH. Original research supporting the 60-day value is not available (J. Covey, NYS DOH, personal communication, 1998).

Boundary Calculation

Because of the complexities of system operation, varying water levels in the reservoirs, and temperature gradients, the delineation of the 60-day travel-time boundary was a difficult undertaking (Klett, 1996). The boundary was calculated using flow data, reservoir-stage data, and bathymetric information for the years 1983–1991. The end points for the calculation are the two locations where water discharges from the Kensico Reservoir (CATLEFF and DEL 18). As discussed in Chapter 6, these points are heavily monitored for compliance purposes.

To be conservative, the boundary calculation represents the 95th percentile case. That is, water originating in areas outside the boundary has a travel time of greater than 60 days in all but 5 percent of the cases. Because of the relatively long residence times experienced by the reservoirs, the time needed for water to travel overland before reaching reservoirs was not considered in the calculation. A map of the Catskill/Delaware watershed showing the 60-day travel-time boundary is presented in Figure 11-5.

Analysis

The appropriateness of a 60-day travel time for source protection must first be framed in the context of what objective is to be achieved. In the absence of precise written documentation justifying the 60-day value, it is assumed that this concept is based on the belief that 60 days will provide sufficient time for decay of microorganisms from a properly treated wastewater, so as to have *de minimus* impact on microbial levels in the source water.[1] In other words, travel time from the WWTP to the reservoirs is considered a barrier to pollution, much like chlorination or other treatment processes. Analyzed in this context, the literature suggests that 60 days may not be appropriate for protozoa, nor indeed for all enteric viruses. This point is elaborated upon below, with particular emphasis on *Giardia* and *Cryptosporidium* (of which the latter is of special concern because its reduction by chlorination alone is minimal).

One way of evaluating the travel time is to consider the log reduction of microorganisms that would occur over that period. This is a convenient way to compare the travel-time barrier with other conventional barriers such as filtration and disinfection. Disinfection criteria under the SWTR require a 3-log removal of *Giardia* and a 4-log removal of enteric viruses. For systems that filter, the Enhanced SWTR will require a 2-log reduction of *Cryptosporidium*.

---

[1] This implies that removal and inactivation of organisms is the primary objective of the 60-day travel-time zone. It should also be noted that the effluents to be released would be diluted with other flows in the watershed, and a policy decision should be made as to whether "credit" for such dilution should be granted to potential dischargers when considering the adequacy of protection thereby afforded to the New York City supply.

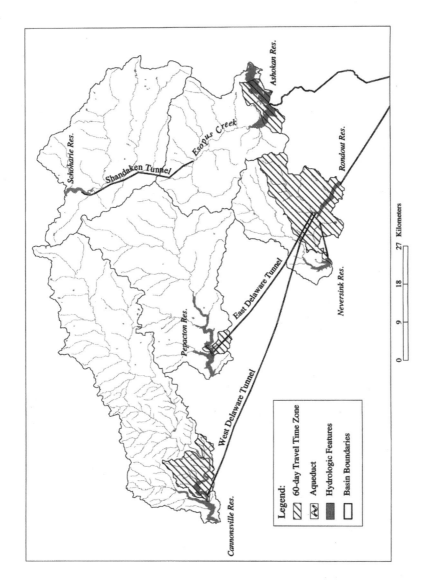

FIGURE 11-5 60-day travel-time zone in the Catskill/Delaware watershed. Courtesy of NYC DEP.

Viruses. There is limited information on viruses with which to evaluate the 60-day travel time. In the case of some viruses, such as adenovirus, 60 days of holding in dechlorinated tap water has been found to produce only a 2-log reduction at 23°C (Enriquez et al., 1995), significantly less than the 4-log reduction required of disinfection processes. The same viruses have greater resistance to disinfection by UV than does poliovirus (for which most disinfection data has been obtained) (Meng and Gerba, 1996), and their resistance to chemical disinfection in water does not appear to have been investigated. These facts suggest that the 60-day travel time may be inadequate to protect against adenovirus.

Giardia. Treated wastewater effluents in the Catskill/Delaware watershed have been found to have average *Giardia* concentrations ranging from less than 1 to over 400 cysts per 100 L (NYC DEP, 1999). These sources have had detectable levels of *Giardia* in 40.8 percent of all samples taken between 1993 and 1998. Over the same time period, Kensico Reservoir effluent locations CATLEFF and DEL 18 have had mean concentrations (substituting zero for measurements below the detection limit) of about 0.2 cysts per 100 L. These values imply that a 1- to 3-log die-off of cysts from the treatment plant effluent discharge to the distribution system would have to occur during the 60-day travel time.

As is the case for viruses, few studies have investigated log removal of cysts over long time periods in conditions similar to those found in the watershed. DeRegnier et al. (1989) suspended *Giardia muris* in river water (Mississippi River at Minneapolis) and lake water at ambient temperatures and monitored viability using propidium iodide and animal infectivity. As measured by infectivity, cysts remained viable at least up to 40 days. It should be noted, however, that the small number of animals did not likely permit measurement of inactivation beyond 1 log. The authors concluded, "*G. muris* cysts suspended in environmental water remained viable for 2 to 3 months, and their survival was enhanced by exposure to low water temperature, despite the fact that the cysts were suspended in the fecal biomass within the sample vial." Other studies have shown that *Giardia* cysts can survive as long as three months in pit latrines and sewage sludge (Deng and Cliver, 1992), but log removal was not calculated. Based on this information, a 60-day travel time for protection against *Giardia* is not supported.

Cryptosporidium. Wastewater effluents discharging into the New York City watershed have substantially lower oocyst levels than do other WWTP effluents across the country. For example, Madore et al. (1994) found that an average wastewater with conventional activated sludge treatment had final effluent oocyst levels of 14,000–396,000/100 L, while NYC DEP studies consistently show concentrations several orders of magnitude lower (Table 11-1). Based on the data in Table 11-1 and the source water oocyst levels measured in Kensico

TABLE 11-1 Wastewater Treatment Plant Effluent Oocyst Levels

	Total *C. parvum* (#/100 L)	
Treatment Plant	Mean	Standard Deviation
Delhi WWTP	0.74	1.76
Hunters Highland WWTP	0.69	3.36
Hobart WWTP	3.25	20.73
Tannersville WWTP	8.57	32.46
Stamford WWTP	2.26	9.85

Source: NYC DEP (1998b).

Reservoir, at least a 1-log (for advanced treated wastewater) to 4-log removal during 60 days might be appropriate benchmarks for *Cryptosporidium*.

Robertson et al. (1992) used the DAPI-PI dye inclusion assay to monitor decay of viability of oocysts held in membrane diffusion chambers in river water under ambient conditions. It has recently been determined that the survival of oocysts in fecal material as measured by vital dyes correlates well with the ability of the oocysts to excyst (Jenkins et al., 1997). Results of the Robertson et al. experiments are shown in Figure 11-6. From these results, 1-log inactivation is estimated to occur at 100 and 180 days for the two strains examined.

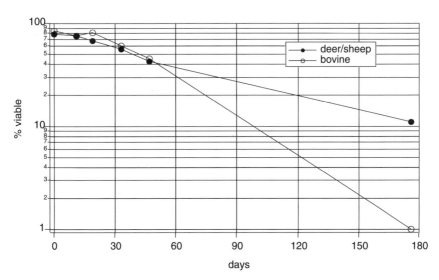

FIGURE 11-6. Inactivation of *C. parvum* oocysts in river water. Source: Reprinted, with permission, from Robertson et al., 1992. © 1992 by the American Society for Microbiology.

These results should be tempered by the observation that vital dyes and excystation appear to be less sensitive indicators of oocyst inactivation by chemical disinfectants than is animal infectivity (Black et al., 1996). That is, greater kill is noted if oocysts are tested for animal infectivity as compared to other oocyst viability methods. Although it is not known whether this same differential would exist for oocysts inactivated over time in surface waters, it may be that dye and excystation tests generally overestimate viability and thus underestimate inactivation. (From a public health point of view, this makes them conservative assays.)

In any event, consideration of both *Giardia* and *Cryptosporidium* information raises questions as to whether a 60-day travel time is sufficiently protective. Studies using the waters under question and sensitive viability assays (animal infectivity or perhaps cell culture in the case of *Cryptosporidium*) should be conducted to establish more site-specific guidelines. If the criterion of no significant adverse impact is used and inactivation is the sole removal mechanism considered (as opposed to dilution and sedimentation), then it would appear that 60 days may be insufficiently conservative, and perhaps a zone of 180 days or more might be required, especially for *Cryptosporidium*. An explicit scientific rationale for such a zone of protection should be developed.

CONCLUSIONS AND RECOMMENDATIONS

1. The upgrades to WWTPs mandated by the MOA should be effective in reducing effluent concentrations of phosphorus, TSS, coliforms, viruses, *Giardia* cysts, and *Cryptosporidium* oocysts. The requirement that these upgrades use BACT is an important component of New York City's watershed management strategy. Because BACT is a "moving target," it is important that these requirements be reevaluated regularly to ensure additional improvements are implemented.

2. Current technologies being used for OSTDS are not adequate and do not represent Best Available Control Technology. Passive systems consisting of a septic tank and drainfield are allowed instead of requiring ATUs, which maximize the destruction and inactivation of microorganisms. Implementation of BACT for OSTDS, including a substantial enforcement effort, could drastically reduce effluent concentrations of *Giardia*, *Crytosporidium*, and viruses in all Catskill/Delaware watersheds, although BACT will not significantly alter OSTDS effluent phosphorus concentrations.

3. Aerobic treatment units should be mandated for new or replacement OSTDS, and enforcement efforts should include annual inspections. This recommendation is especially important for the Kensico watershed, because of its critical location in the water supply and because OSTDS serves the entire population. An evaluation should be made of allowable time to upgrade other OSTDS to BACT. Requiring effluent standards for OSTDS is not absolutely necessary, although a representative sample of ATUs should be performance-monitored to ensure their functionality.

In the future, other technologies may become candidates for BACT. To explore their use, New York State should consider conducting demonstration projects and establishing a rigorous experimental permit program for these technologies (as has been done in Virginia, Florida, Washington, and elsewhere). If ATUs, or some equivalently tested and approved technology, are not required for repairing existing OSTDS and installing new OSTDS, NYC DEP should consider converting users of OSTDS to wastewater treatment plants.

4. Effluent standards for WWTPs are adequate. BACT should be used for all WWTPs whenever possible to counteract declines that occur because of population growth. The implementation of BACT at WWTPs in the Ashokan, Schoharie, Cannonsville, and Pepacton watersheds will significantly reduce contaminant concentrations in plant effluents. BACT for WWTPs will have little present-day impact on effluent concentrations in the Rondout, West Branch, Kensico, and Neversink watersheds. Maximum effort should be made to have the upgrades mandated by the MOA installed as quickly as possible at all existing facilities.

5. To make sure that the Continuous Backwash Upflow Dual Sand Filtration units represent BACT, wastewater treatment plants using this technology should be subject to rigorous long-term monitoring of particle counts and turbidity to verify that equivalency is maintained. This monitoring should determine the effectiveness of the filtration process and backwashing, and it should detect operational problems that may occur, such as clogging of the filter.

6. Waivers that allow placement of OSTDS on slopes of greater than 15 percent should not be permitted. Although slopes between 15 percent and 20 percent are not fundamentally less able to provide adequate treatment, slopes greater than 15 percent can make proper construction of OSTDS more problematic. In addition, such systems may require additional regulatory oversight.

7. The current 60-day value used for siting WWTPs does not appear to be supported by available knowledge. The particular criteria to be achieved by such a barrier must be explicitly defined prior to stipulation of a duration that might be acceptable, and an explicit scientific rationale for such a zone of protection should be developed. Limited research suggests that 60 days may be inadequate for significant inactivation of *Cryptosporidium* oocysts, which are known to be resistant to disinfection. The 60-day limit may not allow adequate protection against *Giardia*, particularly at low temperatures. Sixty days may not be sufficient to substantially reduce virus levels, although their sensitivity to chlorination makes this less problematic than for the protozoans. NYC DEP should pursue further studies of pathogen transport and fate using sensitive viability assays and local source waters to refine this value.

8. Although declines in water quality related to WWTPs and OSTDS will occur with population growth, in most cases 40–100 years will pass before contaminant contributions from WWTPs and OSTDS reach pre-BACT levels. It should be noted that these results are likely to be conservative because the impact analysis did not attempt to quantify any improvements to wastewater quality occurring between the wastewater's point of discharge to a subsurface drainfield and its point of entry into a surface water reservoir. Other factors that may affect the accuracy of these predictions include the fact that no attempt was made to differentiate between seasonal and permanent populations in the watershed, and BACT for OSTDS was assumed to be ATUs.

REFERENCES

Ayres Associates. 1998. Technology Memorandum #7: Technology Assessment of Onsite Wastewater Treatment Systems. Prepared for Monroe County Sanitary Wastewater Master Plan.

Black, E. K., G. R. Finch, R. Taghi-Kilani, and M. Belosevic. 1996. Comparison of assays for *Cryptosporidium parvum* oocysts viability after chemical disinfection. FEMS Microbiology Letters 135: 187–189.

Bott, T. L. 1973. Bacteria and the assessment of water quality. Pp. 61–75 in Biological Methods for the Assessment of Water Quality. ASTM STP 528. Philadelphia, PA: American Society for Testing and Materials.

Deng, M. Y., and D. O. Cliver. 1992. Degradation of *Giardia lamblia* cysts in mixed human and swine wastes. Applied and Environmental Microbiology 58(8):2368–2374.

DeRegnier, D. P., L. Cole, D. G. Schuff, and S. L. Erlandsen. 1989. Viability of *Giardia* cysts suspended in lake, river and tap water. Applied and Environmental Microbiology 55(5):1223–1229.

Enriquez, C. E., C. J. Hurst, and C. P. Gerba. 1995. Survival of enteric Adenoviruses 40 and 41 in tap, sea and wastewater. Water Research 29(11):2548–2553.

Fox, J. 1998. Letter to Ronald Tramontano, New York Department of Health from EPA Region II. Dated April 10, 1998.

Gratz, J. 1998. Letter to Interested Parties. March 20, 1998. Re: NRDC's Letter on Microfiltration. Dated March 20, 1998. Region 2 EPA, New York, N.Y.

Jenkins, M. B., L. J. Anguish, D. D. Bowman, M. J. Walker, and W. C. Ghiorse. 1997. Assessment of a dye permeability assay for determination of inactivation rates of *Cryptosporidium parvum* oocysts. Applied and Environmental Microbiology 63(10):3844–50.

Klett, B. 1996. Delineation of a Sixty-Day Travel Buffer for the Protection of the New York City Water Supply. In New York City Water Supply Studies. Proceedings of a Symposium on Watershed Restoration Management. American Water Resources Association. Herndon, VA.

Madore, M. S., J. B. Rose, C. P. Gerba, M. J. Arrowood, and C. R. Sterling. 1994. Occurrence of *Cryptosporidium* oocysts in sewage effluents and selected surface waters. Journal of Parasitology 73(4): 702–705.

Marx, R., and M. Izeman. 1998. Memorandum to Jeff Gratz, EPA Region 2, on April 21, 1998.

Meng, Q. S., and C. P. Gerba. 1996. Comparative inactivation of enteric adenoviruses, poliovirus and coliphages by ultraviolet irradiation. Water Research 30(11):2665–2668.

New York City Department of Environmental Protection (NYC DEP). 1993. Final Generic Environmental Impact Statement for the Proposed Watershed Regulations for the Protection from Contamination, Degradation, and Pollution of the New York City Water Supply and its Sources. November 1993. Corona, NY: NYC DEP.

NYC DEP. 1998a. DEP Pathogen Studies on *Giardia* spp. and *Cryptosporidium* spp.—Protocol Development for testing equivalency of two technologies to remove pathogens from wastewater effluent. Valhalla, NY: NYC DEP.

NYC DEP. 1998b. DEP Pathogen Studies of *Giardia* spp., *Cryptosporidium* spp., and Enteric Viruses. July 1998. Valhalla, NY: NYC DEP.

NYC DEP. 1999. DEP Pathogen Surveillance Report of *Giardia* spp., *Cryptosporidium* spp., and Enteric Viruses. Semi-Annual Report. Valhalla, NY: NYC DEP.

New York State Department of Health (NYS DOH). 1996. Individual residential wastewater treatment systems design handbook. Albany, NY: NYS DOH.

Robertson, L. J., A. T. Campbell, and H. V. Smith. 1992. Survival of *Cryptosporidium parvum* oocysts under various environmental pressures. Applied and Environmental Microbiology 58:3494–3500.

Sherman, K. 1998. OSTDS Research Coordinator for the Florida Department of Health. Presentation given at the Third NRC Committee Meeting. May 13–16, 1998, Oliverea, New York.

12

Overarching Issues

NEW DIRECTION FOR NEW YORK CITY'S
WATERSHED MANAGEMENT PROGRAM

The principal goal of the Memorandum of Agreement (MOA) is to protect public health in the City of New York by providing the City with safe, potable water while at the same time protecting the rights and needs of the residents of the watershed. A necessary component of any plan to meet this goal involves meeting the provisions set forth in the Safe Drinking Water Act (SDWA) and its amendments. For unfiltered surface water supplies, this legislation and the rules emanating from it involve (1) providing sufficient inactivation to control *Giardia*, (2) maintaining the turbidity and fecal coliform concentrations of the surface water supplies below specified levels, (3) controlling the concentrations of disinfection byproducts (DBPs) in the distribution system below specified levels, (4) maintaining a detectable residual disinfectant concentration "or its equivalent" throughout the distribution system, (5) limiting the total coliform concentration in the distribution system, and (6) controlling corrosion that can lead to high lead and copper concentrations at the consumer's tap.

Nutrients such as phosphorus are addressed in the Clean Water Act and are not a direct public health concern in drinking water. In fact, phosphate is often added to potable water supplies to retard corrosion in distribution networks and household plumbing, thereby reducing lead and copper concentrations. Nutrients can, however, have deleterious effects on drinking water when they stimulate algal and bacterial growth in reservoirs. Increased productivity can lead to the release of organic carbon compounds in the water and, in turn, increase the production of undesirable DBPs when the water is disinfected for potable use.

Future Environmental Protection Agency (EPA) rules are expected to lower the allowable concentrations of several DBPs in drinking water distribution systems. Requirements for the inactivation and/or removal of *Cryptosporidium* oocysts may also be introduced.

Pollutant Priorities

In the committee's opinion, in order to meet the goals enumerated above, four primary contaminants require effective control in the management of the City's watershed. Prioritized by the committee, they are (1) microbial pathogens, (2) natural organic carbon compounds, (3) phosphorus, and (4) turbidity. Toxic compounds such as pesticides, heavy metals, and synthetic organic compounds are also potential pollutants that should not be overlooked. However, their use in the Catskill/Delaware watershed is limited to specific areas. Recent monitoring of toxic compounds in the water supply watersheds has not detected notable concentrations. Current and potential land and resource uses in the region are unlikely to involve industries that have substantial inventories of metals and solvents. Agricultural trends, such as the move from dairy operations to truck crops and organic practices, suggest reductions in pesticide use. For these reasons, toxic compounds are not considered further in this chapter. Rather, specific recommendations for monitoring these compounds are given in Chapter 6.

Microbial pathogens should be the primary focus of the watershed management program for numerous reasons. The MOA was created to comply with the Surface Water Treatment Rule (SWTR), which specifically targets microorganisms for control. These organisms, which can produce mild to life threatening infections, originate from a diverse array of sources. Several are difficult to inactivate by conventional disinfection processes, and some can survive for long periods in aquatic environments. Finally, their transport and fate are difficult to assess, and the effectiveness of common best management practices (BMPs) in reducing their concentrations in stormwater is generally unknown.

Natural organic carbon compounds—measured as dissolved organic carbon (DOC)—are placed second because they are a primary reactant in the formation of most DBPs that are, or are likely to be, regulated. Also, these compounds can serve as carbon and energy sources for microbial growth in the City's distribution system. Their origins in the watershed are diverse and in many cases difficult to manage. Organic carbon will be an important factor in the design and operation of any additional treatment facilities the City may consider.

Third, attention is placed on phosphorus, which has several major impacts on potable water supplies. It can drive algal productivity in surface waters, which is the case in most of the City's water supply reservoirs. Algae, in turn, can generate several other water quality problems for a water supply. Because algae are particles, they scatter light and contribute to turbidity. When algae settle to the bottom waters of a reservoir, their degradation by heterotrophic bacteria can deplete oxygen in these waters and accelerate the release of undesirable contaminants such as iron and manganese from bottom sediments into the water column. The presence of these contaminants can necessitate treatment of the water prior to its potable use. Phytoplankton can release organic taste- and odor-causing substances to surface waters. Algae serve as substrates for bacteria and other organisms that also release organic carbon compounds. This autochthonous DOC is a precursor for DBP formation.

The fourth contaminant considered is turbidity. Turbidity is an indirect and imperfect but easily assessed indicator of the possible presence of other pollutants. Turbidity arising from particles, including algal cells and debris, can protect pathogens from inactivation during disinfection. Particle-reactive pollutants are often associated with turbidity and are transported with sediments in the watershed. Particles can also accumulate in reservoir sediments and have long-term effects. Although the sources of turbidity in a watershed are diverse and intermittent, controls of many of these inputs can be effective.

Current Approaches by New York City

Pathogens

At present New York City provides disinfection with free chlorine to meet EPA's "CT" requirement for the 3-log inactivation of *Giardia* cysts. It is likely that EPA will, in the next few years, require treatment to inactivate *Cryptosporidium* oocysts (EPA, 1998; NYC DEP, 1998a). These organisms are resistant to inactivation by free chlorine. Therefore, substantial changes in the City's disinfection strategy for control of microbial pathogens can be anticipated.

Managers of unfiltered supplies should be at or near the forefront of information on *Cryptosporidium*, other emerging microbial pathogens, and their regulation. A comprehensive framework for controlling pathogens in the watershed should have three primary goals: (1) identifying pathogen sources in the watershed and determining quantitative source terms, (2) determining the effects of management practices on current pathogen concentrations, and (3) directing resources toward the most polluting sources. Reaching these three goals is paramount to demonstrating the role of watershed management in reducing the risk of microbial pathogens to public health. Once its role is established, watershed management can be compared to filtration and other treatment processes to better evaluate its costs and benefits.

The current pathogen monitoring efforts represent a first step toward assessing quantitatively sources of pathogens (goal 1) and the degree to which watershed management improvements can reduce pathogen loading to the reservoirs (goal 2). As discussed in Chapter 6, accomplishing the first goal requires gathering data on the occurrence of pathogenic microorganisms in the watershed and calculating source terms, such as (oo)cyst concentration per area per time, for each catchment. Although some data have been collected as part of the Pathogen Studies and other efforts (Auer et al., 1998a), they have not been systematically analyzed to the extent necessary to evaluate current impacts or future options. For example, the relative contributions of different land uses to overall pathogen loadings have not been determined as has been done for phosphorus, even though some of the necessary concentration and land use data are available. Development of quantitative source terms has not been undertaken.

While wastewater effluents and reservoirs have been subject to regular monitoring for pathogens, there has not been the same regular and systematic monitoring of agricultural and other nonpoint sources (e.g., stormwater, septic systems) and BMPs used to control pathogen concentrations. Available information suggests that most urban stormwater BMPs are only moderately effective in removing bacteria, and no information is available on removal of protozoa. Chapter 9 stresses the urgent need for performance monitoring of agricultural BMPs, suites of BMPs, and Whole Farm Plans to determine their pathogen-removal capabilities.

Finally, existing pathogen monitoring data of nonpoint sources have not been used in a coordinated way with watershed models. Integration of the pathogen monitoring data with watershed models must be accomplished and supported by appropriate validation. Once developed, these models can be coupled with field data and used as an analytical tool to assess the effectiveness of different watershed management options on pathogen concentrations in reservoirs.

To implement this approach, which would place pathogens in a framework commensurate to other contaminants (e.g., phosphorus), additional effort should be made to obtain fundamental information on processes influencing pathogen transport and fate in the watershed. These additional processes include die-off, separation from manure, contributions from subsurface waste-disposal systems, and removal by agricultural and urban stormwater BMPs, among others. Given its prioritization of pollutants, the committee maintains that the level of resources and effort afforded to pathogens be the same as or greater than that currently afforded to phosphorus. The necessary resources should be devoted to expedite development and use of a pathogen management framework that integrates information on sources, life-cycle processes, and BMP removal rates with models of transport and fate.

Natural Organic Carbon Compounds

Natural organic carbon compounds are important because of their role in DBP formation. High levels of DOC in water supplies can reduce flexibility in treating water for pathogens by preventing water suppliers from aggressively disinfecting their water with chlorine. In order to manage DOC effectively, the City must determine the origins of DOC in the water supply and evaluate possible differences among those sources on the formation of DBPs and on other processes.

Autochthonous sources of DOC are amenable to control by limiting the supply of inorganic nutrients to the reservoirs, and New York City is addressing this problem well by controlling phosphorus loading. Allochthonous sources, however, are more difficult to assess and are also very difficult to control. Much of the allochthonous carbon derived from areas high in the watershed is from natural plant tissue (e.g., forests), with less from agricultural, urban, and deep groundwater sources. Closer to New York City, anthropogenic sources (such as agriculture and urban areas) become more important, but because of the high percentage of forest cover in the Catskill/Delaware watershed, the majority of allochthonous DOC likely is derived from widespread natural sources.

Because removal processes for allochthonous DOC are almost all biological, time is the most important factor for reducing concentrations of these recalcitrant compounds. Traditional urban stormwater and agricultural BMPs are not expected to remove DOC to any significant degree. Wetlands have some capacity for controlling DOC, but the percent removal is generally low. Buffer zones can accomplish some removal of allochthonous DOC if they retard its transport, particularly during periods of high DOC loading associated with rain or snow-melt events.

Research efforts have revealed that both allochthonous and autochthonous sources of DOC can contribute to the formation of DBPs in the reservoirs. In the Cannonsville Reservoir, DOC from allochthonous sources was very effective in forming trihalomethane (THM) precursor compounds (Stepczuk et al., 1998a). Similar analyses were conducted within the reservoir open water, and again the DOC was found to be most effective in functioning as THM precursor compounds. In the open water, however, it is not possible to distinguish between allochthonous DOC and the autochthonous DOC produced by phytoplankton decomposition (Stepczuk et al., 1998b). Because inflows purportedly decreased and phytoplankton production increased during the summer, it was assumed that much of the THM precursor compounds was being formed by the algae. That may not be the case, however, because allochthonous DOC decomposes much more slowly in the reservoir than does the DOC from phytoplankton. Therefore, control of DOC (and consequently DBP formation) will not be accomplished by focusing solely on autochthonous sources and reducing nutrient inputs to the reservoirs.

For the past few years, the concentrations of total trihalomethanes (TTHMs) in the City's water have been decreasing; they are well below past and present standards (0.10 and 0.08 mg/L, respectively). However, there is some evidence that the total concentration of five haloacetic acids (HAA5) in the water is close to the new limit of 0.060 mg/L. Under present conditions, the drinking water would exceed the forthcoming Stage 2 maximum contaminant level (MCL) for HAA5 (0.030 mg/L is the placeholder value). In actively managing DOC it is important that the City consider the causes of these trends in DBP levels and also address other DBPs that may be regulated in the future.

Phosphorus

Of the priority pollutants, phosphorus has received the most attention. This may be because it is an important contaminant, because the professional staff have extensive expertise in limnology and eutrophication, and because it may be the least difficult and costly to manage of the four types of priority contaminants. As noted earlier, phosphorus can be present in water in either soluble or particulate forms, with soluble phosphorus being more bioavailable for eutrophication. Because it has been shown that over 95 percent of the phosphorus taken up by phytoplankton in the Cannonsville Reservoir is soluble (Auer et al., 1998b), management efforts are primarily targeted at this form.

Sources of soluble phosphorus in the watersheds are many and diverse. Soluble phosphorus is the primary type of phosphorus emanating from point source inputs such as wastewater treatment plants (WWTPs) and from septic systems with ineffective drainage fields. WWTP sources of soluble phosphorus are adequately addressed by the MOA through its mandated upgrade of all sewage treatment plants. Indications are that all 41 municipal WWTPs in the Catskill/Delaware watershed will receive these upgrades in the next two or three years. The control of soluble phosphorus from on-site sewage treatment and disposal systems (OSTDS) is much less certain. Because there is no current requirement for the use of aerobic treatment units with appropriate sludge disposal (as recommended in Chapter 11), reducing phosphorus loadings from OSTDS is unlikely.

Recent research suggests, however, that point sources are not the primary source of soluble phosphorus in the watershed. Predictions from the Generalized Watershed Loading Function (GWLF) implicate subsurface flow (affected by nonpoint sources) as the greatest contributor of soluble phosphorus (NYC DEP, 1998b). Monitoring data have shown that nonpoint sources contribute significant amounts of soluble phosphorus, comprising about 61 percent of soluble reactive phosphorus between 1992 and 1996 in the Cannonsville watershed (Longabucco and Rafferty, 1998). Urban stormwater is typically 40 percent to 50 percent soluble reactive phosphorus (Schueler, 1995), and similar percentages have been observed for agricultural runoff (Longabucco and Rafferty, 1998). Controlling these sources of soluble phosphorus should be a goal of stormwater and agricultural BMPs. However, the current paucity of performance monitoring of these technologies makes it difficult to know how well these sources are being mitigated.

The situation is more even complex when the contribution of particulate phosphorus to the pool of soluble phosphorus is considered. As much as 50 percent of particulate phosphorus from the Cannonsville watershed may be bioavailable (Auer et al., 1998b). In addition, biogeochemical cycling of phosphorus in reservoir sediments adds to the available pool of dissolved phosphorus. During wet years (such as 1996), extremely high loadings of particulate phosphorus—up to ten times greater than loadings of soluble phosphorus—are possible. Hence, particulate phosphorus cannot be ignored when assessing reservoir health and the potential for eutrophication.

About 90 percent of particulate phosphorus in the Cannonsville watershed comes from nonpoint sources: urban stormwater, agriculture, and forests (Longabucco and Rafferty, 1998). Nonpoint sources are likely to also account for the majority of particulate phosphorus in the other Catskill/Delaware watersheds. Predictions from the GWLF that account for the area and type of land use suggest that much of this particulate phosphorus emanates mainly from forest land and urban areas in all basins except Cannonsville, where agriculture is the primary source (NYC DEP, 1998b). Thus, nonpoint source BMPs must be capable of reducing both particulate and soluble phosphorus loadings, a formidable task.

In summary, available data and model predictions suggest nonpoint sources should be the focus of future management efforts. Although soluble phosphorus may seem, quite logically, to be the priority pollutant, for the reasons stated above, both soluble and particulate phosphorus must be controlled. Improvements in OSTDS design, installation of urban stormwater and agricultural BMPs, and performance monitoring of these technologies are fundamental to meeting these goals.

Turbidity

Phytoplankton growth is a source of turbidity in the water supply reservoirs that is being adequately controlled by the City's efforts to limit phosphorus loading from point and nonpoint inputs. However, most turbidity problems in the Catskill/Delaware watersheds are caused by inorganic sediment from soil and channel erosion mobilized during rainfall, snowmelt, and stormflow events. In undisturbed forests, with the soil surface well protected by leaf litter, most sediment originates within stream channels. As noted earlier, some land and resource uses exceed the inherent resistance and resilience of sites and increase the quantity and rate of soil and channel erosion.

It is not clear what coordinated management efforts have been made to reduce erosion from distributed sources, especially in the Schoharie and Ashokan watersheds where the problem is most acute. In addition, water transfers from the Schoharie to the Ashokan may inadvertently promote channel instability and erosion in the Esopus Creek when releases are made to augment streamflow for recreational use. The operational response to high concentrations of turbidity is to add alum to the Catskill aqueduct just upstream of the Kensico Reservoir. As an alternative, or in addition to current efforts, the committee suggests the following approach to controlling sources of turbidity, some elements of which may already be under way.

First, detailed land cover and land use data should be collected for watersheds with unacceptably high turbidities (e.g., Stony Clove, tributary to the

Esopus Creek at Phoenicia) using high-resolution digital orthophotography. Using land cover/land use, soils, and digital elevation data, a Geographic Information System (GIS)-based model of soil erosion and sediment delivery should be developed. Polygons or grid cells in the resulting soil erosion and sediment delivery layers could then be ranked from the highest to lowest potential contributor of sediment. (Because active management is prohibited in some areas of the Catskill Forest Preserve, the Preserve can be excluded from the ranking procedure. However, erosion control and stormwater management measures—in collaboration with the New York State Department of Environmental Conservation (NYS DEC), the Appalachian Mountain Club, and others—may be needed on heavily used hiking trails and parking areas for recreational access.)

Spatially distributed watershed modeling should be accompanied by the development of a simple, reproducible field assessment technique for soil erosion, channel stability, and sediment transport. Sites identified with the GIS should be visited in priority order to assess field conditions. Preliminary recommendations for erosion, sediment, and stormwater control should be formulated in the field. They could include such practices as (1) water bars on unpaved roads to divert overland flow into adjacent forest before it reaches streams, (2) bioengineering techniques for stream channel and road fill stabilization, (3) crushed stone applied to steep road sections, and (4) revegetation of unstable areas, among others.

To promote successful implementation, the New York City Department of Environmental Protection (NYC DEP) should seek the cooperation of the Catskill Watershed Corporation, the Watershed Agricultural Program, the Watershed Forestry Program, or town officials in contacting landowners to gain permission to undertake erosion and sediment control. Sites selected either at random or on the basis of site characteristics and BMP types should be monitored before, during, and after mitigation activities to validate and refine modeling and assessment procedures, minimize construction impacts, and assess "as-built" performance, respectively. This program of coordinated, incremental, small-scale actions should have positive and cumulative effects on turbidity in the Catskill/Delaware system.

Conclusions and Recommendations for Prioritization
Within the Watershed Management Program

1. The primary focus of the City's efforts in protecting the health of its consumers should be on microbial pathogens in the water supply and on developing means to control them. The strategy for controlling pathogens should have three goals: (1) identifying pathogen sources in the watershed and determining quantitative source terms, (2) determining the effects of management practices on current pathogen concentrations, and (3) directing resources toward the most polluting sources. To accomplish these tasks, fundamental information on pathogen sources, transport, and fate in the watershed must be obtained, and nonpoint source BMPs must be quantitatively assessed for their ability to control pathogens. This is especially important for sources of pathogens such as urban areas, OSTDS, and dairy farms. Finally, greater efforts should be made to link pathogen-monitoring data to disease surveillance activities, as recommended in Chapter 6. Coordination with multiple organizations, such as the Watershed Agricultural Program, the New York City Department of Health, and the Catskill Watershed Corporation, will be necessary to effectively mitigate sources of pathogens in the Catskill/Delaware watershed.

2. After pathogens, New York City should focus efforts on the reduction of DBPs, both present and emerging. Because disinfection via chlorination is likely to continue (even if only to maintain a disinfectant residual following ozonation), control of DBPs will require precursor (organic carbon) removal. NYC DEP has started to characterize and compare sources of DOC in the reservoirs, an activity that should be extended to all Catskill/Delaware reservoirs. Current NYC DEP efforts to reduce phosphorus loading to the reservoirs are an important step in controlling DBP formation. However, it is highly likely that management of allochthonous DOC will be necessary to control DBP formation because of the recalcitrance of allochthonous DOC and its dominance during winter months. It should be noted that watershed management may not be capable of reducing allochthonous DOC levels to meet future DBP regulations. This may necessitate additional treatment. Reliable estimates of the sources of natural organic carbon compounds are critical to evaluating future treatment processes that may be needed to control DBP formation.

3. Efforts to control phosphorus loading to the reservoirs have been strong to date, and a significant body of monitoring data and terrestrial models have been generated to determine the sources of phosphorus. NYC DEP should concentrate on measuring the success of nonpoint source BMPs in reducing both particulate and soluble phosphorus. Nonpoint sources are the logical future focus of phosphorus reduction efforts rather than WWTPs, which are adequately addressed by the MOA upgrades.

4. More effort should be applied to control sources of turbidity in the watersheds rather than downstream in aqueducts and reservoirs. As outlined above, a more comprehensive erosion control program is needed throughout the watershed. Controlling turbidity at its origins has benefits beyond compliance with the SWTR. For example, controlling particulate-phase pollutants that might otherwise reach Kensico and other reservoirs will prolong the service life of the reservoirs and reduce maintenance costs in proportion to sediment accretion rates.

5. New York City should consider using a regular scientific external review/advisory process to expedite the use of new scientific information during watershed management. Whether particular areas of the watershed management program generate and use scientific information is highly variable. For example, phosphorus has been the subject of intense research and monitoring, while pathogens have received much less attention. New scientific information on wastewater treatment had a major role in the upgrades of WWTPs, but almost no role in improving OSTDS (which serve a substantial proportion of the watershed population). Disparities between state-of-the-art science and on-the-ground implementation are also apparent in the MOA's use of setbacks, in the Watershed Agricultural Program, in the Stormwater Pollution Prevention Plans, and in the phosphorus offset pilot program.

6. Integration across programs within the watershed management strategy is needed. One way to accomplish this would be to hold an annual workshop during which all program managers would discuss their accomplishments to date, the challenges and problems, and future goals (perhaps with an external scientific advisory panel). This process could identify gaps and overlap between the programs and could better prioritize actions.

ECONOMIC DEVELOPMENT IN THE WATERSHED REGION

The MOA is the result of a lengthy and complex bargaining process among parties concerned with maintaining the economic and social viability of the watershed region on the one hand and protecting the water quality on the other. As summarized in Chapter 1, the MOA emerged from a sometimes acrimonious negotiation process that extended over many months. Difficult as the negotiation was, the signing of the MOA has allowed New York City and the watershed communities to move from impasse toward realization of their goals of watershed protection and economic growth.

The potential for conflict between these two objectives was acknowledged during the negotiation process and is specifically addressed in the MOA and in the Watershed Protection and Partnership Programs. A substantial financial resource provided by the City under the MOA, the Catskill Fund for the Future, is being used to fund only environmentally sensitive development projects. This $59.7 million dollar fund has the potential to complement the extensive infrastructure upgrades and regulatory actions undertaken in the interest of water quality and to fund income-generating activities that will advance the economic welfare of the region.

Although it is too early to assess the success of the Catskill Fund for the Future, the committee believes that development consistent with maintenance of water quality is a realistic and achievable goal. Success in directing regional development depends upon a clear understanding of the impacts of various activities on water quality as well as having the political will to direct and assist development in accord with current knowledge. Because the relevant science continues to grow and change, it is important for the parties to periodically review existing regulations and policies and to make adjustments in cases where either a significant added benefit to environmental protection can be obtained or controls can be modified without adverse impacts on water quality.

What is the Future of the Catskill/Delaware Region?

The recently completed comprehensive overview of the watershed economy shows that the decade of the 1990s has been a period of economic decline for the Catskill/Delaware watershed (HR&A, 1998a,b). Between 1990 and 1997, the five watershed counties experienced an overall decline in employment of some seven percent, and real wages declined in every sector except finance, insurance, and real estate. For services, which as the largest employment sector accounted for 40 percent of all jobs, the decline in real wages was comparable to the decline for the State as a whole (1.2 vs. 1.8 percent). For the next three largest employment sectors—retail trade, manufacturing, and government, which collectively account for about 42 percent of all jobs—real wages fell more than the respective New York State averages did during the same period. The decline in regional real wages occurred even though real wages in the watershed counties are substantially lower than the State average wages for every Standard Industrial Classification Sector (HR&A, 1998a). This suggests that the labor force of the region is not highly mobile.

External markets have not, for the most part, driven recent regional economic growth. The largest and fastest-growing sector in the watershed, the service sector (including education, healthcare, and social services) serves the local population rather than short-term visitors or persons outside the region. Private sector service jobs, often oriented toward servicing visitors to the region, declined between 1990 and 1997 as the result of the loss of hotel employment opportunities.

One of the goals of the comprehensive economic study is to assist the Catskill Watershed Corporation in defining objectives for the Catskill Fund for the Future in terms of markets to address and products or services to provide. Tourism and recreation, arts and crafts, and specialty manufacturing are among the major activities identified by the study that may hold promise for growth. **Even given possibilities and the impetus provided by the growth-promoting programs under the MOA, the committee sees few signs that rapid increases in economic activity are likely in the region.**

Will New York City's Programs to Promote Economic Development Result in Water Quality Degradation?

Although population growth and increased economic activity may adversely affect water quality, these effects can be offset by careful planning, directed development, more extensive environmental regulation, and improved waste-water management, as provided in the MOA. Such measures will help to maintain high water quality in the West-of-Hudson reservoirs over the next several years, assuming growth rates do not increase substantially. **The committee thinks moderate population growth and a wide range of new economic activities can be accommodated in the watershed without deleterious impacts on water quality as long as development regulations are rigorously implemented and the extensive water quality infrastructure investments now being planned are put in place.**

The role of agriculture in the future of the watershed economy deserves special mention. As discussed in Chapter 9, agriculture is an important contributor to both nutrient and pathogen loadings, and it is an important area of concern for water quality in the region. Agricultural production has been steady during 1990s despite a consolidation of land holdings and an 11 percent decline in agricultural employment between 1990 and 1997. (The agricultural sector accounted for less than 2 percent of watershed employment in 1997.) Dairy production accounts for the bulk of agricultural activity in the watershed, and its economic prospects are tied to programs designed to maintain regional milk prices above unregulated market levels.

Because agriculture is not regulated under the MOA, future agricultural activities will not be held to the same environmental regulations imposed on other activities in the Catskill/Delaware watershed. Rather, the Watershed Agricultural Program is intended to support both the use of environmentally sound management practices and the economic viability of agricultural activities. These are not necessarily opposing goals, because on-farm infrastructure investments can often improve the economic value of the farm enterprise as well as provide replacements for old or obsolete facilities and equipment. The committee recommends that the Watershed Agricultural Program give priority for financial assistance to high-value, low-environmental-impact agricultural activities that could eventually replace activities with inherently high impacts (such as animal-based agriculture).

To summarize, increased population and economic activity could, without appropriate planning, contribute to water quality degradation to varying degrees. However, the committee believes the development incentive programs funded under the MOA are positive contributions to the regional welfare and the deleterious impacts associated with managed growth are small compared to the net benefits that will accrue to regional water quality under the MOA. If the provisions of the MOA are carried out as envisioned, the quality of life in the region will be improved.

Is the Concept of Balancing Development Restrictions and Incentives a Sound Strategy?

The committee believes that the concept of balancing development restrictions and incentives is a reasonable strategy for New York City and possibly other communities. Because the City committed substantial financial resources to help advance environmentally sensitive development, other regulatory, land acquisition, and water quality investment programs that will contribute significantly to protection of its drinking water supply are moving forward.

The Land Acquisition Program, which will seek to solicit 355,000 acres (approximately 30 percent of the Catskill/Delaware watershed), can affect the distribution of new economic activity within the region and alter the competitive prospects of areas within the watershed. Because of the possibility of acquiring such a large percentage of the total land area, the City has reviewed potential and existing recreational uses of City property. The potential for environmentally sensitive uses of City property should be thoroughly explored with the intent of making City ownership of land recognized as an asset that complements and assists regional growth rather than as an inhibitor of regional economic advancement.

In the committee's opinion, the infrastructure improvements (if implemented) and financial assistance provided by the Watershed Protection and Partnership Programs should reduce the adverse impacts of existing development and provide the capacity to accommodate environmentally sustainable growth for several decades. The program expenditures have the potential to benefit the watershed communities by creating job opportunities for the duration of the implementation phase. The lack of strong growth pressures in the Catskill/Delaware region increases the likelihood that assistance through the Catskill Fund for the Future will be effective in shaping future development. The committee recommends a continuing process of program evaluation to ensure that funds are most productively used to meet the goals of environmentally sustainable development.

Does the MOA Adequately Protect Private Property Rights?

Both the Land Acquisition Program and the Watershed Rules and Regulations raise questions about the rights of watershed property owners. **In the committee's opinion, the MOA adequately considers both the private property rights and the economic, social, and political concerns of watershed residents, while allowing a wide range of actions to protect water quality.** The imposition of setbacks along waterbodies within the watershed is similar to measures currently employed in Massachusetts and other states to protect water quality and riparian habitat and to alleviate future flood damage. The setbacks do not involve any right of public access. Although the owner is not compensated, the setback restrictions typically apply to only a narrow portion of land parcels. As such, they are likely to be upheld by courts as reasonable, limited constraints on private land development, with the balance of the property being available for development in accordance with local zoning laws.

Within the Land Acquisition Program, New York City has made significant concessions regarding private property. For example, the City explicitly promised not to exercise its power of eminent domain, promising instead to acquire all land on "willing buyer/willing seller" terms. The Land Acquisition Program excludes from acquisition land with structures and land within the self-defined growth boundaries of towns. The purchase of conservation easements will also soften the effects of the Land Acquisition Program on the use of land for acceptable purposes such as agriculture and recreation.

MOA provisions regarding property rights were made without complete knowledge about the sources of pollution in the watershed. The committee has emphasized the importance of evaluating the relative contributions of pollution sources in the watershed, and it has made recommendations regarding prioritization of pollutants. As new information becomes available, New York City should revisit its land acquisition priorities. In particular, the City should consider how the exclusion of certain lands from acquisition will affect pollutant loadings. The overall water quality impact of the MOA can only be assessed once the requisite monitoring data and information are in place.

BALANCING TREATMENT OPTIONS AND WATERSHED MANAGEMENT

As mentioned in Chapter 1, some perceive watershed management and other treatment options (such as coagulation/filtration) to be mutually exclusive in New York City. Nevertheless, the City's filtration avoidance determination requires that designs for a coagulation/filtration plant be drawn up concurrently with implementation of the watershed protection strategy.

Watershed management and treatment process such as disinfection and coagulation/filtration are examples of barriers that comprise the multiple-barrier approach to providing safe drinking water endorsed by EPA, water supply organizations, and this committee. Other barriers include having the highest-quality source water, using the best available treatment technologies, maintaining a clean distribution system, practicing thorough monitoring and accurate data analyses, having well-trained operators, and maintaining operating equipment. If effectively implemented, the MOA represents the first barrier in a multiple-barrier approach to providing high-quality drinking water to New York City residents.

Why Additional Treatment Options Should Be Considered

Drinking water that meets or exceeds all the requirements of federal, state, and local authorities is, in a regulatory sense, defined as safe to drink. It is, however, *not* risk-free. That is, the concentrations of microbial pathogens and chemicals in drinking water are not zero. Fortunately, each barrier in a multiple-barrier approach can reduce the microbial and chemical risk of drinking water, and these benefits accumulate as additional barriers are placed between pollution sources and water consumers.

Although it presently complies with all federal and state standards for drinking water quality, a time may come when New York City will want or need to add additional barriers for the protection of its drinking water supply. Direct filtration and alternative disinfection, for which risk reduction can be systematically quantified, are the most probable and well-characterized additional treatment options, given the pilot studies already completed on these processes.

Changes in Future Regulations

As discussed in detail in Chapters 3 and 5, future changes to federal regulations may compel New York City to adopt additional treatment processes. The water supply currently appears to be most vulnerable to noncompliance for haloacetic acids (HAAs). When the Stage 2 MCL for HAA5 is promulgated, New York City may not be able to comply unless DBP precursor removal is added to its treatment train. Turbidity may also become a factor within the next ten years. It is possible that negotiations on the long-term Enhanced Surface Water Treatment Rule will lead to a substantially lower allowable turbidity from unfiltered supplies, which might be difficult to meet without some process for particulate control (such as coagulation/filtration). In addition, specific requirements for inactivation and/or removal of *Cryptosporidium* are likely. Finally, EPA's Candidate Contaminant List contains 10 microbial parameters and 30 organic compounds. If any of these should be regulated in the future, having particulate/precursor treatment in place to supplement disinfection would be beneficial for controlling the microbial contaminants, and it could provide a location for addition of an adsorbent to remove organic contaminants.

Uncertainty

Because some sources of pollution within the Catskill/Delaware water have not been quantified or even identified (particularly for microbial pathogens), it is impossible to conclude that the present high water quality will be maintained indefinitely. Uncertainty regarding pollutant sources can be partially ameliorated by adding additional treatment processes to the water supply. Uncertainty regarding sources, transport, and fate is greatest for pathogens, particularly

Cryptosporidium. The Monte Carlo analysis performed by the committee (Table 6-11) suggests that New York City's drinking water may present a risk to consumers that is greater than the acceptable risk level informally suggested by EPA (although the many caveats discussed in Chapter 6 *must* be kept in mind). Because *Cryptosporidium* oocysts are resistant to some disinfectants, the use of alternate disinfectants coupled with coagulation/filtration or some other particulate-control treatment could be used to lower this risk and reduce uncertainty.

The sources of DBP precursors have also not been unequivocally determined in the watershed. The MOA's focused efforts on phosphorus control will have a positive influence on the plankton-generated precursors of DBPs. However, as discussed above, allochthonous DBP precursors are substantially more difficult to control, with streamside buffers having limited effect. Treatment processes such as coagulation and filtration may be necessary to control these organic materials in the future.

Changes in Supply and Demand

Although not a current threat, future demand pressures as well as future supply shortages caused by drought could force New York City to use supplementary sources of drinking water such as the Hudson River. Because of its poor quality compared to water from the Catskill/Delaware watershed, this water is highly likely to require additional treatment beyond chlorination and coagulation with alum (which are currently provided for). The existence of additional treatment facilities for the Catskill/Delaware system would allow Hudson River water to be blended into Catskill/Delaware water with minimal complications.

Summary

New York City and other signatories of the MOA are commended for bolstering the multiple-barrier approach to water supply protection. Although limitations are noted in this report, the MOA is a template for proactive watershed management that, if diligently implemented over the long term, will improve source water quality. The New York City MOA emphasizes the importance of watershed management as a first-line barrier in all water systems, whether they are filtered or not. Box 12-1 highlights recommendations regarding watershed management in New York City that are transferable to other communities across the country.

The committee encourages New York City and managers of all other unfiltered water supplies to be receptive to the possibility of new and additional treatment options. The need may arise because of uncertainties about pollutant sources, the growing scientific understanding of the impacts of pollutants on human health, and changing regulations. For example, New York City may be required to filter its water, *not* because watershed management does not

BOX 12-1
Transferable Conclusions and Recommendations

Although this report addresses the watershed management strategy of New York City, its conclusions and recommendations are relevant to other source water protection programs, whether their source water is filtered or not. This box discusses some of the most global conclusions and recommendations. The reader is referred to specific chapters for more detailed information.

Monitoring (Chapter 6)

Operational monitoring of physical, chemical, and biological parameters should be event-based rather than based on fixed intervals for streamflow, shallow subsurface flow, precipitation, and WWTP effluent. Sampling at fixed intervals will miss significant loadings of pollutants, particularly those in the particulate phase such as microbial pathogens, sediment, and particulate phosphorus. This report strongly encourages the use of performance monitoring to determine whether watershed management is effective over the long run. Examples of such monitoring over a wide range of spatial scales are given.

GIS (Chapter 6)

Land use and land cover should be maintained as separate databases in a GIS that supports watershed management. GIS databases are most useful when they are made readily available to the public.

Public Health and Risk Assessment (Chapter 6)

Active disease surveillance of the consumer population is a component of watershed management for a drinking water supply that is not practiced by all communities. Active disease surveillance can reveal trends in endemic rates of disease and, if enhanced surveillance is practiced, possibly detect outbreaks of disease. These techniques should be supplemented with epidemiological studies in order to determine the proportion of disease attributable to drinking water. Such studies should occur in water utilities where the population is sufficient to provide a statistically powerful evaluation.

Risk assessment is a complementary and powerful tool that can be used to estimate the risk of infection from drinking tap water. The accuracy and significance of risk assessment calculations rely on high-quality data collected from the source water, on accurate estimates of water consumption, and on a determination among relevant stakeholders of an acceptable risk level. Risk assessment can be used to assess the overall impact of watershed management over time.

Nonstructural Protection Strategies (Chapter 7)

A land acquisition program is potentially one of the most successful strategies for source water protection. This report stresses the importance of prioritizing land for acquisition based both on proximity to bodies of water and on land use.

Land use planning is also integral to watershed management. Nonstructural practices such as zoning and public education are sometimes the most effective and least expensive ways to prevent future pollution in water supply watersheds. Their success relies heavily on adequate public participation and effective implementation through enforcement of plan provisions.

TMDLs (Chapter 8)

As demonstrated in New York City, the Clean Water Act's Total Maximum Daily Load program relies heavily on developing accurate models for terrestrial runoff and reservoir water quality. These models in conjunction with the TMDL framework can be powerful tools for determining the relative importance of pollutant sources and directing BMPs to reduce pollutant loading. Adapting, calibrating, and validating such models may be a difficult prospect for many smaller communities.

Pollutant Trading (Chapter 8)

New York City's phosphorus offset pilot program illuminates many important issues that should be considered in the development of effluent trading programs for WWTPs in other communities. First, these programs should clearly define baseline requirements for pollutant reduction beyond which additional reductions can be counted as surplus. The success of these efforts depends largely on the availability and efficacy of monitoring techniques for the pollutant of interest, which should play a role in determining whether effluent trading can be viable. The ratios used in an effluent trading program should reflect the specific goals of the regulatory agency, the unique conditions present in the community, and the characteristics of the pollutant.

continued

BOX 12-1 Continued

Antidegradation (Chapter 8)

This report describes the wide variation among state antidegradation policies and their implementation. The most important issue that all state antidegradation programs should address is how the assimilative capacity of waters will be allocated, a stipulation that is not explicitly stated in most state policies. Depending on the individual regulations used to implement antidegradation, states should bolster their policies to include language about assimilative capacity and to ensure adequate discussions of the social and economic benefits of new discharges and the significance of new discharges in terms of how much assimilative capacity is used.

Additional Treatment (Chapter 8)

This report points out that as future regulations are promulgated, the need for additional treatment processes should be carefully considered. The use of ozone as a sole disinfectant is cautioned against because of the potential creation of biodegradable organic matter that may foul distribution systems. Treatment beyond disinfection alone can provide additional benefits to the consumer. An aggressive watershed management program coupled with effective water treatment would maximize public health protection.

Nonpoint Source Pollution Control (Chapters 9 and 10)

The limitations of urban stormwater BMPs for reducing pollutant concentrations in runoff are applicable to all communities. In particular, the efficacies of individual stormwater BMPs diminish when placed in series, and many BMPs do not have demonstrated performance ratings. For all BMPs discussed, including urban stormwater, forestry, and agricultural

produce substantial and sustained benefits, but to comply with more stringent regulations for DBPs and microbial pathogens. Management of these constituents may be infeasible with the current water supply infrastructure simply because increasing disinfectant concentrations to kill more pathogens *simultaneously* increases the quantity of DBPs. By contrast, coagulation/filtration or its equivalent, when operated properly, is effective against microbial pathogens, particulates, and the organic carbon precursors of DBPs. Furthermore, additional treatment options can reduce risk by predictable amounts. As noted in Chapter 6,

BMPs, performance monitoring is required to demonstrate their effectiveness over the long term. Monitoring techniques should be developed in parallel with the development of new, innovative BMPs.

Multiple recommendations are given on the wise use of setbacks and buffer zones adjacent to bodies of waters, which are a popular type of structural BMP in many settings. In general, buffer zones should not be used as a sole management practice, but should instead be combined with other BMPs to increase overall pollutant removal. The efficacy of buffer zones depends primarily on (1) the establishment of sheet flow upslope of the buffer, (2) the appropriateness of the buffer vegetation, and (3) the regular harvesting of the buffer to remove accumulated pollutants. These management techniques must be applied to setbacks if setbacks are expected to function as buffer zones.

Wastewater Treatment (Chapter 11)

This report suggests that the presence of WWTPs in a water supply watershed will not lead to immediate declines in reservoir water quality if tertiary and quaternary WWTP upgrades, such as microfiltration and continuous backwash upflow dual sand filtration, are put in place and are maintained over the long term. The use of a travel-time criterion for siting WWTPs is cautioned against unless sufficient scientific evidence exists to support the choice of travel time.

Many communities rely on individual septic systems, or OSTDS, in lieu of WWTPs. The most important factors controlling the efficacy of OSTDS are the choice of technology and enforcement activities to reduce failure rates. The committee strongly advocates the use of aerobic treatment systems as BACT and yearly inspections to maintain their ability to remove pollutants. Such requirements are more realistic than setting effluent standards for OSTDS, which would be difficult to comply with. Failure rates for OSTDS can be substantially reduced by siting them on slopes of less than 15 percent.

direct filtration is credited with at least a 100-fold (2-log) removal of *Cryptosporidium* (Nieminski and Ongerth, 1995), suggesting that filtration could lower the risk of waterborne cryptosporidiosis by a factor of 100. It is clear that if other factors controlling water quality remain the same, treatment beyond disinfection alone can enhance public health protection in an unfiltered water supply.

Although its contribution to risk reduction is more difficult to quantify than that of treatment technologies, watershed management is an essential component of a modern water supply system. **Because its watershed management strategy has gained national prominence, the committee encourages New York City to lead in efforts to quantify the contribution of watershed management to overall risk reduction.** For example, New York City may be able to demonstrate the risk-reducing effects of the MOA by comparing the results of regular risk assessment calculations.

The New York City MOA outlines an ambitious and path-breaking program of source water protection, and it is a unique document in the history of water resource management. The program it advances is a prototype of the utmost importance to all water supply managers. The MOA provides for an extraordinary financial and legal commitment from New York City (1) to prevent existing and potential contaminants from reaching the source reservoirs, (2) to monitor a broad range of water quality and drinking water parameters, (3) to conduct new research on public health and water quality, and (4) to promote the economic and social well-being of the Catskill/Delaware watershed communities in an environmentally sustainable manner.

New York City should maintain its standing as a national model for watershed management and sustainable source water protection. The committee is encouraged by the evidence of progress to date in achieving the goals of the MOA. Other public water supplies, whether using filtration or not, have much to learn from these accomplishments.

REFERENCES

Auer, M. T., S. T. Bagley, D. A. Stern, and M. J. Babiera. 1998a. A framework for modeling the fate and transport of *Giardia* and *Cryptosporidium* in surface waters. Journal of Lake and Reservoir Management 14(2-3):393–400.

Auer, M. T., K. A. Tomasoski, M. J. Babiera, M. L. Needham, S. W. Effler, E. M. Owens, and J. M. Hansen. 1998b. Phosphorus bioavailability and P-cycling in Cannonsville Reservoir. Journal of Lake and Reservoir Management 14(2-3):278–289.

Environmental Protection Agency (EPA). 1998. Interim Enhanced Surface Water Treatment Rule (IESWTR), Federal Register 13(24): 69478–69521.

Hamilton, Rabinovitz and Altschuler, Inc. 1998a. West of Hudson Economic Development Study for the Catskill Watershed Corporation. Report 1: Baseline Economic Analysis and Community Assessment, June 22, 1998.

Hamilton, Rabinovitz and Altschuler, Inc. 1998b. West of Hudson Economic Development Strategy. Report 2: Market Sector Assessment & Program Issues Analysis, September 25, 1998.

Longabucco, P., and M. Rafferty. 1998. Analysis of material loading to Cannonsville Reservoir: Advantages of event-based sampling. Journal of Lake and Reservoir Management 14(2-3):197–212.

New York City Department of Environmental Protection (NYC DEP). 1998a. Disinfection Study for the Catskill and Delaware Water Supply Systems. October 30, 1998. Valhalla, NY: NYC DEP.

NYC DEP. 1998b. Calibrate and Verify GWLF Models for Remaining Catskill/Delaware Reservoirs. Valhalla, NY: NYC DEP.

Nieminski, E. C., and J. E. Ongerth. 1995. Removing *Giardia* and *Cryptosporidium* by conventional treatment and direct filtration. J. Amer. Waterworks Assoc. 9:96–106.

Schueler, T. 1995. The importance of imperviousness. Watershed Protection Techniques 1(3):100-112.

Stepczuk, C. L., A. N. Martin, P. Longabucco, J. A. Bloomfield, and S. W. Effler. 1998a. Allochthonous contributions of THM precursors in a eutrophic reservoir. Journal of Lake and Reservoir Management 14(2-3):344–355.

Stepczuk, C. L., A. N. Martin, P. Longabucco, J. A. Bloomfield, and M. T. Auer. 1998b. Spatial and temporal patterns of THM precursors in a eutrophic reservoir. Journal of Lake and Reservoir Management 14(2-3): 356–366.

Appendixes

Appendix A

Abridged Version of the New York City Watershed Memorandum of Agreement (MOA)

LAND ACQUISITION
ARTICLE II

In the Catskill/ Delaware watershed (approximately 1 million acres), New York City owns 6 percent of the land and 20 percent is owned by the state and maintained as a Forest Preserve. To acquire more land, New York City plans to spend $250 million to buy undeveloped land possessing features that are water quality sensitive (e.g., proximity to intakes, streams, and reservoirs). The City will buy from willing sellers only, and will not enforce their power of eminent domain. The agreement allows certain parcels of land to be exempted from outright purchase, but all land is subject to conservation easements. The City will bid at fair market values and continue to pay property taxes on all acquired land. Property in the Catskill/Delaware watershed has been prioritized into five categories (1A, 1B, 2, 3, and 4) for acquisition. The City is required to contact the owners of 350,050 acres.

$10 million of New York City money and $7.5 million of state money is available for land purchase in the Croton watershed.

NEW WATERSHED RULES AND REGULATIONS
ARTICLE III AND ATTACHMENT W

Before the 1997 Memorandum of Agreement, watershed activities were governed by a ruling from 1953. New watershed regulations stipulated in the MOA work in conjunction with existing state and federal regulations and include additional rules unique to the New York City watersheds. An abridged summary is provided for sections with relevance to this report.

Wastewater Treatment Plants (WWTPs)

Minimum technical requirements for all WWTPs:
- phosphorus removal: 1.0 mg/l for <50,000 gallons per day (gpd)
                                 0.5 mg/l for 50,000 - 500,000 gpd
                                 0.2 mg/l for >500,000 gpd
- 99.9% removal and/or inactivation of *Giardia* and enteric viruses

Minimum siting requirements for all WWTPs:
- No building or expansion in phosphorus-restricted basins.
- No building or expansion in coliform-restricted basins.

Requirements for subsurface discharging plants, any travel time:
- Remediate existing and build new WWTPs to include:
  sand filtration
  disinfection for systems > 30,000 gpd
  phosphorus removal
- No part of the seepage unit or absorption field is allowed within 100 feet (ft) of a watercourse or wetland and within 500 ft of a reservoir, reservoir stem, or controlled lake.

Requirements for surface discharging plants, within 60 days travel time:
- No new WWTPs
- Remediate existing WWTPs to include:
  sand filtration
  disinfection
  phosphorus removal
  microfiltration

Requirements for surface discharging plants, outside 60 days travel time:
- No new WWTPs discharging into reservoirs, reservoir stems, controlled lakes, or wetlands
- Build new WWTPs discharging into a watercourse, and remediate all existing WWTPs, to include:
  sand filtration
  disinfection
  phosphorus removal
  microfiltration

Exceptions:
- Can build or expand WWTPs in phosphorus- or coliform-restricted basins if it will eliminate the source of contamination.
- If it will not, *existing plants* in phosphorus-restricted basins can be expanded if a 2:1 offset can be demonstrated.

- Up to 6 new plants in phosphorus-restricted basins (3 West-of-Hudson, 3 East-of-Hudson) are allowed under 5-year pilot program requiring 3:1 phosphorus offsets and a maximum phosphorus concentration of 0.2 mg/l.

Operational Requirements:
- In general, stand-by power units, flood protection, back-up disinfection, two sand filters, flow meter, and an alarm system are required. For subsurface discharge, a pump chamber, an additional area equal to 50% of the absorption field, and percolation and deep hole tests are required.

Subsurface Sewage Treatment Systems

Siting restrictions:
- All new systems require NYC DEP approval.
- For new systems, no part of absorption field can be within 100 ft of watercourse or wetland, or within 300 ft of reservoir, reservoir stem, or controlled lake.
- No new raised systems allowed within 250 ft of a watercourse or wetland, or within 500 ft of a reservoir, reservoir stem, or controlled lake unless site constraints make buffer distances impossible.

Stormwater Controls

Siting of impervious surfaces:
- New impervious surfaces prohibited within 100 ft of a wetland or watercourse, or 300 ft of a reservoir, reservoir stem, or controlled lake.

Exceptions:
- Bridges or water crossings.
- Impervious surfaces used for agricultural activities.
- Roads used to modify and operate WWTPs, public water systems.
- Single family home on existing lot or new lot greater than 5 acres which receives a stormwater permit from the state can build an impervious surface within 100 ft of watercourse or wetland.
- West-of-Hudson, those surfaces located in a village or hamlet or in an area zoned as commercial.
- East-of-Hudson, those surfaces designated as a main street.

Stormwater Pollution Prevention Plans:
- Activities must obtain a general stormwater permit with two sets of technical requirements:
  - *control the first 1/2 inch of run-off*
  - *volume control*

- No permit required for agriculture and silviculture.
- Permits only necessary for areas greater than 5 acres.
- NYC DEP retains final decision-making authority for all activities.
- All plans should include an analysis of the 25-year storm.
- Activities in phosphorus-restricted and coliform-restricted areas require additional analyses.

Hazardous Substances

Siting restrictions:
- No new NYS DEC registered tanks within 100 ft of watercourse or wetland, or within 500 ft of reservoir, reservoir stem, or controlled lake.
- Between 100 and 250 ft of watercourse, owner must:
    - *provide tank registration before installation*
    - *use best management practices to minimize release of substances*
    - *meet all other state requirements for hazardous substances*

Exceptions:
- Hazardous material used at WWTPs, or in connection with a WWTP.
- "Noncomplying regulated activities" exempted, including the replacement of existing facilities.

Petroleum Products

Siting restrictions:
- No new storage tanks requiring registration under State law within 100 ft of watercourse or wetland, or within 500 ft of reservoir, reservoir stem, or controlled lake.
- No new storage tanks not requiring State registration within 25 ft of watercourse or wetland, or within 300 ft of reservoir, reservoir stem, or controlled lake.
- New home heating oil tanks within 100 ft of watercourse or wetland, or within 500 ft of reservoir, reservoir stem, or controlled lake, must be above ground or in the basement.

Exceptions:
- Above buffer distances do not apply to the replacement in kind of existing tanks.
- Agricultural uses of petroleum products are exempted.
- Petroleum products used at or in conjunction with WWTPs are exempted.
- "Noncomplying regulated activities" exempted.

Solid Waste

Siting restrictions:
- No new or expanded solid waste facilities within 250 ft of a watercourse or wetland, or within 1000 ft of a reservoir, reservoir stem, or controlled lake.

Exceptions:
- Recycling facilities (not including oil, batteries).
- Composting facilities.
- Expansion of the landfill in Delaware County.

Agriculture

- No reckless activity that will increase pollution to the water supply.

Pesticides

- No pesticide discharges or storage that would lead to discharge that will impair water quality, according to the standards set in A18-D.

Fertilizers

- No activity is considered "noncomplying."
- Cannot wash fertilizer application machinery into wetland, watercourse, reservoir, reservoir stem, or controlled lake.
- Cannot use water from a reservoir, reservoir stem, or controlled lake to make fertilizer.
- Must use an anti-siphon device when using water from a watercourse to make fertilizer.
- Cannot store fertilizer in the open.

Exceptions:
- Any agricultural activity in compliance with State or Federal laws is exempt.
- Individual, non-commercial uses of fertilizer are exempt.

Snow-Melt Materials

- Do not dispose of snow in wetlands, watercourses, reservoirs, reservoir stems, or controlled lakes if possible.
- Use as little snow-melt materials as possible.
- Store snow-melt materials in structures that will not leak.

WATERSHED PROTECTION AND PARTNERSHIP PROGRAMS
ARTICLE V

The City is investing $270 million for several West-of-Hudson water quality protection and partnership programs and $126 million for East-of-Hudson partnership programs (see Table 7-3). Partners in this effort include the following organizations and programs.

a) *Watershed Agricultural Program's* goal is to refine and demonstrate an environmentally sound approach to farm management. Concerns include nutrients, pathogens, sediment, toxic and organic matter, soil erosion, and pesticides. Leadership is provided by the Watershed Agricultural Council. It is a voluntary program with incentives to farmers and cost sharing for New York City.

b) *Watershed Protection and Partnership Council.* The council is the main forum for discussion that brings together city residents and residents of the watershed. The council will review and assess all watershed protection efforts. The Council will have dispute resolution authority to prevent future arguments from spilling into the courts.

c) *Catskill Watershed Corporation* is a nonprofit organization created to administer $240 million for water quality protection programs in the West-of-Hudson region. The CWC will focus on septic system inspection, stormwater management, environmental education, stream corridor protection, and improved storage of sand, salt, and de-icing materials.

Appendix B

Use Classifications and Water Quality Criteria for New York State

The New York State Department of Environmental Conservation (NYS DEC) is the agency responsible for setting water quality criteria and waterbody use classifications for the New York City drinking water reservoirs. Much of this information is contained in the New York State Codes, Rules, and Regulations, Title 6, Chapter X, Parts 700–705. New York divides water into four main categories for the purpose of classification: fresh surface waters (which includes the New York City drinking water reservoirs), saline surface waters, fresh groundwater, and saline groundwater. Each category is then divided into classes that represent different uses. Water quality criteria (often expressed as concentrations averaged over a certain time period) correspond to these classes for a wide variety of physical, chemical, and biological parameters.

USE CLASSIFICATIONS—FRESH SURFACE WATERS

Eight classes of fresh surface waters are delineated: N, AA-special (AA-S), A-special (A-S), AA, A, B, C, and D. Class N waters are the most pristine and support the greatest number of uses. The best uses of Class N waters are "the enjoyment of water in its natural condition." This phrase is unique to Class N waters. Other uses of Class N waters include use as a drinking water source, bathing, fishing, fish propagation, and recreation. Discharge of sewage and industrial wastes into Class N waters is prohibited, unless the sewage effluent has traveled at least 200 feet through unconsolidated earth. In addition, these waters shall contain no substances that will contribute to eutrophication, nor shall they receive runoff containing such substances. It is not clear which waterbodies in the state of New York carry this classification. None of the drinking water reservoirs of New York City are Class N.

Each class below N supports fewer uses, and restrictions on discharges into those waterbodies become less strict. Classes AA-S, A-S, AA, and A all support use as a drinking water source, culinary or other food processing source, for recreation, for fishing, and for fish propagation. In addition, Class AA-S waters cannot be polluted with solids, oils, sludges, sewage, and other wastes (which is actually a more stringent requirement than for Class N waters). Class AA-S waters shall also contain no nitrogen or phosphorus in amounts that will result in algal growth, weeds, or slimes (similar to the eutrophication requirement for Class N). The main difference between the three remaining "A" classifications concerns their use as a source of drinking water. Class A-S (international boundary waters) and A waters must be treated with coagulation, sedimentation, filtration, and disinfection (or their equivalents) to qualify as sources of drinking water. Class AA waters require only disinfection. All of the reservoirs in the Catskill/Delaware and Croton watersheds are classified as either A or AA. Classes B, C, and D include waters that can be used for recreation and fishing, but not as a source of drinking water. Waterbodies can be reclassified every three years.

WATER QUALITY CRITERIA

For each use classification, there is a list of physical, chemical, and biological parameters that characterize that classification. These criteria are included in Part 703 of the Water Quality Regulations. The criteria fall under three broad categories: health-based assuming the waterbody is a drinking water source, health-based assuming the waterbody contains consumable fish that might bioaccumulate contaminants, and aquatic-based. For many chemicals, the criteria are set equal to their MCLs. For oncogenic chemicals, the criteria are based on a one in a million lifetime cancer risk, with doses based on a 70-kg adult. Criteria can also be based on former regulations, aesthetic considerations, and chemical correlations. The "Basis for Establishment of Standards" is defined in Table 2, Part 703 of the Water Quality Regulations.

Some of the criteria are narrative; e.g., turbidity in all classes of waters must not increase to cause a substantial visible contrast to natural conditions. There are narrative criteria for nitrogen and phosphorus, stating that for all classes there shall be none in amounts that result in the growth of weeds, algae, and slimes that will impair the waters for their best uses. The 20-µg/L guidance value for total phosphorus that is currently used by NYS DEC appears in a subsequent revision of the Water Quality Regulations.

For almost all other important physical, chemical, and biological parameters, there are specific criteria. Dissolved oxygen, pH, Cl, fecal coliforms, and metals are of most concern for New York City. pH must fall between 6.5 and 8.5 for all waters. Most of the other parameters of interest do not differ between the various use classifications. There are apparent differences are between Classes AA, A, and B and Classes C and D for zinc, arsenic, and selenium concentration. For dissolved oxygen, two new use "subclassifications" arise: trout (T) and trout-spawning (TS). Many of the Catskill/Delaware reservoirs have one of these additional subclassifications, shown in parentheses after the main classification, e.g. A(TS). The T and TS subclassifications require higher dissolved oxygen concentrations than the AA, A, B, and C classifications. (T and TS classifications are made by the Department of Fish and Wildlife after observing whether trout are present, and whether, based on water temperature, reproduction is possible.)

Table 1 in Part 703 of the Water Quality Regulations lists almost 160 different chemical substances for which there are numeric criteria. Metals, organics, pesticides, and herbicides are included. For the ammonia and ammonium standard, both health-based and aquatic-based criteria are given. The aquatic-based criteria are considerably more stringent and depend heavily on pH and temperature. Many of the criteria for metals depend on hardness. For substances considered in a group (e.g., dichlorobenzenes), the standard applies to the sum of all substances within that group.

The final table in Part 703 of the regulations, Table 3, gives criteria for point source effluents discharging directly into groundwater. These criteria are typically twice the allowable maximum concentration of the chemical in groundwater, but not always. There are a significant number of chemicals in Table 1 (and consequently Table 3) for which criteria are only given for groundwater and not for fresh or saline surface waters.

REVISIONS TO THE WATER QUALITY REGULATIONS

On October 22, 1993, major revisions to Part 703 the Water Quality Regulations were published by NYS DEC. There are 51 new entries to Table 1, and most contain guidance values rather than criteria. Guidance values may be used where a standard has not yet been established. Table 1 is considerably more complete as a result of the revisions. Chemicals have either criteria or guidance values for all possible water types (fresh surface water, groundwater, etc.). A new addition to Table 1 is the guidance value for total phosphorus of 20 µg/L. This value is based on aesthetic effects for primary and secondary contact recreation (unlike the narrative standard, which is based on eutrophication).

Finally, the revisions contain an additional section dealing exclusively with groundwater—the principal organic contaminant groundwater standard (POC). These criteria are meant to provide protection of groundwater from any organic compound, regardless of whether toxicity data exist.

WATER QUALITY IMPAIRMENT

NYS DEC is responsible for categorizing all waterbodies that are "use-impaired." This information can be found in several documents, including the biennial Clean Water Act 305b, the biennial Clean Water Act 303d report, the triennial NYS DEC Priority Waterbodies List, and numerous NYS DEC publications.

The Priority Waterbodies List (PWL) is the state's basic tool for identifying and organizing impaired waters. The state is divided into drainage basins; those basins that include the New York City reservoirs are the Lower Hudson River basin, the Delaware River basin, and the Mohawk River basin. All available information, including monitoring data, surveys, and public input, are used to evaluate the severity of the water quality impairment. If there is insufficient information for judging a waterbody's use impairment, it is not listed. Fortunately, a great deal of monitoring data exists for the New York City watersheds, and every reservoir and some major tributaries are currently listed.

The Degree of Designated Use Support qualitatively describes water quality. *Threatened* waters have water quality supporting designated uses, with no obvious signs of stress to ecological systems. However, existing or changing land-use patterns may result in restricted use or ecosystem disruption. *Stressed* waters have reduced water quality and designated uses are intermittently or marginally restricted. Natural ecosystems may exhibit adverse changes. *Impaired* waters have water quality and/or habitat characteristics that frequently impair a classified use. This term also applies to water supporting a designated use at a level significantly lower than would be expected. Natural ecosystem function may be disrupted. *Precluded* waters have water quality and/or associated habitat degradation that precludes, eliminates, or does not support a classified use. Natural ecosystem functions may be significantly disrupted.

For each individual waterbody, the PWL includes basic information about the acreage, the location, the use classification, the use impairments, pollutants, pollutant sources, the degree of designated use support, and the resolvability. In some cases the PWL identifies impairments to uses other than the primary use. Depending on the pollutants present, a waterbody may be more impacted for a secondary use than a primary use.

Three reservoirs in the Catskill/Delaware watershed are *threatened* for use as a source of drinking water: Neversink, Rondout, and Pepacton. Nutrients from urban runoff are the primary pollutants at Neversink, while Pepacton is impacted by pathogenic microorganisms from septic systems. Cannonsville, Schoharie, and Ashokan reservoirs are *stressed*, suffering from nutrients in agricultural runoff, silt from construction activities, and nutrients and silt from urban runoff, respectively. In the Croton watershed, West Branch Reservoir is *stressed* due to silt and nutrients derived primarily from urban runoff. The Kensico Reservoir is *threatened* for a variety of uses, with nutrients from urban runoff again as the primary culprit. One major drawback of the PWL is that it does not include a breakdown of all pollutants exceeding the water quality criteria for each waterbody.

The 303d list is required by EPA to identify and rank waterbodies, which may require development of TMDLs. Inclusion on the list, however, does not mean that TMDL calculations will be completed during the following two years. The list simply identifies impaired bodies, describes the primary pollutants, and ranks the waterbody according to its level of impairment. It draws heavily upon the PWL, using an identical format for grouping the waterbodies and describing their condition. Whole sections of the 1996 PWL appear to comprise the bulk of the 1998 303d list.

The 303d list categorizes impaired waterbodies as (1) designated priority for TMDL development, (2) impacted by atmospheric deposition, (3) with fish consumption advisories, (4) closed to shellfish harvesting, (5) showing water quality criteria exceedances supported by monitoring data, and (6) requiring verification of water quality problems. A waterbody can fall into more than one category, but a review of the list shows that very few do. The New York City reservoirs are all designated priority for TMDL development.

The 305b list gives an overall description of the health of waterbodies in New York State. As of 1996, 93 percent of New York's rivers and streams fully supported their designated uses, while only 47 percent of lakes, ponds, and reservoirs support their designated uses. The report identifies nonpoint source pollution as the cause of water quality impairment, with agriculture, urban runoff, and septic systems cited as the most prevalent contributors. Point sources such as wastewater treatment plants are now relatively minor contributors to pollution. However, the report voices the need to upgrade and replace the wastewater treatment infrastructure, which is apparently approaching the limits of its normal life span.

Appendix C

Microbial Risk Assessment Methods

Risk may be defined as the possibility of suffering harm from a hazard. Risk analysis provides the tools by which the magnitude and likelihood of such consequences are evaluated. The objective of risk analysis is to allow a standard to be set to achieve a certain level of public health or ecological protection, or perhaps to balance the costs of achieving such a standard (for example, by application of treatment processes) with the benefits obtained.

The formal risk analysis process can be divided into four discrete components (NRC, 1983):

1. Hazard assessment, in which the amount of a substance discharged from various sources is estimated.

2. Exposure assessment, in which the distribution of a hazardous substance among individuals in the population of concern is estimated. This may involve use of transport or attenuation models.

3. Dose-response assessment, in which the relationship between the level (concentration, duration) of exposure and the likelihood of effects of various degrees of severity is assessed.

4. Risk characterization, in which the information from the previous three elements is integrated.

Most human health risk assessments have been conducted on carcinogens. Such risk assessments are based on chronic, lifetime exposure to a particular agent and almost always rely on evidence obtained from animal tests (such as rats or mice) to quantify the level of exposure below which risk is maintained at an acceptable level. The level of acceptable risk to carcinogens has generally been set in the range of 10^{-6}–10^{-4} (excess cases per lifetime) (Travis et al., 1987), although there are differing actions taken under different federal statutes (Rosenthal et al., 1992).

With infectious microorganisms, for which uniform guidelines do not exist, the process of risk assessment may be conducted in a similar manner (Haas et al., in press). There are several distinguishing aspects of quantitative microbial risk assessment, however, which bear notice:

• Generally adverse outcomes (infections, illnesses) associated with infectious agents may result from single exposures, rather than from chronic long-term exposures, although the duration of exposure will influence the likelihood of the outcome.

• Secondary cases of illness may occur in persons not exposed to the original contaminated material, such as by person-to-person contact.

• In general, human dose-response data are available for a variety of pathogens, and it is not always necessary to rely on animal studies. In fact, the issue of potential host-specificity of certain agents may make extrapolation from animal studies a less generally applicable procedure (without extensive validation).

• Because the doses of microorganisms that may provoke adverse outcomes are very small (in principle one competent organism has the potential to act in this matter), the statistics of small particles adds an element of variability that it not present with many chemicals (where even for very potent carcinogens, an average daily dose of millions or more molecules may be needed for an adverse effect to occur). This variability must be accounted for in the dose-response model and in the estimation of exposure (and is superimposed on other sources of variability, such as in the analytical method itself).

• There is also the possibility (although its importance has not been conclusively demonstrated) that populations exposed to pathogens over time may change in sensitivity (either to become more or less sensitized to the particular agent).

The process is illustrated schematically in Figure C-1. From the dose-response curve, the risk corresponding to a particular pathogen concentration and water ingestion rate (volume per day of water ingested without treatment, such as boiling, which might inactivate pathogens) can be determined. If this is above an acceptable level, then the concentration of the pathogen in the water supply may be regarded as too high. In initial formulations of the Surface Water Treatment Rule, an annual risk of 1 infection per 10,000 persons was regarded as being acceptable (Regli et al., 1988) although there is some question about the appropriateness of this value (Haas, 1996).

The use of this approach is attractive in that it has the potential to impute risks far lower than might be readily detectable by epidemiological surveillance in a routine manner. Furthermore, it may be applied to a series of pathogen monitoring results to assess any trends in the implied risk.

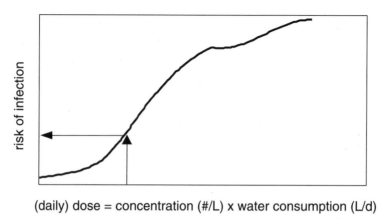

(daily) dose = concentration (#/L) x water consumption (L/d)

FIGURE C-1. Use of dose-response to estimate microbial risk.

There are a number of uncertainties that need to be considered with this approach. First, the ingestion of water is variable from person to person [in common EPA risk assessments, a default value of 2 L/person/day is generally used (Cohrssen and Covello, 1989) emanating from the NRC report *Drinking Water and Health* (NRC, 1977)]. It is now recognized that use of a population distribution may be more appropriate (Roseberry and Burmaster, 1992). Second, the methods for pathogen measurement (and especially determination of viability and infectivity of pathogens in environmental samples) have limitations, although it is far from clear whether particular methods may err on the positive or negative side. Third, existing methods to detect pathogenic microbes in water tend to be time consuming, and so use of this approach on a rapid or real-time basis to assess water quality is not realistic. Rather, this approach is more suited for periodic retrospective assessment of long-term trends. Finally, the dose-response curve used in this approach has typically been obtained on a defined population (generally "normal, healthy" individuals) and may not be reflective of differential susceptibilities in other segments of the population (Gerba et al., 1996).

Risk assessment may be particularly beneficial in conjunction with an epidemiological surveillance program. If the dose-response curve is extrapolated to low levels of risk, it is possible to ascertain the level of infection implied by a particularly low level of exposure. At the 1 in 10,000 risk level, surveillance of many thousands of persons over a period of time would be necessary to ascertain a positive effect. If pathogen-monitoring data are available, the imputed risk from the dose-relationship can be compared to the bounds from surveillance to assess consistency.

With pathogens, it is possible to perform risk assessments on a frequent basis, perhaps monthly or quarterly, as new pathogen monitoring data becomes available. It is also possible to use the monitoring data in conjunction with dose-response information and perhaps other data (such as the presence of excess turbidity in a finished water) to define levels of contamination where particular public health actions must be taken. Limitations on method efficiency and rapidity may, however, limit the present ability to use microbial risk assessment in such a real-time manner.

REFERENCES

Cohrssen, J. J., and V. T. Covello. 1989. Risk Analysis: A Guide to Principles and Methods for Analyzing Health and Environmental Risks. Springfield, VA: National Technical Information Service, U.S. Department of Commerce.

Gerba, C. P., J. B. Rose, and C. N. Haas. 1996. Sensitive Populations: Who is at the Greatest Risk? International Journal of Food Microbiology 30(1–2):113–123.

Haas, C. N. 1996. Viewpoint: Acceptable Microbial Risk. Journal of the American Water Works Association 88(12):8.

Haas, C. N., J. B. Rose, and C. P. Gerba. In press. Quantitative Microbial Risk Assessment. New York: Wiley.

National Research Council (NRC). 1977. Drinking Water and Health. Washington, DC: National Academy Press.

NRC. 1983. Risk Assessment in the Federal Government: Managing the Process. Washington, DC: National Academy Press.

Regli, S., A. Amirtharajah, B. Borup, C. Hibler, J. Hoff, and R. Tobin. 1988. Panel Discussion on the Implications of Regulatory Changes for Water Treatment in the United States. Advances in *Giardia* Research:275.

Roseberry, A. M., and D. E. Burmaster. 1992. Log-Normal Distributions for Water Intake by Children and Adults. Risk Analysis 12(1):99–104.

Rosenthal, A., G. M. Gray, and J. D. Graham. 1992. Legislating Acceptable Cancer Risk from Exposure to Toxic Chemicals. Ecology Law Quarterly 19:269–363.

Travis, C.C., S. A. Richter, E. A. C. Crouch, R. Wilson, and E. D. Klema. 1987. Cancer Risk Management. Environmental Science and Technology 21(5):415–420.

Appendix D

Analysis of Wastewater Treatment Plants and Onsite Sewage Treatment and Disposal Systems: The Impact Index

This appendix describes the methods used creating the Impact Index found in Chapter 11. The Impact Index shows the relative pollutant loadings of wastewater treatment plants (WWTPs) and on-site sewage treatment and disposal systems (OSTDS) in individual reservoir basins. For each West-of-Hudson reservoir, Kensico Reservoir, and West Branch Reservoir, a worksheet similar to the one presented below was created. An explanation of each worksheet entry is given below, followed by the eight worksheets completed for each reservoir.

Impact Index Worksheet

Known Permanent Watershed Population[1]
Permitted Number of Sewage Treatment Plants[2]
Permitted Wastewater flow from WWTPs[3]

Discharger   Daily Flow

Parameter	Input from WWTPs Now[4]	Input w/ upgrades[5]	Input 2010 w/growth[6]
Phosphorus (lbs/day)			
Giardia Cysts (10^6/day)			
Fecal Coliforms (10^6/day)			
Viruses (10^6/day)			
Total Suspended Solids (lbs/day)			
Cryptosporidium Oocysts (10^6/day)			

Known Number of OSTDS[7]
Estimated Wastewater flow from OSTDS[8]
Estimated Percent of Population served by OSTDS[9]
Estimated Population served by OSTDS (@2.5 persons/home)[10]

Parameter	Input from OSTDS Now	Input w/ upgrades	Input 2010 w/ growth
Phosphorus (lbs/day)	11	17	23
Giardia Cysts (10^6/day)	12	18	24
Fecal Coliforms (10^6/day)	13	19	25
Viruses (10^6/day)	14	20	26
Total Suspended Solids (lbs/day)	15	21	27
Cryptosporidium Oocysts (10^6/day)	16	22	28

Notes

1. Population from NYC DEP (1993), and based on 1990 Census values.

2. Permitted number of WWTPs from Table 5-4.

3. Permitted flow from WWTPs from Table 5-4.

4. Obtained from NYC DEP (1993 EIS), Table NC-2 "Existing, Expanded and New WWTPs Summary of No Action/Action Impact on Pollutant Loadings by District and Drainage Basin Totals."

5. Obtained from NYC DEP (1993 EIS), Table NC-2 "Existing, Expanded and New WWTPs Summary of No Action/Action Impact on Pollutant Loadings by District and Drainage Basin Totals."

6. Calculated by applying the growth rate observed for the period 1990 to 1996 to the period from 1990 to 2010. The observed annual growth rate of 0.25 percent for the West-of-Hudson basin resulted in a total population increase of 5 percent in 2010. The observed annual growth rate of 1.06 percent for the East-of-Hudson basin resulted in a total population increase of 21.2 percent for the Kensico and West Branch basins in 2010.

7. Septic system numbers from NYC DEP (1993, Table VIII F-3).

8. Based on 50 gallons of wastewater per day per person (Chanlett, 1979).

9. Derived by dividing the estimated population served by OSTDS (#10) by the known permanent watershed population (#1). Values slightly exceeding 100 percent appears to reflect the fact that many of the WWTPs serve schools and not residential subdivisions, while values greatly exceeding 100 percent may be attributable to large numbers of seasonal residences.

10. Calculated by multiplying the known number of OSTDS (#7) by 2.5 residents. This assumes that there is one OSTDS per residence, 2.5 occupants per residence, and that only residences (not commercial facilities) are served by OSTDS. 2.5 occupants per household is the figure for Florida derived from the 1990 Census values for population (12,937,926) divided by the number of households (5,138,360).

11. This is the average value for total phosphorus (0.0082 lb/day) found in effluent from conventional septic tank and drainfield systems as reported by Ayres (1993). It has been multiplied by the estimated population served by OSTDS (#10).

12. Assumes 6.4 percent of the population using OSTDS (#10) is positive for *Giardia*. This is based on the average of two CDC studies of prevalence of parasites in stool specimens which reported *Giardia* was found in 7.2 percent of samples examined in 1987 and 5.6 percent of samples examined in 1991 (Kappus et al., 1994). The number of positive residents is then multiplied by the maximum number of *Giardia* cysts excreted per day by an individual (9×10^6 cysts reported by Knight, 1980). This has been multiplied by 0.05 to reflect reported failure rates (defined as surfacing sewage) for conventional systems. This assumes that the remaining 95 percent of the systems are functional and remove all *Giardia* cysts.

13. This is the value for fecal coliform bacteria (2.7 MPN × 10⁶/100mL) found in effluent from conventional septic tank and drainfield systems as reported by Canter and Knox (1986). It has been multiplied by 0.05 of the total daily flow from OSTDS (#8) to reflect reported failure rates (defined as surfacing sewage) for conventional systems. This assumes that the remaining 95 percent of the systems are functional and remove all fecal coliforms.

14. This is the value for viruses (22 units/L) found in effluent from conventional septic tank and drainfield systems as reported by Anderson et al. (1991). It has been multiplied by 0.05 of the total daily flow from OSTDS (#8) to reflect reported failure rates (defined as surfacing sewage) for conventional systems. A similar process is applied to the remaining 95 percent of functional systems using the value for viruses (0.002 units/L) found 24 inches beneath the drainfield of conventional systems by Anderson et al. (1991).

15. Based on negligible value for total suspended solids (mg/L) found in effluent from conventional septic tank and drainfield systems (K. Sherman, FL DOH, personal communication, 1998).

16. Assumes 0.21 percent of the population using OSTDS (#10) is positive for *Cryptosporidium*. This is based on the average of two CDC studies of prevalence of parasites in stool specimens which reported *Cryptosporidium* was found in 0.23% of 216,275 stools examined in 1987 and 0.20 percent of 178,786 samples examined in 1991 (Kappus et al., 1994). The number of positive residents is then multiplied by the average number of *Cryptosporidium* oocysts excreted per day by each infected individual (1 × 10⁷, which is the midpoint of the range found in Goodgame et al., 1993). This has been multiplied by 0.05 to reflect reported failure rates (defined as surfacing sewage) for conventional systems. This assumes that the remaining 95 percent of the systems are functional and remove all *Cryptosporidium* oocysts.

17. This is the average value for total phosphorus (0.0014 lb/day) found in effluent from aerobic treatment systems as reported by Ayres (1993). It has been multiplied by the estimated population served by OSTDS (#10).

18. This assumes that BACT (aerobic treatment unit) removes virtually all *Giardia* cysts and that the failure rate, with a stringent regulatory structure, would be negligible.

19. This reflects the fact that BACT (aerobic treatment unit) removes virtually all fecal coliform bacteria and the failure rate, with a stringent regulatory structure, would be negligible.

20. This is the value for viruses (0.002 units/L) found in effluent 24 inches beneath the drainfield of conventional septic tank and drainfield systems as reported by Anderson et al. (1991). It has been multiplied by the total daily flow from OSTDS (#8) and assumes that all systems would be functional with adequate regulatory oversight.

21. Based on negligible value for total suspended solids (mg/L) found in effluent from BACT (aerobic treatment unit) (K. Sherman, FL DOH, personal communication, 1998).

22. This assumes that functional BACT (aerobic treatment unit) removes virtually all *Cryptosporidium* oocysts and that the failure rate, with a stringent regulatory structure, would be negligible.

23. This is the value from #17 with the growth factor described in #6 applied.
24. This is the value from #18 with the growth factor described in #6 applied.
25. This is the value from #19 with the growth factor described in #6 applied.
26. This is the value from #20 with the growth factor described in #6 applied.
27. This is the value from #21 with the growth factor described in #6 applied.
28. This is the value from #22 with the growth factor described in #6 applied.

RESERVOIR: Kensico

Known Permanent Watershed Population	4,432
Permitted Number of Wastewater Treatment Plants	0
Permitted Wastewater flow from WWTPs (gallons/day)	0

Known Number of OSTDS	580
Estimated Wastewater flow from OSTDS (gallons/day)	29,000
Estimated Percent of Population served by OSTDS	31
Estimated Population served by OSTDS (@2.5 persons/home)	1,450

Parameter	Input from OSTDS Now	Input w/ upgrades	Input 2010 w/ growth
Phosphorus (lbs/day)	11.9	2.0	2.4
Giardia Cysts (10^6/day)	42	0	0
Fecal Coliforms (10^6/day)	146,770	0	0
Viruses (10^6/day)	0.1	0	0
Total Suspended Solids (lbs/day)	0	0	0
Cryptosporidium Oocysts (10^6/day)	0.2	0	0

RESERVOIR: West Branch

Known Permanent Watershed Population	3,210
Permitted Number of Wastewater Treatment Plants	1
Permitted Wastewater flow from WWTPs (gallons/day)	20,000

Discharger	Daily Flow
Clear Pool Camp	20,000

Parameter	Input from WWTPs Now	Input w/ upgrades	Input 2010 w/ growth
Phosphorus (lbs/day)	0.8	0.2	0.3
Giardia Cysts (10^6/day)	13.6	13.6	16.3
Fecal Coliforms (10^6/day)	302.8	302.8	363.4
Viruses (10^6/day)	0.4	0.4	0.5
Total Suspended Solids (lbs/day)	3.3	1.7	2.0
Cryptosporidium Oocysts (10^6/day)	1.2	1.2	1.4

Known Number of OSTDS	1,777
Estimated Wastewater flow from OSTDS (gallons/day)	222,125
Estimated Percent of Population served by OSTDS	138
Estimated Population served by OSTDS (@2.5 persons/home)	4,443

Parameter	Input from OSTDS Now	Input w/ upgrades	Input 2010 w/ growth
Phosphorus (lbs/day)	36.5	6.1	7.3
Giardia Cysts (10^6/day)	128	0	0
Fecal Coliforms (10^6/day)	1,139,501	0	0
Viruses (10^6/day)	0.9	0	0
Total Suspended Solids (lbs/day)	0	0	0
Cryptosporidium Oocysts (10^6/day)	0.5	0	0

RESERVOIR: Ashokan

Known Permanent Watershed Population	14,294
Permitted Number of Wastewater Treatment Plants	6
Permitted Wastewater flow from WWTPs (gallons/day)	605,826

Discharger	Daily Flow
Belleayre Mt. Ski Center	29,000
Camp Timberlake	34,000
Onteora Schools	27,000
Mountainside Rest	3,076
Pine Hill WWTP	500,000
Rotron, Inc.	12,750

Parameter	Input from WWTPs Now	Input w/ upgrades	Input 2010 w/ growth
Phosphorus (lbs/day)	2.0	1.3	1.4
Giardia Cysts (10^6/day)	113.4	113.4	120.2
Fecal Coliforms (10^6/day)	1070.3	1070.3	1134.5
Viruses (10^6/day)	3.5	3.5	3.7
Total Suspended Solids (lbs/day)	5.3	5.3	5.6
Cryptosporidium Oocysts (10^6/day)	11.2	11.2	11.9

Known Number of OSTDS	5,360
Estimated Wastewater flow from OSTDS (gallons/day)	670,000
Estimated Percent of Population served by OSTDS	94
Estimated Population served by OSTDS (@2.5 persons/home)	13,400

Parameter	Input from OSTDS Now	Input w/ upgrades	Input 2010 w/ growth
Phosphorus (lbs/day)	110.0	18.3	19.4
Giardia Cysts (10^6/day)	386	0	0
Fecal Coliforms (10^6/day)	3,437,100	0	0
Viruses (10^6/day)	2.8	0	0
Total Suspended Solids (lbs/day)	0	0	0
Cryptosporidium Oocysts (10^6/day)	1.4	0	0

RESERVOIR: Schoharie

Known Permanent Watershed Population   17,662
Permitted Number of Wastewater Treatment Plants   19
Permitted Wastewater flow from WWTPs (gallons/day)   1,751,113

Discharger	Daily Flow
Camp Loyaltown, Inc.	21,000
Colonel Chair Estates	30,000
Crystal Pond Townhouse	36,000
Elka Park	10,000
Forester Motor Lodge	3,900
Frog House Restaurant	1,788
Golden Acre Farms	9,200
Grand George WWTP	500,000
Harriman Lodge	20,000
Hunter Highlands	80,000
Latvian Church Camp	7,000
Liftside at Hunter Mountain	81,000
Mountain View Estates	13,000
Ron-De-Voo Restaurant	1,000
Snowtime	120,000
Tannersville WWTP	800,000
Thompson House Inc.	4,775
Whistle Tree Development	12,450

Parameter	Input from WWTPs Now	Input w/ upgrades	Input 2010 w/ growth
Phosphorus (lbs/day)	12.7	4.5	4.8
Giardia Cysts (10^6/day)	536.9	299.3	317.3
Fecal Coliforms (10^6/day)	3,718.4	2,567.8	2,721.9
Viruses (10^6/day)	16.5	9.3	9.9
Total Suspended Solids (lbs/day)	29.7	10.7	11.3
Cryptosporidium Oocysts (10^6/day)	53.0	30.8	32.6

Known Number of OSTDS   7,869
Estimated Wastewater flow from OSTDS (gallons/day)   98,363
Estimated Percent of Population served by OSTDS   111
Estimated Population served by OSTDS (@2.5 persons/home)   19,673

Parameter	Input from OSTDS Now	Input w/ upgrades	Input 2010 w/ growth
Phosphorus (lbs/day)	161.7	26.9	28.5
Giardia Cysts (10^6/day)	567	0	0
Fecal Coliforms (10^6/day)	504,602	0	0
Viruses (10^6/day)	0.4	0	0
Total Suspended Solids (lbs/day)	0	0	0
Cryptosporidium Oocysts (10^6/day)	2.1	0	0

RESERVOIR: Cannonsville

Known Permanent Watershed Population   17,000
Permitted Number of Wastewater Treatment Plants   7
Permitted Wastewater flow from WWTPs (gallons/day)   2,000,000

Discharger	Daily Flow
Delaware-BOCES	2,500
Delhi	515,000
SEVA Institute	7,800
Allen Center	20,000
Stamford WTP	500,000
Village of Hobart	160,000
Walton WTP	1,170,000

Parameter	Input from WWTPs Now	Input w/ upgrades	Input 2010 w/ growth
Phosphorus (lbs/day)	86.6	4.3	4.6
Giardia Cysts (10^6/day)	69,658.6	22,847.3	24,217.8
Fecal Coliforms (10^6/day)	1,471.9	1,324.0	1,403.4
Viruses (10^6/day)	253.3	82.2	87.1
Total Suspended Solids (lbs/day)	287.3	6.3	6.7
Cryptosporidium Oocysts (10^6/day)	6,473.3	308.5	327

Known Number of OSTDS

8,388

Estimated Wastewater flow from OSTDS (gallons/day)

1,048,500

Estimated Percent of Population served by OSTDS

123

Estimated Population served by OSTDS (@2.5 persons/home)

20,970

Parameter	Input from OSTDS Now	Input w/ upgrades	Input 2010 w/ growth
Phosphorus (lbs/day)	172.3	28.7	30.4
Giardia Cysts (10^6/day)	604	0	0
Fecal Coliforms (10^6/day)	5,378,805	0	0
Viruses (10^6/day)	4.4	0	0
Total Suspended Solids (lbs/day)	0	0	0
Cryptosporidium Oocysts (10^6/day)	2.2	0	0

RESERVOIR: Pepacton

Known Permanent Watershed Population

15,622

Permitted Number of Wastewater Treatment Plants

6

Permitted Wastewater flow from WWTPs (gallons/day)

579,400

Discharger	Daily Flow
Camp NuBar	12,500
Camp Tai Chi	7,500
Margeretville WWTP	400,000
Mountainside Farms, Inc.	49,800
Regis Hotel	9,600
Roxbury Run Village	100,000

Parameter	Input from WWTPs Now	Input w/ upgrades	Input 2010 w/ growth
Phosphorus (lbs/day)	8.6	1.5	1.6
Giardia Cysts (10^6/day)	227.2	82.1	87
Fecal Coliforms (10^6/day)	2,396.3	407.2	431.6
Viruses (10^6/day)	7.0	2.6	2.8
Total Suspended Solids (lbs/day)	21.9	3.3	3.5
Cryptosporidium Oocysts (10^6/day)	23.5	8.9	9.4

Known Number of OSTDS

6,417

Estimated Wastewater flow from OSTDS (gallons/day)

802,150

Estimated Percent of Population served by OSTDS

103

Estimated Population served by OSTDS (@2.5 persons/home)

16,043

Parameter	Input from OSTDS Now	Input w/ upgrades	Input 2010 w/ growth
Phosphorus (lbs/day)	131.9	22.0	23.3
Giardia Cysts (10^6/day)	462	0	0
Fecal Coliforms (10^6/day)	4,115,029	0	0
Viruses (10^6/day)	3.4	0	0
Total Suspended Solids (lbs/day)	0	0	0
Cryptosporidium Oocysts (10^6/day)	1.7	0	0

RESERVOIR: Rondout

Known Permanent Watershed Population	4,170
Permitted Number of Wastewater Treatment Plants	1
Permitted Wastewater flow from WWTPs (gallons/day)	180,000

Discharger	Daily Flow
Grahamsville WWTP	180,000

Parameter	Input from WWTPs Now	Input w/ upgrades	Input 2010 w/ growth
Phosphorus (lbs/day)	0.3	0.3	0.3
Giardia Cysts (10^6/day)	12.7	12.7	13.5
Fecal Coliforms (10^6/day)	33.5	33.5	35.5
Viruses (10^6/day)	0.4	0.4	0.4
Total Suspended Solids (lbs/day)	0.1	0.1	0.1
Cryptosporidium Oocysts (10^6/day)	1.5	1.5	1.6

Known Number of OSTDS	1,414
Estimated Wastewater flow from OSTDS (gallons/day)	176,750
Estimated Percent of Population served by OSTDS	85
Estimated Population served by OSTDS (@2.5 persons/home)	3,535

Parameter	Input from OSTDS Now	Input w/ upgrades	Input 2010 w/ growth
Phosphorus (lbs/day)	29.1	4.8	5.1
Giardia Cysts (10^6/day)	102	0	0
Fecal Coliforms (10^6/day)	906,728	0	0
Viruses (10^6/day)	0.7	0	0
Total Suspended Solids (lbs/day)	0	0	0
Cryptosporidium Oocysts (10^6/day)	0.4	0	0

RESERVOIR: Neversink

Known Permanent Watershed Population	1,557
Permitted Number of Wastewater Treatment Plants	0
Permitted Wastewater flow from WWTPs (gallons/day)	0

Known Number of OSTDS	822
Estimated Wastewater flow from OSTDS (gallons/day)	102,750
Estimated Percent of Population served by OSTDS	132
Estimated Population served by OSTDS (@2.5 persons/home)	2,055

Parameter	Input from WWTPs Now	Input w/ upgrades	Input 2010 w/ growth
Phosphorus (lbs/day)	16.9	2.8	3.0
Giardia Cysts (10^6/day)	59	0	0
Fecal Coliforms (10^6/day)	527,108	0	0
Viruses (10^6/day)	0.4	0	0
Total Suspended Solids (lbs/day)	0	0	0
Cryptosporidium Oocysts (10^6/day)	0.2	0	0

REFERENCES

Anderson, D. L., A. L. Lewis, and K. M. Sherman. 1991. Human enterovirus monitoring at on-site sewage disposal systems in Florida. Pp. 94-104 In On-Site Wastewater Treatment, Proceedings of the Sixth National Symposium on Individual and Small Community Sewage Systems. St. Joseph, MI: Amer. Soc. Ag. Engineers.

Ayres, A. 1993. Onsite Sewage Disposal Systems Research in Florida: An Evaluation of Current OSDS Practices in Florida. Tallahassee: Florida Department of Health and Rehabilitative Services.

Canter, L. W., and R. C. Knox. 1986. Septic Tank System Effects on Groundwater Quality. Chelsea, MI: Lewis Publishers Inc.

Chanlett, E. T. 1979. Environmental Protection. New York, NY: McGraw-Hill.

Goodgame, R. W., R. M. Genta, A. C. White, and C. L. Chappell. 1993. Intensity of infection in AIDS-associated cryptosporidiosis. Journal of Infectious Diseases 167:704–709.

Kappus, K. D., R. G. Lundgren, Jr., D. D. Juranek, J. M. Roberts, and H. C. Spencer. 1994. Intestinal parasitism in the United States: Update on a continuing problem. Am. J. Trop. Med. Hyg. 506:705–713.

Knight, R. 1980. Epidemiology and transmission of giardiasis. Transactions of the Royal Society of Tropical Medicine and Hygiene 74(4):433–435.

Marx, R., and E. Goldstein. 1993. A guide to New York City's reservoirs and their watersheds. New York, NY: Natural Resource Defense Council.

New York City Department of Environmental Protection (NYC DEP). 1993. Final Generic Environmental Impact Statement for the Proposed Watershed Regulations for the Protection from Contamination, Degradation, and Pollution of the New York City Water Supply and its Sources. November 1993. Corona, NY: NYC DEP.

Appendix E

Acronyms

AIDS	acquired immunodeficiency syndrome
BACT	best available control technology
BDL	below-detection-limit
BMP	best management practice
BOD	biological oxygen demand
CBUDSF	continuous backwash upflow dual sand filtration
CDC	Centers for Disease Control and Prevention
CFU	colony forming units
CREP	Conservation Reserve Enhancement Program
CWA	Clean Water Act
CWC	Catskill Watershed Corporation
CWT	Coalition of Watershed Towns
DAF	dissolved air flotation
DBP	disinfection byproduct
D/DBP	disinfectants/disinfection byproducts
DO	dissolved oxygen
DOC	dissolved organic carbon
DOM	dissolved organic matter
ECL	Environmental Conservation Law
ECM	event mean concentration
EIS	environmental impact statement
EOH	East-of-Hudson
EPA	Environmental Protection Agency
ESWTR	Enhanced Surface Water Treatment Rule
FAD	filtration avoidance determination

GAC	granulated activated carbon
GIS	Geographic Information System
GWLF	generalized watershed loading function
HAA	haloacetic acid
HAAFP	haloacetic acid formulation potential
HIV	human immunodeficiency virus
HPC	heterotrophic plate count
ICR	Information Collection Rule
IESWTR	Interim ESWTR
IR	inactivation ratio
LA	load allocation
LCR	Lead and Copper Rule
LT1 ESWTR	Long-term 1 ESWTR
LT2 ESWTR	Long-term 2 ESWTR
MCL	maximum contaminant level
MCLG	maximum contaminant level goal
MDC	Metropolitan District Commission
MIP	Model Implementation Program
MOA	Memorandum of Agreement
MOS	margin of safety
MRDL	maximum residual disinfectant level
MRDLG	maximum residual disinfectant level goal
MTBE	methyl tert-butyl ether
MWRA	Massachusetts Water Resources Authority
NOM	natural organic matter
NPDES	National Pollutant Discharge Elimination System
NPS	nonpoint source
NSF	National Sanitation Foundation
NTU	nephelometric turbidity unit
NYC DEP	New York City Department of Environmental Protection
NYC DOH	New York City Department of Health
NYS DEC	New York State Department of Environmental Conservation
NYS DOH	New York State Department of Health
NYS DOS	New York State Department of State
OSTDS	on-site sewage treatment and disposal system
PP	particulate phosphorus
QA/QC	quality assurance/quality control
SDWA	Safe Drinking Water Act
SEQR	State Environmental Quality Review
SPDES	State Pollutant Discharge Elimination System
SPPP	stormwater pollution prevention plan
SRP	soluble reactive phosphorus
SS	suspended solids

SWAP	source water assessment program
SWTR	Surface Water Treatment Rule
TCR	Total Coliform Rule
THM	trihalomethane
THMFP	trihalomethane formulation potential
TMDL	total maximum daily load
TOC	total organic carbon
TP	total phosphorus
TSS	total suspended solids
TTHM	total trihalomethane
TVA	Tennessee Valley Authority
WAC	Watershed Agricultural Council
WAP	Watershed Agricultural Program
WBDR	West Branch of the Delaware River
WFP	Watershed Forestry Program
WLA	waste load allocation
WOH	West-of-Hudson
WRI	Water Resources Institute
WWTP	wastewater treatment plant

Appendix F

Biographical Information

Charles R. O'Melia, *Chair*, is the Abel Wolman Professor of Environmental Engineering in the Department of Geography and Environmental Engineering at the Johns Hopkins University. He is a member of the National Academy of Engineering and formerly served on the Water Science and Technology Board. His professional experience includes positions at Hazen & Sawyer Engineers, University of Michigan, Georgia Institute of Technology, Harvard University, and the University of North Carolina, Chapel Hill. His research interests are in aquatic chemistry, environmental fate and transport, predictive modeling of natural systems, and theory of water and wastewater treatment. He received a B.C.E. from Manhattan College and an M.S.E. and Ph.D. in sanitary engineering from the University of Michigan.

Max J. Pfeffer, *Vice-Chair*, is an associate professor in the Department of Rural Sociology at Cornell University. His research has focused on the social aspects of agriculture, the environment, and development planning. He has done recent work on the social dimensions of watershed planning within the New York City watershed and works on natural resource management in Central America. Dr. Pfeffer was a member of the NRC's Watershed Management Committee. He received his Ph.D. in sociology from the University of Wisconsin in 1986.

Paul K. Barten is an associate professor of forest resources at the University of Massachusetts at Amherst and lecturer in hydrology and watershed management at Yale University. Dr. Barten received his Ph.D. (1988) and M.S. (1985) in hydrology and watershed management from the University of Minnesota. A native of the Catskill Mountain region, he holds undergraduate degrees in forestry from SUNY Environmental Science and Forestry (1983) and the New York State Ranger School (1977). His research focuses on hydrological processes and land use impacts on watersheds, forests, and wetlands. He is co-chair of the Quabbin (Boston water supply) Science and Technical Advisory Committee.

G. Edward Dickey is a consultant to public and private organizations interested in water policy and infrastructure development and management. He also is adjunct professor of economics at Loyola College in Maryland. Dr. Dickey retired from federal service in 1998 after a career in water resources planning and project development. In his last position as Chief of the Planning Division of the U.S. Army Corps of Engineers, he directed the Corps' nationwide water resources planning programs and its small project programs. In his prior positions as Deputy Assistant Secretary of the Army and Acting Assistant Secretary of the Army (Civil Works), he provided leadership and policy direction for all army civil works activities including the Section 404 regulatory program. He received his B.A. in Political Economy from the Johns Hopkins University and his M.A. and Ph.D. from Northwestern University.

Margot W. Garcia received her Ph.D. in watershed management from the University of Arizona in 1980. She is an associate professor in the Department of Urban Studies and Planning, Virginia Commonwealth University, where she teaches courses in environmental planning and citizen participation. Previously, she worked for the U.S. Forest Service doing land use planning, computer modeling, and public involvement. Dr. Garcia was a member of the Steering Committee of Water Quality 2000, a coalition of 85 national organizations which proposed a national water policy; one of its main points was to call for a watershed approach to clean water. She has given invited talks and written policy papers on watershed management since 1989 and was a member of the NRC's Watershed Management Committee.

Charles N. Haas earned a B.S. in biology and an M.S. in environmental engineering from the Illinois Institute of Technology and a Ph.D. in environmental engineering from the University of Illinois. He is the Betz Chair Professor of Environmental Engineering at Drexel University. He was formerly a professor and acting chair in the Department of Environmental Engineering at the Illinois Institute of Technology. His areas of research involve microbial and chemical risk assessment, hazardous waste processing, industrial wastewater treatment, waste recovery, and water and wastewater disinfection processes. Dr. Haas has chaired a number of professional conferences and workshops, has served as a member of several advisory panels to the Environmental Protection Agency, and is currently on an advisory committee to the Philadelphia Department of Health.

Richard G. Hunter is the Deputy State Health Officer with the Florida Department of Health in Tallahassee, Florida, where he is responsible for the day-to-day management of all public health programs. He has been involved with numerous issues relating to the impact of land use on the environment. These include long-term studies on the impacts of development on the near-shore waters of the Florida Keys, a research project relating to nutrient input from dairies and septic systems in the Suwannee River watershed, regulation of land disposal of wastewater treatment residuals, and a study of *Giardia* contamination of water wells by septic systems. Dr. Hunter received his Ph.D. in environmental health from the University of Oklahoma Health Sciences Center in 1988.

R. Richard Lowrance is an ecologist with the U.S. Department of Agriculture, Agricultural Research Service. He received his Ph.D. in ecology from the University of Georgia in 1981. His research involves analyzing and modeling the ability of buffer zones to protect surface water and groundwater from nonpoint source pollution derived from agriculture. Recently, he chaired a scientific panel for the Chesapeake Bay Program to evaluate the water quality functions of riparian forest buffer systems in the Chesapeake Bay watershed. Dr. Lowrance served on the NRC Workshop on Monitoring and Managing Natural Resources for the Office of International Affairs in 1990.

Christine L. Moe is an assistant professor in the Department of Epidemiology, University of North Carolina, Chapel Hill, and an adjunct professor in the Division of Environmental and Occupational Health at the Emory University School of Public Health. She received her Ph.D. in environmental sciences and engineering from the University of North Carolina and has done extensive laboratory and field research on waterborne transmission of infectious agents and diagnosis and epidemiology of enteric virus infections. Dr. Moe served on the NRC Committee to Evaluate the Viability of Augmenting Potable Water Supplies with Reclaimed Water.

Cynthia L. Paulson is a vice president and Central North Manager at Brown and Caldwell in Denver, Colorado. Her experience is in watershed and water quality planning and assessment, including the evaluation of point and nonpoint source pollutant impacts on receiving waters and performance of wasteload allocations. She receive her Ph.D. in environmental engineering from the University of Colorado in 1993. Dr. Paulson has served on the Water Science and Technology Board's Committee on U.S.G.S. Water Resources Research.

Rutherford H. Platt is a professor of geography and planning law at the University of Massachusetts at Amherst. He received his Ph.D. in geography from the University of Chicago and also holds a J.D. from the University of Chicago Law School. He served as assistant director and staff attorney for the Open Lands

Project, Inc., Chicago, and is a member of the Illinois bar. He has served on other NRC committees including the Committee on Flood Insurance Studies, the Committee on Water Resources Research Review, the Committee on a Levee Policy for the National Flood Insurance Program, and the Committee on Managing Coastal Erosion. He chaired the NRC Committees on Options to Preserve the Cape Hatteras Lighthouse and Flood Control Alternatives in the American River Basin.

Jerald L. Schnoor is a University of Iowa Foundation Distinguished Professor of Environmental Engineering and co-directs the Center for Global and Regional Environmental Research. A new member of the National Academy of Engineering, he received his Ph.D. in environmental health engineering from the University of Texas. His research interests are in mathematical modeling of water quality, aquatic chemistry, and impact of carbon emissions on global change. He has research projects in aquatic effects modeling of acid precipitation, global change and biogeochemistry, groundwater and hazardous wastes, and exposure risk assessment modeling. He is the editor of four books and the author of *Environmental Modeling*. Dr. Schnoor is also the associate editor of *Environmental Science & Technology*.

Thomas R. Schueler is the executive director of the Center for Watershed Protection, a nonprofit organization devoted to the protection, restoration, and stewardship of the nation's watersheds. As Chief of the Anacostia Restoration Team at the Metropolitan Washington Council of Governments, he directed the restoration of the Anacostia watershed. Over the past 15 years, he has pioneered new designs for stormwater ponds, wetlands, and filtering systems, developed new methods for urban watershed planning, demonstrated new techniques for stream restoration, sediment control, riparian reforestation, and stormwater retrofits, and developed new approaches to use impervious cover as a management tool for watershed planning. He received his B.S. in environmental science from the George Washington University.

James M. Symons, a member of the National Academy of Engineering, is presently retired and holds the position of the Cullen Distinguished Professor Emeritus of Civil Engineering at the University of Houston. His primary research focus was disinfection of drinking water, specializing in the formation of disinfection byproducts. Other areas of research included source water control, treatment for removal of organic, inorganic, and particulate contaminants, and distributed water quality. His career includes 20 years in the federal government, the U.S. Public Health Service and the Environmental Protection Agency. Dr. Symons received his B.C.E. in Civil Engineering from Cornell University in 1954, and the S.M. and Sc.D. degrees in Sanitary Engineering from MIT in 1955 and 1957.

Robert G. Wetzel is the Bishop Professor of Biological Sciences at the University of Alabama in Tuscaloosa. His research interests include the physiology and ecology of bacteria, algae, and higher aquatic plants; biogeochemical cycling in fresh waters; and functional roles of organic compounds and detritus in aquatic ecosystems. His prior professional experience includes positions as professor at Michigan State University, Erlander National Professor of the Institute of Limnology of Uppsala University in Sweden, and professor at the University of Michigan. Dr. Wetzel is an elected member of the Royal Danish Academy of Sciences and the American Academy of Arts and Sciences. He earned a B.Sc. and M.Sc. from the University of Michigan and a Ph.D. from the University of California at Davis.

STAFF

Laura J. Ehlers is a senior staff officer of the National Research Council's Water Science and Technology Board and the study director for this report. She is also the study director for the WSTB Committee on Environmental Remediation at Naval Facilities, the Committee on Riparian Zone Functioning, and the Committee on Bioavailability of Contaminants in Soils and Sediment. Dr. Ehlers received her B.S. in biology and engineering and applied science from the California Institute of Technology, and her M.S.E. and Ph.D. degrees in environmental engineering from the Johns Hopkins University.

Ellen A. De Guzman is a senior project assistant at the National Research Council's Water Science and Technology Board. She received a B.A. from the University of the Philippines and is currently majoring in economics from the University of Maryland University College. Her previous NRC reports include *Valuing Groundwater, Innovations in Ground Water and Soil Cleanup, Issues in Potable Reuse, Improving the American River Flood Frequency Analyses,* and *New Directions in Water Resources Planning for the U.S. Army Corps of Engineers.*